5G New Radio

5G New Radio

A Beam-based Air Interface

Edited by
Mihai Enescu

This edition first published 2020
© 2020 John Wiley & Sons Ltd

The right of Mihai Enescu to be identified as the editor of this work has been asserted in accordance with law.

Registered Offices
John Wiley & Sons, Inc., 111 River Street, Hoboken, NJ 07030, USA
John Wiley & Sons Ltd, The Atrium, Southern Gate, Chichester, West Sussex, PO19 8SQ, UK

Editorial Office
The Atrium, Southern Gate, Chichester, West Sussex, PO19 8SQ, UK

For details of our global editorial offices, customer services, and more information about Wiley products visit us at www.wiley.com.

Wiley also publishes its books in a variety of electronic formats and by print-on-demand. Some content that appears in standard print versions of this book may not be available in other formats.

Library of Congress Cataloging-in-Publication Data applied for

Hardback ISBN: 9781119582380

Cover Design: Wiley
Cover Image: © ixpert/Shutterstock

Set in 9.5/12.5pt STIXTwoText by SPi Global, Chennai, India

Printed and bound by CPI Group (UK) Ltd, Croydon, CR0 4YY

10 9 8 7 6 5 4 3 2 1

Contents

List of Contributors *xiii*
Preface *xv*
Acknowledgments *xvii*
Abbreviations *xix*

1   **Introduction and Background** *1*
Mihai Enescu and Karri Ranta-aho
1.1   Why 5G? *1*
1.2   Requirements and Targets *2*
1.2.1   System Requirements *3*
1.2.2   5G Spectrum *7*
1.3   Technology Components and Design Considerations *10*
1.3.1   Waveform *12*
1.3.2   Multiple Access *13*
1.3.3   Scalable/Multi Numerology *13*
1.3.3.1   Motivation for Multiple Numerologies *13*
1.3.3.2   5G NR Numerologies *13*
1.3.4   Multi-antenna *17*
1.3.5   Interworking with LTE and Other Technologies *18*
1.3.6   5G Beam Based Technologies Across Release 15 and Release 16 *19*
1.3.6.1   Integrated Access and Backhaul *19*
1.3.6.2   NR Operation on Unlicensed Frequency Bands (NR-U) *20*
1.3.6.3   Ultra-Reliable and Low Latency Communications *21*
1.3.6.4   Vehicular-to-everything (V2X) *21*
1.3.6.5   Positioning *22*
1.3.6.6   System Enhancements *22*

2   **Network Architecture and NR Radio Protocols** *25*
Dawid Koziol and Helka-Liina Määttänen
2.1   Architecture Overview *25*
2.2   Core Network Architecture *26*
2.2.1   Overview *26*
2.2.2   Service Request Procedure *29*

2.3   Radio Access Network   *31*
2.3.1   NR Standalone RAN Architecture   *31*
2.3.2   Additional Architectural Options   *32*
2.3.3   CU-DU and UP-CP Split   *37*
2.4   NR Radio Interface Protocols   *41*
2.4.1   Overall Protocol Structure   *41*
2.4.2   Main Functions of NR Radio Protocols   *44*
2.4.3   SDAP Layer   *47*
2.4.4   PDCP Layer   *47*
2.4.4.1   PDCP Packet Transmission   *48*
2.4.4.2   PDCP Duplication   *49*
2.4.4.3   Access Stratum (AS) Security   *50*
2.4.4.4   Robust Header Compression (ROHC)   *50*
2.4.5   RLC   *50*
2.4.5.1   Segmentation and Concatenation   *51*
2.4.5.2   RLC Reordering   *51*
2.4.5.3   ARQ Retransmissions and Status Reporting   *52*
2.4.6   MAC Protocol   *53*
2.4.6.1   Overview   *53*
2.4.6.2   Multiplexing and Demultiplexing   *53*
2.4.6.3   Logical Channel Prioritization   *54*
2.4.6.4   Hybrid Automatic Repeat Request (HARQ)   *57*
2.4.6.5   BWP Operation   *58*
2.4.6.6   Scheduling Request   *60*
2.4.6.7   Semi Persistent Scheduling and Configured Grants   *60*
2.4.6.8   Discontinuous Reception (DRX)   *60*
2.4.6.9   Buffer Status Reports   *62*
2.4.6.10   Timing Advance Operation   *62*
2.4.6.11   MAC Control Elements   *63*
2.4.7   Radio Resource Control (RRC)   *67*
2.4.7.1   Overview   *67*
2.4.7.2   RRC State Machine   *68*
2.4.7.3   Cells, Cell Groups, and Signaling Radio Bearers   *70*
2.4.7.4   System Information   *71*
2.4.7.5   Unified Access Control (UAC)   *78*
2.4.7.6   Connection Control   *79*
2.4.7.7   NAS Information Transfer   *87*
2.4.7.8   UE Assistance Information   *87*
2.4.7.9   RRC PDU Structure   *89*

3   **PHY Layer**   *95*
           Mihai Enescu, Youngsoo Yuk, Fred Vook, Karri Ranta-aho, Jorma Kaikkonen,
           Sami Hakola, Emad Farag, Stephen Grant, and Alexandros Manolakos
3.1   Introduction (Mihai Enescu, Nokia Bell Labs, Finland)   *95*
3.2   NR Waveforms (Youngsoo Yuk, Nokia Bell Labs, Korea)   *96*

3.2.1   Advanced CP-OFDM Waveforms for Multi-Service Support   *96*
3.2.2   Low PAPR Waveform for Coverage Enhancement   *102*
3.2.3   Considerations on the Waveform for above 52.6 GHz   *104*
3.3   Antenna Architectures in 5G (Fred Vook, Nokia Bell Labs, USA)   *105*
3.3.1   Beamforming   *105*
3.3.2   Antenna Array Architectures   *108*
3.3.3   Antenna Panels   *110*
3.3.4   Antenna Virtualization   *111*
3.3.5   Antenna Ports   *113*
3.3.6   Beamforming for a Beam-Based Air Interface   *115*
3.4   Frame Structure and Resource Allocation (Karri Ranta-aho, Nokia Bell Labs, Finland)   *115*
3.4.1   Resource Grid   *115*
3.4.2   Data Scheduling and HARQ   *118*
3.4.3   Frequency Domain Resource Allocation and Bandwidth Part   *119*
3.4.4   Time Domain Resource Allocation   *123*
3.5   Synchronization Signals and Broadcast Channels in NR Beam-Based System (Jorma Kaikkonen, Sami Hakola, Nokia Bell Labs, Finland)   *125*
3.5.1   SS/PBCH Block   *125*
3.5.2   Synchronization Signal Structure   *126*
3.5.3   Broadcast Channels   *128*
3.5.3.1 PBCH   *128*
3.5.3.2 SIB1   *129*
3.5.3.3 Delivery of Other Broadcast Information and Support of Beamforming   *135*
3.6   Physical Random Access Channel (PRACH) (Emad Farag, Nokia Bell Labs, USA)   *139*
3.6.1   Introduction   *139*
3.6.2   Preamble Sequence   *140*
3.6.2.1 Useful Properties of Zhadoff-Chu Sequences   *140*
3.6.2.2 Unrestricted Preamble Sequences   *142*
3.6.2.3 Restricted Preamble Sequences   *144*
3.6.3   Preamble Formats   *147*
3.6.3.1 Long Sequence Preamble Formats   *148*
3.6.3.2 Short Sequence Preamble Formats   *149*
3.6.4   PRACH Occasion   *150*
3.6.5   PRACH Baseband Signal Generation   *155*
3.7   CSI-RS (Stephen Grant, Ericsson, USA)   *159*
3.7.1   Overview   *159*
3.7.1.1 CSI-RS Use Cases   *159*
3.7.1.2 Key Differences with LTE   *161*
3.7.2   Physical Layer Design   *162*
3.7.2.1 Mapping to Physical Resources   *162*
3.7.2.2 Antenna Port Mapping   *167*
3.7.2.3 Sequence Generation and Mapping   *167*
3.7.2.4 Time Domain Behavior   *168*

3.7.2.5   Multiplexing with Other Signals   *169*
3.7.3   Zero Power CSI-RS   *170*
3.7.4   Interference Measurement Resources (CSI-IM)   *170*
3.7.5   CSI-RS Resource Sets   *171*
3.7.5.1   CSI-RS for Tracking   *171*
3.7.5.2   CSI-RS for L1-RSRP Measurement   *173*
3.8   PDSCH and PUSCH DM-RS, Qualcomm Technologies, Inc. (Alexandros Manolakos, Qualcomm Technologies, Inc, USA)   *176*
3.8.1   Overview   *176*
3.8.1.1   What Is DM-RS Used for?   *176*
3.8.1.2   Key Differences from LTE   *176*
3.8.2   Physical Layer Design   *178*
3.8.2.1   Mapping to Physical Resources   *178*
3.8.2.2   Default DM-RS Pattern for PDSCH and PUSCH   *189*
3.8.2.3   Sequence Generation and Scrambling   *193*
3.8.3   Procedures and Signaling   *200*
3.8.3.1   Physical Resource Block Bundling   *200*
3.8.3.2   DM-RS to PDSCH and PUSCH EPRE Ratio   *205*
3.8.3.3   Antenna Port DCI Signaling   *207*
3.8.3.4   Quasi-Colocation Considerations for DM-RS of PDSCH   *209*
3.9   Phrase- Tracking RS (Youngsoo Yuk, Nokia Bell Labs, Korea)   *210*
3.9.1   Phase Noise and its Modeling   *210*
3.9.1.1   Phase Noise in mm-Wave Frequency and its Impact to OFDM System   *210*
3.9.1.2   Principles of Oscillator Design and Practical Phase Noise Modeling   *211*
3.9.2   Principle of Phase Noise Compensation   *216*
3.9.3   NR PT-RS Structure and Procedures   *221*
3.9.3.1   PT-RS Design for Downlink   *221*
3.9.3.2   PT-RS Design for Uplink CP-OFDM   *224*
3.9.3.3   PT-RS Design for Uplink DFT-s-OFDM   *225*
3.10   SRS (Stephen Grant, Ericsson, USA)   *228*
3.10.1   Overview   *228*
3.10.1.1   SRS Use Cases   *228*
3.10.1.2   Key Differences with LTE   *229*
3.10.2   Physical Layer Design   *230*
3.10.2.1   Mapping to Physical Resources   *230*
3.10.2.2   Antenna Port Mapping   *237*
3.10.2.3   Sequence Generation and Mapping   *239*
3.10.2.4   Multiplexing with Other UL Signals   *243*
3.10.3   SRS Resource Sets   *244*
3.10.3.1   SRS for Downlink CSI Acquisition for Reciprocity-Based Operation   *244*
3.10.3.2   SRS for Uplink CSI Acquisition   *245*
3.10.3.3   SRS for Uplink Beam Management   *246*
3.11   Power Control (Mihai Enescu, Nokia Bell Labs, Finland)   *246*
3.12   DL and UL Transmission Framework (Mihai Enescu, Nokia, Karri Ranta-aho, Nokia Bell Labs, Finland)   *249*

3.12.1   Downlink Transmission Schemes for PDSCH   *249*
3.12.2   Downlink Transmit Processing   *250*
3.12.2.1 PHY Processing for PDSCH   *250*
3.12.2.2 PHY Processing for PDCCH   *251*
3.12.3   Uplink Transmission Schemes for PUSCH   *254*
3.12.3.1 Codebook Based UL Transmission   *254*
3.12.3.2 Non-Codebook Based UL Transmission   *255*
3.12.4   Uplink Transmit Processings   *255*
3.12.4.1 PHY Processing for PUSCH   *255*
3.12.5   Bandwidth Adaptation   *256*
3.12.5.1 Overview   *256*
3.12.5.2 Support for Narrow-Band UE in a Wide-Band Cell   *257*
3.12.5.3 Saving Battery with Bandwidth Adaptation   *257*
3.12.5.4 Spectrum Flexibility   *260*
3.12.6   Radio Network Temporary Identifiers (RNTI)   *260*

4   Main Radio Interface Related System Procedures   *261*
          Jorma Kaikkonen, Sami Hakola, Emad Farag, Mihai Enescu, Claes Tidestav,
          Juha Karjalainen, Timo Koskela, Sebastian Faxér, Dawid Koziol, and Helka-Liina
          Määttänen
4.1   Initial Access (Jorma Kaikkonen, Sami Hakola, Nokia Bell Labs, Finland, Emad
          Farag, Nokia Bell Labs, USA)   *261*
4.1.1   Cell Search   *261*
4.1.1.1   SS/PBCH Block Time Pattern   *262*
4.1.1.2   Initial Cell Selection Related Assistance Information   *265*
4.1.2   Random Access   *265*
4.1.2.1   Introduction   *265*
4.1.2.2   Higher Layer Random Access Procedures   *266*
4.1.2.3   Random Access Use Cases   *274*
4.1.2.4   Physical Layer Random Access Procedures   *274*
4.1.2.5   RACH in Release 16   *283*
4.2   Beam Management (Mihai Enescu, Nokia Bell Labs, Finland, Claes Tidestav,
          Ericsson, Sweden, Sami Hakola, Juha Karjalainen, Nokia Bell Labs,
          Finland)   *287*
4.2.1   Introduction to Beam Management   *287*
4.2.2   Beam Management Procedures   *289*
4.2.2.1   Beamwidths   *291*
4.2.2.2   Beam Determination   *291*
4.2.3   Beam Indication Framework for DL Quasi Co-location and TCI States   *296*
4.2.3.1   QCL   *296*
4.2.3.2   TCI Framework   *297*
4.2.4   Beam Indication Framework for UL Transmission   *303*
4.2.4.1   SRS Configurations   *305*
4.2.4.2   Signaling Options for SRS Used for UL Beam Management   *306*
4.2.4.3   Beam Reporting from a UE with Multiple Panels   *306*

4.2.5   Reporting of L1-RSRP   *307*
4.2.6   Beam Failure Detection and Recovery   *312*
4.2.6.1   Overview   *312*
4.2.6.2   Beam Failure Detection   *313*
4.2.6.3   New Candidate Beam Selection   *314*
4.2.6.4   Recovery Request and Response   *315*
4.2.6.5   Completion of BFR Procedure   *316*
4.3   CSI Framework (Sebastian Faxér, Ericsson, Sweden)   *317*
4.3.1   Reporting and Resource Settings   *318*
4.3.2   Reporting Configurations and CSI Reporting Content   *323*
4.3.2.1   The Different CSI Parameters   *323*
4.3.2.2   CSI-RS Resource Indicator (CRI)   *323*
4.3.2.3   SSB Resource Indicator   *324*
4.3.2.4   Precoder Matrix Indicator (PMI) and Rank Indicator (RI)   *324*
4.3.2.5   Channel Quality Indicator (CQI)   *325*
4.3.2.6   Layer Indicator (LI)   *327*
4.3.2.7   Layer-1 Reference Signal Received Power (L1-RSRP)   *327*
4.3.2.8   Reporting Quantities   *327*
4.3.2.9   Frequency-Granularity   *331*
4.3.2.10   Measurement Restriction of Channel and Interference   *332*
4.3.2.11   Codebook Configuration   *333*
4.3.2.12   NZP CSI-RS Based Interference Measurement   *333*
4.3.3   Triggering/Activation of CSI Reports and CSI-RS   *334*
4.3.3.1   Aperiodic CSI-RS/IM and CSI Reporting   *334*
4.3.3.2   Semi-Persistent CSI-RS/IM and CSI Reporting   *335*
4.3.4   UCI Encoding   *337*
4.3.4.1   Collision Rules and Priority Order   *338*
4.3.4.2   Partial CSI Omission for PUSCH-Based CSI   *339*
4.3.5   CSI Processing Criteria   *340*
4.3.6   CSI Timeline Requirement   *341*
4.3.7   Codebook-Based Feedback   *344*
4.3.7.1   Motivation for the Use of DFT Codebooks   *346*
4.3.7.2   DL Type I Codebook   *349*
4.3.7.3   DL Type II Codebook   *352*
4.4   Radio Link Monitoring (Claes Tidestav, Ericsson, Sweden, Dawid Koziol, Nokia
            Bell Labs, Poland)   *356*
4.4.1   Causes of Radio Link Failure   *357*
4.4.1.1   Physical Layer Problem   *357*
4.4.1.2   Random Access Failure   *363*
4.4.1.3   RLC Failure   *364*
4.4.2   Actions After RLF   *365*
4.4.2.1   RLF in MCG   *365*
4.4.2.2   RLF in SCG   *368*

4.4.3   Relation Between RLM/RLF and BFR   *368*
4.5   Radio Resource Management (RRM) and Mobility (Helka-Liina Määttänen, Ericsson, Finland, Dawid Koziol, Nokia Bell Labs, Poland, Claes Tidestav, Ericsson, Sweden)   *370*
4.5.1   Introduction   *370*
4.5.2   UE Mobility Measurements   *371*
4.5.2.1   NR Mobility Measurement Quantities   *372*
4.5.2.2   SS/PBCH Block Measurement Timing Configuration (SMTC)   *374*
4.5.2.3   SS/PBCH Block Transmission in Frequency Domain   *376*
4.5.3   Connected Mode Mobility   *376*
4.5.3.1   Overview of RRM Measurements   *378*
4.5.3.2   Measurement Configuration   *378*
4.5.3.3   Performing RRM Measurements   *383*
4.5.3.4   Handover Procedure   *384*
4.5.4   Idle and Inactive Mode Mobility   *388*
4.5.4.1   Introduction   *388*
4.5.4.2   Cell Selection and Reselection   *389*
4.5.4.3   Location Registration Udate   *393*
4.5.4.4   Division of IDLE Mode Tasks between NAS and AS Layers   *396*

5   **Performance Characteristics of 5G New Radio**   *397*
          Fred Vook
5.1   Introduction   *397*
5.2   Sub-6 GHz: Codebook-Based MIMO in NR   *398*
5.2.1   Antenna Array Configurations   *398*
5.2.2   System Modeling   *399*
5.2.3   Downlink CSI Feedback and MIMO Transmission Schemes   *399*
5.2.4   Traffic Models and Massive MIMO   *401*
5.2.5   Performance in Full Buffer Traffic   *401*
5.2.6   Performance in Bursty (FTP) Traffic   *404*
5.2.7   Performance of NR Type II CSI   *411*
5.3   NR MIMO Performance in mmWave Bands   *413*
5.4   Concluding Remarks   *416*

6   **UE Features**   *419*
          Mihai Enescu
6.1   Reference Signals   *422*
6.1.1   DM-RS   *422*
6.1.2   CSI-RS   *423*
6.1.3   PT-RS   *424*
6.1.4   SRS   *424*
6.1.5   TRS   *425*
6.1.6   Beam Management   *426*

6.1.7   TCI and QCL   *428*
6.1.8   Beam Failure Detection   *428*
6.1.9   RLM   *429*
6.1.10   CSI Framework   *429*

References   *433*
Index   *437*

List of Contributors

Mihai Enescu
Nokia Bell Labs, Finland

Emad Farag
Nokia Bell Labs, USA

Sebastian Faxér
Ericsson, Sweden

Stephen Grant
Ericsson, USA

Sami Hakola
Nokia Bell Labs, Finland

Jorma Kaikkonen
Nokia Bell Labs, Finland

Juha Karjalainen
Nokia Bell Labs, Finland

Timo Koskela
Nokia Bell Labs, Finland

Dawid Koziol
Nokia Bell Labs, Poland

Helka-Liina Määttänen
Ericsson, Finland

Alexandros Manolakos
Qualcomm Technologies, Inc., USA

Karri Ranta-aho
Nokia Bell Labs, Finland

Claes Tidestav
Ericsson, Sweden

Fred Vook
Nokia Bell Labs, USA

Youngsoo Yuk
Nokia Bell Labs, Korea

Preface

5G New Radio is perhaps the most awaited technology of 2020 promising not only to improve the mobile broadband transmission speed and capacity but also to revolutionize our lives and important industries, allowing for more reliable and fast communications, and for massive machine-type communications. One new development of this technology is the operation in millimeter wave, in carrier frequencies above 3.5 GHz, while the backbone of the design relies on multiple antenna transmission and reception. Indeed, multiantenna transmission is not a development area anymore but becomes a central part of the system design, and this was one of the main motivations for the introduction of this book. The complexity of the 5G specification is higher compared to legacy wireless technologies such as LTE. Understanding the design and functionality of 5G is way harder by reading solely the specification text; hence, the motivation to connect the dots of the existing specifications and explaining in a more formal way how the system works and how the specified technologies are functioning.

The book is written by active participants in the standardization, experts from top companies who have been proposing some of the specified techniques, and have been part of the standardization discussions. Our intention was to present the physical layer structure and procedures by looking at the system design also from the perspective of multiantenna/beam operation. Because of the vast amount of information and its complexity, the book does not cover some of the physical aspects, which might be more beam agnostic, such as channel coding. However, in addition to the physical layer part, the book has a thorough presentation of the related network architecture and NR radio protocols, information which we believe complements the other physical layer aspects. From a timeline perspective of the specification, the book aims at converging what is known in the industry as of Release 15; however, some of the Release 16 developments are also mentioned in various areas of the design.

The structure of this book is as follows.

Chapter 1 takes a general view on the introduction of 5G and its technology components. We present the system requirements and targets and elaborate on the available 5G spectrum. A brief overview of the 5G technology components is presented, with a more in-depth description following in the rest of the book. In this chapter, we touch some of the higher level aspects related to waveforms, multiple access, scalable/multiple numerologies, and multiantenna architectures, interworking with other technologies. Release 16 beam-based technologies are mentioned as well, such as integrated access and backhaul (IAB), NR operation in unlicensed frequency bands (NR-U), ultrareliable and low-latency communications (URLLC), vehicular-to-everything (V2X), positioning and other system enhancements in the area of two-step RACH, cross-link interference (CLI) mitigation, UE power saving, mobility enhancements, multiantenna technologies, and the support for above 52 GHz frequencies.

Chapter 2 provides an overview on 5G core (5GC) network and functionalities by presenting the specified logical network nodes that are the standardized building blocks for the 5GC and by giving an example of a signaling flow between the nodes for connection establishment. We also go through the different dual-connectivity options and related terminologies.

Physical layer components are described in Chapter 3. Waveforms are described first, being at the foundation of the system design. Throughout the description, we are explaining the selected waveforms and the reasons behind these choices. Antenna architectures are described next, highlighting the main possible antenna array architectures: analog, digital, and hybrid. Antenna panels, virtualization, and ports are also described. Frame structures and resource allocation strategies are introduced. These are followed by the detailed description of synchronization signals and broadcast channels that are complemented by the random access channels. The reference symbols of NR are as follows, with a detailed description of reference symbols used for channel sounding in both downlink and uplink, demodulation, and phrase tracking. The chapter closes with a description of power control and a downlink and uplink transmission framework.

The main radio interface-related system procedures are described in Chapter 4. The chapter starts with initial access and continues with one of the core elements of 5G physical layer: beam management. Issues such as beam management procedures, signals quasi-colocation, and beam failure are described in detail. The CSI framework presents not only the ways in which CSI resources are configured and used in the reporting procedure but also details on codebook configurations and operation. Radio link monitoring procedures, radio resource management, and mobility are also covered in Chapter 4.

In standardization, the introduction of system components is usually done based also on performance. In Chapter 5, we look into selected system performance aspects, the overall system performance characteristics being never the less quite large. We are considering some key components based on MIMO, below and above 6 GHz operation, which translated into the utilization of advanced codebook design as well as on the beam-based operation.

Chapter 6 ends the book with a presentation of the UE features. All the UE functionalities used by the 3GPP 5G specifications are defined as UE features and these are mandatory and optional, or mandatory with optional capability. In this chapter, we are trying to give a glimpse of some of the mandatory physical layer parameters.

Acknowledgments

This book is the result of the fruitful cooperation of a team of experts from several companies. The dedication and professionalism are hereby gratefully acknowledged, as well as the support of their companies.

5G New Radio is one of the most complex standards to date, which would have not been possible without the contribution of all the entities participating in the standardization process. Without their work and contribution to the standardization process, this book would have not been possible. We hereby provide our deep thanks to all our colleagues participating in the 3GPP discussions. The help provided by ETSI is also gratefully acknowledged.

Finally, this book is intended to present the 5G New Radio physical layer from a multi-antenna perspective; however, the reader should refer to the technical specifications published by 3GPP. Any views expressed in this book are those of the authors and do not necessarily reflect the views of their companies.

Abbreviations

3GPP	Third Generation Partnership Project
ACK	Acknowledgment
ACLR	Adjacent channel leakage ratio
ADC	Analog-to-Digital Converter
BPCH	Broadcast Physical Channel
BSR	Buffer Status Report
BWP	Bandwidth part
BPSK	Binary Phase Shift Keying
BS	Base Station
CBG	Code block group
CCDF	Complementary cumulative distribution function
CCP	Common Control Plane
CDD	Cyclic delay diversity
CDMA	Code Division Multiple Access
CMOS	Complementary Metal-Oxide-Semiconductor.
CN	Core Network
CP	Cyclic prefix
CQI	Channel quality indicator
CPU	CSI processing unit
CRB	Common resource block
CRC	Cyclic redundancy check
C-RNTI	Cell Radio-Network Temporary Identifier
CRI	CSI-RS Resource Indicator
CRS	Common Reference Signals
CSI	Channel state information
CSI-RS	Channel state information reference signal
CS-RNTI	Configured scheduling RNTI
CSI-RSRP	CSI reference signal received power
CSI-RSRQ	CSI reference signal received quality
CSI-SINR	CSI signal-to-noise and interference ratio
CU-DU	Central Unit - Distributed Unit
CW	Codeword
DAC	Digital-to-Analog Converter

DC	Dual Connectivity
DCI	Downlink control information
DFT	Discrete Fourier Transform
DL	Downlink
DM-RS	Dedicated demodulation reference signals
DRX	Discontinuous Reception
DSP	Digital Signal Processing
eMBB	Extreme Mobile Broadband
EIRP	Effective Isotropic Radiated Power
EN-DC	E-UTRA NR Dual-Connectivity
EPC	Evolved Packet Core
EPRE	Energy per resource element
E-UTRA	Evolved UTRA
EVM	Error Vector Magnitude
FDD	Frequency Division Duplex
FDM	Frequency Division Multiplexing
FDMA	frequency division multiple access
FFT	fast Fourier transform
FR1	Frequency Range 1
FR2	Frequency Range 2
FTP	File Transfer Protocol
GaAs	Gallium arsenide
GaN	Gallium nitride
GNSS	Global navigation satellite system
GTP	GPRS Tunneling Protocol
HARQ	hybrid automatic repeat request
HO	Hand Over
IAB	Integrated Access and Backhaul
IM	Interference measurement
IMS	IP Multimedia Subsystem
ITU	International Telecommunications Union
ITU-R	International Telecommunications Union-Radiocommunications Sector
IoT	Internet of Things
KPI	Key Performance Indicator
L1-RSRP	Layer 1 reference signal received power
LCID	Logical Channel Index
LI	Layer Indicator
LOS	Line of Sight
LTE	Long Term Evolution
LSB	Least Significant Bit
MAC	Medium Access Control
MBMS	Multimedia Broadcast Multicast Service
MCS	Modulation and coding scheme
MIB	Master Information Block
MCL	Maximum coupling loss

MTC	Machine Type Communications
mMTC	Massive Machine Type Communications
MIMO	Multiple Input Multiple Output
MMSE	Minimum mean square error
MMTEL	multimedia telephony service
MN	Master Node
MSB	Most Significant Bit
MU-MIMO	Multi-User MIMO
NAK	Negative Acknowledgment
NB-IoT	Narrow Band Internet of Things
NR-DC	New Radio Dual Connectivity
NSA	non-standalone
NR	New Radio
NR-U	New Radio Unlicensed
NZP	Non zero power
NW	Network
OFDM	Orthogonal Frequency Division Multiplexing
OFDMA	Orthogonal Frequency Division Multiplexing Access
PA	Power Amplifier
PAPR	peak-to-average power ratio
PBCH	Physical Broadcast Channel
PCID	Physical Cell Identity
PDCCH	Physical downlink control channel
PDCP	Packet Data Convergence Protocol
PDSCH	Physical downlink shared channel
PDU	Protocol Data Unit
PHY	Physical Layer
PLL	Phase-Locked Loop
PLMN	public land mobile network
PMI	Precoding-Matrix Indicator
PN	Phase Noise
PRACH	Physical Random-Access Channel
PRB	Physical Resource Block
P-RNTI	Paging RNTI
LSB	Least Significant Bit
PSK	Phase Shift Keying
PSS	Primary Synchronisation signal
PUCCH	Physical uplink control channel
PUSCH	Physical uplink shared channel
PBCH	Physical Broadcast Channel
PCID	Physical Cell Identifier
PDCP	Packet Data Convergence Protocol
PHY	Physical Layer
PMI	Precoding Matrix Indicator
PRACH	physical random access channel

PRB	Physical resource block
PRG	Precoding resource block group
PT-RS	Phase-tracking reference signal
QAM	Quadrature Amplitude Modulation
QCI	QoS Class Identifier
QCL	Quasi co-location
QoS	quality of service
QPSK	Quadrature Phase Shift Keying
RACH	random access channel
RAN	Radio Access Network
RA-RNTI	Random Access RNTI
RAT	Radio Access Technology
RB	Resource block
RE	Resource Element
RBG	Resource block group
RDI	DRB mapping Indication
RF	Radio Frequency
RI	Rank Indicator
RIT	Radio Interface Technology
RLC	Radio Link Control
RMSI	Remaining Minimum System Information
RNTI	Radio-Network Temporary Identifier
RIV	Resource indicator value
RS	Reference signal
RRC	radio resource control
RRM	Radio Resource Management
SA	standalone
SIB	System Information Block
SLNR	Signal-to-leakage-and-noise ratio
SNR	Signal to Noise Ratio
SSB	Synchronization Signal Block
SSS	Secondary Synchronization Signal
SFN	Single Frequency network
SLIV	Start and length indicator value
SI	System Information
SiGe	Silicon Germanium
SR	Scheduling Request
SRIT	Set of Radio Interface Technologies
SRS	Sounding reference signal
SS	Synchronisation signal
SS/PBCH	Synchronisation signal Physical Broadcast Channel
SSS	Secondary Synchronisation signal
SS-RSRP	SS reference signal received power
SS-RSRQ	SS reference signal received quality
SS-SINR	SS signal-to-noise and interference ratio

SRB	Signaling radio bearers
SU-MIMO	Single-User MIMO
SDL	Supplemental downlink
SVD	Singular Value Decomposition
SUL	Supplemental uplink
TB	Transport Block
TR	Technical Report
TCI	Transmission Configuration Indicator
TCP	Transmission Control Protocol
TDD	Time Division Duplex
TDM	Time division multiplexing
TD-RA	Time domain resource assignment
TPC	Transmit Power Control
TPMI	Transmit Precoding Indicator
TTI	Transmit Time Interval
TX	Transmitter
UCI	uplink control information
UE	User equipment
UL	Uplink
UMTS	Universal Mobile Telecommunications System
UP-CP	User Plane – Control Plane
URLLC	Ultra Reliable and Low Latency Communications
UTC	Universal Time
VRB	Virtual Resource Block
V2X	Vehicular-to-everything
WG	Working group

1

Introduction and Background

Mihai Enescu and Karri Ranta-aho

Nokia Bell Labs, Finland

1.1 Why 5G?

It is perhaps a reasonable question to ask why we need 5G? Was in fact 4G/LTE and its evolution sufficient? Looking at LTE we note that indeed, we identify a good set of advanced technical components (speaking from the perspective of physical layer): it has multiple antennas scaling to the possibility to use massive MIMO (Multiple Input Multiple Output), it supports emerging connectivity techniques such as device-to-device, NB-IoT, Ultra Reliable and Low Latency Communications (URLLC). So why 5G? Every decade seems to bring a new wave of technology, this was the case with the arrival of 3G and 4G. From this basic perspective, the arrival of 5G strengthens this "rule." Is there a need for 5G, especially given the fact that LTE and LTE-pro are capable of delivering high data rates and flexible technology? Yes, there is a need for 5G and there are a couple of reasons which we will elaborate next.

Like any evolving system, LTE has reached the limit of being a complicated design, its limitations coming from the fact that the basic design conceived a decade ago cannot be updated further in a simple way and cannot be kept easily compatible with the earlier equipment. The evolution of 3GPP generations has maintained the rule that all new devices are able to function in the older networks and vice versa without loss of functionality. As some of the components introduced at the very beginning of 4G need complicated solutions in order to be optimized, new functionality must be introduced so that compatibility with the old functionality is maintained. To give a simple example, the very first LTE release was designed based on Common Reference Signals (CRSs), up to four ports, which are transmitted all the time, hence they are always ON. However, the system cannot turn these reference signals OFF, nor can scale them when the number of antennas is increasing at the transmitter without causing complications to existing device base. While CRS were a good design at its time, it was also unscalable in the future, by scalability meaning, for example, increasing the number of ports as the number of transmit antennas increases. Perhaps one of the most important lessons learned from LTE is indeed the need for a more flexible and scalable system toward the future. This is an intrinsic characteristic of 5G NR design and we are covering perhaps the most important new development over LTE, the multi-antenna perspective, in this book.

Another reason for 5G is the need for harvesting the spectrum where cellular technology has not been used so far. Spectrum is a costly resource, not easy to handle in a global environment, especially keeping in mind the cost of both devices and network equipment. 4G/LTE has been deployed in the low bands and with these occupied, there is an obvious need and choice to evolve to the areas of unoccupied spectrum to obtain more capacity for the system in general. Most of the first deployments rely on mid band of 3.5/4.5 GHz as well as 28/39 GHz, bands that are mostly unoccupied by LTE in the lower end, and lacking ability for LTE to be even deployed on in on the higher end. While moving to higher frequency bands not earlier used for cellular radio transmissions is not an easy step, it comes with the good news that the antenna form factor gets smaller as the carrier frequency gets higher. However, a lot of other challenges are arising from the fact that the pathloss on higher frequencies is also higher, and the power amplifier technologies need to evolve to support sufficient transmit powers with reasonable energy efficiency in a consumer device price point. As a first step, the 5G NR is designed to operate on bands starting from under 1 GHz and reaching up to 52 GHz, with more design focus to come in the area of 52 GHz and beyond.

The need for this book came from the fact that it seems necessary to link the dots of the existing 5G specifications into a cursive description of what the system is about and why it has been created the way it is. This book has been written by standardization experts who have been and currently are participating actively in 3GPP discussions. Most of them have also been involved in designing the 4G system with the area of expertise reaching back to 2G and 3G. We hope that the description herein will help the reader into getting a better understanding of how the 5G physical layer and related higher layer components are functioning and especially why they were designed to function that way.

While the existing 3GPP standards describe what has been agreed in the meetings, in this book we want to take this one step further and also present the context of some of the discussions and various alternatives which have been discussed but, in the end, not adopted as a standardized solutions.

1.2 Requirements and Targets

The 5G specification work was preceded by a study whose objective [1] was to identify and develop technology components needed for new radio (NR) systems being able to use any spectrum band ranging at least up to 100 GHz. The goal was to achieve a single technical framework addressing all usage scenarios, requirements, and deployment scenarios defined in TR38.913 [2]. This TR was defined to ensure that the ITU-R IMT-2020 requirements [3] will be covered, and additional industry needs are also taken into account in the technical work. Naturally the requirements were not developed in a vacuum but based on the practical understanding of what should also be achievable in real life without the need to violate the laws of physics. The new radio access technology (RAT) was to be designed for inherent forward compatibility to allow specification in multiple phases and cater for the needs of the future that today we can't imagine. This design requirement was also learning from LTE, which had a somewhat rigid setup of always-own full-band CRS and fairly fixed carrier bandwidth structure even if the original LTE design has regardless of these things proven itself to be very capable of integrating new features over a number of subsequent

3GPP releases. The New Radio study item contained technical features needed to meet these objectives set in the said TR38.913 [2]. In addition to very demanding performance requirements the additional functional requirements include:

- Tight interworking between the new RAT and LTE
- Interworking with non-3GPP systems
- Operation in licensed bands (paired and unpaired), and licensed assisted operations in unlicensed bands.
- Stand-alone operation in unlicensed bands.

One of the focus areas for the initial work related to defining a fundamental physical layer signal structure for new RAT, the key topics in this area including waveform, frame structure(s), and channel coding scheme(s).

1.2.1 System Requirements

In order to understand how wireless technologies are being created and especially labeled, it is necessary to discuss briefly about IMT-2020. There is indeed a basic question: who is certifying that a particular technology can be called 5G and based on what criteria is this done?

The International Telecommunications Union (ITU), based in Geneva, Switzerland, is a United Nations agency responsible for issues related to the communications technology. The areas of ITU cover Radio communications (ITU-R), Standardization (ITU-T), Development (ITU-D). In order to have a unified certification system on a global scale, technologies must be certified by ITU-R. This is based on ITU-R-crafted requirements, for a technology being labeled as 5G, this needs to pass these ITU requirements. The first NR release is already done, however, in order for this to be called a 5G technology, this design must pass the ITU-R requirements, this evaluation being done during the IMT-2020 discussions. According to the IMT-2020 vision, a 5G technology needs to reach requirements presented in Figure 1.1 [4].

The area of 5G requirements is crafted around the three main development cases: Extreme Mobile Broadband (eMBB), URLLC, and Massive Machine Type Communications (mMTC). eMBB capabilities relate to the maximum achievable data rate in terms of peak data rate (20 Gbps) but also in terms of coverage (1 Gbps), a high speed environment where quality of service (QoS) needs to be met for 500 kmph, an energy efficient operation at least similar to 4G and the possibility for delivering services in dense areas where the total traffic across the coverage area is of 1000 Mbps m^{-2}. URLLC has its own requirements of a very tight latency (round trip time) of 1 ms while the eMBB target of mobility needs also to be met. Finally, mMTC targets a total number of devices per unit area of 10 000 devices per km^2.

This vision was worked further, and to some extent in parallel with the 3GPP requirement and evaluation environment definition, and a more thorough requirement document with a very detailed evaluation methodology document were later produced by ITU.R. [5, 6] (summarized in Table 1.1). In some cases, the requirement name alone is insufficient to understand what is required, as what matters is how the detailed definition and the test setup have been defined. E.g. the user experienced data rate refers to a cell edge user defined as a 5th

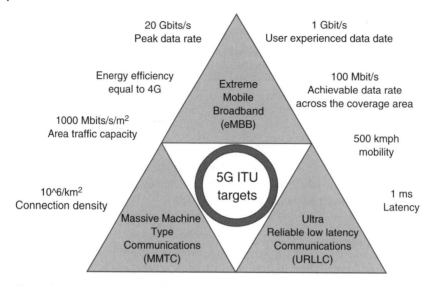

Figure 1.1 ITU-R vision for 5G [4].

percentile point in the cumulative distribution function in a setup where there are 10 users in each cell of the system all of them having data to be scheduled. This leads to 100% loading in all cells all the time generating both worst case own cell loading and worst possible other cell interference. The simulation scenarios for evaluation are defined in fine detail in [6].

The 3GPP RAN meeting #70 in December 2015 opened a study on scenarios and requirements for Next Generation Access Technologies [7]. The draft Technical Report of the System Information (SI) [2] released in March 2016 consisted of a large number of requirements that were used as a basis for the physical layer design work. The RAN#71 (in March 2016) also approved a new study on New Radio (NR) Access Technology that tasked the Radio Access Network (RAN) WGs "...*to develop an NR access technology to meet a broad range of use cases including enhanced mobile broadband, massive MTC, critical MTC, and additional requirements defined during the RAN requirements study*" [8].

The Requirements and Scenarios Technical Report [2] listed a large number of requirements to which the physical layer of the 5G New Radio is to provide solutions, which are summarized in the list below:

- A very *diverse set of deployments* ranging from Indoor Hotspot to Extreme Rural coverage
- A *wide range of spectrum bands* up to 100 GHz *and bandwidths* up to 1 GHz
- Wide range of *device speeds*, up to 500 kmph
- *Ultra-deep indoor coverage* with tentative target of 164 dB MCL
- *D2D/V2V links*
- Target *peak rate of 20 Gbps in downlink and 10 Gbps in uplink*
- Significantly *improved system capacity, user data rates and spectral efficiency* over LTE
- Target *C-plane latency* of 10 ms
- Target *U-plane latency* of 4 ms for mobile broadband, and 0.5 ms for *ultra low latency* communication
- Target *mobility*-incurred *connection interruption of 0 ms*

Table 1.1 ITU-R requirements for IMT-2020 [5].

Requirement name	Minimum requirement
Peak data rate	DL: 20 Gbps, UL: 10 Gbps
Peak spectral efficiency	DL: 30 bps Hz^{-1}, UL: 15 bps Hz^{-1}
User experienced data rate	DL: 100 Mbps, UL: 50 Mbps
5th percentile user spectral efficiency	Indoor hotspot – DL: 0.3 bps Hz^{-1}, UL 0.21 bps Hz^{-1} Dense urban – DL: 0.225 bps Hz^{-1}, UL 0.15 bps Hz^{-1} Rural – DL: 0.12 bps Hz^{-1}, UL 0.045 bps Hz^{-1}
Average spectral efficiency	Indoor hotspot – DL: 9 bps Hz^{-1}, UL 6.75 bps Hz^{-1} Dense urban – DL: 7.8 bps Hz^{-1}, UL 5.4 bps Hz^{-1} Rural – DL: 3.3 bps Hz^{-1}, UL 1.6 bps Hz^{-1}
Area traffic capacity	10 Mbps m^{-2}
User plane latency	4 ms for eMBB, 1 ms for URLLC
Control plane latency	20 ms requirement, 10 ms encouraged
Connection density	1 000 000 devices km^{-2}
Energy efficiency	Network energy efficiency is the capability of a RIT/SRIT to minimize the radio access network energy consumption in relation to the traffic capacity provided. Device energy efficiency is the capability of the RIT/SRIT to minimize the power consumed by the device modem in relation to the traffic characteristics.
Reliability	The minimum requirement for the reliability is 99.999% success probability of transmitting a layer 2 PDU (protocol data unit) of 32 bytes within 1 ms in channel quality of coverage edge for the Urban Macro-URLLC test environment
Mobility link spectral efficiency	Indoor hotspot 10 kmph–1.5 bps Hz^{-1} Dense urban 30 kmph–1.12 bps Hz^{-1} Rural 120 kmph–0.8 bps Hz^{-1} Rural 500 kmph–0.45 bps Hz^{-1}
Mobility interruption time	0 ms
Bandwidth	At least 100 MHz and up to 1 GHz in higher frequency bands

- Target *reliability* of delivering a packet in 1 ms with 1-10^{-5} reliability
- Tentative target *User Equipment (UE) battery life of 15 years* for *massive MTC* type terminals
- Improved *UE energy efficiency* while providing much better MBB data rate
- Improved *network energy efficiency*
- Target connection *density of 1 million devices km^{-2}*
- Tight *interworking with LTE*
- Connectivity through *multiple transmission points*
- Operator-controlled *sidelink (device-to-device) operation*

Looking at the above list of diverse requirements and deployment environments envisioned for New Radio system, they cannot be met with a single monolithic physical layer design – but rather by a physical layer design that is able to adapt to different deployments and use cases, all these in a common operation framework.

At least the following implications to the physical layer design were derived from the requirement list presented in Section 2.1:

- Both frequency division duplex (FDD) and time division duplex (TDD) need to be supported to be able to deploy the New Radio on all bands from sub-1 GHz up to 100 GHz.
- The network topologies will depend on the operating carrier frequency, a possible split being roughly as listed in the following points.
- The ultra-deep indoor coverage operation of mMTC would be applicable for low band frequencies, e.g. below 1 GHz, while high band frequencies could potentially support a different category of mMTC applications such as household appliances, this latter category not having the same coverage requirements as the low band mMTC.
- There may not be a need to optimize FDD for high bands (e.g. above 3 GHz) due to the fact that most known band plans above 3 GHz are either TDD or unidirectional.
- A single numerology for all cases as in LTE is not sufficient to support very wide bandwidth and relatively small coverage cells closer to 100 GHz band range, and large coverage cells around traditional cellular bands and bandwidths.
- The same multicarrier OFDMA waveform suitable e.g. for eMBB use cases up to around 40 GHz may not be optimal for, e.g. narrow-band mMTC type links and eMBB links above 40 GHz. A single-carrier waveform should be considered for these cases.
- A single fixed slot duration is not sufficient for meeting both ultra-reliability requirements as well as extreme rural and deep indoor coverage.
- Operation on unlicensed frequencies calls for Listen Before Talk or similar interference management technique.
- Wireless relay may have special implications to control channel design, beam detection, reference signals, and interference management.
- D2D/V2V sets specific requirements to accommodate for the time misalignment and has special implications to control channel and reference signal design.
- Network energy efficiency requirement needs to be addressed with minimizing the network transmission in no-data TTIs (e.g. synchronization signals, CRS, Physical Broadcast Channel [PBCH]).
- Consider enabling UEs that do not support the full system BW in order natively to enable low-cost IoT devices.
- Develop advanced Massive MIMO framework that scales from low band digital beamforming/MIMO to high-band analogue/hybrid beamforming with very large number of antenna elements.

Some of the above implications can be summarized in Table 1.2 as follows:

Initially the need for supporting all features across the full frequency range was debated, partially because the design characteristics needed for the support of high order antenna arrays was not yet known. Once the base design started to form, the desire to be able to deploy all features on all frequency ranges was becoming more evident, and the solutions being designed were more and more obviously agnostic to frequency bands. The one exception to the norm was some specific design details supporting analogue component of the beamformers that is essentially just applicable for high frequency bands.

Table 1.2 Summary of 5G NR technology components.

	Below 2 GHz	2–6 GHz	3–40 GHz	20–100 GHz
Network topology	Wide area deployment	Urban area deployment	Small cell deployment	Small and ultra-dense small cells
Technologies	eMBB, mMTC, URLLC		eMBB (URLLC)	eMBB
Waveform	OFDMA based, consider contention based and/or non-orthogonal access		OFDMA based	ZT-DFT-s-OFDM/ Null CP-single carrier
Frame structure	Unified and flexible across deployments			
Multi antenna	Digital BF		Digital/analogue/ hybrid BF	Analogue/ hybrid BF
	Unified feedback framework, base station processing transparent to the UE, CSI feedback agnostic to base station processing			
Mobility	Decreasing with the increase of frequency band operation and with decreasing cell size			

1.2.2 5G Spectrum

Spectrum availability is the corner stone for 5G NR development and deployment. The large amount of use cases as described in the previous section, implies a more efficient utilization of the existing spectrum as well as the identification of new spectrum for cellular use. LTE and other legacy cellular radio deployments are operating in relatively low frequency bands, these bands fitting well for eMBB services and mMTC deployments. This is because the radio propagation characteristics in low bands allow for a good compromise between coverage for a given base station as well as capacity through being able to use the same frequency resource with a neighboring base station without requiring a very large distance between the two base stations [29]. The established volume market has also led to the availability of simple, low cost radio implementations for those frequency bands. 5G deployments are likely to benefit from the refarming of this spectrum range, hence in time, at least some of the bands today known as 2G, 3G, or LTE bands will be migrated to 5G use. However, higher frequency bands not used by cellular and not accessible to the earlier generations due to technology limitations also become a deployment possibility with NR – if such new bands can be identified for cellular use.

3GPP does not have control over frequency allocations or bands, and it does not specify a particular frequency band unless the band is available or becoming available for cellular use in some part of the world, but the 3GPP system design in itself is band agnostic up to a point – there are some limitations to how high or how low the 3GPP-defined technologies can practically go, as well as what bandwidths the different generations support. ITU-R, regional and national regulators are identifying a new spectrum, while 3GPP is defining these bands in order to create the necessary radio frequency (RF) requirements which are band specific. In Tables 1.3 and 1.4 we present the 3GPP defined bands as of summer 2019. There are several types of identified spectrum: paired bands used for FDD, unpaired bands used for TDD, unpaired bands used for SUL (Supplemental uplink) or SDL (Supplemental

Table 1.3 Spectrum bands for low-to-mid frequency (FR1).

Operating band	Total spectrum (MHz)	Uplink (MHz)	Downlink (MHz)	Duplex mode
n1	2 × 60	1920–1980	2110–2170	FDD
n2	2 × 60	1850–1910	1930–1990	FDD
n3	2 × 75	1710–1785	1805–1880	FDD
n5	2 × 25	824–849	869–894	FDD
n7	2 × 70	2500–2570	2620–2690	FDD
n8	2 × 35	880–915	925–960	FDD
n12	2 × 17	699–716	729–746	FDD
n20	2 × 30	832–862	791–821	FDD
n25	2 × 65	1850–1915	1930–1995	FDD
n28	2 × 45	703–748	758–803	FDD
n34	15	2010–2025	2010–2025	TDD
n38	50	2570–2620	2570–2620	TDD
n39	40	1880–1920	1880–1920	TDD
n40	100	2300–2400	2300–2400	TDD
n41	194	2496–2690	2496–2690	TDD
n50	85	1432–1517	1432–1517	TDD
n51	5	1427–1432	1427–1432	TDD
n66	70 + 90	1710–1780	2110–2200	FDD
n70	15 + 25	1695–1710	1995–2200	FDD
n71	2 × 35	663–698	617–652	FDD
n74	2 × 43	1427–1470	1475–1518	FDD
n75	1 × 85	—	1432–1517	SDL
n76	1 × 5	—	1427–1432	SDL
n77	900	3300–4200	3300–4200	TDD
n78	500	3300–3800	3300–3800	TDD
n79	600	4400–5000	4400–5000	TDD
n80	1 × 75	1710–1785	—	SUL
n81	1 × 35	880–915	—	SUL
n82	1 × 30	832–862	—	SUL
n83	1 × 45	703–748	—	SUL
n84	1 × 60	1920–1980	—	SUL
n86	1 × 70	1710–1780	—	SUL

Table 1.4 Spectrum bands for high frequency (FR2).

Operating band	Total spectrum	Uplink (GHz)	Downlink (GHz)	Duplex mode
n257	3 GHz	26.5–29.5	26.5–29.5	TDD
n258	3.25 GHz	24.25–27.5	24.25–27.5	TDD
n260	3 GHz	37–40	37–40	TDD
n261	850 MHz	27.5–28.35	27.5–28.35	TDD

Downlink), although one may observe that as of today none of the SUL bands are actually uplink only bands due to frequency regulation as in all cases the SUL band is actually the uplink portion of an FDD band, and the SUL designation is driven by a specific deployment idea within a band combination.

The 3GPP specifications are written in an agnostic way with respect to the carrier frequency, yet, the introduction of FR1 (frequency range 1) and FR2 (frequency range 2) terminology was needed to differentiate the cases where particular frequency range dependent operation is needed. There is no strong border line in terms of FR1 and FR2, but FR1 is mainly referring to the operation in bands below 6 GHz (this was extended to 7.125 GHz in 2019 for future compatibility) while FR2 is referring to the operation in bands in the range of 24.25–52.6 GHz. The background of these frequency ranges is related to the candidate frequency ranges for IMT use that the World Radio Conference (WRC) 2015 agreed to be studied for possible IMT identification. The WRC had a number of distinct frequency blocks that 3GPP lumped together. The later extension of the FR1 upper edge was to accommodate for a potential new unlicensed frequency block 5.925–7.125 GHz. For future compatibility 3GPP has already started to look at frequencies above 52.6 GHz as well as frequencies between FR1 and FR2. The >52.6 GHz frequency range will very likely get introduced as a new Frequency Range with some additional technology solutions to better optimize the radio system for these frequencies, while it is somewhat unclear if a new FR would be introduced between 7.125 and 24.25 GHz, or if FR1 and FR2 would be extended to close the gap. This will at least partially depend on what additional frequencies will be identified for cellular use in the future. The detailed operating bands and spectrum allocations for FR1 and FR2 are depicted in Figures 1.2 and 1.3 and also summarized in Tables 1.3 and 1.4.

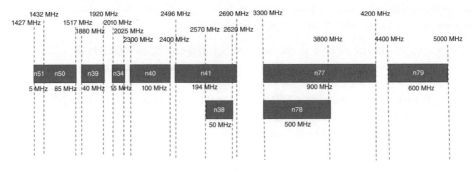

Figure 1.2 Spectrum bands for low-to-mid frequency (FR1).

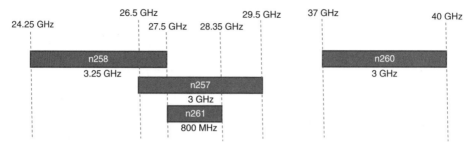

Figure 1.3 Spectrum bands for low-to-mid frequency (FR2).

1.3 Technology Components and Design Considerations

In its very beginning, the very first objective of the 5G was to identify and develop technology components needed for new radio systems being able to use any spectrum band ranging at least up to 100 GHz. The goal was to achieve a single technical framework addressing all usage scenarios, requirements, and deployment scenarios defined in TR38.913 [2], which was developed to also meet all the ITU-R requirements for IMT-2020 in ITU_R M.2410 [5]. One of the focus areas for the initial work related to defining fundamental physical layer signal structure for new RAT, the key topics including waveform, frame structure(s), and channel coding scheme(s). As one of the new physical layer building blocks for the new 5G, support for beam-based common control plane (CCP) in a massive MIMO framework was an important component. In LTE, dedicated control and data channels can be beamformed but support for common control plane beamforming was not provided, something which seems mandatory in the new system when moving to higher carrier frequencies.

The basic LTE Release 8 has been evolving with every release, adding in time new technologies and addressing new use cases. 5G NR has started with an ambitious plan of delivering a way more advanced basic package of techniques from its very first release (Release 15). Certainly, it is not feasible to have everything from the beginning, from both the perspective of not having enough time to create the specifications for the proposed techniques but also from the practical perspective of delivering products based on such techniques. Ideally, the evolution of the specification should go hand in hand with the market demand. Nevertheless, in the first stage of 5G discussions, a broad set of technology components and design guidelines have been on the design table, these being able to address the system requirements we have been discussed in a previous section.

- Large component *carrier bandwidth* is supported, ranging up to 400 MHz on FR2.
- *Multiple numerologies* are supported, since the system scales into very high carrier frequencies.
- While the *waveform is based on OFDM*, additional functionalities such as DFT-S-OFDM are supported. In a dedicated chapter of the book we are discussing also about the waveform variants which have been discussed and are considered in fact in 5G NR.
- A *flexible frame structure* is used to support the downlink and uplink of various techniques in FDD/TDD duplex arrangement. In addition, sidelink, and integrated access and backhaul (IAB) link are being considered as evolutionary steps on top of Release 15.

- All these are being supported for *standalone and non-standalone (NR anchored to an LTE carrier) operation.*
- *Enhanced massive MIMO* covering the analog/digital/hybrid beamforming is at the center of the physical layer design.
- New forms of access, such as *UL-grant free transmission* are supported.

The 5G NR has been designed in a forward compatible manner. Indeed, for example LTE suffers in various areas from the lack of flexibility, something the NR avoids already in its core design. The design principles building around forward compatibility have been considering:

- *Minimizing the transmission of always-on signals:* In LTE, the transmission of the always-on CRS is somehow problematic as even when a cell is empty, that particular cell transmits these signals and hence creates interference. The UEs' expectation for full bandwidth CRS also leads to rigidity of the utilized bandwidth. During the LTE evolution there have been solutions to alleviate this problem, such as turning off some of the CRS ports, in case the cell operates with four CRS ports, or developing techniques at the UE in order to mitigate the amount of interference created by the CRS. NR reference signals are designed mainly in a UE centric way while all of them have the possibility to be turned ON/OFF as needed by the system when there is no UE in the cell actually needing those signals.
- *All the signals and channels are confined within configurable time/frequency/spatial resources:* Such modular design, where we can imagine "tiles" of time/frequency/spatial resources containing different signals, possibly transmitted even with different numerologies, and multiplexed together in a seamless way. In legacy technologies it has been often difficult to handle signals which are transmitted in a wideband fashion and they need to be multiplexed with signals from other technologies which utilize a narrowband transmission. When such conflict happens, the only choice is to blank, in the overlapping area, one of the signals, typically the wideband one, and hence allow for a performance degradation of that channel. This is one situation NR avoids in its basic design.
- *Allowing for blanked resource which can be used flexibly in case of backward compatibility issues:* In every system it is important that the very first UEs designed for the system are able to understand such blanked resources which in future releases can be used by other techniques, not envisioned in the very first stage of the design. In other words, what is a blank resource for a Rel 15 UE may be a useful transmission for a later release UE.
- *User-specific bandwidth, waveform, and numerology:* Finally, in earlier cellular systems, the cell configuration is common to all UEs, and as a general principle, it does not vary in time. In NR, only the initial boot-strap needed for the UE to access the system is common. After that, each UE can operate with its own cell bandwidth (UE-specific bandwidth part [BWP]), uplink waveform can be selected between the CP-OFDM and DFT-S-OFDM, while in DL this is not possible not because the system flexibility wouldn't allow, but at the time of writing, there is only one DL waveform in the specifications. Also, the UE's subcarrier spacing and CP can be individually configured. Different combinations can exist at the same cell and one UE does not need to care of what some other UE is configured with. This flexibility by and large is not necessary and all UEs in the cell operate

with the same basic configuration for natural reasons, but it offers a natural growth path as whatever the current UEs are capable of doing does not restrict what the future UEs can do in the same system.

1.3.1 Waveform

Following the experience of LTE which is utilizing OFDM in downlink and DFT-S-OFDM in UL, the new radio system has considered the OFDM-based waveforms at least as a reference in its early design stage. Several design requirements were steering the waveform selection: the applicability of the technologies in a wider spectrum of carrier frequency and the range of deployment services identified in TR 38.913, including:

- Enhanced mobile broadband (eMBB)
- Massive machine-type-communications (mMTC)
- Ultra reliable and low latency communications (URLLC)

In such a mix of scenarios, it was clear that one needs to combine the choice of waveform with the choice of numerology in order to meet the requirements. Indeed, for short symbol durations one needs to utilize larger subcarrier spacings when targeting low latency services while for long symbol durations, smaller subcarrier spacings are a good choice for high delay spreads and multicast/broadcast services. As often it is the case, a compromise was needed in order to converge to a limited amount of choices which are feasible for implementation.

The targeted frequency bands are playing their role in the waveform selection. For FR1 and FR2 operation, waveforms of the OFDM family are a good choice in this frequency range, due to high flexibility in multiple access, good MIMO compatibility and low complex fast Fourier transform (FFT)-based implementations with scalar equalization. NR Release 15 is supporting CP-OFDM waveform for up to 52.6 GHz for both DL and UL transmission, while this is complemented by the utilization of DFT-S-OFDM based waveform in the uplink. CP-OFDM waveform can be used for a single-stream and multi-stream transmissions, while DFT-S-OFDM based waveform is limited to a single stream transmission (targeting for link budget limited cases). The multiplexing of different air interface configuration is mainly done in a frequency division multiple access (FDMA) manner which means that time or frequency domain filtering could be applied over parts of the carrier bandwidth in order to improve isolation between sub-band and to allow asynchronous communication or user specific numerology. A DL/UL symmetric utilization of CP-OFDMA is a good choice in providing the best capacity solution and localizing the waveform well in time, in order to achieve high capacity and hence allowing the utilization of higher order modulations. The efficient in-band multiplexing of different non-orthogonal waveforms/numerologies is defined by the in-band emission requirements which leads to good spectrum efficiency. Such implementation may be achieved by filtering, the requirements being specified by the standard while they are met in an implementation specific way by each manufacturer. Good spectral localization is driven by controlling efficiently the in-band (intra-carrier) adjacent channel leakage ratio (ACLR).

1.3.2 Multiple Access

The basic NR multiple access concept is TDM/FDM, where different users may be separated to transmit or receive by scheduling them to utilize a given frequency resource at different times. This concept of slotted access was used already in many 2G systems. In addition, the OFDM/DFT-S-OFDM system allows for scheduling different users to use a different frequency portion of the cell's frequency resource at the same time. This is familiar e.g. from LTE. Using a more formal terminology, the mechanism the UE uses to initially access the network is called *contention based* access, where different UEs' transmission attempts may collide with each other as the basic random access uses this access mechanism. Also, the uplink transmissions with configured grant, where the UE may start transmitting at any given time on pre-determined resources can be used in this way by configuring multiple UEs with the same resources to use when they have something to send. All other access mechanisms in NR are essentially *contention free*, the uplink and downlink transmissions are scheduled on resources orthogonal to those used for some other users.

Having the radio resources of the network being controlled by a dynamic scheduler it is something which legacy technologies are utilizing, this being the case of LTE. 5G NR has adopted also this strategy as it proves to be the most reliable way in meeting the users' QoS requirements, adapting to the radio conditions, etc. Indeed, the three families of applications considered in 5G, the eMBB, mMTC, and URLLC, are having diverse sets of requirements and these are best met based on network controlled scheduled resources.

1.3.3 Scalable/Multi Numerology

1.3.3.1 Motivation for Multiple Numerologies

The different 5G NR scenarios discussed so far in this chapter are having a direct connection to the 5G NR numerology design, as discussed in Table 1.5

In addition to different scenarios, TR 38.913 [9] describes the key performance indicators for the next generation access technologies. Many of them have a direct connection to the new radio numerology design. Table 1.6 summarizes the most important KPIs from the new radio numerology point of view.

Based on the 5G new radio scenarios and KPIs, there was a need for a common numerology framework to cover the full frequency range up to 100 GHz and support for all the envisioned environments/deployments and scenarios. One single numerology is clearly not sufficient to fulfill all the requirements, while a common framework significantly reduces not just the specification effort, but implementation effort as well, while maximizing the hardware commonality for different scenarios.

1.3.3.2 5G NR Numerologies

It is advantageous from an implementation point of view if the different options of scalable new radio numerology can be obtained utilizing the same common base clock. Thus, the need for scalability leads to all the 5G new radio sampling rates and corresponding subcarrier spacing being 2^N from the one base sampling rate.

Table 1.5 Scenarios impacting 5G new radio numerology.

Scenario	Requirements set for 5G new radio numerology
Support for different usage scenarios including • Enhanced mobile broadband • Massive machine-type-communications • Ultra reliable and low latency communications	• Support scalable numerology w.r.t. scenario.
Support for carrier frequencies up-to 100 GHz	• Support scalable numerology (e.g. the used bandwidth and subcarrier spacing, subframe length) w.r.t. carrier frequency: ■ Spectrum below 6 GHz is rather fragmented and composed of a mixture of bands for paired and unpaired operation. ■ As an outcome of WRC-15*, frequency bands between 24.25 and 86 GHz for IMT/5G, are studied for ITU-R WRC-19. ■ 28 GHz band (27.5–29.5 GHz), excluded from ITU studies, may have potential to become a de-facto 5G band in certain countries. ■ Other bands beyond WRC-19 may be identified in later WRCs or regional/national bodies • Co-existence and tight interworking between the new RAT and LTE require specific consideration especially in TDD scenario. Even though backward compatibility to LTE is not required, it is beneficial to utilize similar symbol timings especially on lower carrier frequencies, where LTE and new RAT may exist on adjacent carriers. One possible option for co-existence could be partial 5G re-use of LTE spectrum with MBMS subframes. Also in-band coexistence with LTE-M/NB-IoT for low frequencies (<2–3 GHz) should be supported. • Phase noise requires specific consideration in high carrier frequencies. It can be alleviated through larger subcarrier spacing and phase noise estimation and compensation.
Different deployment scenarios: • Indoor hotspot • Dense urban • Rural • Urban macro • High speed	• Support for scalable numerology w.r.t environment. • Smaller cell sizes imply lower delay spreads and hence allow shortening cyclic prefix (or other corresponding guard time) length. • Required CP duration may be assumed to be in the order of ~2 and ~4 μs, for the FDD and TDD WA cases respectively, with typical WA cell sizes. • Shorter TDD CP of ~2 μs is sufficient for dense urban deployments, given the typical values of excess delay spread in urban macrocells (e.g. 1.9 μs for the ITU Urban Macro cell channel model) with cell range limited below 1 km. • Even shorter CP of ~1 μs or smaller can be considered for small cells and in higher carrier frequencies.
Support for massive MIMO	• Support for RF beam switching within a subframe.

Table 1.6 Key performance indicators impacting the new radio numerology.

Key performance indicator	Definition [2]	Requirements set for 5G new radio numerology
Peak data rate Peak spectral efficiency	Obtained when all assignable radio resources for the corresponding link direction are utilized	Support for scalable numerology to minimize overhead and to maximize peak data rate/spectral efficiency in different scenarios.
Control plane latency	The time to move from a battery efficient state to start of continuous data transfer.	Support short subframe length(s)
User plane latency	The time it takes to successfully deliver an application layer packet from the radio protocol layer 2/3 SDU ingress point to egress point.	Support short subframe length(s)
Coverage	Definitions for mMTC type deep indoor coverage and long-distance communication coverage are still under discussion	Support long/concatenated subframe length(s)
Mobility	The maximum user speed at which a defined QoS can be achieved (in kmph). The target for mobility target should be 500 kmph.	Subcarrier spacing needs to be decided so that the influence of the Doppler effect can be managed with minimal performance degradation

Further, it i even more beneficial to reuse the same common base clock with LTE. Co-existence with LTE-TDD indicates that numerology evolvement from LTE and for example symbol level alignment between the new radio and LTE may be preferred especially with lower carrier frequencies. Thus, it is advantageous to be able to synthesize the new radio *sampling* rates utilizing the same common clock with LTE and to align the symbol length with LTE-TDD. Furthermore, to facilitate efficient implementation of multi-RAT base stations and multi-RAT fronthaul interfaces, using the same base clock is considered beneficial, hence the 5G new radio base clock rate is 2^N of the base sampling rate used in LTE.

TR38.913 requires at least dual-connectivity type of aggregation of LTE and 5G new radio data flows: "*Considering high performing inter-RAT mobility and aggregation of data flows via at least dual connectivity between LTE and new RAT. This shall be supported for both collocated and non-collocated site deployments*" [2]. The support of DC level aggregation of data flows in L2 benefits from synchronized sub-frame boundaries of the two RATs. In order to support common L2 that is seen as a good enabler for joint scheduling and efficient implementation, the 5G new radio subframe length is defined as integer division of 1 ms corresponding to LTE subframe. The key motivation for this was to allow for a phased introduction of NR, as the DC setup allowed for the radio-to-core network as well as the UE-to-core network protocols as they were, and the core network itself agnostic to the fact that an NR booster was added to the LTE architecture, and only in a later phase introduces

Table 1.7 5G new radio 5G PHY numerologies for OFDM based waveforms.

Spectrum band	<6 GHz			3...40 GHz	20...100 GHz	
Maximum carrier bandwidth (MHz)	5	20	100	200	400	1600
Waveform	OFDM/SC-FDMA (/UF-OFDM)			OFDM	OFDM	ZT-S-OFDM
Clock rate (Mchip s^{-1})	Up to 30.72	30.72	122.88	245.76	491.52	1966.08
Sub-carrier spacing (kHz)	3.75	15	60	120	240	960
Ts (μs)	266.67	66.67	16.67	8.335	4.17	1.04
Maximum (I)FFT size[a]	16–2048	2048	2048	2048	2048	2048
# symbols per subframe	FFS[b]	7	7	14	28	120
Sub-frame length (μs)	FFS[b]	500	125	125	125	125
CP (μs)	19.2	4.8	1.2	0.6	0.3	—
CP overhead (%)	6.7	6.7	6.7	6.7	6.7	—

a) Implementation issue, not to be standardized, however standard needs to support efficient implementation. (I)FFT size higher than 2 K FFT is not seen feasible from implementation point of view.
b) The length of the sub-frame is to be defined.

a full-blown 5G core network and the corresponding network and UE protocols that are needed to come with it.

Larger subcarrier spacing of NR offers increased robustness to the higher phase noise as well as an option for keeping the FFT size reasonable when carrier bandwidths are increased. The amount of available bandwidth increases as a function of increasing carrier frequency, leading also to larger carrier bandwidths. Carriers up to 100 MHz may be available at the conventional <6 GHz cellular spectrum area for eMBB communication. For lower frequencies, the 15 kHz subcarrier spacing is a feasible baseline for large microcell deployments needing the long CP. For 20–60 GHz frequency range, carrier bandwidths ~200 MHz or more can be expected, whereas for >60 GHz they are expected to be even larger, possibly over 1 GHz. In order to keep the system design and implementation complexity in feasible limits, fully flexibly adjustable numerology is not preferred due to high number of cases to define requirements for implementation and testing. Hence, a limited set of possible numerologies covering the different spectrum areas and scenarios is more practical.

For example, the supported 5G new radio PHY numerologies and waveforms is presented in Table 1.7. In addition to the parameters listed in the table, also Physical Resource Block (PRB) size was assumed to be scalable in a similar manner. It can be seen that the proposal was not too far from what was eventually specified for NR Release 15, while some significant differences can be observed as well. One reason for many of the original proposals having a lot of common ground was an extensive research collaboration under several public–private–partnership research frame works as well as some of the need for LTE compatibility related aspects, as explained earlier, being universally common.

1.3.4 Multi-antenna

Larger antenna arrays in terms of number of antenna elements are required at higher carrier frequencies to compensate increasing path loss between transmitter and receiver. As the carrier frequency increases, a single antenna becomes smaller and captures a smaller surface area of the radio propagation sphere, or conversely a fixed area (panel) can accommodate a larger number of physical antennas. On the other hand, the number of transceiver units cannot easily be increased mainly due to cost and energy consumption of the transceiver units (power hungry converters). Thus, from a technology point of view it can be noted that the NR designed to operate across carrier frequencies up to 100 GHz will need to support a variety of different transceiver and antenna system architectures at the BS side. These architectures can be categorized at a high level into fully digital, hybrid, and fully analogue beamforming systems. The architecture types with their characteristics are depicted at a high level in Figure 1.4.

Typically, fully digital transceiver architectures can be expected to be utilized in the lower end of the 5G spectrum where number of antennas is limited, and bandwidth is relatively low, hybrid architectures may be utilized in higher spectrum bands to keep the number of transceiver units in a cost and power efficient realm while providing support for a higher number of antennas. Fully analogue beamforming architectures are the extreme simplification of the hybrid setup with just a single transceiver unit and could be used to minimize the product cost in the very high end of the new radio spectrum. Furthermore, it is expected that a similar tendency in transceiver and antenna architecture can be considered for the UE side as well as in the base station. However, it is to be noted that architecture selection is vendor specific and thus no strict applicable carrier frequency range can be defined for different architectures. More detailed description of such techniques is described in further sections of this book.

Beamforming techniques are thus the centerpiece of the 5G NR design. When operating in lower carrier frequencies common control plane signaling (referred here as downlink synchronization signal(s), system information transmission in broadcast manner and paging, and random access channel [RACH] in uplink) may be based on wide sector beams at BS for transmission and reception. On the other hand, as discussed above, coverage enhancement will be particularly important at higher carrier frequencies, where the

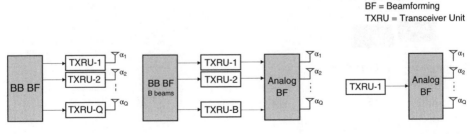

Figure 1.4 Examples of different transceiver and antenna system architectures: left: fully digital, middle: hybrid, right: fully analogue architecture.

deployments will tend to be mostly coverage limited due to the poor path loss conditions. Severe coverage limited situations pose difficulty that can be overcome with massive MIMO: a sector-beam type of approach for the common control signaling may not be feasible since there wouldn't be sufficient gain to achieve the needed cell radius. Therefore, to improve the cell radius a sweeping narrow beam approach is essentially a requirement for the common control. In a beam-based common control scheme the gNB transmits downlink and receives common control signals and channels via narrow beams in sweeping manner throughout the cell. In general, common channel beamforming can be used by the system when needed in all frequencies to improve coverage and cell detection.

The content of this book tries to focus on the utilization of beam-based operation in 5G NR. Naturally, the whole design, starting from initial access to transmitting and receiving data, must take this into account. The utilization of beams is generating quite a new system design approach, as one needs to encompass the challenges accounting for such operation, these ranging from achieving a reasonable latency when accessing the system, to an efficient power consumption for the UE, etc.

1.3.5 Interworking with LTE and Other Technologies

One core requirement was tight interworking with LTE. The way this was realized was a dual connectivity solution with the LTE carrier as the anchor cell to which the NR carrier is attached as a secondary cell, and the user plane data flows are connected at the Packet Data Convergence Protocol (PDCP) layer. The 3GPP supported architecture options allowing for either of the two radios to act as the master node, and the core network connection can be toward the 4G Evolved Packet Core (EPC) core as well as the 5G core network. The primary architecture that was designed to enable fast NR roll-out is leveraging the existing LTE deployments and 4G EPC core network by attaching the NR cell as a secondary cell to the existing LTE/EPC deployment. Figure 1.5 depicts the high-level view of this architecture option, while the full set of architecture options are discussed in more detail in Chapter 2.

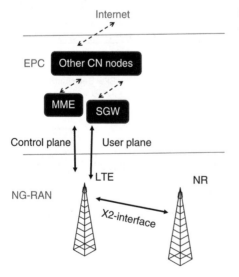

Figure 1.5 LTE-NR tight interworking architecture optimized for fast roll-out.

In this LTE-NR Dual Connectivity setup the UE is connected to both the LTE and the NR base stations and two radios operate independently on physical and Medium Access Control (MAC) layer and the data flow split is made in the PDCP layer. The user plane data flow can also be routed directly from the EPC to the NR base station rather than via the LTE base station.

An LTE-NR Dual Connectivity deployment can later be evolved to a stand-alone NR system with more traditional inter-system mobility between the NR and LTE, and the deployment can appear as a stand-alone NR system for the UEs supporting that and LTE-NR DC deployment for the UEs that do not support the stand-alone NR mode of operation. The NR Release 15 supports inter-technology handovers to (and from) LTE, while Release 16 is working to introduce NR-to-UMTS handovers (but not the other way around), and the mobility back to NR is based on cell reselection after the call has been released. NR – Wi-Fi interworking has no special support in the NR Release 15 specifications, and if the UE supports them both, the two technologies operate independently of each other with no traffic steering support, although it is likely that some interworking features will be specified in future releases of the 3GPP specifications.

1.3.6   5G Beam Based Technologies Across Release 15 and Release 16

The first release of 5G NR, Release 15, introduces basic system functionality, the main components being mentioned in the previous chapter. To state these again, the system is based on OFDM in downlink and on OFDM and SC-FDMA in uplink, operates with multiple numerologies across a vast amount of carrier frequencies spanning from 600 MHz up to 52 GHz, while MIMO functionality is possible in any deployment, in fact the MIMO functionality is a basic functionality on all bands, and in deployments above 3.5 GHz beamforming is assumed to be commonplace as everything happens through "beams." In the first release, the system design has focused more on the traditional single transmission point operation, that is the UE connects to a transmission point (cell) and the data exchange takes place in this way. Additionally, the dual connectivity to LTE and NR is also possible, and in future, UE connection to multiple 5G transmission points either in a dual connectivity setup, or as a single cell's multi-panel configuration similar to multi-point MIMO. 5G intends to serve a large amount of application areas, so called verticals, and hence technology components are added on top of the main functionality designed in Release 15. In the following we briefly describe the upcoming technologies.

1.3.6.1   Integrated Access and Backhaul

IAB has the scope of enabling an integrated solution for providing access and backhaul links over the same carrier frequency. Indeed, 5G NR promises highly increased DL and UL availability, however, such massive data communication may be too expensive to deploy using traditional wired or dedicated point-to-point wireless backhaul links. IAB is a practical answer to such questions and such technology is facilitated by the existence of massive MIMO as a basic building block of the system. An example of IAB deployment is presented in Figure 1.6 where the IAB nodes can multiplex the access and backhaul data traffic in both time, frequency and/or space.

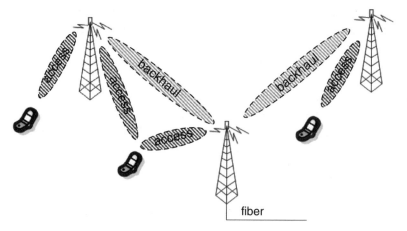

Figure 1.6 Illustration of integrated access and backhaul.

Initial access is a first area where IAB is expected to need particular enhancements as there is a need for orthogonal SSB to the SSB used for UEs. RACH design is also one area of development as there is expected to be a need for extending the RACH occasions and periodicities for backhaul RACH resources w.r.t the RACH resources. Resource multiplexing is a key area of design as we area dealing with both backhaul and access links.

1.3.6.2 NR Operation on Unlicensed Frequency Bands (NR-U)

Spectrum and its availability are critical enablers of communication technologies. We have previously mentioned in this chapter the spectrum utilization in 5G, however, while the licensed spectrum remains superior from the QoS point of view as it provides seamless coverage and highest reliability and spectral efficiency, it is of critical importance to provide 5G NR functionality in the area of unlicensed spectrum as well, this being possible by either with so called Licensed Assisted Access (LAA) carriers that are paired with anchor carrier on a licensed band, or with a fully stand-alone deployments in unlicensed spectrum. Thus, 3GPP targets a single global solution for access to unlicensed spectrum, the considered operation being in the bands of 5 and 6 GHz, but some band and regional differences in how the unlicensed operation is allowed requires some flexibility in the solutions to be specified. Co-existence with other technologies occupying the same unlicensed spectrum is of critical importance and governed by specific regulations, like for example in 5 GHz where there are already Wi-Fi deployments.

Channel access mechanisms are of critical importance in NR-U. Initial access is an area where the baseline specification of Release 15 needs to be extended as NR-U has its own channel access particularities and the number of candidate SS Blocks might need to be increased in order to lead to reliable discovery. Other particularities of the physical layer and procedures relate to the design of the frame structure and of the discovery reference signal, the possible extension of the physical random access channel (PRACH) formats to support the minimum bandwidth requirement given by the regulations. The DL and UL control channel design as well as the transmission of PDSCH and PUSCH are areas of investigation as well. The functions of radio link monitoring (RLM)/radio link failure (RLF) may need to be extended due to reduced transmission opportunities for DL signals.

1.3.6.3   Ultra-Reliable and Low Latency Communications

URLLC provides a key platform for various application areas from which we name a few: electrical power distribution, factory automation, transport industry where the focus is on the information exchange between a UE supporting V2X and a V2X application server, augmented reality, virtual reality (entertainment industry). The common denominator of these enumerated application areas are the need for high reliability for various degrees of latency bounds ranging from 1 ms air interface delay for AR/VR to 15 ms in the case of power distribution. One of the main goals of 5G NR was to provide native support for URLLC. In this respect, Release 15 contains several basic URLLC enablers such as slot structure facilitating low latency as well as other components enabling improved reliability. URLLC targets the operation across all frequency bands, including the beam-based operation in FR2. However, the main focus is on enhancing the control channel design, enhancing the uplink control information (UCI) operation, improving PUSCH for both grant based and configured grant (grant free) operation, improving the scheduling/hybrid automatic repeat request (HARQ) in the direction of low latency support.

1.3.6.4   Vehicular-to-everything (V2X)

3GPP/LTE has already a tradition in providing a platform to automotive industry as the support of V2V already exists in LTE since Release 14 2016, while further integration in the cellular environment has happened in V2X. The range of applications for V2X is quite vast, out of the identified 25 use cases we highlight the following:

Extended Sensors is perhaps one application area with a potential immediate impact for a large number of drivers. In this case the vehicle collects information based on installed sensors and exchanges this information based on V2X application servers. The use of extended sensors enable other application areas such as *advanced driving* which can be semi or fully automated and where the exchange for data between vehicles can lead to collision avoidance and improved traffic efficiency.

Vehicle Platooning implies a dynamically formed group of vehicles moving together. The lead vehicle stands as the main source of information for the rest of the vehicles from the group. Obviously, the latency of the communication between the vehicles is rather small, while the reliability needs to be also high.

The V2X communication links are a bit different from the communication of the UE with the gNB. This technology enables the communication of the devices directly between each other, this being called a sidelink. Such sidelink communication can be unicast (a connection between two UEs) or sidelink groupcast (a one-to-many sidelink message) or sidelink broadcast (a one-to-whoever can hear sidelink message). V2X operation is envisioned for both FR1 and FR2. While no particular design is attempted or FR2 with respect to the beam management part, it is expected that the basic Release 15 "beam based" functionality will be used.

Some of the physical layer expected specification areas include the synchronization mechanism, the resource allocation strategies, HARQ, channel state information (CSI) acquisition, and power control.

1.3.6.5 Positioning

Providing accurate positioning to the 5G NR UEs is of key importance in today's world. Accurate positioning is possible based on many technologies such as GNSS or radio based (LTE and Wi-Fi networks, etc.), however, 5G NR has the benefit of supporting a large array of antennas at the base station as well as indoor and outdoor deployments facilitating the UEs localization in a wide range of scenarios. Positioning techniques need to operate in all NR deployments, hence in higher carriers as well. Accurate positioning boils down to the transmission of reliable reference signals whose measurements are providing a very good input in determining the location of the device. While Release 15 5G design has defined a good set of reference signals, these have not been considering the support for positioning techniques, hence one needs to investigate if additional reference signals design is needed from the perspective of utilizing both the downlink and the uplink reference signals.

1.3.6.6 System Enhancements

1.3.6.6.1 2 Step RACH Low latency initial access is a high incentive for any radio system. While the current 5G design is based on a four steps RACH operation, it is believed that two step RACH can bring potential benefits for channel access in several application areas such as mMTC, URLLC, and eMBB and for both licensed and unlicensed operation.

1.3.6.6.2 Cross Link Interference Mitigation (CLI) Cross link interference (CLI) is an inherent problem of unsynchronized TDD operation and in some circumstances can require particular attention when handling communication beams. Indeed, in high carrier frequencies, advanced MIMO techniques can be utilized to mitigate cross-link interference.

1.3.6.6.3 UE Power Saving Efficient battery life is an undisputed essential component of any device. Release 15 NR consists of power saving mechanisms such as Discontinuous Reception (DRX) and BWP adaptation. Ultimate power saving mechanisms are targeting the adaptation/finetuning of UE operation on a wide range of characteristics such as the dynamic adaptation of the MIMO layers, the efficient scheduling procedure, the efficient utilization of the control channel.

1.3.6.6.4 Mobility Enhancements Release 15 NR design is based on a basic mobility mechanism where a source gNB triggers handover by sending HO request to target gNB and after receiving ACK from the target gNB, the source gNB initiates handover by sending HO command with target cell configuration. The UE sends PRACH to the target cell after radio resource control (RRC) reconfiguration is applied with target cell configuration. In high carriers of FR2, due to the need of beam sweeping, the HO can incur larger interruption time compared to lower carrier frequency deployments or to the LTE. The sensitivity to beam, due to beam blockage or UE rotation, is another reason for looking into enhancements in the mobility area. Identifying handover solutions to achieve high handover performance with 0 ms interruption, low latency, and high reliability is of great importance.

1.3.6.6.5 *MIMO Enhancements* MIMO is a basic system component of 5G NR, not only a capacity enabler. MIMO techniques are in a continuous evolution, either from the perspective of increasing system capacity where CSI feedback is a critical component, or from the perspective of increasing beam management reliability in higher carriers. The basic NR MIMO framework allows for an easy introduction of new MIMO enhancements without disrupting the MIMO operation with existing device base, both through network implementation innovation as well as additional improvements requiring also UE support.

1.3.6.6.6 *Support for Above 52 GHz Frequencies* While 5G NR is promises to harvest a wide range of spectrum, Release 15 5G design fits carrier frequencies up to 52.6 GHz. Frequencies above 52.6 GHz are very likely requiring a new DL waveform solution likely in the form of low peak-to-average power ratio (PAPR) single carrier operation, something which is already possible in UL. The propagation characteristics in these bands are more challenging due to high atmospheric absorption, lower power amplifier efficiency, and strong power spectral density regulatory requirements. However, the incentive is also high as potential for wider bandwidths, this fitting the need for a wide range of use cases such as V2X, IAB, NR licensed and unlicensed, and even non-terrestrial operations.

2

Network Architecture and NR Radio Protocols

Dawid Koziol[1] and Helka-Liina Määttänen[2]

[1]*Nokia Bell Labs, Poland*
[2]*Ericsson, Finland*

2.1 Architecture Overview

The first phase of new radio (NR) standard, which was finalized in December 2017, is called a non-standalone (NSA) NR as it is based on dual connectivity between an LTE radio access technology (RAT) and the NR RAT, known also as E-UTRA – NR Dual Connectivity (EN-DC). In EN-DC operation, the NR RAT connectivity may be added to the User Equipment (UE) after the UE has initiated a connection to LTE. The second phase of NR standard is called the standalone (SA) NR and it enables UEs to access directly the NR RAT and to have only NR connectivity, with or without dual connectivity with another Radio Access Network (RAN) node. Specification for the standalone NR was finished in June 2018. On top of those two options, additional architectural possibilities were also specified. They introduce further multi radio dual connectivity (MR-DC) alternatives between an LTE RAT and the NR RAT, but differ in detail from EN-DC option mentioned above (details are provided in subsequent section). A dual connectivity operation involving two nodes of NR RAT was also finalized in March 2019. The principle of dual connectivity means that the UE is connected to two nodes simultaneously and it is a concept already introduced in LTE Release 12. There are several options for the dual connectivity which are further discussed in the section. Figure 2.1 gives an overview for two principles where the master node (MN), which is initiating and controlling the dual connectivity for the UE, may be connected to LTE core network (CN) or 5G core network.

In Section 2.2 we give the overview on 5G core (5GC) network and functionalities by presenting the specified logical network nodes which are the standardized building blocks for the 5GC and by giving an example of a signaling flow between the nodes for connection establishment. In Section 2.3 we go through the different dual connectivity options and related terminology.

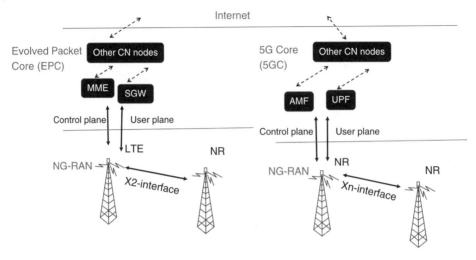

Figure 2.1 Overview of MR-DC connectivity principles with EPC and 5GC.

2.2   Core Network Architecture

2.2.1   Overview

The overall 5G Core Network (5GC) architecture is presented in Figure 2.2 where different blocks represent logical nodes of the network while the connecting lines represent interfaces between them.

Each interface has an associated protocol which is used to exchange data between the involved nodes in a way understandable for both sides. The figure mentions three such protocols – Non-Access Stratum (NAS) protocol, Next Generation Application Part (NGAP),

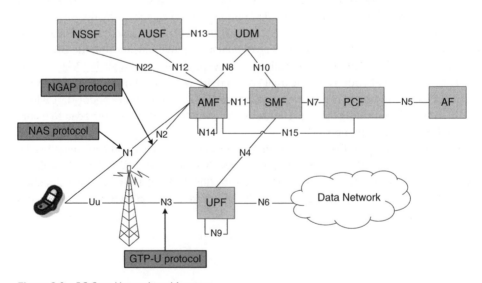

Figure 2.2   5G Core Network architecture.

and GPRS tunneling protocol user plane (GTP-U) protocol. The NAS protocol is responsible for exchanging core network messages between the CN and the UE. The NGAP protocol is responsible for exchanging control messages between RAN and Access Management Function (AMF) via the N2 interface, for example, during the Service Request procedure described later in this section. While the NGAP is for control plane message exchange, the GTP-U protocol is used for user data delivery between RAN and User Plane Function (UPF). The functions depicted in the figure are shortly explained below.

Functions not belonging to 5GC with which 5GC interacts:

- UE represents 5G system connectivity capability of the end user's device such as smartphone, wireless modem, smart meter, a sensor in the factory, or any other 5G-capable device.
- RAN which is capable of connecting to 5G Core Network (5GC, 5G CN). This includes NR and E-UTRA (LTE) networks.
- Data Network (DN), e.g. operator services, Internet access, or 3rd party services.

5GC functions:

- Authentication Server Function (AUSF) supports authentication for 3GPP access and untrusted non-3GPP access (e.g. to WLAN).
- Access and Mobility Management Function (AMF) is mainly responsible for management of UE's registration, connection, and mobility related procedures. It also supports functions such as lawful interception, participates in access authorization and authentication, provides transport of SMS messages, manages location services for regulatory services. It interacts with both UE (N1 interface) and RAN (N2 interface) to accomplish its tasks.
- Network Slice Selection Function (NSSF) provides slice management functionality.
- Policy Control Function (PCF) ensures policy and charging control for service data flows (SDFs) (e.g. provides authorized Quality of Service [QoS] for a SDF), protocol data unit (PDU) Session related policy control and event reporting.
- Session Management Function (SMF) main responsibility is establishment, modification, and release of UE data sessions. It is also responsible, e.g. for UE IP address allocation and management, charging data collection and for roaming functionality.
- Unified Data Management (UDM) performs functions related to security (generation of authentication credentials), subscription management (e.g. access authorization, storage, and management of privacy-protected subscription identifiers) as well as SMS management services.
- Application Function (AF) interacts with the 3GPP Core Network in order to provide additional network services, e.g. interacting with the policy framework for policy control.
- UPF is in general responsible for delivery of user data between data network and the UE (via RAN). As such, it hosts functions similar to those of a traditional IP router such as packet routing and forwarding, packet inspection, QoS enforcement, traffic usage reporting, etc.

We will not go into the details of each of the functions, but the interactions between different nodes will be presented by showing an exemplary signaling flow for a Service Request procedure. This procedure is triggered by the UE for requesting a connection establishment,

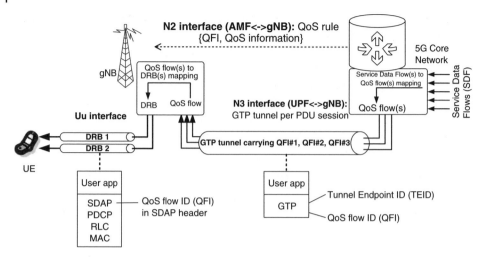

Figure 2.3 Overall QoS architecture in 5G system.

e.g. due to data arriving at the UE buffer, in order to be able to deliver the data to the recipient using Internet access via Next Generation Radio Access Network (NG-RAN) and 5GC. However, before proceeding with this, it is worth having a look at Figure 2.3 which explains concepts such as PDU session, data flow, and radio bearers, and shows interdependencies between them. Further details on these and other functions related to user plane data transmission are discussed in Section 2.4, especially the part related to user plane protocols.

In order to run a service or an application in a device that is connected to 5G network, data needs to be exchanged between the device and the other end point of the service/application. For this exchange of data, a logical connection called PDU session needs to be established. The data flow within this logical connection is called a service data flow (SDF). It is usually a flow of IP packets related to the service/application running in the device.

As there can be several applications running on a device, several SDFs may be routed between 3GPP CN and Data Network for a UE. These SDF(s) are mapped in the Core Network to QoS flows based on their QoS requirements and the rules defined in 5GC. Packets which are supposed to receive the same QoS treatment are marked with the same QoS flow identifier (QFI). Subsequently, QoS flows are mapped to a PDU session, which has to be first established between the UE and CN. In case there is an existing PDU session which can carry a certain QoS flow, it may be reused. Each PDU session of the UE is carried using a single transport connection, i.e. GTP tunnel between UPF and RAN. In the tunnel, each packet can be differentiated in the transport network based on the QFI marking, thus in this network segment, the achieved QoS granularity is a QoS flow. In RAN, QoS flows are extracted from the PDU session and based on their QoS requirements, SDAP layer maps them to DRBs, which are established between the gNB and the UE on Uu interface. Hence, on the radio interface, the granularity of QoS differentiation is DRB, which can carry one or more QoS flows. Similar rules apply to traffic delivery in uplink (UL) direction. With that knowledge, it may be easier to understand the procedures described in the subsequent paragraphs.

2.2.2 Service Request Procedure

The Service Request Procedure assumes the UE has an established PDU sessions, that is, the logical connection for a service or an application. The PDU session is established by a PDU session establishment procedure which is not shown here. After a PDU session is established, the user plane connection within the session may be activated or inactivated depending on the ongoing data flow's activity.

The Service Request procedure is presented in Figure 2.4. The Service Request procedure is initiated at the UE when uplink data arrives at the UE's buffer or when the UE is paged

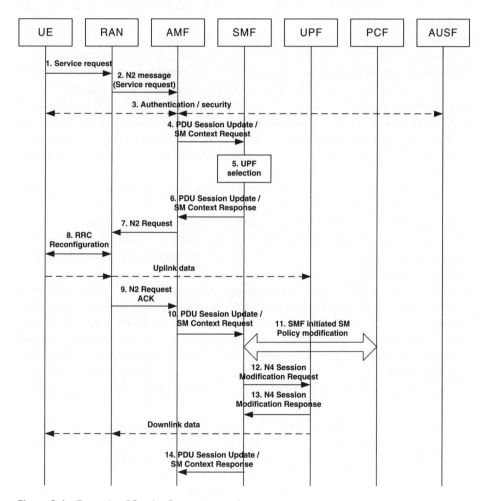

Figure 2.4 Example of Service Request procedure.

by the network. It allows the UE to establish a connection with the network, so that data can be exchanged. The procedure comprises the following steps:

Steps 1 and 2: UE sends a Service Request message to the RAN node it is connected to, which forwards it to the AMF function. It contains information such as a list of PDU sessions whose user plane connections are to be activated, security parameters, UE's identity in the 5G system (5G-S-TMSI), UE's selected PLMN and establishment cause (e.g. emergency call, response to paging, UE originated data connection).

Step 3: Based on the information included in the Service Request, the AMF sends a message to RAN containing information about UE's security context and potential mobility related parameters informing RAN about certain mobility restrictions the UE is subject to or including a list of recommended cells, Tracking Areas, etc. The AMF may interact with AUSF function in this step for example to authenticate the UE.

Step 4: The AMF identifies the PDU sessions for which user plane connections should be activated and contacts the SMF which maintains UE data sessions with a request to activate the user plane connections.

Step 5: The SMF chooses an UPF which should serve the request sent by the UE. The choice is based on UE's location in the network and additional UPF Selection Criteria used in the network. In this example, it is assumed that UPF function serving as PDU Session Anchor is directly connected to RAN, i.e. so called Intermediate UPF is not used.

Step 6: The SMF replies to the AMF that it accepted the activation of the PDU sessions indicated previously by the AMF. The reply indicates tunnel identifier which should be used by RAN to forward the data from the UE to the UPF. Additionally, it may also contain the information which the AMF should forward to RAN (e.g. QFI, QoS profile, slice related information, etc.).

Step 7: The AMF sends N2 Request message to RAN including information received from the SMF relating the user plane parameters of the session (QoS info, CN Tunnel ID) as well as additional control information such as UE Radio Capability Information, Radio Resource Control (RRC) Inactive Assistance Information, security context, mobility restrictions applicable to the UE, maximum UE's bit rate based on its subscription, tracing requirements, etc.

Step 8: RAN performs RRC connection reconfiguration (more about the procedure can be found in Section 2.4.6) leveraging the information received from AMF as well as additional information it might have received from the UE. How the UE is reconfigured by RRC, depends on the UE's capabilities as well as on the QoS information for all the QoS Flows of the PDU Sessions whose user plane connections are activated and current configuration of DRBs. User Plane security is also established on radio interface in this step. From this moment data may be sent in uplink from the UE to RAN and from RAN to UPF.

Step 9: RAN replies to the request received from AMF with N2 Request Ack message containing information such as Access Network (AN) Tunnel Info, list of accepted and/or rejected QoS Flows for the PDU Sessions whose user plane connections are activated.

Step 10: AMF contacts SMF again and provides information about the Access Type and RAT Type which it derived based on the Global RAN Node ID associated with N2 interface. Additionally, AMF may include, e.g. information about UE's time zone.

Step 11: In this step SMF may initiate the change of Session Management Policy, e.g. based on the fact that UE's location/time zone has changed, and PCF may provide updated policies to SMF.

Step 12: SMF provides UPF with at least the information about AN Tunnel Info allowing UPF to start transmitting downlink data to RAN.

Step 13: UPF confirms requested session modification from Step 12 and starts transmitting described downlink (DL) data to RAN which forwards it to the UE.

Step 14: SMF sends PDU Session Update/SM Context Response message to confirm the request received in Step 10.

It has to be noted that the described procedure is just an example and the outcome of the procedure can be different, e.g. the requested session may be rejected by RAN in case it is not possible to meet the requested QoS requirements. Depending on the network deployment, additional nodes may be involved in it as well, e.g. Intermediate UPF(s). However, the example presents well how different nodes are involved even in the basic procedures such as connection establishment and how much network signaling needs to exchanged for a single call or data connection to be established.

2.3 Radio Access Network

2.3.1 NR Standalone RAN Architecture

The baseline architecture in the NG-RAN, which is shown in Figure 2.5, follows the architecture known form E-UTRAN, i.e. LTE system.

Base stations in 5G networks are called gNBs to distinguish them from eNBs utilized in LTE. Similarly to eNBs in LTE, gNBs are network access points for user devices. They

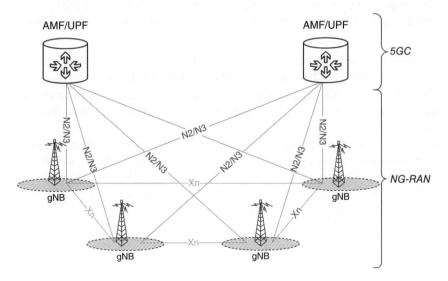

Figure 2.5 NG-RAN architecture for NR standalone deployment.

interact with the UE on one side and with the 5G Core Network (5GC) on the other side and their main functions include:

- Radio Resource Management (RRM) which includes sub-functions such as radio bearer control, radio admission control, connection mobility control, allocation of resources to UEs in uplink and downlink directions (i.e. scheduling), measurements, and measurement reporting configuration
- Transmission of downlink data toward the UEs and reception of uplink data from the UEs
- Security functions such as encryption and integrity protection of transferred data
- Routing and ensuring QoS for user data and control information sent toward the 5GC
- Connection management including connection setup, release, and mobility
- Broadcasting of system information
- Locating UEs in the radio network by sending paging messages
- QoS Flow management and mapping to DRBs for downlink data

gNBs may be inter-connected with each other using Xn interface, which serves similar purpose as X2 interface in LTE. Its two main functions include:

- Exchange of control information and data forwarding between neighboring gNBs to support UE mobility
- Procedures and data exchange between two nodes to support UE dual-connectivity

Additionally, Xn interface is utilized also for other purposes such as:

- Neighboring cell information exchange
- RAN paging for UEs in RRC Inactive state
- Indication of cell activation/deactivation to neighboring gNBs for the sake of energy saving in RAN
- Coordination of radio resources usage, e.g. to minimize mutual interference
- Functions supporting user data exchange between the gNBs such as flow control, fast retransmission and sharing assistance information, e.g. related to radio conditions of the UE.

gNBs are connected to the Core Network (CN) using two interfaces: N2 and N3, which are used to interact with AMF and UPF respectively. The purpose of these nodes was described in more detail in the preceding section. Figure 2.6, in turn, presents simplified functional split between the NG-RAN nodes and main nodes of the 5GC as described also in 3GPP TS 38.300 [10].

The architecture which involves only gNBs, which are connected to a 5G Core Network is a so-called standalone NR deployment. In the simplest words, it is an equivalent of LTE network where eNBs are connected to Evolved Packet Core (EPC). However, this is not the only possible network architecture, which was standardized in 3GPP Rel-15 and subsequent section provides additional insight into this aspect.

2.3.2 Additional Architectural Options

When browsing through 3GPP TS 38.300 [10], which is a technical specification providing overall description of NR and NG-RAN, the trained eye will spot that the figure presenting

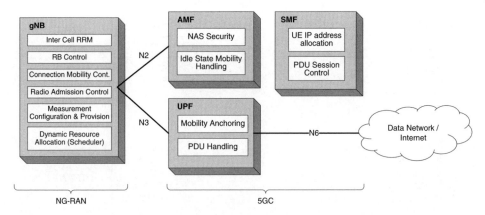

Figure 2.6 Functional split between NG-RAN and 5GC nodes.

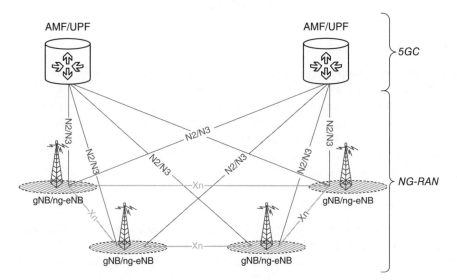

Figure 2.7 Full NG-RAN architecture.

NG-RAN architecture included there differs from the one presented in Figure 2.5 which, indeed, is incomplete. The missing pieces are RAN nodes called ng-eNBs, which are now shown in Figure 2.7 presenting the more complete version of NG-RAN architecture.

The ng-eNB stands for next generation eNB and it represents the possibility of enhancing LTE eNBs to support N2, N3, and Xn interfaces to enable their connectivity to 5GC and to other NG-RAN nodes including other ng-eNBs and gNBs. In other words, ng-eNBs, are RAN nodes which on network side support interfaces associated with 5G network while reusing 4G/LTE air interface (i.e. E-UTRA) for connectivity with the UEs. This architectural option is also known as eLTE.

On top of the two options mentioned already, which are so called standalone options where the UE connects to a single network node at a time, 3GPP specified multiple NSA architectures, i.e. the ones which require the UE to be simultaneously connected to

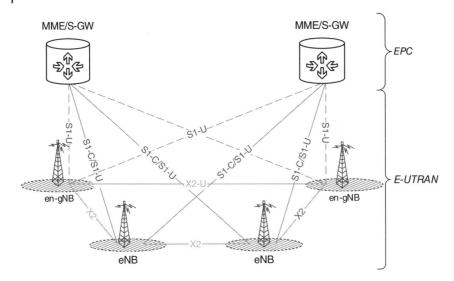

Figure 2.8 EN-DC architecture.

two RAN nodes, each supporting different RAT. Such operation is commonly named as Multi-RAT Dual Connectivity (MR-DC). In MR-DC, one node is always acting as a Master Node (MN) and the other one as a Secondary Node (SN). MN is the node responsible for controlling the connection on radio side with the UE and with the Core Network on network interface side and hence it is responsible for functions such as, e.g. connection establishment and maintenance, passing control messages between the UE and the Core Network, etc. User plane data on the other hand, may be provided to and from the UE via both MN and SN, which allows achieving higher throughputs than in case of a single-node operation. The following alternatives of MR-DC are supported:

- EN-DC where eNB is acting as a Master Node and gNB is acting as a Secondary Node and both network nodes as well as the UE are connected to EPC (Figure 2.8). The gNB in this case is denoted as en-gNB to underline the fact that it is not operating in the standalone mode.
- NG-RAN E-UTRA – NR Dual Connectivity (NGEN-DC) where eNB is acting as a Master Node and gNB is acting as a Secondary Node and both network nodes as well as the UE are connected to 5GC (Figure 2.9). The eNB in this case is denoted as ng-eNB to underline the fact that it is connected to 5GC and not to EPC.
- NR – E-UTRA Dual Connectivity (NE-DC) where gNB is acting as a Master Node and eNB is acting as a Secondary Node and both network nodes as well as the UE are connected to 5GC (Figure 2.10). The eNB in this case is denoted as ng-eNB to underline the fact that it inter-connects with other RAN nodes using Xn interface and not X2 interface.

All those architectural options may seem confusing at first, but there is a certain logic behind introducing them and clear differences allowing to identify quickly with which option we deal based on:

- which RAN node (eNB or gNB) is acting as a Master Node and which as a Secondary Node

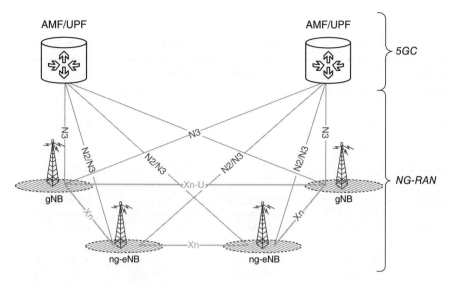

Figure 2.9 NGEN-DC architecture.

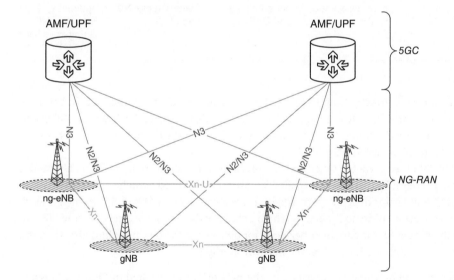

Figure 2.10 NE-DC architecture.

- which Core Network (EPC or 5GC) Master Node has a control plane connection to.

Those principles, together with some other main characteristics of each architectural option, are presented in Table 2.1. The table contains also one additional option not presented on the figures above, i.e. NR-DC. NR-DC represents Dual Connectivity where both of the nodes the UE has connection with are gNBs and the Core Network that is used is 5GC.

One additional clarification worth adding is that in all of the Dual-Connectivity options, data may flow between the RAN nodes and the Core Network in two ways:

Table 2.1 Summary of different Dual Connectivity architecture options.

	EN-DC	NGEN-DC	NE-DC	NR-DC
Master node	eNB	eNB (ng-eNB)	gNB	gNB
Secondary node	gNB (en-gNB)	gNB	eNB (ng-eNB)	gNB
Interface used between MN and SN	X2	Xn	Xn	Xn
Core Network used	EPC	5GC	5GC	5GC
Control Plane connection between CN and RAN	S1-C interface between MN and MME	N2 interface between ng-eNB and AMF	N2 interface between gNB and AMF	N2 interface between gNB and AMF
User Plane connection between CN and RAN	S1-U interface between MN and S-GW and optionally between SN and S-GW	N3 interface between ng-eNB and UPF and optionally between gNB and UPF	N3 interface between gNB and UPF and optionally between ng-eNB and UPF	N3 interface between gNB acting as an MN and UPF and optionally between gNB acting as an SN and UPF

- All data is sent directly from CN to RAN node using S1-U or N3 interface; in this case the radio bearer is called as Master Cell Group (MCG) bearer for bearers established directly between CN and MN or Secondary Cell Group (SCG) bearer for bearers established directly between CN and SN
- Part of data is sent via the other node involved in Dual Connectivity by using UPF of X2 or Xn interface. In this case, a so called split bearer is utilized, which can either be MCG or SCG split bearer depending on whether it is terminated in MN or SN on the network side. The terminating node decides which data packets to provide directly to the UE and which of them to route to the other node for delivery to the UE. In uplink direction, the split is done by the UE based on the network configuration.

Different kind of bearers established by the network are presented in Figure 2.11. Having read all of this information on architectural options, the question which probably arises in the minds of many readers is: why are all of those actually needed? The short answer to such a question would be: they are not for a single network. However, considering that each of the operators may have different 4G to 5G migration plans, there was a need for a standard to specify tools supporting these plans as much as possible. There was a common understanding between the operators involved in 3GPP work that the first step on the migration path is by leveraging existing LTE deployments and enhancing their capacity by tight inter-working with NR system utilized to provide a secondary connection to the UE. That is the reason for which EN-DC architecture was specified as a first one, even before NR standalone option. Thanks to that, operators do not have to deploy 5GC in order to

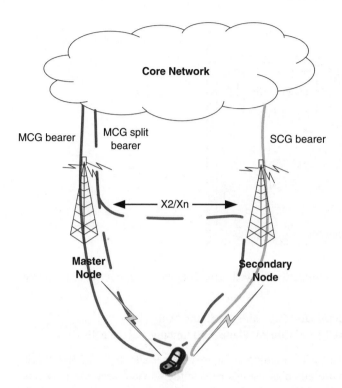

Figure 2.11 Different kind of bearers established by the network.

enhance capacity of their networks and the development effort required from the UE vendors is lower, since they do not have to support full functionality of the NR system (e.g. IDLE mode procedures, control connection with NG-RAN and 5GC, etc.) and may reuse existing LTE modems to a large extent. However, operators may have different plans on how to migrate their networks further and many migration paths were considered during the study phase on NR system as can be seen in 3GPP Technical Report 38.801 [11]. Once an operator deploys 5GC, it now has many possibilities on how to upgrade its RAN network. It may, e.g. choose to either allow its eNBs to remain under EPC jurisdiction or upgrade them and connect to new 5GC. In any case, it will be able to leverage dual connectivity feature, either via NE-DC or via NGEN-DC. This is of course under the assumption that UEs would support all of the options, which is rather improbable as it would likely be non-cost efficient. Thus, we will all find out which of the architecture options will eventually make it into the real-world network deployments and that will probably happen in not so distant future.

2.3.3 CU-DU and UP-CP Split

In Section 2.3.1, it was mentioned that NR architecture is reminiscent of the architecture which was used in LTE. However, this is only partially true. In fact, when looking at the architecture on the high level, nothing seems to have changed with respect to 4G systems. On the other hand, the cloud networking revolution did not remain unnoticed in 3GPP and

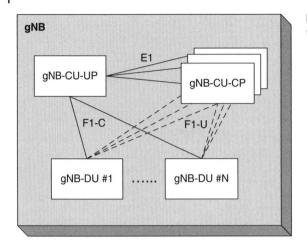

Figure 2.12 gNB architecture with CU-DU split and UP-CP separation.

had an impact on NG-RAN architecture as well. This is apparent in two main principles introduced for NR:

- splitting gNB into Centralized Unit (CU) and Distributed Unit (DU)
- separation of User Plane (UP) and Control Plane (CP) functions of the gNB-CU

The resulting gNB architecture is presented in Figure 2.12, see also 3GPP TS 38.401 [12].

As can be seen, what was presented as a single gNB in the previous sections, may actually consist of multiple logical and physical nodes. First level of separation is between gNB's Centralized Unit (gNB-CU) and gNB's Distributed Units (gNB-DUs), which are interconnected using F1 interface. gNB-CU is hosting RRC, SDAP, and Packet Data Convergence Protocol (PDCP) layers of the overall protocol stack (please see Section 2.4.1) while gNB-DU hosts Radio Link Control (RLC), Medium Access Control (MAC), and Physical layer PHY layers and may provide multiple cells (similarly as a single base station in LTE system). The protocol stack architecture with CU-DU split is presented in Figure 2.13.

The operation of gNB-DUs is partly controlled by the gNB-CU and control messages are exchanged using F1 protocol. The number of gNB-DUs under one gNB-CU is limited to 2^{36}. However, this is only a theoretical number allowed by the specifications and in real deployments it will be rather lower, although volumes going into thousands, tens of thousands, or even hundreds of thousands are anticipated.

The other dimension, in which the split of the gNB into multiple logical nodes can be implemented, is by separating control and user planes. As can be seen in Figure 2.12, such separation is taking place only in the gNB-CU part and multiple gNB-CU-UP functions can be connected to a single gNB-CU-CP function. Furthermore, gNB-DU may be under the control of a single gNB-CU-CP at a time, but on the other hand it may have F1-U connections to multiple gNB-CU-UPs. E1 interface which inter-connects gNB-CU-CP with gNB-CU-UP has only a control plane part, since there is no user data flowing between those

Figure 2.13 Protocol stack architecture with CU-DU split.

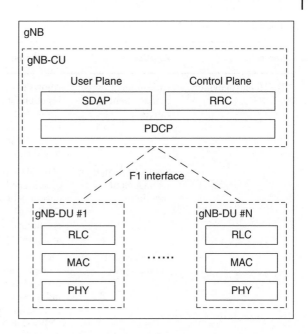

two functions. The most important functions of the F1 and E1 interfaces are summarized in Table 2.2.

Similarly, as in the case of multiple architecture options, the reader may wonder why the split gNB architecture was introduced. NR system supports operation in the frequency bands much higher than those used by LTE. Decreased cell ranges in these frequency bands result in the need of densifying the networks, which means more base stations need to be deployed. At the same time, cells with larger coverage will be deployed to provide seamless mobility and control signaling robustness. Macro cells deployed for coverage will co-exist with abundance of small cells providing additional capacity, which calls for an extra multi-cell coordination possibility to, e.g. enable collaboration between cells for efficient interference control, coordinated multi-point techniques, etc. This can be facilitated by having the centralized gNB node responsible for those tasks which require higher level of coordination. On the other hand, there are services requiring extra-low latency for data delivery, which means that UP functions need to be deployed closer to the user, which is enabled by the CP-UP separation where gNB-CU-UP function may be located closer to the user than its control plane related counter-part. In general, all of the functions mentioned above may be deployed flexibly by an operator according to the needs in the network in a certain geographical area. It is also important to remember that the separation mentioned in this chapter is not mandatory and there is a possibility to implement and deploy so called standalone gNBs as well.

Table 2.2 Summary of F1 and E1 protocol functions.

Protocol	Function	Description
F1-C (control plane)	F1 interface management function	Allows to manage F1 interface by providing functionalities such as interface setup, configuration update, error indication, etc. Using this function, gNB-DU may also inform gNB-CU about the status of its cells (activated/deactivated) or indicate that it is overloaded.
	System Information management function	gNB-DU is responsible for scheduling of all the broadcasted System Information and for the encoding of MIB and SIB1 (see Section 2.4.6.4). On the other hand, gNB-CU is responsible for encoding of all other System Information Blocks and providing them to gNB-DU for transmission.
	F1 UE context management function	The main purpose of this function is establishment, modification, and release of the overall UE context as well as of signaling and data radio bearers of the UE. The establishment of the context and bearers is always requested by the gNB-CU and gNB-DU may either accept or reject it based on its admission control criteria, e.g. depending on the radio resources being available or not.
	RRC message transfer function	Since the gNB-DU is the node having direct connectivity to the UE while RRC protocol is hosted in the gNB-CU, RRC messages need to be transferred between gNB-DU and gNB-CU and this function provides that functionality.
	Paging function	gNB-DU is the node which is responsible for scheduling and sending paging information. However, there are certain parameters which are required to calculate properly the Paging Occasion and Paging Frame of the UE (see Section 3.5.3.2) and this function can be used by the gNB-CU to deliver those to gNB-DU.
	Warning messages information transfer function	This function supports sending of warning messages related to Public Warning Systems, i.e. cooperates with PWS on Core Network side and encodes the warning message into proper System Information message and provides it to gNB-DU for transmission over the radio interface.
F1-U (user plane)	Transfer of user data	This function allows to transfer user data between gNB-CU and gNB-DU.
	Flow control function	This function allows to control the downlink user data flow, e.g. by gNB-DU providing feedback to gNB-CU about the delivered or non-delivered PDCP PDUs, information about the currently desired buffer size or data rate for a specific UE data radio bearer or about the radio link quality of the UE.

(Continued)

Table 2.2 (Continued)

Protocol	Function	Description
E1 (control plane)	E1 interface management function	Allows to manage E1 interface by providing functionalities such as interface setup, configuration update, error indication, interface reset, etc. It also allows gNB-CU-CP to provide information such as supported cells, PLMNs, network slices and QoS to the gNB-CU-UP. On the other hand, gNB-CU-UP may inform gNB-CU-CP about its capacity or about being in overload condition.
	E1 bearer context management function	This function is used to setup and modify the QoS-flow to DRB mapping configuration based on the QoS flow establishment/modification requests arriving from the Core Network to gNB-CU-CP. It is also used to provide security related information to the gNB-CU-UP while gNB-CU-UP may use it to inform gNB-CU-CP about events related to user data activity such as packet arriving for the UE which is in RRC Inactive state (gNB may then trigger UE paging procedure) or data inactivity timer for the UE having expired (upon which gNB-CU-CP maÿ, e.g. release the connection with the UE)

2.4 NR Radio Interface Protocols

2.4.1 Overall Protocol Structure

Each telecommunications system is characterized by its own, specific protocol structure. Even though the functions performed by the particular protocols in each system depend on the design and its main traits, for example whether the system is wired or wireless or whether it operates with frequency division duplex or time division duplex, all of them can still be mapped to a reference model developed over 40 years ago. This model is known as the Open Systems Interconnection reference model, commonly known as the Other System Information (OSI) model or the seven-layer model, contained in the ITU-T norm X.200 [13]. The seven layers and their functions can be shortly characterized as follows:

1. *Physical layer*: Responsible for physical properties of carried data/signals (represented by bits)
2. *Data link layer*: Ensures correct reception/transmission of data between network entities
3. *Network layer*: Responsible for packet forwarding and routing between network entities
4. *Transport layer*: Provides means for reliable data transfer taking into consideration a required performance
5. *Session layer* Manages sessions (data exchanges) between separated applications

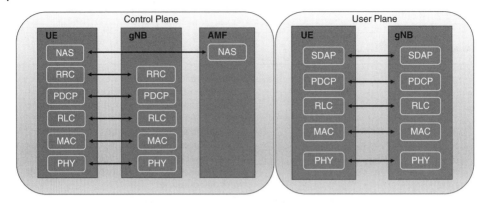

Figure 2.14   Control plane and User plane protocol stack in NR system.

6. *Presentation layer*: Provides for the representation of information originating from application layer
7. *Application layer:* Represents end user application

Two different protocol stacks exist in NR system: control plane (CP) protocol stack, carrying control information between UE and gNB or Core Network, and user plane (UP) protocol stack, carrying data between different applications in the UE and in the network. The two protocols stacks are presented in Figure 2.14.

The reference to the OSI model is recalled here, since very often, when speaking of radio protocols of NR system, they are referred to as Layer 1 or Layer 2 protocols. While the physical layer of NR directly maps to the Layer 1 of the OSI model, the NR layers on top of the physical Layer could be characterized as Layer 2 or Layer 3. Indeed, one may also come upon a term "Layer 2/3 protocols", since some of the functions of the protocols in question cannot be directly mapped to OSI model. However, the OSI nomenclature of service-access-points (SAP) and services offered to higher layer protocol is used in 3GPP standards and for NR it can be described as follows:

- The physical layer offers to the MAC sublayer *transport channels*
- The MAC sublayer offers to the RLC sublayer *logical channels*
- The RLC sublayer offers to the PDCP sublayer *RLC channels*
- The PDCP sublayer offers to the SDAP sublayer *radio bearers*
- The SDAP sublayer offers to 5GC *QoS flows*

Details of the different data channels, bearers, and flows are described further down in this chapter. The reason for naming the data pipes differently for each layer is to be able to distinguish directly which part of the protocol stack is being discussed. It becomes more intuitive when presenting the data flow from top layer to bottom layer following an example of downlink data flow presented in Figure 2.15.

The main functionality of each protocol layer depicted in the figure is shortly explained below:

- SDAP sublayer maps QoS flows on the radio bearers in a many-to-one manner (i.e. multiple QoS flows can be mapped onto a single radio bearer)

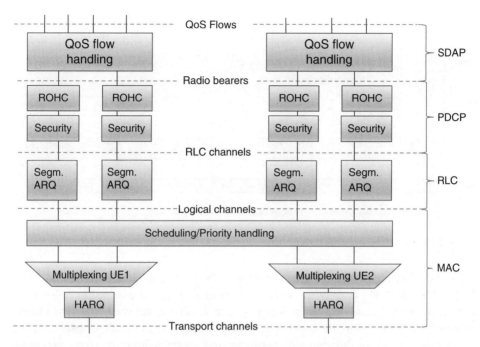

Figure 2.15 Data flow example through Layer 2 protocol stack in NR system.

- PDCP sublayer processes the radio bearers and maps them in a one-to-one manner to RLC channels (i.e. a single radio bearer corresponds to a single RLC channel)
  ○ The PDCP layer also handles Robust Header Compression (ROHC) and security for the data packets
- RLC sublayer processes RLC channels and maps them in a one-to-one manner to logical channels (i.e. a single RLC channel corresponds to a single logical channel)
  ○ The RLC is responsible for segmentation and Automatic Repeat Request (ARQ) functionality
- MAC sublayer processes logical channels and maps them to transport channels in a many-to-one manner (i.e. multiple logical channels can be mapped to a single transport channel)
  ○ MAC provides many different functions including those related to data handling such as scheduling, multiplexing, and Hybrid Automatic Repeat Request (HARQ)
- PHY layer processes transport channels and maps them to physical channels in a one-to-one manner (i.e. a single transport channel is mapped onto a single physical channel)

It can be seen that, for downlink transmission, gNB has parallel processing tracks for multiple UEs and operations at different layers are applied separately for traffic from each UE. The only function spanning across data streams from different UEs is scheduling and priority handling, since the scheduler needs to be able to divide available radio resources between all the connected UEs. In uplink direction, the UEs apply very similar processing of the data to be transmitted with the difference that they do not perform scheduling, but just utilize the grants issued by the gNB. They still need to perform traffic prioritization by applying Logical Channel Prioritization (LCP) procedure described further in this chapter.

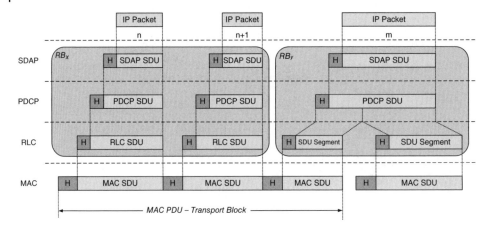

Figure 2.16 Data flow example through Layer 2 protocol stack in NR system.

The data flow can also be presented on data packet level as done in Figure 16 extracted from 3GPP TS 38.300 [10]. When a service is needed, a logical connection called PDU session between UE and data network (see Section 2.2.1) is established. The PDU session may be IPv4, IPv6 or Ethernet and UE may establish more than one PDU session with a single data network or different data networks over single or multiple access networks. The outer IP header of an IP packet has the end point address at UE or data network. This forms a tunnel for the PDU session that is visualized in Figure 2.16. Each packet that is output of a given layer is called PDU of that layer which then serves as service data unit (SDU) for the receiving layer. For example, SDAP PDU comprises the SDAP SDU (IP packet) and the SDAP header.

In this example, there are three QoS flows arriving from IP layer, which are then mapped to two radio bearers by SDAP layer. After being processed by PDCP and RLC sublayers, logical channels arriving at MAC sublayer can be multiplexed onto a single transport block (TB). Different transport blocks can be then mapped to different or the same physical channels.

Since the information belonging to control plane needs to be carried between UE and gNB on the same medium (air interface), most of the sub-layers will be common for user and control planes and the functions provided by each sub-layer in both protocol stacks will be very similar. Despite the clear division into control and user plane in the NR system, the control functions and data delivery functions very often run side by side, e.g. a single MAC PDU can contain both data belonging to user data related logical channels and MAC Control Elements (CEs) responsible for functions such as power control or beam management. Then again, in uplink direction, physical layer Channel State Information (CSI) reports, or other uplink control information (UCI), can be sent via both dedicated control channel (PUCCH) and shared data channel (PUSCH).

2.4.2 Main Functions of NR Radio Protocols

Table 2.3 introduces shortly the main functions of each NR radio protocol layer.

Table 2.3 Summary of the main functions of various NR radio protocol layer's.

Protocol layer	Description and functions
Physical layer (PHY)	○ On PDSCH and PUSCH, i.e. downlink and uplink shared channels, PHY layer performs all the actions required to turn the transport block generated by the MAC layer into a physical signal on the air interface, i.e. ■ CRC attachment to transport blocks or code blocks ■ Segmentation of transport blocks into code blocks ■ Channel coding ■ Processing of HARQ on physical layer ■ Rate matching ■ Scrambling ■ Mapping to assigned resources and antenna ports ■ Signal modulation and waveform generation according to the configured numerology ○ PDCCH and PUCCH, i.e. downlink and uplink control channels can be used to: ■ schedule transmissions in uplink and downlink ■ notify the UE about the utilized slot format ■ provide power control commands ■ switch UE's active Bandwidth Part (BWP) ■ perform random access procedure ■ generate reference signals used for Layer 1 and Layer 3 measurements ■ provide HARQ-ACK/NACK feedback ■ report Channel State Information (CSI) ■ send Scheduling Request to the gNB
Medium Access Control (MAC)	○ Maps logical channels onto transport channels ○ Multiplexes and demultiplexes MAC SDUs belonging to one or different logical channels into/from transport blocks delivered to/from the physical layer on transport channels ○ Is used for scheduling information reporting ○ Performs error correction through Hybrid Automatic Repeat Request (HARQ) mechanism ○ Handles priority between different UEs and different types of data (logical channels) within a single UE ○ Adds padding bits
Radio Link Control (RLC)	RLC sub-layer supports three transport modes: Transparent Mode (TM), Unacknowledged Mode (UM) and Acknowledged Mode (AM). The functions provided by this protocol depend on the utilized transport mode: ○ Sequence numbering, independent of the one in PDCP (AM and UM [for the latter, only in case of segmentation]) ○ Error correction through Automatic Repeat Request (ARQ) mechanism (AM only), i.e. retransmission of RLC SDUs or segments based on RLC status reports ○ Segmentation (AM and UM) and re-segmentation (AM only) of RLC SDUs to match the provided grant size ○ Reassembly of SDU from segments (AM and UM) ○ Detection of duplicated PDUs (AM only) ○ Discarding of some RLC SDUs, e.g. the ones received outside of reassembly window or based on indication from PDCP layer (AM and UM) ○ RLC re-establishment, e.g. upon UE handover ○ Protocol error detection (AM only)

(Continued)

Table 2.3 (Continued)

Protocol layer	Description and functions
Packet Data Convergence Protocol (PDCP)	Functions of PDCP sub-layer depend mainly on whether the processed data unit belongs to control plane (CP) or user plane (UP): ○ Sequence Numbering of PDCP PDUs (both UP and CP) ○ Header compression and decompression using Robust Header Compression mechanism (ROHC) (UP only) ○ Reordering and duplicated PDCP PDU detection (both UP and CP) ○ In-order delivery of PDCP SDUs to higher layers (both UP and CP) ○ PDCP PDU routing for so called split bearers in case of operating in Dual Connectivity mode (UP only) ○ Retransmission of PDCP SDUs (UP only) ○ Security functions such as ciphering, deciphering, and integrity protection (both UP and CP) ○ PDCP SDU discard (UP only) ○ PDCP re-establishment and data recovery for RLC AM, e.g. after handover (UP only) ○ PDCP status reporting for ARQ operation in RLC AM (UP only) ○ Duplication of PDCP PDUs to enhance reliability of transmission ○ Duplicate discard indication to RLC layer after receiving confirmation of delivery of the other copy of PDCP PDU (both UP and CP)
Service Data Adaptation Protocol (SDAP)	This is the only protocol, which was not present in LTE system and is applied only when the base station is connected to 5GC. The main functions of SDAP entity, configured per PDU session in the UE are: ○ Mapping between a QoS flow and a data radio bearer ○ Marking QoS flow ID (QFI) in both DL and UL packets
Radio Resource Control (RRC)	○ Broadcast of System Information ○ Controlling UEs access to the network, e.g. depending on the network load ○ Paging of the UE based on the request from 5GC or NG-RAN ○ Establishment, maintenance, and release of an RRC connection between the UE and NG-RAN ○ Security functions, e.g. security keys management ○ Establishment, configuration, maintenance, and release of Signaling Radio Bearers (SRBs) and Data Radio Bearers (DRBs) ○ Mobility functions for UEs in both IDLE and CONNECTED states including mobility within NR and between different Radio Access Technologies (e.g. between NR and LTE) ○ QoS management functions by providing relevant radio resources configuration ○ UE measurement reporting and control of the reporting ○ Detection of and recovery from radio link failure ○ NAS message transfer between Core Network and the UE
Non Access Stratum (NAS)	This is the only protocol in the protocol structure depicted in Figure 14 which goes beyond air interface (or otherwise – Access Stratum). Its termination points are UE on one side and Access Management Function (AMF) residing in the Core Network on the other side. It is mainly used to provide the Core Network control over the UE by utilizing the following functions: ○ managing mobility of the UE including procedures such as authentication, identification, generic UE configuration update and security mode control procedures ○ session management for establishment and maintenance of data connectivity between the UE and the data network ○ NAS transport procedure for support of transport of messages related to applications such as, e.g. SMS, LTE Positioning Protocol (LPP), 5G System Session Management

More details about each protocol layer and its functions are provided in dedicated sub-sections, an exception being made for the NAS protocol. Considering that the focus of this book is on radio interface aspects, NAS protocol has not received a dedicated sub-section. However, since Access Stratum (AS) and NAS operate next to each other in the UE, there is a need for them to interact on some occasions. Thus, NAS protocol or NAS responsibilities will be mentioned on several occasions, e.g. when describing RRC protocol or IDLE mode mobility in further chapters. In case a reader is interested in more details about NAS protocol and the related procedures, she or he is welcome to refer to 3GPP TS 24.501 [14], which describes this topic in detail.

2.4.3 SDAP Layer

The main functionality of the SDAP layer is to map the QoS flows of each PDU session to radio bearers as is visible in Figure 2.15. The QoS flow is a 5GC specific feature thus the SDAP layer is applied when 5GC is connected to NG-RAN and the EPC does not apply SDAP.

In MR-DC with 5GC, the network may host up to two SDAP protocol entities for each individual PDU session of a UE, one for MN and another one for SN. The MN decides which QoS flows are assigned to the SDAP entity in the SN. If the SDAP entity at SN cannot host a given QoS flow any longer, the SN informs the MN and the MN cannot reject the request to tear down a QoS flow.

At the UE, one SDAP entity is configured per a PDU session. The RRC configuration of the UE maps QoS flows identities (QFIs) to a bearer and is allowed to map more than one QoS flow to one DRB. However, in UL, only one QoS flow is mapped onto only one DRB at a time. The RRC configuration does not have to map all QoS flows to a bearer, and if the configuration does not exist for a QoS flow, the mapping is based on stored information and if that is not available either, the SDAP entity maps the QoS flow to a default bearer. Consequently, within the bearers that have a mapping to QoS flows of this PDU session, one bearer needs to be configured as a default bearer.

The encapsulated SDAP header can contain different QoS markings. The header includes a Reflective QoS flow to DRB mapping Indication (RDI), a Reflective QoS Indication (RQI) and the flow ID. The reflective mapping means that the mapping applied by the UE in UL reflects the mapping applied by the network in DL. The RRC configuration also tells whether an SDAP header is present for the UL or DL data flow mapped to a bearer.

SDAP entity at the UE may also use an End-Marker control PDU to indicate that it stops the mapping of the SDAP SDU of the QoS flow indicated by the QFI to the DRB on which the End-Marker PDU is transmitted.

2.4.4 PDCP Layer

The data and control signaling between gNB and UE goes via radio bearers and the bearer types, or categories, are named accordingly. The DRBs carry user plane data between UE and gNB and the signaling radio bearers (SRBs) transfer RRC and NAS messages.

A bearer can further be a split or a non-split bearer as depicted in Figure 2.17. A concept of splitting a DRB at PDCP layer developed for LTE dual connectivity is reused in NR. It means that the data packets belonging to that bearer may be routed via the MN connection or SN

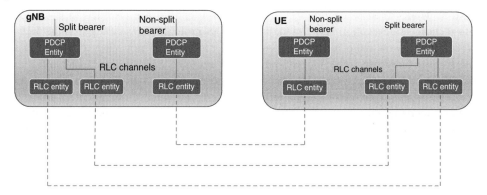

Figure 2.17 PDCP protocol with a split and a non-split bearer.

connection. The receiving PDCP entity has to then reorder the packets received via the two connections such that they are forwarded in a correct order to higher layers. In NR, there are more variations of dual connectivity options, as presented in Section 2.3, and more bearer types and thus a larger number of possible user plane configuration options. As opposed to LTE, in standalone NR and MR-DC architectures, not only a DRB, but also an SRB can be a split bearer. To avoid high impact to the UE, it was decided to simplify the bearer types from UE perspective by applying a concept of bearer harmonization. The UE sees only three types of bearers: MN terminated, SN terminated or a split bearer. The UE does not see or care whether the bearer is terminated at LTE or NR RAN node, which increases the network deployment flexibility while minimizing the risk of market fragmentation by having a lot of different types of UEs implemented with respect to which bearer types they support.

2.4.4.1 PDCP Packet Transmission

The PDCP entity produces PDCP data PDUs and control PDUs. PDCP data PDUs contain RRC messages or IP, Ethernet, or unstructured packets as all of them are user plane data from the PDCP layer point of view. There are two control PDCP PDUs, one is for ROHC related feedback and the other one is a PDCP status report. The PDCP status report includes an indication of successful delivery of PDCP PDUs. Each PDCP PDU is associated with a sequence number (SN) and the status report includes the SN of the first missing PDCP PDU and a bitmap of which subsequent SNs are successfully received. It is not specified when the network should send status reports to the UE while the UE should send the status report at PDCP re-establishment or data recovery if the bearer has been configured such that the status reporting is required. A PDCP window example is presented in Figure 2.18.

At the receiving side, the PDCP entity performs reordering and in-order delivery of data packets to upper layers unless out of order delivery is configured for the bearer. For the in-order delivery, the receiving PDCP entity maintains a *T-reordering* timer and performs reordering based on SNs while the timer is running. When the timer expires, it delivers the packets received so far to upper layers regardless of missing PDCP PDUs. In LTE, PDCP always delivers packets in order and in NR the option of out of order delivery was added. This may increase data rate as the PDCP entity does not wait for the T-*reordering timer* to expire before it delivers the packets with higher sequence numbers in case some previous packets are missing.

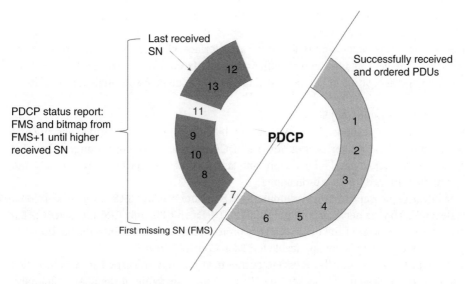

Figure 2.18 PDCP window example.

In order to avoid receiving PDCP packets with the same sequence number, the transmitting PDCP entity may have a maximum of half of the window of the PDCP SN space in use. Thus, the maximum SN value determines limit for maximum data rate. As the receiving entity may need to keep as many packets as half of the maximum SN space at the receiver buffer, the SN space also sets the receiving end buffer size requirement.

At the transmitting side of the PDCP, the PDCP entity discards an SDU if it is confirmed by lower layers as delivered or if its *discardTimer* expires. The *discardTimer* is a mechanism introduced to allow higher layer transport protocols such as TCP to adjust their rate better. Further, it avoids delivering a packet if it is delayed more than indicated by the QoS requirements of the service. The value for the T-*reordering* and *discardTimer* are set by RRC for the UE.

Retransmissions of the PDCP PDUs are only made by the UE during PDCP re-establishment and during PDCP data recovery. At those events, the UE resends all the PDUs that have not been acknowledged by lower layers. The PDCP re-establishment may be requested by the network while modifying the bearer configuration. The retransmissions happen typically at bearer setup after handover to another cell. During the handover process, there may be PDCP PDUs that are lost and the retransmission operation allows the handover to be seamless for the upper layer services. It means that, for example, the TCP layer does not receive a peak of lost TCP packets every time a handover occurs. In LTE, the PDCP re-establishment is only supported for DRBs but NR supports it for both DRBs and SRBs.

2.4.4.2 PDCP Duplication
NR supports PDCP packet duplication which may be configured for a bearer by the network using RRC protocol. Once the duplication is configured for a data bearer, it may be dynamically activated and deactivated using a dedicated MAC CE. For SRBs, once duplication is configured, it is always active and cannot be dynamically deactivated. Duplication

relies on adding a secondary RLC entity for the bearer which is supposed to use duplication. Subsequently PDCP layer provides two copies of the same PDCP PDU to RLC layer. In the end, the same packet is sent twice on the air interface, which increases the reliability of its delivery. The packets may be sent either on two different carriers within the same gNB using Carrier Aggregation or can be sent via two distinct gNBs using Dual-Connectivity.

2.4.4.3   Access Stratum (AS) Security

The PDCP entity is responsible for AS security by performing ciphering/deciphering and integrity protection and verification for the PDCP packets. At the beginning of the RRC connection, AS security and related ciphering keys are negotiated using RRC protocol and they are then utilized for data protection.

The integrity protection is always applied to PDCP Data PDUs of SRBs while it is optional (i.e. depending on the network configuration) for DRBs. In LTE, only RRC messages (SRBs) can be integrity protected and in NR integrity protection for DRBs was introduced. Integrity protection or ciphering is not applicable to PDCP Control PDUs.

For integrity protection, the message authentication code is carried in a field called MAC-I. At transmission of a packet, the UE computes the value of the MAC-I field and at reception it calculates a code called X-MAC. If the X-MAC corresponds to the received MAC-I, integrity protection is verified successfully.

When security is activated, ciphering is applied to the data PDUs for both DRBs and SBRs. Ciphering is applied to the MAC-I and the data part of the PDCP Data PDU except the SDAP header and also to the SDAP Control PDU, if included in the PDCP SDU.

2.4.4.4   Robust Header Compression (ROHC)

ROHC protocol performs header compression and decompression. There are nine different header profiles which can be compressed by ROHC which correspond to types of network and transport protocols utilized at higher layers (e.g. IP, IP/UDP/IP/UDP/RTP, etc.). The most popular use case for header compression in mobile networks is Voice over IP (VoIP) where it is estimated that headers relating to a single VoIP packet can be compressed from 40 bytes down to even a single byte.

2.4.5   RLC

The RLC layer processes the PDCP PDUs (or also referred to as RLC SDUs) and produces the RLC PDUs that are sent to MAC layer. The RLC layer has three different modes: Transparent Mode (TM), Unacknowledged Mode (UM), and Acknowledged Mode (AM), and the functionality that the RLC layer provides is specific to different modes. The transparent mode means that the RLC entity directly maps an RLC SDU to an RLC PDU without any modification (even the RLC header is not added in this case). The most important functions the RLC entity in UM mode provides is segmentation and concatenation and reordering of out-of-sequence PDUs. In RLC AM, the main additional function on top of the UM mode is to provide reliability through ARQ retransmissions. Table 2.4 compares the different transmission modes of RLC.

Both RLC AM and UM modes can be configured for all data bearer types (MCG, SCG, and split bearers) depending on the traffic type and service requirements. For example, UM

Table 2.4 Comparison of different RLC modes.

RLC transmission mode	Description and functions
Transparent mode (TM)	• Directly maps an RLC SDU to an RLC PDU • No RLC header applied
Unacknowledged Mode (UM)	• Segmentation and concatenation of RLC SDUs • RLC header with SN applied to the RLC SDU • Reordering of out-of-sequence PDUs • Duplicate discarding
Acknowledged Mode (AM)	In addition to the UM functionality: • Retransmission of Data PDUs using ARQ • Re-segmentation of ARQ-retransmissions • Status reporting • Polling for status reports

mode fits better services delivered over UDP transport protocol for which lossless transmission is not required or applicable (e.g. live TV broadcast, low latency data, voice) while AM mode, thanks to its lossless delivery mechanism (ARQ), is more appropriate for applications utilizing TCP protocol requiring lossless data delivery (e.g. web browsing, file download). For SRBs other than SRB0, the AM mode is applied in order to provide reliable delivery of RRC and NAS messages. For the SRB0, which is used for paging and broadcast system information, TM mode is applied.

2.4.5.1 Segmentation and Concatenation

Segmentation and concatenation are performed by the RLC layer by dividing and combining parts of PDCP PDUs. The purpose is to provide the MAC layer with appropriately sized packets that match to the packet size requested by the MAC layer in order to avoid zero-padding. For example, when MAC layer requests a certain size packet and RLC layer provides smaller RLC PDU, the MAC layer adds padding bits to fill up the MAC PDU.

In RLC AM, the RLC layer is further able to re-segment an RLC PDU in case of RLC layer re-transmissions. This is needed in case the physical layer transport block size (TBS) on which the RLC PDU is sent, which then translates to different packet size requested by MAC layer, varies between the original transmissions and for the retransmission.

An RLC header is appended to each RLC PDU for UM and AM modes. The header content depends on the RLC mode and the configured RLC sequence number space as well as whether re-segmentation or concatenation is used.

2.4.5.2 RLC Reordering

The RLC reordering functions in a similar way as the PDCP reordering. The receiving entity starts the RLC entity specific reordering timer called *t-reassembly*, configured by RRC, when the PDU arrives with a sequence number that is larger than expected, i.e. larger than the last received sequence number plus one. For PDCP, the corresponding timer is called *t-reordering*. Upon expiry of the timer, the missing PDU will be considered lost and the SDUs associated with the PDU will be discarded. The timer is stopped when all of the PDUs up to the last received PDU are in order and can thus be delivered to the higher layers.

2.4.5.3 ARQ Retransmissions and Status Reporting

For RLC AM, the transmitting entity retransmits missing RLC PDUs based on RLC status reports received from the RLC receiver entity. The receiver entity, both at UE and network side, sends a status report when polled by the transmit entity by toggling a poll bit the header of the PDU. In uplink direction, the transmitting RLC entity is at the UE and the operation is controlled by network by RRC configuration. The poll bit can be configured to be toggled by the UE every X PDUs or after every X kB. In DL, network can set the poll bit in any PDU for the UE to send a status report. The RLC status report will be triggered at the UE also upon expiry of the *t-reassembly* timer. Like PDCP status report, the RLC status report includes information on which RLC PDUs are successfully delivered. Figure 2.19 and Figure 2.20 gives examples of RLC ARQ operation in DL and UL, respectively. In an example presented in Figure 2.19, UE starts the timer *T-reassembly* after not receiving RLC packets in a correct order, that is, SN = 3 is received but SN = 2 is missing. The timer *T-reassembly* expires when the packet with SN = 2 is not received during the time set by the corresponding RRC parameter. The UE sends a status report reflecting the correctly received RLC packets when the timer expires. Additionally, the network may poll the UE for a status report. Here, the network decides to poll the UE by toggling the status report bit in the PDU with SN = 9. After receiving that PDU, UE would send the status report.

In uplink direction, RRC configures the number of PDUs or a number of kilobytes after which the UE should toggle the status report poll bit in the PDU. In an example shown in Figure 2.20, the poll bit is toggled in PDU with SN = 4 and SN = 8 and the gNB sends the status report accordingly, reflecting the correctly received RLC packets.

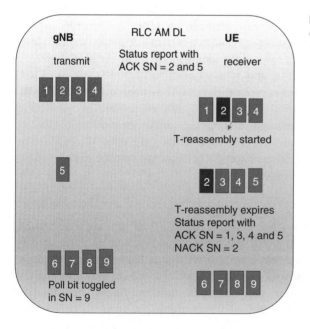

Figure 2.19 RLC AM operation example in DL.

Figure 2.20 RLC AM operation
example in UL.

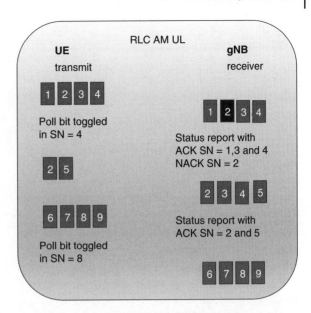

For uplink, RRC also configures a maximum number of allowed retransmissions. When a packet is retransmitted the maximum allowed times, the RLC entity informs upper layers that the limit is reached and upper layers declare radio link failure (see Section 4.4 for details).

It should also be mentioned that, as opposed to HARQ, there is no possibility for the RLC layer to do any type of soft-combining. This is due to the fact that the RLC sequence number would be needed to map different soft buffer sequences and for that the RLC PDU would first need to be decoded (in order to read the sequence number). The consequence of this is that there is no soft-combining gain in RLC.

2.4.6 MAC Protocol

2.4.6.1 Overview

MAC is a protocol layer above the physical layer and below the RLC layer. In MR-DC, the UE is configured with two MAC entities: one MAC entity for the MCG and one MAC entity for the SCG. Figure 2.21 depicts the MAC entity and its main functions. The MAC control function is responsible for configuring and steering other functions of the MAC entity according to the specifications and given RRC parameters. Within the MCG or SCG MAC entity, there is a separate HARQ entity for each serving cell belonging to the cell group when carrier aggregation is configured in the UE.

The MAC entity is responsible for the handling of logical and transport channels and provides mapping between those two types of channels. Table 2.5 lists the specified channels and their main functions.

2.4.6.2 Multiplexing and Demultiplexing

The MAC entity multiplexes MAC SDUs (from Common Control Channel [CCCH], Dedicated Control Channel [DCCH], and Dedicated Traffic Channel [DTCH]) and the MAC

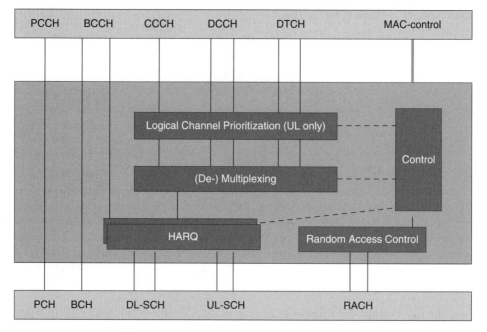

Figure 2.21 MAC entity.

CEs into MAC PDUs. A MAC PDU consists of several MAC subPDUs, which in turn consist of a MAC subheader and one of: MAC SDU, MAC CE or padding. Figure 2.22 shows an example of the MAC multiplexing operation.

The short version of the MAC subheader consists of just a reserved field and logical channel ID. The long version has additionally a Format (F) field that has a length of 1 bit and a 8 or 16 bit Length (L) field which describes the length of the MAC SDU or the MAC CE. The F bit tells whether the L field is 8 or 16 bit long.

The messages from Paging Control Channel (PCCH) and Broadcast Control Channel (BCCH), that is paging and system information, fit directly to the physical layer resources reserved for those. For these, a transparent MAC is used which is a MAC PDU that solely consists of a MAC SDU that has size aligned to a physical layer transmission block (see Section 6.1.4 of 3GPP TS 38.321 [15]).

2.4.6.3 Logical Channel Prioritization

The logical channel prioritization (LCP) procedure is used to make sure the data is sent by the UE according to its priority. The LCP of NR follows the same principle as in LTE and is controlled by three main parameters which are configured by RRC for each logical channel (LCH): a priority value, a prioritized bit rate (PBR) and a bucket size duration (BSD). Additional restrictions related to allowed sub-carrier spacing (SCS), configured grant type, serving cells, and maximum PUSCH duration can be configured by RRC as well for a logical channel.

LCP procedure is applied whenever new UL transmission is performed. The logical channel with highest priority is selected among applicable logical channels which are those that

Table 2.5 Main functions of logical and transport channels in NR.

Channel name	Channel type	Description
Broadcast Control Channel (BCCH)	Logical	BCCH carries the system information from the network to the UE, i.e. RRC messages for MIB, SIB1, and SystemInformation message. The MIB message is transmitted via direct mapping between BCCH and BCH while the rest of the system information is mapped from BCCH to DL-SCH. There is no signaling radio bearer related to system information as the UE is reading system information while in IDLE mode and bearers are not configured for the IDLE mode UE. These RRC messages that use BCCH are delivered as RLC SDUs (TM).
Common Control Channel (CCCH)	Logical	CCCH is used for sending such RRC messages as RRC Setup, RRC Reject (in downlink) and RRC Setup Request, RRC Resume Request, RRC Re-establishment, RRC System Info Request (in uplink).
Dedicated Control Channel (DCCH)	Logical	DCCH is used to carry such RRC messages as RRCReconfiguration, RRC resume, RRC release or UE capability Enquiry (in downlink) while for uplink, some examples are RRC Reconfiguration Complete, RRC Setup Complete, RRC Resume Complete.
Dedicated Traffic Channel (DTCH)	Logical	DTCH is for dedicated traffic thus carries data radio bearers.
Paging Control Channel (PCCH)	Logical	PCCH which is directly mapped to PCH carriers the paging message.
Broadcast Channel (BCH)	Transport	BCH is used to deliver MIB to the UE.
Downlink Shared Channel (DL-SCH)	Transport	DL-SCH is used to carry user data and some of the control information (e.g. System Information messages) in downlink.
Paging Channel (PCH)	Transport	PCH is used for delivering paging messages to the UE.
Uplink Shared Channel (UL-SCH)	Transport	UL-SCH is used to carry user data in uplink.
Random Access Channel (RACH)	Transport	RACH channel is used for random access procedure.

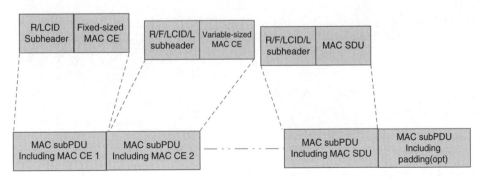

Figure 2.22 (De-) multiplexing in MAC.

fulfill all configured restrictions for the UL grant and have the configured PBR value more than 0 kbits. The amount of data that is allocated for the selected logical channel depends on:

(1) how much data resources have been allocated in the past for this logical channel,
(2) the PBR and BSD values
(3) the time elapsed (T) from the last time the procedure was run.

The larger the PBR, the more resources can be allocated to an LCH, taking into account the restriction set by BSD as well as how much resources had been allocated to this logical channel in the past. The PBR can be set to infinity which means that all available resources are allocated to this LCH as long as it has data to be sent. The SRBs always have PBR as infinity.

In addition to PBR, each LCH maintains a counter called Bj which is used to control the possible starvation of lower priority LCH. In the beginning, the Bj of each channel is set to 0. Each time when LCP is about to be executed by the UE, the Bj is set to PBR x T to mark the amount of prioritized data for the channel as set by the PBR and the time when the procedure was run previously. If the product of PBR x T is larger than BSD, Bj is set to BSD. That is, the BSD sets the maximum amount of pending data allowed for a LCH. The value of Bj is decreased every time the radio resources are allocated for the LCH. If the value of Bj gets negative for an LCH, the data from this LCH does not receive resources allocation. If there are resources left after considering the highest priority LCH among the applicable LCHs, the remaining resources are allocated in strict decreasing priority order regardless of Bj, PBR, or BSD.

Figure 2.23 gives an example of the principle of how data is allocated from logical channels to the available resources. In the example, as the LCH 1 has PBR set to infinity, typical to signaling bearers, all available data from LCH 1 is mapped to the MAC PDU. Next, data

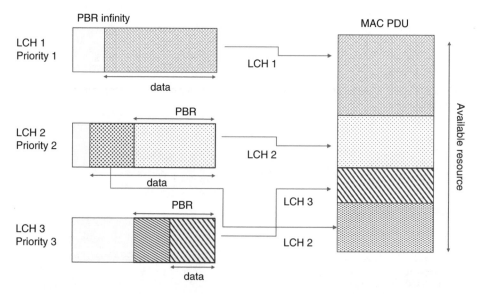

Figure 2.23 Logical Channel Prioritization example.

from LCH 2 is mapped to the MAC PDU according to PBR. The PBR is important in order not to starve a lower priority LCH like LCH 3 in this example. After allocating PBR amount of data from LCH 2, the data from LCH 3 has a chance to be allocated with resources. Afterwards, as there are still resources left in the MAC PDU, the rest of the data from LCH 2 is allocated.

This is the basic operation assuming each LCH had the value Bj larger than 0 at the start of LCP procedure. If the Bj of LCH 2 would have been negative, but Bj of LCH 3 positive, data from LCH 3 would have been allocated in the MAC PDU before any data from LCH 2. In the second round when MAC entity checks if there are still resources in the MAC PDU to be allocated, the Bj value is not checked but data is allocated to the remaining resources in the logical channel's priority order.

2.4.6.4 Hybrid Automatic Repeat Request (HARQ)

HARQ is a mechanism providing error correction for data transmissions. It is a function spanning across MAC and PHY layers. HARQ is described from physical layer perspective in Section 3.4.2 while this section focuses on MAC aspects.

A HARQ entity maintains a number of parallel HARQ processes associated with a HARQ process identifier, separately for uplink and for downlink, as depicted in Figure 2.24. In DL, the MAC entity includes HARQ entity for each serving cell. Depending on whether the physical layer is configured with spatial multiplexing or not, the related HARQ processes support only one Transport Block (TB) or up to two TBs. The HARQ process handles the data transmission for that HARQ process according to information received from the HARQ entity. The HARQ information consists of New Data Indicator (NDI), Transport Block size (TBS), Redundancy Version (RV), and HARQ process ID.

In DL, the HARQ entity allocates the TBs received from the physical layer along with the associated HARQ information to the HARQ process this data belongs to. Depending on whether the decoding was successful or not, the HARQ process either forwards the MAC

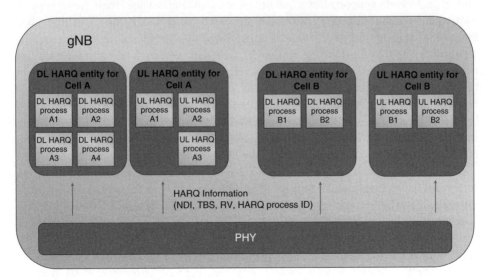

Figure 2.24   Example of HARQ entities and processes in a gNB.

PDUs for demultiplexing or updates the data in the soft buffer as well as generates the corresponding acknowledgments.

A dedicated HARQ process is used for SIB1 and the System Information message carrying the other SIBs. For these, the HARQ process performs soft combining but does not generate acknowledgments as they are broadcast data obtained by the UE via BCCH.

In UL, the MAC entity includes HARQ entity for each serving cell with configured uplink. The HARQ entity handles the UL transmission based on UL grants received. The UL grant may be received via Physical Dedicated Control Channel (PDCCH), in a random access response, or configured semi persistently by RRC. The HARQ information is also received from lower layers. The HARQ entity identifies HARQ process associated with the grant and instructs the HARQ process to trigger a retransmission or delivers a MAC PDU, UL grant and HARQ information in order to generate a new transmission. Each HARQ process supports one TB and is associated with a HARQ buffer from which the retransmission is generated. The HARQ entity is responsible for flushing (i.e. emptying) the HARQ buffer of the process.

As explained in Section 3.4.2, in NR, the specifications guarantee the minimum time before HARQ-ACK can be sent to the UE, so that the UE processing timeline can be met. The retransmissions are not fixed to a subframe and are dynamically scheduled by the Downlink Control Information (DCI) in any subframe. This means that the number of allowed HARQ re-transmissions is not configured but rather scheduled.

For UL transmissions, the network schedules a new (re)transmission for the specific HARQ ID with an NDI which signals to the UE if the last transmission failed or not. This implies that a HARQ process ID is allocated to a specific TB until feedback in the DCI indicates a new transmission.

2.4.6.5 BWP Operation

The DL bandwidth supported by a cell might be considerably wider than the total bandwidth that a UE can support. Furthermore, a power consumption of the UE configured to use a wide bandwidth increases while it might not be justified by the throughput requirements of the application running on the UE. Different services might also have varying requirements when it comes to the required reliability and latency and hence the UE may be configured to operate using different subcarrier spacing settings. To tackle such different needs of the UEs and their applications, while allowing the cells to be deployed with wide bandwidth, a concept of a Bandwidth Part (BWP) was introduced. The BWP is a subset of the total cell bandwidth and the network may configure and adjust the width, frequency domain location and subcarrier spacing of the BWP the UE operates in.

When the BWP operation (also referred to as BWP adaptation) is needed, the UE can be configured with a maximum of four BWPs per serving cell. Note that this is subject to the capability of the UE and it is very likely that in initial NR deployments, UEs will be capable of supporting only a single BWP. In Rel-15, even when more than one BWP is configured, there can be only one active BWP for a UE at a time and the UE needs to monitor PDCCH only on the active BWP of a cell. The network may order the UE to switch its active BWP either using RRC Reconfiguration message or by sending a DL assignment or UL grant via PDCCH. For unpaired spectrum the BWP switching is common for UL and DL, i.e. both

UL and DL BWPs are switched with a single command. For paired spectrum, i.e. in FDD mode, BWP is switched separately for UL and for DL.

A BWP can further be configured as "initial BWP", "first active BWP" or "default BWP". The initial BWP is referred to by BWP-Id = 0 and it is used for initial access to the PCell. It is configured by SIB1 which the UE acquires before accessing the cell from RRC IDLE state. The initial BWP can then be updated by dedicated RRC signaling for further operation.

The first active BWP (UL or DL) can be configured for the UE by dedicated signaling when configuring a UE with SCells or with an additional cell group (SCG) and, for spCells (in MCG and in SCG), upon reconfiguration with sync (i.e. handover). First active BWPs are the ones which are automatically active upon the RRC configuration for all active cells of the UE or upon an activation of the SCells.

The UE may also be configured with a *BWPInactivityTimer* and a default BWP. The MAC specifies how the timer is started and restarted. For example, the timer is started when the UE switches the active BWP or when the UE receives dedicated data on the active BWP. If the timer expires, the UE falls back to the default BWP. Since default BWP is expected to have narrower bandwidth than the BWP used for data reception and transmission, the UE can save its battery power during the data inactivity periods. If default BWP is not configured, the UE uses the initial BWP as a default BWP.

Figure 2.25 gives an example of the BWP operation.

BWP switching may also happen when the UE initiates Random Access procedure. If there are no Physical Random Access Channel (PRACH) occasions configured for the active BWP, the UE autonomously switches to the initial UL BWP. If the PRACH is initiated for SpCell, the DL BWP has to "match" the UL BWP. This means that, if the UE switches to the initial UL BWP, it also switches to the initial DL BWP. If it does not need to switch the UL BWP, i.e. the currently active UL BWP has PRACH occasions configured, it checks that the DL BWP-Id matches that of the UL BWP-Id. If it does not, the UE switches to DL BWP that

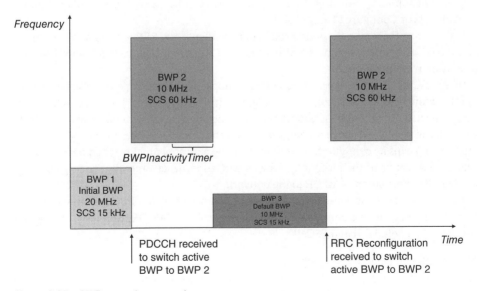

Figure 2.25 BWP operation example.

does have the same BWP-Id as UL BWP. The network configures the UL BWP with PRACH occasions only if the matching DL BWP has SS/PBCH block associated to it.

If the UE receives PDCCH with BWP switch command, it is left to UE implementation whether to stop an ongoing random access procedure and restart it after switching the BWP or to ignore the PDCCH for BWP switching.

2.4.6.6 Scheduling Request

Scheduling request (SR) is used by the UE for requesting UL resources for a new transmission. The SR is sent by the UE on dedicated resources which are configured to the UE by the gNB. Each SR configuration consists of a set of PUCCH resources across different BWPs and cells. Each SR also corresponds to one or more logical channels which means that UE is able to make the query for UL resources for a selected logical channel. For a logical channel at most one PUCCH resource per BWP is configured.

SR is then sent on PUCCH resource on an active BWP. For the same SR configuration, UE may send up to a maximum number of SR requests and the maximum value is configured by network. UE may also be configured with an SR prohibit timer which prevents UE from sending multiple SRs for same SR configuration, which correspond to one or more logical channels, consecutively.

An SR cannot be sent during a measurement gap or when UL-SCH is scheduled.

2.4.6.7 Semi Persistent Scheduling and Configured Grants

The idea of the semi-persistent scheduling (SPS) and Configured Grants (CG) is to allocate the UE with periodic DL assignment (SPS) or UL grant (CG) to serve a certain kind of traffic type which has a defined interval between when the packets have to be received/transmitted. VoIP was the service which inspired defining SPS in the past, but it may be used for any service with periodic data. Voice data flow, or any other QCI type, can be mapped to a bearer which can be mapped to a logical channel ID. By network implementation or by utilizing the LCP, the bearer can be steered to use SPS downlink assignments or Configured Grants for uplink.

In downlink, the semi-persistent resources are configured by RRC with a given periodicity and the DL assignment scrambled with a specific CS-RNTI is then used for activation and deactivation of the resource.

In uplink, there are two types of CG configurations, Type 1 and Type 2. In Type 1, the UL grant is configured by RRC and, once configured, it is always active. Type 2 CG is first configured by RRC, but it has to be additionally activated by PDCCH scrambled with CS-RNTI before it can be utilized. Once not required, it may be dynamically deactivated via PDCCH, but the UE keeps its configuration, so that it can be activated quickly when needed again. The UE also confirms the reception of the L1 activation/deactivation by sending a configured grant confirmation MAC CE to the network.

The resources are configured per serving cell and per BWP and in Rel-15 multiple configurations can be active only on different serving cells. In MR-DC, SPS resources can be configured on both Pcell and PSCell.

2.4.6.8 Discontinuous Reception (DRX)

When the UE is in RRC CONNECTED mode, the network may configure a UE with so called Discontinuous Reception (DRX) configuration, which tells the UE when it should monitor its control channel (PDCCH). Without DRX being configured, UE needs to monitor

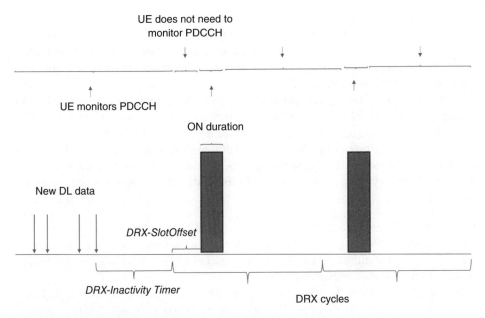

Figure 2.26 Basic DRX operation.

PDCCH continuously, which causes high UE battery consumption and DRX allows power savings in the UE. How exactly the UE monitors PDCCH is controlled by various timers configured by the network.

Within a DRX configuration, UE is provided with a setting of time interval called "DRX short cycle" or "DRX long cycle". For simplicity we call it "DRX cycle" in the following description as the rules of operation are similar. Within each "DRX cycle" there is an "ON duration" during which the MAC entity of the UE needs to monitor PDCCH. The *drx-onDurationTimer* describes the duration of the "ON duration" and *drx-SlotOffset* defines the delay from the start of the "DRX cycle" before starting the *drx-onDurationTimer*. The "DRX cycle" is started if UE has not received a new DL assignment or UL grant for a period of time defined by *drx-InactivityTimer* which is started/restarted every time PDCCH indicates a new transmission. Figure 2.26 depicts this basic operation.

The other DRX timers are related to HARQ and corresponding retransmissions. These timers are per HARQ process and also affect when the UE needs to monitor PDCCH and when it can save battery. An example of the usage of HARQ related timers is the following:

- When a MAC PDU is received in a configured downlink assignment, the *drx-HARQ-RTT-TimerDL* is started for the corresponding HARQ process in the first symbol after the end of the corresponding transmission carrying the DL HARQ feedback. If neither *drx-InactivityTimer* nor *drx-onDurationTimer* are running, the UE does not have to monitor PDCCH. If *drx-HARQ-RTT-TimerDL* expires and if the data of the corresponding HARQ process was not successfully decoded the *drx-RetransmissionTimerDL* is started and UE needs to monitor PDCCH.

The UE may also receive a DRX Command MAC CE or a Long DRX Command MAC CE. When the UE receives either of the MAC CEs, it starts using short DRX cycle or long DRX cycle depending on which MAC CE was received and whether short DRX was configured for the UE. If the UE is not configured with short DRX cycle and receives a DRX Command MAC CE, UE starts the long DRX cycle. The reason to have both short and long DRX is to balance between power saving in the UE and the delay related to delivering uplink or downlink data when it becomes available.

For further power consumption optimizations, the network may configure the UE to limit CSI reporting to only ON durations of the DRX. However, the MAC entity transmits HARQ feedback and aperiodic SRS when such is expected regardless of PDCCH monitoring.

It is worth noting that in EN-DC, separate DRX configurations are provided for MCG and SCG, but to allow for maximum power consumption optimizations at the UE, Master Node and Secondary Node can exchange and coordinate DRX configurations they provide to the UE.

2.4.6.9 Buffer Status Reports

The buffer status reporting is used in uplink to provide the base station with information about the amount of uplink data a UE currently has in its buffer. This helps the network to know how much uplink resources to allocate for the UE.

The logical channels of the UE are grouped by RRC into logical channel groups (LCG). The BSR reports the buffer status per LCG by using the long or short BSR format. The short format allows reporting for just one of the LCGs while the long format allows reporting the status of each of the configured LCGs.

There are three types of BSR: Regular BSR, Periodic BSR, and Padding BSR. A Regular BSR is reported in three different situations:

- When data becomes available for a logical channel which belongs to an LCG when none of the logical channels had any UL data, i.e. the buffer was previously empty for that LCG.
- When data becomes available for a logical channel which has higher priority than any other logical channel in any LCG.
- When the *retxBSR-Timer* expires and there is still data in the logical channel buffer.

The triggering of Regular BSR based on the *retxBSR-Timer* is similar to the Periodic BSR which is triggered when *periodicBSR-Timer* expires. The difference is that the Periodic BSR is triggered regardless type of the amount of data in the buffer. It is up to the network to choose which BSR reporting is configured for the UE. For Regular and Periodic BSR, UE chooses short BSR if it has data available only for one LCG and otherwise, the long BSR is used.

The Padding BSR is triggered when the number of padding bits in a MAC PDU allows to fit a BSR report in the MAC PDU instead of the padding bits. In this case, the UE may simply utilize the remaining space from the grant to provide some useful information to the gNB as the Padding BSR is sent using the padding bit space in a MAC PDU. In addition to long and short formats, there are Truncated Short and Long BSR formats which report the status of one LCG but indicate that there are more data in other buffers. The truncated BSR versions give more options for the BSR reporting when the size of the BSR report needs to fit to the available space for it.

2.4.6.10 Timing Advance Operation

Timing advance (TA) is an offset, which the UE is using to advance its UL transmission in relation to the time where it receives DL transmissions from the base station. It is then the offset at the UE between the start of a received downlink subframe and a transmitted uplink subframe. It is needed to ensure that the downlink and uplink subframes are synchronized at the gNB. For uplink timing advance operation, the UE can assume certain cells to be time aligned (e.g. the cells provided on adjacent frequency channels). Group of these cells is called timing advance group (TAG). If the group contains a SpCell, it is called a primary TAG (PTAG) and otherwise it is called a secondary TAG (STAG). RRC configures the UE with a timer value called *timeAlingmentTimer* for each of these groups which tells how long MAC entity of the UE can assume the cells to be uplink time aligned.

The UE receives the initial timing advance value in a random access response message, see Section 4.1.2.2.1, and starts the timer. Timing advance value can be updated with a MAC CE giving a new TA value and the *timeAlingmentTimer* is then restarted. If the timer expires, the UE releases UL configurations for the cells in the group and needs to perform Random Access procedure before further uplink transmissions within these cells are possible.

2.4.6.11 MAC Control Elements

In NR, many MAC CEs were introduced which do not have a corresponding MAC CE in LTE. For example, MAC CEs in NR are an essential part of the CSI framework. Table 2.6 lists the NR MAC CEs related to reference signal resource set activation/deactivation.

As an example, the detailed format for the "SP CSI-RS / CSI-IM Resource Set Activation / Deactivation MAC CE" is presented. Description for the RRC configuration for the NZP CSI-RS and CSI-IM resources and resource sets can be found in Sections 3.7 and 4.3.1. Figure 2.27 shows the fields of this MAC CE. The first field is A/D field which defines whether this MAC CE is to activate (the bit in the field is set to 1), or deactivate (the bit in the field is set to 0) the resources indicated in the further fields of the MAC CE. The MAC CE indicates NZP CSI-RS resource set ID and optionally the CS-IM set ID to be activated

Table 2.6 NR MAC CEs related to reference signal resource set activation/deactivation.

MAC CE name	Description
SP CSI-RS/CSI-IM Resource Set Activation/Deactivation MAC CE (see also Section 3.7 and Section 4.3.1)	A UE can be configured with lists of semi persistent NZP CSI-RS and CSI-IM resources sets. This MAC CE activates or deactivates an NZP CSI-RS resource set and optionally also a CSI-IM resource set. If NZP CSI-RS resource set is activated by the MAC CE, the MAC CE also gives TCI state ID for each resource of the NZP CSI-RS resource set.
SP SRS Activation/Deactivation MAC CE (see also Section 3.10)	A UE can be configured by RRC signaling with up to 16 semi persistent SRS resource sets per UL BWP and serving cell. This MAC CE activates or deactivates one of the semi-persistent resource sets with the 4 bit SP SRS Resource set ID field for an indicated UL BWP. The MAC CE also gives the spatial relationship for each SPS resource of the activated SPS resource set with an NZP CSI-RS resource index, SSB index or SPS resource index.
SP ZP CSI-RS Resource Set Activation/Deactivation MAC CE (see also Section 3.7.2)	In PDSCH configuration for a BWP, UE is given a list of ZP CSI-RS resource sets. This MAC CE activates or deactivates a ZP CSI-RS resource set for a given BWP of a cell.

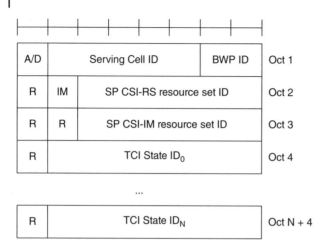

A/D		Serving Cell ID		BWP ID	Oct 1
R	IM	SP CSI-RS resource set ID			Oct 2
R	R	SP CSI-IM resource set ID			Oct 3
R		TCI State ID_0			Oct 4

...

R	TCI State ID_N	Oct N + 4

Figure 2.27 SP CSI-RS / CSI-IM Resource Set Activation/Deactivation MAC CE. Source 3GPP TS 38.321 [15].

or deactivated and the serving cell and BWP in which the resource set belongs to. The octet containing the CSI-IM field is present only if the bit in the "IM" field is set to 1. Additionally, this MAC CE indicates a Transmission Configuration Indication (TCI) state for each resource within the activated CSI-RS or CSI-IM resource sets. These octets are not present if A/D field is set to 0, which is when the MAC CE just deactivates the indicated resource sets.

The TCI framework, as well as the related concept of DL Quasi co-location (QCL), is explained in detail in Section 4.2.3. In short, the TCI defines pairs of reference signals for QCL indication. The TCI describes which reference signals are used as QCL source, and what QCL properties can be derived from each reference signal. The QCL in short is the relation between two reference signals at the UE receiver. When the relation between two reference signals is known at the UE, the UE can use the similarities like delay spread, Doppler spread, Doppler shift, and average delay when forming the corresponding channel estimations.

In NR, there are further TCI state or spatial relation related MAC CEs which are listed in Table 2.7. As explained in Section 4.2.3, the amount of options for a QCL relation for a reference signal is huge and the relation can be pointed to a UE with RRC configuration, DCI or MAC CE or a combination thereof. First two MAC CEs in Table 2.7 indicate directly the TCI state (QCL) for a channel (for the reference signal belonging to the channel as UE makes channel estimate out of that to actually receive the data part of the channel), similarly as the "SP CSI-RS/CSI-IM Resource Set Activation/Deactivation MAC CE" when activating a resource set. The last MAC CE of Table 2.7, "TCI States Activation / Deactivation for UE-specific PDSCH MAC CE", does not directly signal TCI state or QCL of a reference signal for a UE but is used to map the TCI states configured in the UE by RRC signaling into DCI codepoints. This is necessary, since the UE can be configured with up to 128 TCI states by RRC while it is desirable to keep the DCI size as small as possible. Hence the MAC CE is used to down-select from up to 128 TCI states to up to eight TCI states, which can subsequently be selected directly by a DCI to indicate the UE with the exact TCI state assumption valid for a given channel at a given time.

Table 2.7 TCI state or spatial relation related MAC CEs.

MAC CE name	Description
TCI State Indication for UE-specific PDCCH MAC CE (see also Section 3.12.2.2 and 4.2.3)	This MAC CE indicates a TCI state for a UE specific PDCCH. RRC may configure at most maxNrofTCI-StatesPDCCH (64) per BWP. The MAC CE indicates cell ID, ResourceSet ID and the TCI state ID that UE should assume for the PDCCH with this ResourceSet ID.
PUCCH spatial relation Activation/Deactivation MAC CE (see also Section 3.4.4 and 4.2.3)	This MAC CE can be used to indicate one of up to 8 spatial relations to be used for an indicated PUCCH resource. UE can be configured with options on which signal to assume as spatial relationship as well as power control related assumptions.
TCI States Activation/Deactivation for UE-specific PDSCH MAC CE (see also Section 3.12.2.1 and 4.2.3)	This MAC CE works as a selector to map codepoints in a DCI transmission configuration field to TCI states configured by RRC. RRC may configure up to maxNrofTCI-States (128) for the UE when UE receives RRC configuration or reconfiguration for a serving cell. UE receives a list of TCI states for each PDSCH configuration per BWP. MAC CE is used to select active TCI states out of all configured TCI states for an indicated BWP by a bitmap where bit value 1 means the TCI state is activated and 0 means the TCI state is deactivated. The codepoints for TCI in the DCI then refer to the activated TCI states. When UE is scheduled, the DCI selects one of the activated TCI states as the TCI to be assumed for the PDSCH reception.

Table 2.8 CSI related NR MAC CEs.

MAC CE name	Description
SP CSI reporting on PUCCH Activation/Deactivation MAC CE (see also Section 4.3.3)	Within the CSI framework configurations, UE can be configured with a list of CSI report configurations that define options on reporting periodicity and offset and PUCCH resource to be used for reporting. Essentially, the MAC CE selects a subset of PUCCH resources and reporting periodicity and offset for CSI reporting for a BWP.
Aperiodic CSI Trigger State Subselection MAC CE (see also Section 4.3.3)	UE can be configured by the network with a list of aperiodic trigger states for aperiodic CSI reporting. A trigger state points to reference signals to be used for channel and interference measurement which can be combination of CSI-RS, SSB, and CSI-IM as well as a configuration on which CSI to report. DCI is used to indicate to the UE an aperiodic trigger state to be used for reporting CSI from the set of the configured trigger states. The maximum number of configured trigger states is 128 while DCI can only indicate one out of 63 states. Hence, when more than 63 states are configured, MAC CE is used to select which of the configured trigger states are mapped to the DCI codepoints.

Table 2.9 Other NR MAC CEs.

MAC CE name	Description
Buffer status report (see also Section 2.4.6.9)	This MAC CE is used for reporting amount of UL data UE has in its buffers.
	The buffer status is calculated per logical channel group according to the procedures specified in 3GPPT TS 38.322 [16] and 3GPP TS 38.323 [17]. A short fixed-size MAC CE can report BSR for one LCG while the long MAC CE can report buffer status for multiple LCGs and has a variable size. The accuracy and maximum value of the BRS depends on the bit length of the buffer size field. 3GPP TS 38.321 [15] specifies 5- and 8-bit quantization of the buffer volume. In addition, there are truncated MAC CE versions for Padding BSR.
	In MR-DC, the BSR configuration, triggering, and reporting are independently performed per cell group.
	For split bearers, the PDCP data is considered in BSR in the cell group(s) configured by RRC.
Timing Advance Command MAC CE (Section 2.4.6.10)	This MAC CE is used for timing advance value adjustment after RACH.
	It comprises a single octet containing TAG Identity (TAG ID) and timing advance command (0-63).
C-RNTI MAC CE (see also Section 4.1.2.2.1)	This MAC CE is used in RACH procedure where Msg3 contains C-RNTI MAC CE if the UE triggering the CBRA procedure is in CONNECTED Mode with a C-RNTI.
	C-RNTI MAC CE has a fixed size of 16 bits.
UE Contention Resolution Identity MAC CE (see also Section 4.1.2.2.1)	This MAC CE is used in RACH procedure where Msg3 contains UE Contention Resolution Identity MAC CE if the UE does not have a C-RNTI.
	The MAC CE contains the first 48 bits of the UL CCCH SDU and has a fixed 48-bit size.
SCell Activation/Deactivation MAC CE	This MAC CE is used if a UE is configured with carrier aggregation in order to activate or deactivate SCells configured for the UE.
	It has a size of one or four octets and indicates the status (activated/deactivated) of up to seven or up to 31 SCells.
DRX Command MAC CE (see also Section 2.4.6.7)	This MAC CE is used to send UE into CONNECTED mode DRX. It has a fixed size of 0 bits, the UE recognizes the MAC CE type from the LCID subheader.
Long DRX Command MAC CE (see also Section 2.4.6.8)	This MAC CE is used to send UE into CONNECTED mode DRX. It has a fixed size of 0 bits, the UE recognizes the MAC CE type from the LCID subheader.
Duplication Activation/Deactivation MAC CE (see also Section 2.4.4.2)	This MAC CE is used for activation/deactivation of PDCP duplication for a given DRB for which the duplication has been configured.

Table 2.9 (Continued)

MAC CE name	Description
	It has a fixed size and consists of a single octet. Each bit in the octet is a field indicating the activation (bit set to 1) or deactivation (bit set to 0) status of the PDCP duplication of DRB i where i is the ascending order of the DRB ID among the DRBs configured with PDCP duplication and with RLC entity(ies) associated with this MAC entity.
Single Entry PHR MAC CE	This MAC CE is used for power headroom reporting by the UE to inform the network about difference between the nominal UE maximum transmit power and the estimated UL power needed. It has a fixed size of 2 octets.
Multiple Entry PHR MAC CE	This MAC CE is used for power headroom reporting when the UE is configured with UL carrier aggregation. It has flexible size that depends on configured number of SCells.
Configured Grant Confirmation MAC CE (see also Section 2.4.6.8)	This MAC CE is used by the UE to confirm a configured grant activation or deactivation when the related DCI is received. It has a fixed size of 0 bits, the UE recognizes the MAC CE type from the LCID subheader.
Recommended bit rate MAC CE	A bit rate recommendation message from the gNB to the UE or a bit rate recommendation query message from the UE to the gNB for a logical channel.

As explained in Section 4.3.3, there is a MAC CE involvement in aperiodic CSI operation. This MAC CE, "Aperiodic CSI Trigger State Subselection MAC CE", is similar to the "TCI States Activation / Deactivation for UE-specific PDSCH" MAC CE as it does not indicate anything directly for the UE to assume but down-selects from the maximum of 128 "aperiodic trigger states" to up to 63 to be further signaled to the UE by a DCI. The aperiodic trigger state is as concept similar to TCI-state as it is a list of assumptions UE should operate according to when a given "aperiodic trigger state" is indicated to the UE. The CSI related NR MAC CEs are summarized in Table 2.8.

Table 2.9 lists other MAC CEs which are not related to reference signals or CSI.

Additional information (e.g. format) about all MAC CEs can be found in Sections 5.18 and 6.1.3 of 3GPP TS 38.321 [15].

2.4.7 Radio Resource Control (RRC)

2.4.7.1 Overview

In Section 2.4.1, the main functions of RRC protocol were mentioned. In general, this is the protocol which is responsible for managing the radio resources of the network in the most efficient way achievable. It is used to configure or reconfigure all other protocols from the air interface stack, starting from PHY through MAC, RLC and up to PDCP and SDAP. It may be used to configure UEs in all mobility states, i.e. Idle, Inactive, and Connected and virtually every feature existing in NR system, which involves UE, requires RRC signaling to work. The main role of the RRC protocol is no different than in LTE or even earlier generation mobile systems, the main differences being:

- Introduction of On-demand System Information
- Introduction of RRC Inactive State
- Necessity to handle beams, e.g. via additional configuration or by taking them into consideration during mobility
- Necessity to handle BWPs and mixed numerologies

Other than that, the number of parameters which are configurable with RRC is much higher than in the case of previous generations of 3GPP system. This is to a very large extent thanks to significantly much more complex and flexible physical layer, which now supports higher bit rates, higher reliability, lower latency and, what is probably even more impactful, higher frequency ranges and beams. This chapter will present an overview of the RRC protocol.

2.4.7.2 RRC State Machine

There are three RRC states in which a UE may be at a certain time: RRC_IDLE, RRC_INACTIVE, and RRC_CONNECTED. The RRC state can be otherwise called a "connection state", because the state UE is in depends mostly on its data activity, i.e. whether there is an established connection between the UE and the network and whether there is an ongoing data exchange between them. In addition, RRC states are very often called "mobility states", since each RRC state is characterized by the mobility procedures which are performed by a UE and a network in a certain state. This, together with other attributes of each RRC state, is summarized in Table 2.10.

There are many reasons for which different RRC states exist, but the main one is the necessity to manage the power consumption of the UE efficiently. Firstly, in the RRC_CONNECTED state, the UE needs to constantly monitor control channels for scheduling grants while there is no reason to do that in case the UE does not have any data to send or receive. Furthermore, mobility in RRC_CONNECTED state is network controlled and generates many signaling messages on the air interface and in the RAN and sometimes even Core Network. Thus, it makes sense to keep the UE in RRC_CONNECTED state only in case data needs to be transmitted between the UE and the network and otherwise let the UE perform mobility procedures which do not require network involvement. To understand why third RRC state was introduced, it is worth noting that applications running over contemporary LTE networks such as, e.g. instant messaging or web browsing, generate short lived, but frequent connections. An exemplary traffic from a real network can be summarized as follows:

- average packet call is very short (200–2000 kB)
- up to 500 call establishments per day per UE
- connection establishment overhead is relatively high (~150 bytes to send first data byte)

As can be seen from the above data, an overhead related to connection establishment compared to the data exchanged during a packet call is significant already in LTE networks. The design target behind 5G system was to support 10 000 times more traffic than in 4G

Table 2.10 Main attributes of the RRC states in NR system.

	RRC_IDLE	RRC_INACTIVE	RRC_CONNECTED
Core Network connection state (CM state)	CM-IDLE	CM-CONNECTED	CM-CONNECTED
RRC connection state	Not established	Established, but suspended	Established
Mobility control	UE controlled mobility based on network configuration	Same as in RRC_IDLE	Network controlled mobility
Measurements and mobility procedures	UE performs neighboring cell measurements and may perform cell selection and cell reselection	Same as in RRC_IDLE	UE performs neighboring cell measurements and reports measurements to serving gNB. Additionally, UE performs channel quality measurements and reports them to the network.
Data exchange	Only control signaling can be exchanged between the UE and a serving cell (single one), e.g. UE may read broadcast System Information (SI) or send control messages in order to establish a connection or send SI request. No user data exchange possible.	Same as in RRC_IDLE	In addition to acquiring SI, unicast data transfer is possible including both control signaling and user data. UE may be configured with additional serving cells using Carrier Aggregation or Dual-Connectivity features for increased bandwidth.
UE location knowledge on the network side	UE position is known on a per Tracking Area or Tracking Area list granularity by the Core Network. UE performs Tracking Area update procedure periodically and when crossing Tracking Areas boundaries.	UE position is known on a per RAN-based Notification Area (RNA) by the RAN network. UE performs RNA updates periodically and when moving outside the configured RNA.	UE position is known on a per cell basis by the RAN network. UE performs handovers initiated and controlled by the network when moving to other cells.
UE reachability by the network	UE monitors Paging channel for paging initiated by Core Network and monitors Short Messages for SI modification and Public Warning System notifications.	UE monitors a Paging channel for paging initiated by Core Network or RAN network and monitors Short Messages for SI modification and Public Warning System notifications.	UE monitors control channels to determine whether there is data scheduled for it.

(Continued)

Table 2.10 (Continued)

	RRC_IDLE	RRC_INACTIVE	RRC_CONNECTED
UE power saving	The most power efficient state for the UE. A UE specific IDLE mode DRX may be configured by upper layers.	A UE specific IDLE mode DRX may be configured by upper layers or by RRC layer.	The UE may be configured with a UE specific CONNECTED mode DRX by RRC layer.
UE context storage	No Access Stratum (AS) context[a] stored by the UE.	The UE stores the UE Inactive AS context.	The UE stores the AS context.

a) AS context can be most easily defined as the configuration provided to the UE by the network using RRC signaling. In RRC_CONNECTED mode, UE operates according to the parameters as present in its AS context. UE Inactive AS context is the AS context which is stored by the UE when the connection is suspended (UE transitions from RRC_CONNECTED to RRC_INACTIVE state) and restored when the connection is resumed (UE transitions back from RRC_INACTIVE to RRC_CONNECTED state).

systems and that would inevitably lead to signaling storms. Second important reason for introducing RRC_INACTIVE state is the support for massive MTC use case where only very short packets are sent during a single connection. In such case, the battery consumption and signaling overhead of establishing the connection would be often bigger than that of sending actual user data. Finally, there are applications (e.g. sensors generating alarms), which require low latency communications and low power consumption at the same time (i.e. the UEs cannot be constantly kept in RRC_CONNECTED state) and in this case the latency of establishing the connection from scratch would not be acceptable. In the RRC_INACTIVE state, the connection can be resumed quickly thanks to, among others, the fact that the UE context is stored in RAN network and therefore there is no need to involve Core Network in its retrieval which is required when moving from RRC_IDLE to RRC_CONNECTED state.

Figure 2.28 presents the UE state machine and state transitions possible between different RRC states considering a UE capable of both E-UTRA (LTE) and NR communications.

It has to be noted that EUTRA RRC_INACTIVE state is only supported in case E-UTRA connected to 5GC. It can also be seen that the mobility between NR and E-UTRA is possible in both RRC_IDLE and RRC_INACTIVE as well as in RRC_CONNECTED state. More information about IDLE and CONNECTED state mobility is given in Section 4.5 while subsequent sections of this chapter will provide some more information about RRC connection management procedures.

2.4.7.3 Cells, Cell Groups, and Signaling Radio Bearers
Before delving into the details of various functions of RRC protocol, it is worth explaining some of the basic terms related to cells and cell groups which can be very often met throughout RRC and other 3GPP technical specifications as well as in this book.

- *MCG*: A group of cells comprising at least a PCell and optionally SCells. In case of Dual-Connectivity MCG is the group of cells belonging to the gNB (or eNB in case of

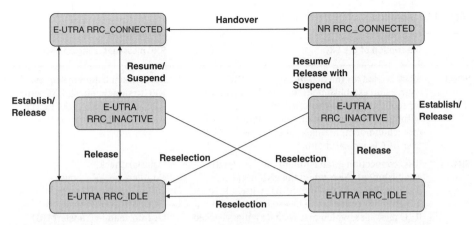

Figure 2.28 UE state machine and state transitions between NR/5GC, E-UTRA/EPC, and E-UTRA/5GC.

EN-DC operation) where the UE performed RRC connection establishment procedure and can be changed via handover procedure.

- *SCG*: A group of cells added as serving cells for the UE by its Master gNB when configuring dual connectivity. It comprises at least PSCell and optionally SCells.
- *Primary cell (PCell):* A cell belonging to the MCG on which the UE performed initial connection establishment procedure or which became the primary cell through handover procedure.
- *Primary SCG cell (PSCell):* The SCG cell in which the UE performs random access when SCG is added for the UE (i.e. UE is configured with dual connectivity operation).
- *Special cell (Spcell)*: Common term for referring to either of PCell or PSCell.
- *Secondary cell (SCell)*: An additional cell belonging to either the MCG or SCG, which is configured by Master gNB or Secondary gNB to provide UE with additional bandwidth (i.e. when UE is configured with Carrier Aggregation in either MgNB or SgNB).
- *Serving cell*: Common term for referring to either of PCell, PSCell, or SCell.

RRC messages which are going to be described throughout this section, are sent on radio bearers called signaling radio bearers (SRBs), as opposed to DRBs used for application data transmissions. There are four SRBs defined in NR, which are summarized in Table 2.11.

It is worth noting that split bearer concept may be applied to both SRB1 and SRB2 in all MR-DC cases. SRB0 and SRB3 cannot be configured as split bearers.

2.4.7.4 System Information
System Information (SI) is information delivered to the UE and which is necessary for the UE to operate in the NR system or to support additional features. The details of how SI is provided to the UE on the physical layer are given in Section 3.5.3 while in this section the focus is on higher layer aspects of delivering SI and what its contents are.

In contrast to the previous generations of the system, in NR, SI can be delivered using both broadcast and unicast transmission from gNB to UEs. In NR, SI is divided into three types of information (summarized also in Table 2.12):

Table 2.11 Signaling Radio Bearers in NR description.

	Description of the SRB	RRC messages using an SRB
SRB0	SRB0 is used for the first messages of RRC connection establishment, re-establishment, and for requesting system information. SRB0 is used without security activation and neither integrity protection nor ciphering applies for SRB0. SRB0 uses CCCH logical channel.	RRCReestablishmentRequest, RRCResumeRequest, RRCsetup, RRCSetupRequest, RRCReject, RRCSystemInfoRequest
SRB1	RRC connection establishment involves SRB1 establishment. Also, when the gNB decides to push the UE to RRC IDLE, SRB1 (with at least integrity protection) shall be used. SRB1 is for RRC messages, which may include a piggybacked NAS message, as well as for NAS messages prior to the establishment of SRB2. SRB1 uses DCCH logical channel.	CounterCheck, CounterCheckResponse, DLInformationTransfer, ULInformationTransfer, ULInformationTransferMRDC, FailureInformation, Location-MeasurementIndication, MeasurementReport, MobilityFromNRCommand, RRCReestablishment, RRCReestablishmentComplete, RRCReconfiguration, RRCReconfigurationComplete, RRCRelease, RRCResume, RRCResumeComplete, RRCSetup, RRCSetupComplete, SCGFailureInformation, SecurityModeCommand, SecurityModeComplete, SecurityModeFailure, UEAssistanceInformation, UECapabilityEnquiry, UECapabilityInformation
SRB2	SRB2 may only be configured by the network after security activation and is used for NAS messages. Only UEs having an SRB2 established can be moved to RRC INACTIVE state. SRB2 uses DCCH logical channel, but has a lower-priority than SRB1.	DLInformationTransfer, ULInformationTransfer
SRB3	SRB3 is used for specific RRC messages when UE is operating in EN-DC mode. The decision to establish SRB3 is taken by the SN, which provides the SRB3 configuration using an SN RRC message. SRB3 establishment and release can be done during Secondary Node Addition and Secondary Node Change procedures. SRB3 may be used to send SN RRC Reconfiguration, SN RRC Reconfiguration Complete and SN Measurement Report messages directly between SN and the UE (i.e. not via MN). SRB3 can only be used in procedures where MN's involvement is not necessary. SRB3 is of higher scheduling priority than all DRBs.	FailureInformation, MeasurementReport, RRCReconfiguration, RRCReconfigurationComplete

Table 2.12 Main characteristics and contents of System Information in NR system.

	MIB	SIB1	Other SI
Contents	MIB contains basic cell configuration and information required for the UE to acquire SIB1, in particular: • six most significant bits of System Frame Number (the remaining 4 bits are encoded in the PBCH transport block) • Subcarrier spacing which is to be assumed by the UE for the sake of SIB1 acquisition, Msg2 and Msg4 reception during initial access, for paging and for receiving broadcast SI messages • SSB subcarrier offset parameter, which allows the UE to determine the location of the SSB with respect to the overall resource block grid • Position of the first DMRS to be used by the UE for DL and UL • Configuration of Common Control Resource Set (CORESET) and common search space, which are required by the UE to receive SIB1 • Cell barred indication, which is used by the UE to determine whether it is possible to camp on the cell. Barring indication is used, e.g. to prevent UE from camping on the NR cells which are to be used only in dual connectivity mode.	Contains information about other SI available in the cell, information relevant for the UE to evaluate whether it may access the cell and part of the information required to perform RRC IDLE mode mobility procedures, in particular: • parameters defining the minimum quality level a cell needs to meet to be selectable by the UE • list of supported PLMNs and additional PLMN related information (e.g. NR cell global identities (NR-CGIs), Tracking Area Codes (TAC), RAN area codes (RANAC), etc.) • parameters for connection establishment failure control (e.g. how many times the UE is allowed to attempt to access the cell before having to refrain from further attempts) • information about other SI available in the cell such as: the list of available SIBs, their mapping on SI messages, how are SI messages provided (e.g. are they currently broadcast or not), how to perform request for SI messages available on-demand, what is the schedule (periodicity and SI-window) of the SI, etc.	There are nine SIBs (SIB1 to SIB9) defined in NR Rel-15. Additional SIBs have the following purpose and contents: • SIB2 – contains information required by the UE to perform cell re-selection which is common for intra-frequency, inter-frequency and/or inter-RAT cell re-selection, e.g. parameters used to calculate cell ranking, number of SS blocks to be considered for cell measurement • SIB3 – contains information about specific intra-frequency neighboring cells such as cells with specific re-selection parameters or blacklisted cells • SIB4 – contains additional cell-reselection related parameters relevant only for inter-frequency cell reselection, e.g. SSB Measurement Timing configuration specific to a certain frequency layer, subcarrier spacing of an SSB on another frequency, etc. It may also contain cell specific information for inter-frequency cells (similar as the one in SIB3 for intra-frequency cells)

(Continued)

Table 2.12 (Continued)

	MIB	SIB1	Other SI
	• Indication of whether the UE is allowed to reselect to another cell on the same frequency in case the highest ranked cell is barred. This indication can be used, e.g. in case all the NR cells on a certain frequency are to be used in dual connectivity mode (e.g. in early NR deployments using EN-DC)	• basic radio configuration of the cell which is common to all UEs in the cell, e.g. downlink and uplink common channels configuration and SSB configuration • information about whether a cell supports emergency calls and eCall service over IMS • configuration of Unified Access Class barring configuration which defines which UE categories may or may not access the cell currently	• SIB5 – contains additional cell-reselection related parameters relevant only for inter-RAT cell reselection (i.e. cell reselection from NR to E-UTRA). It may also contain cell specific information for E-UTRA cells (similar as the one in SIB3/SIB4 for intra- and inter-frequency cells) • SIB6 – contains an Earthquake and Tsunami Warning System (ETWS) primary notification • SIB7 – contains an ETWS secondary notification • SIB8 – contains a Commercial Mobile Alert System (CMAS) notification • SIB9 – contains information related to GPS time and Coordinated Universal Time (UTC) which the UE may use to determine UTC, GPS, and local time. How the information is used by the UE is up to the UE implementation. It may be, e.g. used to assist GPS initialization or to synchronize UE clock or applications in the UE.
Physical channel	PBCH	PDSCH	PDSCH
Periodicity	80 ms with repetitions made within 80 ms	160 ms with repetitions within 160 ms	SIBs can be grouped in System Information messages which have a configurable periodicity between 80 ms and 5.12 s (between 8 and 512 radio frames)
Broadcast or unicast delivery	Broadcast	Broadcast	Broadcast or unicast

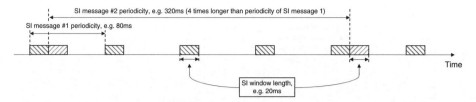

Figure 2.29 An example of scheduling of SI broadcast transmissions using SI windows.

- *Minimum System Information (MSI)*: Comprises the information necessary for the UE to start operating in NR system, i.e. select a cell to camp on where an initial RRC connection can be established. It consists of Master Information Block (MIB) and System Information Block 1 (SIB1)
- *Remaining Minimum System Information (RMSI):* An alternative term referring to SIB1 (i.e. it is a part of MSI which comprises SIB1)
- *OSI*: All other SIBs existing in NR which in Rel-15 of NR standard means SIB2 through SIB9

Other SIBs, as mentioned above, are mapped onto SI messages which can be either broadcasted or delivered with unicast signaling. When a certain SIB is required by the UE, it first checks SIB1 to know whether the SIB is available in the cell and whether it is currently broadcast. In case it is broadcast, the UE may derive its schedule based on the configuration in SIB1 as well. The basic concept of SI delivery via broadcast remained unchanged from LTE and is presented in Figure 2.29.

Broadcast SI messages are delivered in so called SI windows with a specified periodicity. The periodicity defines how often the network provides a certain SI message, while the SI window specifies the time during which the SI message can be scheduled and thus the time when the UE willing to receive that message, needs to monitor PDCCH channel for SI scheduling information. Some additional details about SI scheduling can be found in Section 3.5.3.

In case a SIB is indicated to be available in the cell, but the current status of the SI message it is contained in is set to "not broadcasting", the UE may request the message to be provided via on-demand System Information mechanism. There are two different on-demand SI mechanisms defined for NR: message 1 (i.e. RACH preamble) based and message 3 (i.e. RRC based). Which of them to utilize depends on the network configuration as presented in Figure 2.30.

Message 1 based mechanism is used in case SIB1 indicates that at least one of the available SI messages is not being broadcast at the moment and at the same time SIB1 provides configuration for message-1 based SI request which includes:

- Configuration of RACH occasions dedicated to SI request based on message 1
- Indication of which RACH resource (e.g. RACH preamble) should be used to request a delivery of a certain SI message

After receiving a RACH preamble configured for SI request, the gNB may change the status of the requested SI message to "broadcast" and start broadcasting it according to the schedule as provided in SIB1. It then sends the confirmation of the request to the UE

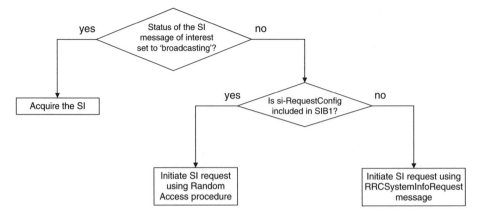

Figure 2.30 Choice of the System Information request mechanism depending on the network configuration.

(with message 2), which subsequently starts SI message acquisition according to the rules presented in Figure 2.29. The gNB keeps on broadcasting the requested SI message at least until the end of current BCCH modification period (i.e. a minimum period for which the SI broadcast does not change in the cell).

In case, for a certain SI message available in the cell, its status is advertised as 'not broadcasting' and at the same time configuration of message 1 based SI request is not provided in SIB1, the UE determines that it should utilize message 3 based SI request mechanism. For this purpose, it needs to send an RRC message called RRCSystemInfoRequest on the CCCH with a default configuration specified in 3GPP TS 38.331 [18]. The content of the message is the list of SI messages UE would like to receive and which are currently not broadcast by the cell. After receiving the request from the UE, gNB confirms reception of the message (by sending message 4) and starts broadcasting the requested messages at least until the end of the current BCCH modification period similarly as in the case of message 1 based mechanism. Both mechanisms require the UE to perform Random Access procedure and this is why message 1–4 nomenclature is used above. The main differentiators between the two procedures are:

- a message from which a gNB gets to know the requested SI messages, which is message 1 (RACH preamble) or message 3 (RRCSystemInfoRequest) respectively
- how many messages are exchanged between the UE and the gNB before requested SI is provided (two and four respectively).

It might seem that message 1 based mechanism is more efficient as it requires less messages to be sent over the air interface. However, the final choice may depend on the particular network deployment and there are also some other aspects to consider. Table 2.13 contains a comparison of message 1 and message 3 based on-demand SI request procedures.

As can be seen from Table 2.13, the choice of the mechanism to use is not that straightforward as might have seemed initially. It depends on many factors such as capabilities of the UEs in the network, services provided by the operator, feasibility of grouping SI messages which may be required to be commonly received by the UEs, required capacity for

Table 2.13 Comparison of message 1 and message 3 based SI request mechanisms.

Message 1 based on-demand SI request	Message 3 based on-demand SI request
<ul><li>Only two messages exchanged between the UE and the gNB before requested SI is provided.</li><li>Requires RACH preambles to be reserved specifically for SI request purpose which limits the number of preambles available for cell access and beam failure recovery purposes.</li><li>The granularity of the request depends on the network configuration and is a trade-off between number of preambles reserved for SI request, e.g.:<ul><li>○ Two preambles may be assigned for SI request so that it is possible to request SI message #1 and SI message #2 separately, but that means 2 preambles cannot be used for other purposes. It also means that in case the UE is interested in receiving both SI message #1 and #2, it needs to send two separate SI request messages.</li><li>○ One preamble may be assigned for the request of both SI message #1 and #2, but in this case even though the UE is interested to receive only one of those, both of them need to be provided by the gNB which is not able to determine which one of them is of real interest to the UE.</li></ul></li></ul>	<ul><li>Four messages exchanged between the UE and the gNB before requested SI is provided.</li><li>Does not require RACH preambles to be reserved specifically for the purpose of SI request. The UE utilizes the preambles from the set of Contention Based Random Access preambles (i.e. the same ones as used for other purposes, e.g. initial cell access).</li><li>The UE may always request exactly the SI messages it is interested in receiving in a single SI request message (e.g. only a single one or any set of SI messages) and is not constrained by the network configuration in that area.</li></ul>

initial access, etc. The rationale for introducing on-demand mechanism for SI delivery is to increase radio resources efficiency in the network. Whenever a certain SI message is not needed by the UEs currently present in the cell or when the up to date version of the message was acquired by all the UEs, the gNB may use the resources, which would otherwise be occupied by SI, for unicast transmission of user data instead. The gain in the efficiency is even more significant considering that, in NR, beam sweeping needs to be performed on the network side to ensure delivery of the SI in the full cell coverage. It means that exactly the same SI contents need to be repeated over each SS/PBCH block in the cell to make sure it is receivable by each UE in the cell regardless of which beam it currently uses for SI reception. This is explained in more detail in Section 3.5.3.

When discussing the ways to deliver SI to the UEs, it is worth mentioning that for RRC Connected UEs, it is possible to provide System Information also with dedicated signaling, i.e. using RRCReconfguration message. Using this method, a gNB may provide a UE with an updated version of SIB1 and the SIBs carrying Earthquake and Tsunami Warning System (ETWS) and Commercial Mobile Alert System (CMAS) notifications, i.e. SIB6, SIB7, and SIB8. This feature is especially useful in case a UE is configured with multiple BWPs which do not overlap with an initial BWP which is normally used to provide SIB1 and System Information modification notifications. If the traditional, broadcast delivery of SI was to be supported, then the gNB would have to configure common search space for monitoring system information or paging in each of the UE's BWP or switch all the UEs to the BWP with a CSS configured before sending SI update notifications. With the dedicated SIB delivery, this can be avoided – UEs are not required to monitor CSS for SI change notifications in other BWPs and the network has to make sure that the relevant information is updated in the UE whenever required.

Last, but not least aspect of SI delivery which distinguishes NR system from previous generations systems, is that SIB's validity may cross the boundary of the cell in NR. For each SIB, except for SIB1 which is always cell-specific, gNB may indicate whether it is cell-specific or SI-area specific. In the latter case, UE is not required to re-acquire the SIB when reselecting to another cell if the new cell belongs to the same SI area identified by the identifier provided in SIB1. This saves the battery of the UEs, especially in case they move around the cells belonging to the same SI area.

2.4.7.5 Unified Access Control (UAC)

RRC protocol's foremost function is connection control. However, another important one, which, at least in LTE, was always applied before the actual connection was established, is access control. Thanks to that function, it is possible for the UE to evaluate whether it is allowed to make an attempt to connect to a network via a certain cell at a specific moment. The feature is designed in a way allowing an operator to control how this evaluation is done, which may depend e.g. on the operator's needs, subscribers' profiles, services available in the network or instantaneous state of the network (e.g. current load, failures in some areas, etc.). In general, the feature is used to control congestion in the network. The Unified Access Control (UAC) steps are presented in Figure 2.31.

The details of the particular steps of the procedure are as follows:

1. For each access attempt a NAS layer in the UE determines UE's Access Category (AC) which may either be standardized (e.g. Mobile Originated (MO) signaling, SMS, MMTEL, (multimedia telephony service) voice/video, MO data, etc.) or defined by an operator (in that case its definition needs to be pre-provisioned in the UE). In addition, the UE is also configured with an Access Identity (AI) which may e.g. refer to a UE which is configured for Multimedia Priority Service (MPS) or Mission Critical Service (MCS). The introduction of Access Identities allows to differentiate the same type of attempts to be treated differently for different UEs, e.g. an operator may configure barring for calls originating from "normal" users, but at the same time it would like to allow them to be established without any disruption by policemen, firefighters or its own field engineers.

2. When needed (e.g. network overload), SIB1 may provide UAC barring parameters applicable to each Access Category (AC) and Access Identity (AI): a barring factor denoting a probability and barring time (specifying the time for which the UE refrains from subsequent access attempts for a barred access attempt).

3. When a UE would like to perform access attempt for which barring parameters are broadcast by the network, it proceeds in the following way:
   a. It randomly draws the number from 0 to 1 and compares it with the broadcast barring factor.
   b. If the drawn number is lower than barring factor, the UE is allowed to make an access attempt and it initiates initial access procedure.

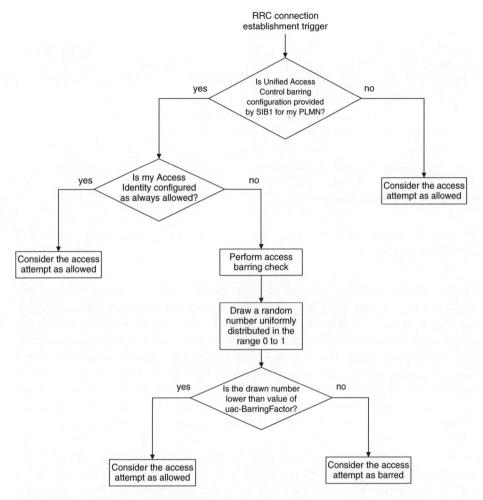

Figure 2.31 Unified Access Control procedure executed in the UE.

c. Otherwise, the UE should wait for time calculated based on the barring time indicated by the network before it will be allowed to attempt to access the network again.

The mechanism is very similar to the one known from LTE, the difference being that Access Identity was introduced. What is also new is that access control is now applied not only for UEs in RRC_IDLE state, but also to new session requests for UEs in RRC_CONNECTED state. "New session request" in this sense refers to events such as new MMTEL voice or video session, sending of SMS, new PDU session establishment, existing PDU session modification and service request to re-establish the user plane for an existing PDU session.

2.4.7.6 Connection Control

2.4.7.6.1 Introduction There are numerous RRC messages and procedures which are used to establish, maintain and release the connection between the UE and the network and

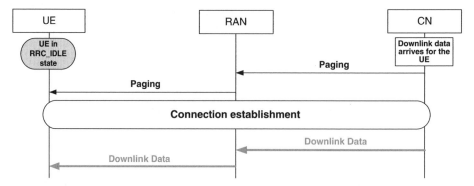

Figure 2.32 CN-initiated Paging.

this section aims at presenting them one by one by simulating an exemplary life-cycle of the UE's connection.

2.4.7.6.2 Paging In the 5G system, similarly as for earlier generations of mobile systems, connection may be established either due to UE data becoming available on the Core Network side or at the UE side itself. How the connection is being established is very similar in both cases, the main difference being that in the former case, before the actual connection establishment begins, the network initiates a procedure called Paging. Since in the RRC_IDLE state, UE's position is known at maximum with Tracking Area granularity as explained in Section 4.5.4, the Core Network does not know to which RAN node to route the UE's data. It firstly needs to identify the exact gNB under the coverage of which UE currently is and this is done by using CN-initiated Paging procedure (Figure 2.32).

In RRC_INACTIVE state, UE's position is known by the network on a RAN Notification Area level, which may cover multiple gNBs. Since from CN perspective the UE is still in a connected state, the CN does not send Paging, but simply forwards user data to the last known gNB that has served the UE, which initiates RAN-initiated Paging procedure (Figure 2.33). Even though it is not shown in the figure, RAN paging may require forwarding Paging message also to other gNBs within the RNA of the UE.

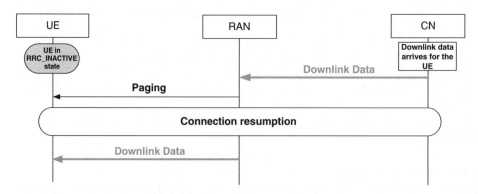

Figure 2.33 RAN-initiated Paging.

Paging procedure is also used to notify the UEs about the modifications of the System Information and about presence of ETWS/CMAS indications. From the perspective of radio interface, the procedures are virtually the same and can be realized either with Paging messages sent over PCCH or with so called Short Messages sent directly over PDCCH with physical layer signaling message called DCI. In RRC_IDLE and RRC_INACTIVE states the UE is monitoring for these notifications in paging channels. To save the UE battery, the UE is required to monitor for paging in only a single Paging Occasion (PO) per its Idle mode DRX cycle. A PO is a set of PDCCH monitoring occasions, which can consist of multiple time slots where paging DCI can be sent. The PO is determined by the UE based on UE's identity (usually 5G-S-TMSI is used for this purpose) and additional parameters signaled by the network such as, e.g. DRX configuration. More information about how the PO is deduced can be found in Section 3.5.3. It is also worth noting that in case of multi-beam operation the same paging message should be repeated over all of the beams, so that the UE may choose any beam it prefers for its reception.

2.4.7.6.3 *RRC Connection Setup*

When a UE needs to establish an RRC connection, e.g. in response to paging reception or due to the need to send uplink data, it triggers RRC Setup procedure. Figure 2.34 presents the message flow for RRC Setup triggered by CN-initiated Paging.

The procedure comprises the following steps:

1. Firstly, an AMF in CN sends Paging message to RAN network including information such as, e.g. UE ID and assistance information allowing RAN to calculate proper POs where Paging notification for a particular UE should be sent (i.e. subframes where the UE monitors for paging messages destined for it, see Section 3.5.3.2).
2. Each of the receiving gNBs sends the Paging message on Uu interface in the POs calculated based on the information provided by the Paging message from AMF. This message

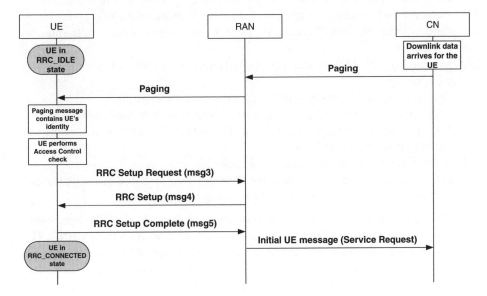

Figure 2.34 RRC Connection establishment triggered by receiving Paging message.

is different from the one provided from AMF. It may contain up to 32 paging records, i.e. identities of the UEs which are paged. Multiple UEs may be paged using a single Paging message on Uu interface and a single PO may be monitored by multiple UEs.

3. When a UE receives a Paging message containing its identity, it attempts to establish the RRC connection by sending RRCSetupRequest message. From that point in time, as far as RAN network side is concerned, the connection establishment procedure is virtually the same as for connection attempt triggered by data arriving from the application at the UE side. RRCSetupRequest message contains UE identity, which for a UE already registered in the network is 5G-S-TMSI (or to be 100% truthful, its 39 rightmost bits) and a connection establishment cause, which provides an information about the reason for establishing the connection, e.g. emergency call, voice/video call establishment, etc. It also determines whether the call is being established due to data originating in the UE or due to paging.

4. The gNB may reply with either RRCReject or RRCSetup message depending on whether the call is to be allowed or not.

   a. In case the setup request is rejected, e.g. due to congestion, gNB additionally provides a time (from 1 to 16 seconds) during which the UE should not attempt subsequent connection establishments (i.e. it should refrain from sending another RRCSetupRequest messages in the same cell) unless the reason to do that is either to respond to paging from the network or to initiate an emergency call. Afterwards the procedure ends.

   b. In case the setup request is accepted, gNB sends RRCSetup message which contains SRB1 configuration (no other radio bearers are established at this stage) and configuration of the spCell which allows the UE to receive SRB1 at minimum (e.g. PDCCH and PDSCH channels configuration) as well as MAC and physical layer's configuration parameters which are common for the whole cell group. For the moment, since the detailed UE capabilities are not yet known by the network, the provided configuration is based on settings of the parameters which are mandatorily supported by all the UEs which support NR system. They can be however reconfigured by the network at a later stage.

5. After receiving the RRCSetup message, the UE replies with a RRCSetupComplete message containing:

   a. The identity of the PLMN to which the UE performs an access to

   b. UE's identity, if available (i.e. 5G-S-TMSI's missing part i.e. its 9 leftmost bits)

   c. NAS message to be forwarded from the gNB to the Core Network to finalize the connection establishment on NAS

   d. AMF to which the UE is registered (if available)

   e. List of S-NSSAI(s) which identify Network Slices UE is subscribed to and which were configured in the UE by the Core Network using NAS protocol.

The RRCSetupComplete message confirms successful completion of the RRC connection establishment procedure and from that point in time the UE is in a RRC_CONNECTED state. After this step, the basic RRC connection establishment ends. The connection establishment might happen only for the sake of the UE sending a single NAS message to the Core Network, e.g. when UE needs to update its Tracking Area due to mobility in the network.

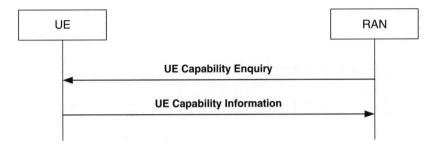

Figure 2.35 UE capability Enquiry message.

However, in most of the cases the purpose is to exchange user data between the UE and the network. The network deduces that based on the establishment cause indicated by the UE in the RRCSetupRequest message and the network's further behavior depends on that. In the first case, it might release the connection right away, but in the second one, it will have to establish DRBs and configure all other radio protocols considering factors such as service to be provided, required QoS parameters (reliability, latency, bit rate), UE capabilities, current network load, available radio resources, etc. Before the network is able to do that, it has to understand what options, with respect to configuring this particular UE, the network has, i.e. what the UE's capabilities are. The next sub-section introduces a procedure allowing this.

2.4.7.6.4 *UE Capability Enquiry* The UE capability Enquiry procedure, allowing the network to obtain information about UE's capabilities, is presented in Figure 2.35.

The network sends a UECapabilityEnquiry message and the UE replies with a UECapabilityInformation in which the UE's capabilities are included, i.e. information about additional (non-mandatory) features and parameters supported for each of the radio protocol layers as well as for measurements and mobility. Such information allows the network to reconfigure the UE in a way which allows an efficient usage of both network and UE resources, e.g. by taking advantage of more sophisticated beam management features or possibility to configure additional carriers with Carrier Aggregation for increased bandwidth. The UE also informs the network about security algorithms it supports, so that security can be established between the network and the UE, which normally is the next step of the RRC connection establishment procedure.

2.4.7.6.5 *Initial Security Activation* Security is activated between the network and the UE for the sake of data encryption and integrity protection. Initial Security activation (Figure 2.36) is done by the network by sending a SecurityModeCommand message to the UE and the UE replying with a SecurityModeComplete message. The former message contains information about the ciphering and integrity protection algorithms the UE should apply to its radio bearers while the latter is a simple message confirming that the configuration was applied successfully. In case the SecurityModeCommand message does not pass security check on the UE side, the UE replies with SecurityModeFailure message as presented in Figure 2.37.

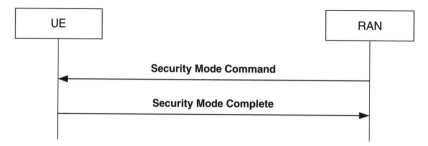

Figure 2.36 Initial AS Security Activation procedure (success).

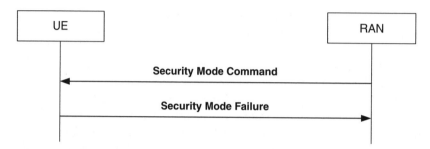

Figure 2.37 Initial AS Security Activation procedure (failure).

2.4.7.6.6 *RRC Connection Reconfiguration* The RRC Connection Reconfiguration (Figure 2.38) is performed by the network sending a RRCReconfiguration message and the UE confirming its successful reception and new configuration application by replying with a RRCReconfigurationComplete message.

In case the UE is not able to comply with the configuration provided by the network, RRC reconfiguration failure event happens and the UE initiates RRC Reestablishment procedure (Figure 2.39).

In general, the RRCReconfiguration message is used to modify an existing RRC connection. The network uses this message, e.g. to:

- Establish, modify or release radio bearers
- Perform reconfiguration with sync (e.g. for the purpose of handover)
- Set up, modify or release measurements (e.g. for Layer 3 mobility or for Channel State Estimation)

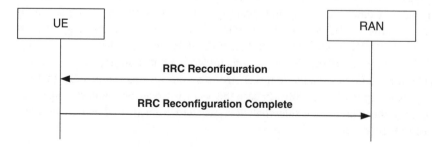

Figure 2.38 Successful RRC reconfiguration.

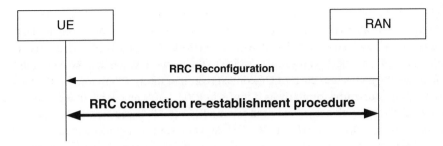

Figure 2.39   RRC reconfiguration failure.

- Add, modify or release SCells and cell groups
- Transfer NAS dedicated information from the Network to the UE.

In practice, virtually each parameter of any of PHY, MAC, RLC, PDCP, or SDAP layers may be reconfigured using the RRCReconfiguration message, so it would be impossible to list them all here. It would not be an exaggeration to say that almost each procedure applied by the UE in RRC_CONNECTED state and described in this book, requires configuration with the RRCReconfiguration message prior to its execution. It has to be also noted that even though the RRCReconfiguration message is described here in the context of connection establishment, it may be sent by the network many times during the course of an on-going RRC connection in reaction to events such as, e.g. UE mobility, new service request or change in the surrounding radio environment.

2.4.7.6.7   RRC Connection Release   After a certain time during which there was no data activity from the UE (i.e. the UE did not transmit or receive any data), the network may decide to release the RRC connection using a RRCRelease message (Figure 2.40).

In NR, the RRCRelease message may be used to transit the UE into one of the two non-active states: RRC_INACTIVE or RRC_IDLE. When the RRCRelease message is used by the network to definitely end the connection with the UE, the UE discards completely its current configuration by releasing all security keys, radio resources, RLC, MAC, PDCP, and SDAP entities and all of its radio bearers. Also, the gNB removes the UE context from its memory. When there is again data to be sent to or from the UE, the UE needs to go through the whole RRC connection establishment procedure from scratch.

The RRCRelease message may also indicate that the UE is supposed to move to an RRC_INACTIVE state and, in this case, it will contain parameters relating to suspend configuration. Upon receiving the RRCRelease message with the suspend configuration,

Figure 2.40   RRC Release procedure.

the UE stores its current configuration received from the gNB previously (e.g. with the RRCReconfiguration message), its current security keys and information about its last serving cell. Afterwards, it suspends all of its SRBs and DRBs, except the SRB0 to leave itself a possibility to send a connection resume message to the network when data arrives in its buffer. Similarly, the gNB stores current UE context (configuration). From now on, the UE performs mobility procedures similarly as if it was in RRC_IDLE mode, i.e. based mainly on cell reselections, but with occasional need to send RNA (RAN-based Notification Area) update to the network. Once the UE is in RRC_INACTIVE state, it cannot send or receive user data to/from the network. Therefore, when data arrives at the UE or UE receives a Paging message (meaning that the network has data to send to the UE), it initiates RRC Resume procedure by sending RRCResumeRequest message, which, among others, contains the UE's identifier assigned by the gNB when suspending the connection (so called I-RNTI, included in the RRCRelease message). Based on this identifier, the gNB receiving the request may deduce what the UE's last serving gNB was and may retrieve the UE's context, i.e. all the information about the UE and its configuration allowing the new gNB to resume the previous connection. Afterwards, it sends a RRCResume message to the UE, which may optionally include a new configuration of UE's radio bearers, cell groups or measurements in case the network wants to modify the one currently stored in the UE. After the successful reception of this message, UE is again in RRC_CONNECTED state, it sends the RRCResumeComplete message to the network and may again transmit and receive data. The successful RRC Resume procedure is presented in Figure 2.41 for the case where the connection is resumed in the gNB different than the one receiving the request from the UE.

Figure 2.40 above presents the intended steps of the RRC Resume procedure where the connection is resumed without any issues. However, it may also have different course of actions triggered by some unexpected situations happening in the network, network congestion, etc. For example, in case the new gNB is not able to retrieve the UE's context from the last serving gNB successfully, but is able to accommodate UE's connection, it may fall back to RRC Setup procedure (Figure 2.42).

On the other hand, if the network is not able to serve the UE's request to resume the connection (e.g. due to network congestion), it may reply with a RRCReject message (similarly as during initial RRC connection setup) as can be seen on Figure 2.43.

RRC Resume procedure takes yet another course in case the resume request is sent by the UE for the sake of RNA update procedure (see Section 4.5.4). In this case, the anchor gNB

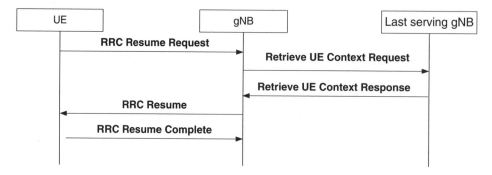

Figure 2.41 Successful RRC Resume procedure.

Figure 2.42 RRC Resume procedure with fall back to RRC Setup.

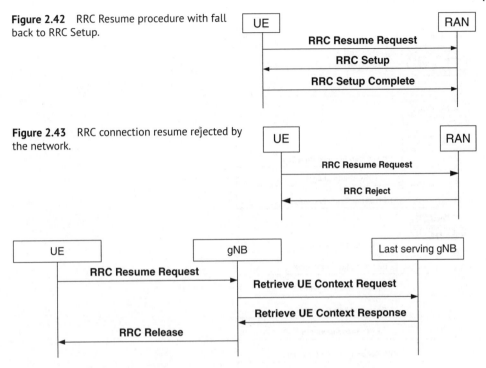

Figure 2.43 RRC connection resume rejected by the network.

Figure 2.44 RRC Resume procedure triggered for the sake of the RNA update.

has to be updated, but there is no subsequent data exchange and hence the UE is sent back to RRC INACTIVE state, as presented in Figure 2.44.

2.4.7.6.8 Exemplary RRC connection's Life-Cycle The procedures described above can be seen as building blocks which are utilized by the UE and by the network to control the connection over time. An exemplary life-cycle of the RRC connection using these building blocks is presented in Figure 2.45. It shows the RRC messages exchange from detection of DL data at the network side and establishing the connection, through suspending the connection, and releasing it in the end.

2.4.7.7 NAS Information Transfer

The NAS protocol allows the UE to exchange control messages with the Core Network. NAS relies on RRC protocol for the delivery of its messages and there are two dedicated RRC messages specified for this purpose: DLInformationTransfer and ULInformationTransfer. Their names are very straightforward and their sole content is a NAS message, which is provided to the RRC layer from the NAS layer and is sent in a way transparent to the RRC or to the AS in general.

2.4.7.8 UE Assistance Information

The purpose of the UEAssistanceInformation message is for the UE to be able to provide the network with an additional information assisting the network to configure radio parameters in the UE. Based on the network configuration, the UE may send a UEAssistanceInformation message during an on-going RRC Connection for two purposes:

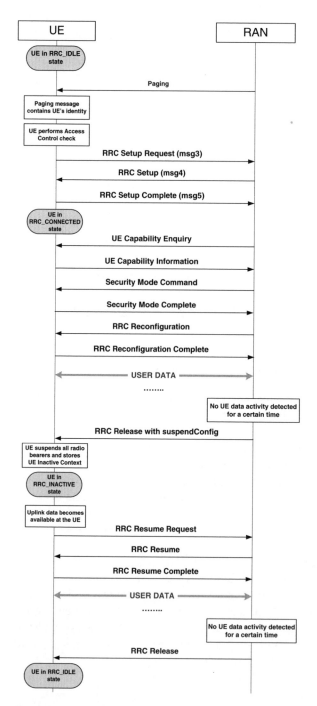

Figure 2.45 Exemplary life-cycle of the UE's RRC connection.

- For the UE to be able to suggest the network to adjust its Connected DRX setting, see Section 2.4.6.8. It may happen that the Connected DRX is configured in such a way that causes the data generated by applications running on the UE to be buffered for a long time before the UE is able to send them. By modifying the time when DRX "ON time" is occurring, the time the data spends in the UE's buffer can be minimized together with the delay the data experiences between the UE and gNB in general.
- For the UE to be able to inform the gNB about potential overheating problems it experiences. In a normal situation, the gNB may provide the UE with a preferred configuration in terms of the number of configured frequency layers, component carriers, cells, MIMO, etc. as long as they are compliant with the UE capabilities, which the UE advertised earlier. However, some configurations, together with external factors such as simultaneous Bluetooth or WiFi connection, processor heavy application running on the UE or simply high environment temperature, may lead to UE experiencing an overheating situation, which in extreme cases could lead to (permanent) device malfunction. When such a situation occurs, the UE may decide to inform the network about the overheating problem and indicate its preference to address it by reducing the configuration of one of the following parameters:
  - ○ maximum aggregated bandwidth across all of its configured UL or DL carriers on either Frequency Range 1 (FR1) or FR2
  - ○ maximum number of configured Component Carriers in either UL or DL
  - ○ maximum number of MIMO layers of each serving cell configured in either FR1 or FR2

Even though the UE Assistance Information message serves only those two purposes at the moment, it is very likely it will be extended in the future when new use cases or features are introduced (e.g. NR V2X), similarly as in LTE.

2.4.7.9 RRC PDU Structure

3GPP TS 38.331 [18], i.e. RRC protocol specification, defines both the UE procedures and UE behavior upon reception of various RRC messages as well as the structure and contents thereof. When analyzing how the system works, it is crucial to understand how to interpret the contents of the described RRC messages and this chapter shortly explains this aspect using selected RRC messages and Information Elements (IEs) as examples.

The contents of RRC PDU are described using abstract syntax notation one (ASN.1), which is specified by ITU-T within Recommendations X.680 [19] and X.681 [20]. There are different types of encoding rules available for ASN.1 determining how the RRC message is translated into bits when it is transferred over the air interface available for ASN.1. The one used by RRC is called Packet Encoding Rules unaligned (PER unaligned) and is specified in ITU-T Recommendation X.691 [21].

ASN.1 syntax reminds that of some programming languages and many data types are common for those. Hence, they are easy to understand for anyone who had at least a brief experience with programming in the past. ASN.1 allows having a standardized RRC parameters representation and at the same time it allows companies implementing the protocol stack to use whichever programming language they prefer without posing the risk that the RRC messages will not be understood between two end devices developed by different vendors, e.g. because of using different programming environment, languages, and tools. At the same time, it is easy to understand for the human being trying to read and interpret the specifications without using any additional specialized tools.

The structures used by RRC protocols described using ASN.1 can be in general divided into two types: messages and Information Elements (IEs). Messages are the structures which are sent between the UE and the gNB and they in turn contain many fields (or parameters), which are defined using Information Element structures. Many of the RRC messages were already mentioned throughout this chapter when describing different RRC procedures. Here, the representation of SIB1 message using ASN.1 notation is depicted as an example, copied directly from 3GPP TS 38.331 [18]:

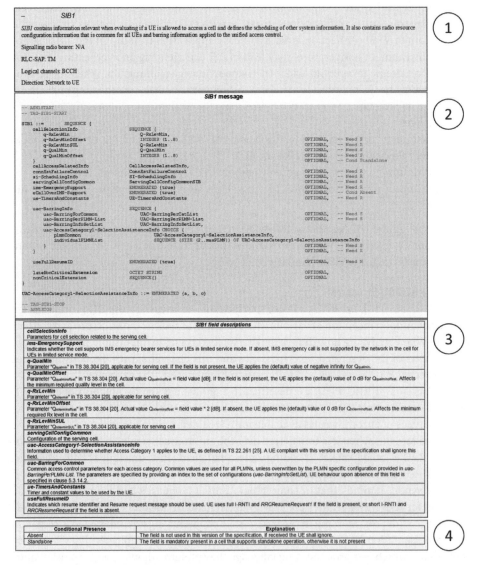

Below are the explanations of each of the parts of message definition, following the color-coding and numbering in the figure.

1. An overall description of the message purpose and contents. It also contains information about the signaling bearer used for the message delivery (if applicable, which is not the case for broadcast messages), utilized RLC mode (Transparent Mode in this case), logical channel to which the message is mapped (Broadcast Control Channel for SIB1) and the direction in which the message is sent.

2. Each of the fields (or almost each) is explained in the table below the message's or Information Element's structure. The field description may contain information about the meaning of the field, interpretation of the field or its allowable values, references to other specifications where the parameter is used or defined, etc.

3. Structure of the message using ASN.1 notation. This is a sequence of parameters (called also "fields") of various types. The type of the field can either be ASN.1 built-in type, e.g. INTEGER or ENUMERATED, or be an Information Element specified in RRC specifications. Information Elements may be either defined within particular Information Element's or message's structure (e.g. UAC-AccessCategory1-SelectionAssistanceInfo in this example) or be defined outside of it, as a separate Information Element, e.g. CellAccessRelatedInfo. As can be seen, some fields can also be defined as a set of other parameters introduced with a SEQUENCE type in ASN.1, e.g. cellSelectionInfo. Another possibility is to define a field as a CHOICE structure containing a list of parameters out of which a single one is chosen by the UE or the network when sending the message, e.g. uac-AccessCategory1-SelectionAssistanceInfo.

   Each parameter can either be mandatory or optional. In the latter case, the OPTIONAL keyword is put on the right side of the parameter. As can be seen, the only mandatory field of SIB1 message is cellAccessRelatedInfo. Next to most of optionality markers, in gray, there are so called Need codes, which specify how the UE should interpret the absence of the optional field (see Table 2.14). Some of the optional fields may also have a "Cond" keyword (standing for "conditional presence") followed by the condition name. The last two fields of the message (lateNonCriticalExtension and nonCriticalExtension) have a purpose of providing extensibility of the message in the future releases of the standard. Thanks to them, new contents can be added to the message in a backwards compatible way, i.e. in such a way that will not cause UEs compliant with the older versions of the specifications to treat the message as erroneous. They will still be able to decode the part of the message before the extension marks and will simply ignore the contents inside the extension markers. New UEs, on the other hand, will be able to interpret the new content and will be able to take advantage of the new functionality it provides. When browsing through the latest RRC specifications, one can already notice some extension markers being used. This is due to the fact that Rel-15 of NR specification, firstly released for NSA operation, was then complemented to support Standalone operation. Some of the corrections, which had to be introduced after approving initial versions of the specifications, also required using extension markers in order to keep backwards compatibility with the UEs which might have already implemented previous versions of RRC specifications.

4. Below the fields descriptions, the conditional presence markers are explained. Usually, the situations or conditions under which the field must, can or cannot be signaled are indicated here.

Table 2.14 Meaning of the NEED codes.

NEED code	Meaning
Need S	*S = Specified* Used for fields, whose field description or a related procedure specifies the UE behavior upon receiving a message with the field being absent.
Need M	*M = Maintain* Used for fields that are stored by the UE i.e. not one-shot. Upon receiving a message with the field absent, the UE maintains the current value.
Need N	*N = No action* Used for fields that are not stored and whose presence causes a one-time action by the UE. Upon receiving message with the field absent, the UE takes no action.
Need R	*R = Release* Used for fields that are stored by the UE i.e. not one-shot. Upon receiving a message with the field absent, the UE releases the current value.

Source: 3GPP TS 38.331

As another example of ASN.1 notation, the definition of one of the Information Elements is provided below.

– CSI-ResourceConfig

The IE *CSI-ResourceConfig* defines a group of one or more *NZP-CSI-RS-ResourceSet*, *CSI-IM-ResourceSet* and/or *CSI-SSB-ResourceSet*.

CSI-ResourceConfig information element

```
-- ASN1START
-- TAG-CSI-RESOURCECONFIG-START

CSI-ResourceConfig ::=        SEQUENCE {
    csi-ResourceConfigId        CSI-ResourceConfigId,
    csi-RS-ResourceSetList      CHOICE {
        nzp-CSI-RS-SSB              SEQUENCE {
            nzp-CSI-RS-ResourceSetList  SEQUENCE (SIZE (1..maxNrofNZP-CSI-RS-ResourceSetsPerConfig)) OF NZP-CSI-RS-ResourceSetId
                                                                                    OPTIONAL, -- Need R
            csi-SSB-ResourceSetList     SEQUENCE (SIZE (1..maxNrofCSI-SSB-ResourceSetsPerConfig)) OF CSI-SSB-ResourceSetId
                                                                                    OPTIONAL  -- Need R
        },
        csi-IM-ResourceSetList      SEQUENCE (SIZE (1..maxNrofCSI-IM-ResourceSetsPerConfig)) OF CSI-IM-ResourceSetId
    },

    bwp-Id                      BWP-Id,
    resourceType                ENUMERATED { aperiodic, semiPersistent, periodic },
    ...
}

-- TAG-CSI-RESOURCECONFIG-STOP
-- ASN1STOP
```

CSI-ResourceConfig field descriptions
bwp-Id
The DL BWP which the CSI-RS associated with this CSI-ResourceConfig are located in (see TS 38.214 [19], clause 5.2.1.2
csi-ResourceConfigId
Used in CSI-ReportConfig to refer to an instance of CSI-ResourceConfig
csi-RS-ResourceSetList
Contains up to maxNrofNZP-CSI-RS-ResourceSetsPerConfig resource sets if ResourceConfigType is 'aperiodic' and 1 otherwise (see TS 38.214 [19], clause 5.2.1.2)
csi-SSB-ResourceSetList
List of SSB resources used for beam measurement and reporting in a resource set (see TS 38.214 [19], section FFS_Section)
resourceType
Time domain behavior of resource configuration (see TS 38.214 [19], clause 5.2.1.2). It does not apply to resources provided in the csi-SSB-ResourceSetList.

As can be seen, Information Elements are defined in a very similar way as messages. They also do contain mandatory and optional fields of the same types. This particular IE defines the CSI-RS resource configuration, which is one of the elements used for channel state estimation or for beam management. In the field descriptions, there are a lot of references to 3GPP TS 38.214 [22], which is one the physical layer specifications of NR. At the end of the IE's definition we can see an ellipsis, which is yet another type of extension marker used in

ASN.1 allowing for the IE to be extended with new fields or new values for the existing fields in future versions of the specifications developed in the next releases of 3GPP standard.

2.4.7.9.1 *Readers Take* It is worth pointing out that, in general, the protocol structure and main functions of the protocols are similar to those known from the previous generation system, i.e. LTE. However, the support of higher bit rates, higher frequencies, services requiring ultra-reliable and short latency data transfer (URLLC) as well as for new cloudified RAN deployments demanded for certain modifications and optimizations. These are rather obvious and clear when looking at physical layer design but are definitely not limited to it, which is proven throughout this chapter. Some notable examples of enhancements in the RAN architecture and Layer 2/3 protocols domains include:

- Cloudified RAN deployments are facilitated by introducing F1 interface allowing to split the gNB into Centralized and Distributed Units and by and introducing E1 interface separating the gNB Centralized Unit into control plane and user plane functions.
- Specification of standalone and various Multi-RAT Dual Connectivity architectural options provides operators with flexibility in choosing an adequate 4G to 5G migration path.
- User plane data processing was streamlined by removal of duplicate functions, e.g. there is no reordering in RLC (only in PDCP in NR while in both RLC and PDCP in LTE), there is no concatenation in RLC in NR (only multiplexing in MAC while in LTE concatenation takes place in RLC followed by multiplexing in MAC), Sequence Number is added in RLC UM in NR only in case RLC SDU needs to be segmented.
- Introduction of PDCP duplication to enhance reliability for URLLC services (even though similar enhancement was eventually introduced for LTE as well).
- New functions in MAC layer were introduced such as support for multiple numerologies, BWPs and beam recovery. MAC participates also in CSI related procedures.
- A new SDAP protocol used for QoS flow to DRB mapping was introduced.
- New control plane functions were introduced in RRC such as on demand System Information delivery, RRC INACTIVE state, beam and BWPs handling, Unified Access Control.

3

PHY Layer

Mihai Enescu[1], Youngsoo Yuk[2], Fred Vook[3], Karri Ranta-aho[1], Jorma Kaikkonen[1], Sami Hakola[1], Emad Farag[3], Stephen Grant[4], and Alexandros Manolakos[5]

[1]*Nokia Bell Labs, Finland*
[2]*Nokia Bell Labs, Korea*
[3]*Nokia Bell Labs, USA*
[4]*Ericsson, USA*
[5]*Qualcomm Technologies, Inc.*

3.1 Introduction (Mihai Enescu, Nokia Bell Labs, Finland)

In the following two chapters we are following on the 5G new radio (NR) physical layer description as well as on the defined procedures utilizing the PHY elements. We are starting this chapter with two main design elements: the 5G waveform and the antenna architectures. These two are basic components of the new system especially from the perspective of defining a unified framework across the operating frequency bands, hence we tackle the challenges in selecting these technical components as well as describing the specified framework. Following the introduction of these two design points, we describe the basic system design components such as frame structure and resource allocation, synchronization signals, and broadcast channels, the physical random access channel (PRACH). The 5G NR reference signals have their dedicated chapters where Channel–state information reference signals (CSI-RS), DeModulation Reference Signals (DM-RS), phase tracking reference signal (PT-RS) and Sounding reference signals (SRS) are described in detail, including their downlink (DL) and uplink (UL) characteristics (where applicable). A short description on power control is also provided and the chapter is closed by the description of the DL and UL transmission framework.

With respect to the reference signals, it is worth already mentioning here the convention on the port numbering which is used in 5G NR system.

The following antenna ports are defined for the uplink:

- Antenna ports starting with 0 for DM-RS for physical uplink shared channel (PUSCH)
- Antenna ports starting with 1000 for SRS, PUSCH
- Antenna ports starting with 2000 for physical uplink control channel (PUCCH)
- Antenna port 4000 for PRACH

The following antenna ports are defined for the downlink:

- Antenna ports starting with 1000 for physical downlink shared channel (PDSCH)
- Antenna ports starting with 2000 for physical downlink control channel (PDCCH)
- Antenna ports starting with 3000 for CSI-RS
- Antenna ports starting with 4000 for SS/PBCH block transmission

3.2   NR Waveforms (Youngsoo Yuk, Nokia Bell Labs, Korea)

3.2.1   Advanced CP-OFDM Waveforms for Multi-Service Support

The introduction of new services and spectrum bands, as the main drivers of 5G system, have brought new challenges into the design of 5G waveform. 5G Radio Access is expected to handle a very wide range of use cases and requirements, including, among others, enhanced mobile broadband (eMBB), massive machine type communications (mMTC) and ultra-reliable/low latency communications (URLLC). For mobile broadband, 5G system is expected to fulfill the demand of exponentially increasing data traffic and to allow people and machines to be served with gigabit data rates with very low latency.

In order to reach to the goal of 5G system design, a number of new requirements have been considered as design principles of 5G waveform:

- High spectrum utilization
- Flexible numerology support
- Multi-service support: low latency and high reliability
- Coverage enhancement

For high spectrum utilization, the guard band (GB) between carrier bandwidth should be minimized. In LTE, 90% of the carrier bandwidth can be used for data transmission (referred as Transmission bandwidth configuration) and 5% of the spectrum at each edge should be reserved as guard band. In NR, the goal is to improve the overall spectral efficiency more than 90%, such that up to 98% of the bandwidth can be used for data transmission.

OFDM- based waveforms are considered as the most suitable choice for the downlink, uplink, and relay and backhaul transmissions in 5G NR. Waveforms from the OFDM family (including their single-carrier frequency division multiple access [SC-FDMA] and time/frequency domain variants) are a good choice over the whole frequency range, due to high flexibility in multiple access, good MIMO compatibility and low complexity FFT-based implementations with scalar equalization. Despite of OFDM's good properties for a strong candidate for waveform of 5G, there are well-known drawbacks of OFDM waveforms. A most well-known problem of OFDM is high peak-to-average power ratio (PAPR) due to the random addition of subcarriers in the time domain. This is critical in uplink coverage extension; hence single carrier waveform is preferable. In LTE uplink, DFT-S-OFDM, so-called SC-FDMA, has been used for uplink waveform by adding DFT operation before the OFDM modulation, and the data is transformed into time domain which results in low PAPR property.

Another problem of OFDM is its high out of band emissions (OOBE). The rectangular pulse waveform of traditional OFDM leads to undesirable magnitude responses that suffer from large side lobes in the frequency domain. The large out-of-band radiation causes severe interference when heterogeneous traffic types are simultaneously served by sharing the unified resource grid. Further, to avoid the interference, OFDM waveform imposes generous guard bands and it severely deteriorates the spectral efficiency. Finally, to prevent inter-carrier interference (ICI), OFDM signals should be tightly synchronized.

For 5G multi-service support, waveforms should provide the time/frequency localization property to mix various services having different requirements. In time domain, a fixed cyclic prefix (CP) to mitigate inter-symbol interference (ISI) is set to cover maximum delay spread and synchronization margin resulting in a fixed overhead not fully optimized in the transmission channels. Perfect time synchronization across different frequency resources is required to avoid ICI between the frequency allocations. Classical CP-OFDM is limited to support different numerologies with various frequency/time scaling due to ICI when different frequency scaling (subcarrier spacing) is applied to different sub-bands.

Figure 3.1 is illustrating the example of the multi-service support in a carrier bandwidth. In this example, three user equipments (UEs) are considered, each UE utilizing its own frequency resources. To support low latency service for UE2, shorter OFDM symbol length with larger subcarrier spacing (e.g. 60 kHz) can be used while UE1 and UE3 are served by normal eMBB service with 15 kHz subcarrier spacing. In this case, to serve UEs with different service requirements, different numerology should be supported in a carrier bandwidth, and the waveform should be capable of well-localizing the sub-band by reducing the interference outside the sub-band. Also, in uplink, such frequency/time localization may allow asynchronous transmission from different UEs.

Figure 3.2 is illustrating how ICI is generated when different numerologies are applied in adjacent physical resource blocks (PRBs). The orthogonality of the OFDM subcarriers may be broken because the different zero crossing points are not aligned for different numerologies in each frequency bin. Though a guard band can reduce the strong ICI, the residual interference at the transmitter degrades the signal quality regardless of the signal-to-noise and interference ratio (SINR) of the received signal.

To overcome these shortcomings of OFDM, new waveform technologies to satisfy 5G requirements have been proposed and evaluated [23]. The considered waveforms were

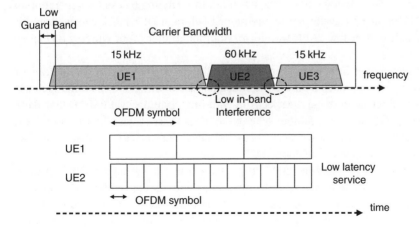

Figure 3.1 Illustration of multi-service support of 5G new radio.

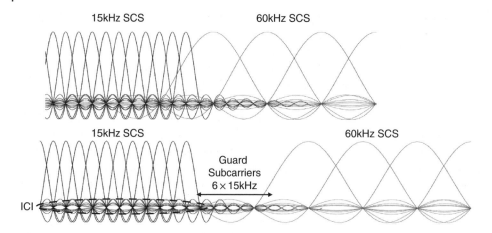

Figure 3.2 Illustration of ICI among OFDM subcarriers with different numerologies.

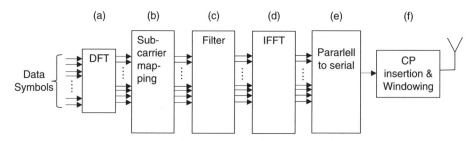

Figure 3.3 Generation of enhanced OFDM based waveforms.

designed with the goal of satisfying following properties: robustness against ISI/ICI and synchronization error, well-localized pulse responses in time/frequency domain, suitability for short burst transmission, low out-of-band power radiation, and, etc.

Most common approach to enhance OFDM waveform is to apply spectral shaping with additional filtering or windowing. Figure 3.3 shows the building blocks for generation of enhanced OFDM based waveforms. As mentioned before, a DFT block can be used for generating single-carrier low PAPR waveform. Additional filter (c) or window (f) can be added.

Two types of schemes have been proposed according to where to apply the filtering or the windowing. One approach is to use the subband-wise filtering and the other is to use subcarrier-wise filtering to shape the spectrum with finer granularity. FBMC (Filter Bank Multi-Carrier) [24] is one of the famous schemes to use subcarrier-wise filtering and it was one of strong candidates for NR waveform because of its superior spectral efficiency as well as reducing ISI resulting in the deletion of CP.

Despite such key advantages of FBMC, there were a few drawbacks which had to be considered. The first problem was the low compatibility with the conventional OFDM techniques. Many component technologies such as modulation and MIMO techniques should have been redesigned for FBMC. Therefore, backward compatibility was the most challenging issue. Second, because of overlap and sum structure of FBMC, a longer transition time is required. For short transmit symbols, the overhead caused by transition time can be larger than overhead of CP. Therefore, FBMC is not adequate waveform for short burst traffic signals. Due to such drawbacks, finally FBMC has not been considered as a candidate waveform of NR, and CP-OFDM-based waveform was agreed as baseline waveform for 5G.

Various advanced CP-OFDM waveforms applying subband-wise spectral shaping have been proposed during the release-14 NR study in 3GPP RAN WG1 such as WOLA (Weighted OverLap-Add) [23], filtered-OFDM or universal filtered (UF)-OFDM [25]. Each waveform provided the improvement in OOBE over conventional CP-OFDM with the additional complexity. The trade-off between complexity and performance gain was the key decision criteria for NR waveform design. The proposed schemes have their own advantages as well as drawbacks, and the best waveform option may be differently selected according to the implementation options and the use cases to be considered. Table 3.1 provides the comparison of the various schemes in terms of performance and their complexity.

Also, from the 3GPP RAN4 studies, the feasibility of the communication between UE and a gNB with the different implementation without specifying any filtering or windowing schemes under certain requirements has been considered.

Figure 3.4 shows the evaluation results of the receiver performance when the desired signal (15 kHz of subcarrier spacing) is transmitted with the interference signal (60 kHz subcarrier spacing) adjacent with guard band (GB). From the evaluation, with certain guard band (e.g. 0.5 or 1 PRB), various schemes show the performance gain of interference suppression over CP-OFDM.

$$64\text{-QAM}, R = 3/4 \ 256\text{-QAM}, R = 3/4$$

Figure 3.5 shows the evaluation results on error vector magnitude (EVM) performance when different TX and RX transceiver options are used [27]. Left figure shows the evaluation results when CP-OFDM is transmitted but different RX processing schemes (WOLA, f-OFDM, UF-OFDM and CP-OFDM) and parametrizations are applied, while the right figure shows the results when the same processing schemes are applied to both TX and RX. As shown in the figure, the EVM performance is rather impacted by the RX schemes only, and the matched operation in TX an RX is not necessary. Thus, the TX waveform can be transparent to the receiver as long as enough interference suppression is applied at the receiver.

Table 3.1 Comparison of various candidate waveforms for NR.

	Implementation	Spectral efficiency	Out-of-band radiation	Short burst transmission	Complexity	Backward compatibility
CP-OFDM	Baseline	Low	High	Yes	Low	Yes
UF-OFDM	Sub-band filtering	Moderate	Low	Yes	High	Yes
f-OFDM	Sub-band filtering	Moderate	Low	Yes	High	Yes
FBMC	Subcarrier filtering	High	Very Low	No	High	No
WLOA	Windowing	Moderate	Reduced	Yes	Moderate	Yes

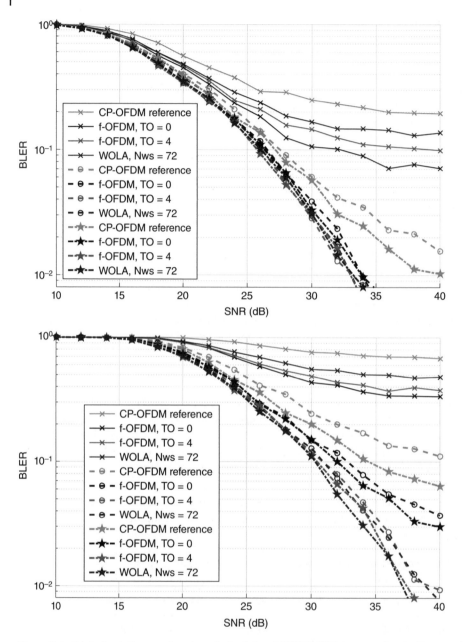

Figure 3.4 Link simulation results when desired signal (15 kHz SCS) is transmitted adjacent with the interfering signal (60 kHz SCS) (solid line has 0 GB, Dashed line has 6 SC GB (90 kHz) and Dot-Dashed line 12 SC GB (180 kHz)) [26].

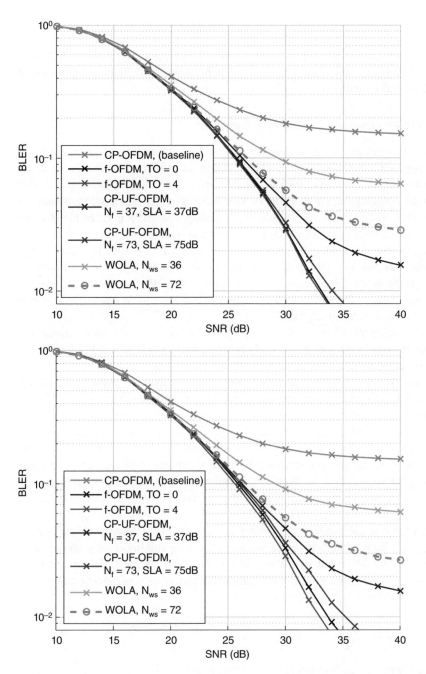

Figure 3.5 Rx unit performance comparison between matched transceiver with asynchronous interfering signal on both sides of the desired signal. Left hand side: Different Rx waveform processing, CP-OFDM Tx, right hand side: matched Tx and Rx waveform processing.

From the study results, though the principle of the enhanced waveform has been widely agreed, 3GPP has finally decided the application of the specific schemes to be considered as implementation instead of defining as a part of physical layer specification. In 3GPP RAN4, additional requirements on in-band emission and in-band sensitivity have specified as gNB and UE radio requirements [28, 29].

3.2.2 Low PAPR Waveform for Coverage Enhancement

Another aspect considered in system design is coverage enhancement, and low PAPR is one of the key requirements. OFDM has high PAPR due to the random addition of subcarriers in the time domain. As a result of such high peaks, the power amplifier at the transmitter operates in the nonlinear region causing a distortion and unwanted emission outside the bandwidth. Thus, single-carrier based waveforms having low PAPR gets the benefit over multi-carrier-based waveform like CP-OFDM in terms of coverage enhancement.

DFT-S-OFDM has been used for uplink waveform of LTE because uplink coverage enhancement was the key performance factor, however, DFT-S-OFDM is limited in providing higher spectral efficiency compared to CP-OFDM. DFT-S-OFDM waveform can be simply implemented by adding DFT operation before CP-OFDM generator, and they can share many common building blocks.

During NR study period, various new single-carrier waveforms have been proposed as complementary waveforms for uplink such as single carrier QAM, zero-tail/head DFT-S-OFDM [30], unique word-OFDM, etc.

The zero-tail DFT-s-OFDM signals may overcome the limitations imposed by a fixed-length CP. Specifically, to decouple the radio numerology from the radio channel characteristics by replacing the fixed length CP with a set of very low power samples (named zero-tail), which are part of the inverse fast Fourier transform (IFFT) output. The usage of zero-tail signals allows decoupling of the radio numerology from the expected channel characteristics, which brings corresponding benefits in terms of overhead/block-error rate (BLER). Also, the zero-tail can be used as a transition time between different carriers or different beams in FR2. The Zero-tail DFT-s-OFDM signal generation diagram is presented in Figure 3.6

However, such non-CP based signal cannot co-exist with CP-OFDM waveform due to different timing. Finally, thanks to the advantage of the compatibility of the waveform generation with CP-OFDM, in NR uplink in addition to CP-OFDM, as baseline waveform, CP-based DFT-s-OFDM can be used as the complementary waveform for coverage-limited use case.

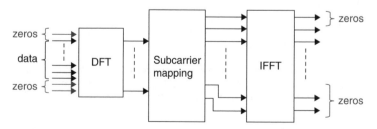

Figure 3.6 Zero-tail DFT-s-OFDM signal generation.

In NR specification, the DFT spreading operation is specified as "transform-precoding," and it can be applied for PUSCH and PUCCH format 3/4. For PUSCH, only single layer transmission with localized resource allocation is applied.

For Msg3 PUSCH transmission, which is a PUSCH transmission scheduled by Random Access Response (RAR) UL grant or a PUSCH scheduled by downlink control information (DCI) format 0_0 scrambled by TC-RNTI, the UE shall apply the transform precoding to the Msg3 transmission when the higher layer parameter *msg3-transformPrecoder in RACH-ConfigCommon* is set to "enabled."

For PUSCH transmission except Msg3 PUSCH, transform precoding is enabled, when

- *transformPrecoder* in *PUSCH-Config* (or *configuredGrantConfig*) is set to "enabled," or
- *transformPrecoder* is not configured in *PUCCH-Config* but *msg3-transformPrecoder in RACH-ConfigCommon* is enabled.

In NR, when transform precoding is enabled, $\pi/2$-BPSK modulation can be used to both PUSCH and PUCCH transmission for further PAPR reduction. In this case, additional spectral shaping is also applicable for further PAPR reduction.

Figure 3.7 shows the comparison of the PAPR when $\pi/2$-BPSK modulation with/without spectral shaping are applied. When transform precoding is enabled, the modulation and coding scheme (MCS) table to use can be configured by the higher layer parameter *mcs-TableTransformPrecoder*. If *mcs-TableTransformPrecoder* is not set to "qam256," the UE can be configured to use $\pi/2$-BPSK modulation by the higher-layer parameter *tp-pi2BPSK*.

Any spectrum shaping can be applied as long as it fulfills the requirement on EVM equalizer spectrum flatness specified in TS38.101-1 [28] and TS38.101-2 [31].

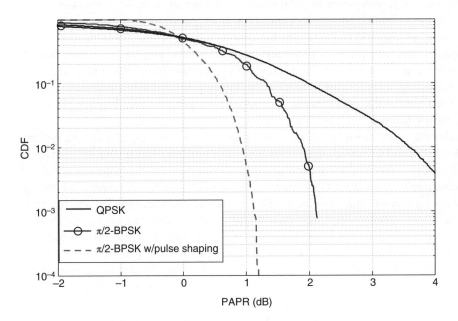

Figure 3.7 PAPR reduction with $\pi/2$-BPSK modulation and spectral shaping.

3.2.3 Considerations on the Waveform for above 52.6 GHz

The frequency bands beyond 52.6 GHz contain large spectrum allocations and support higher capacity and higher bit rate use cases than lower mmW bands. The main challenges are increased path loss and less efficient RF components. The larger number of antenna elements can be accommodated in the same form factor as in lower mmW bands thus compensating for the pathloss and providing comparable LoS coverage. Also, a dense network deployment is needed in dense urban and indoor environments. Considering these benefits and limitations, specific enhancements to NR must be considered to operate effectively at these higher bands.

The system design for above 52.6 GHz has not been part of the first phase of NR 5G design, however, 3GPP is considering such deployments and has crafted the objectives of the study are as following [32]:

- identify target spectrum ranges, including survey on global spectrum availability, regulatory requirements, channelization, and licensing regimes,
- identify potential use cases and deployment scenarios
- identify NR design requirements and considerations on top of regulatory requirements

The study has captured the following considerations and requirements for the waveform design for NR beyond 52.6 GHz. [33]

○ *Power efficiency of power amplifier (PA)*: PA efficiency at higher frequency is expected to degrade and low PAPR waveforms designed to minimize PA backoff and maximize efficiency should be considered. At higher frequencies and especially in millimeter wave frequencies, output power per transistor as well as power added efficiency decrease. Waveform with high power back-off to support EVM and out-of-band emission (OBE) requirements could dramatically reduce PA efficiency even further.

○ *Dynamic range of ADC and DAC*: The increase in PA back-off also affects the other device requirements, for example, the dynamic range of the ADC and DAC. A higher Tx DAC effective number of bits (ENOB) is required to accommodate higher PAPR, and extra oversampling in the baseband DSP, and Tx DAC may be needed to accommodate wider channel bandwidth. All of these are impacted by waveform, and therefore should be carefully evaluated.

○ *Modulated signal accuracy and out-of-band emission*: Power amplifier is designed and adjusted to meet RF requirements, such as spectral mask emission (SEM), adjacent channel leakage ratio (ACLR), in-band emission (IBE) and OBE requirements, and EVM requirements. Proper RF requirements are needed to determine appropriate in-band signal quality characteristics, minimize adjacent channel interference and impact to signals in adjacent channels. Occupied signal bandwidth and guardband for a given channel bandwidth in high carrier frequencies above 52.6 GHz require further investigation.

○ *Complexity and performance of waveform*: Given the high data rate and high sampling rates the system is expected to operate, the complexity and performance tradeoff for waveform generation/modulation and reception/demodulation should be considered.

○ *Spectrum flexibility of waveform*: Use cases and frequency allocations by various government bodies may require various bandwidth to be supported. Therefore, flexibility to support different system bandwidth should be considered in the design.

o *Robustness to frequency offset and phase noise*: Carrier frequency offset and phase noise is much higher in spectrum beyond 52.6 GHz because of imperfection of PA and crystal oscillator is more severe than that of lower bands. In addition, Doppler shift/spread is also larger with the carrier frequency increasing. As a result, robustness on frequency offset and phase noise is one of the key requirements for systems operating on bands above 52.6 GHz. Increasing the subcarrier spacing for CP-OFDM waveform to better cope with increased phase noise could be investigated. For other potential waveforms impact from phase noise and ability to robustly handle phase noise should be investigated.

o *Feature re-usability and design commonality with existing NR specification*: It would be good to be able to support features for FR1 and FR2 as defined in NR with minimal change (if possible) and support a common design structure that could support various use cases. To that extent further considerations of using an integer ratio between clock rates of NR below and NR above 52.6 GHz should be investigated. One possibility to achieve this would be to maintain the NR numerology scaling principle but extend to higher numerologies, i.e. $\Delta f = 2^\mu \times 15$ kHz with an appropriate range of possible integer values for μ.

As following work, new release-17 study item may be continued with the focus of the physical layer design aspect. Waveform design is considered as one of key study objectives, including the introduction of low PAPR waveform for both downlink and uplink.

3.3 Antenna Architectures in 5G (Fred Vook, Nokia Bell Labs, USA)

In this section, we discuss various aspects of how the antenna arrays that are being used in 5G NR systems are implemented and configured for creating a "beam-based air interface." We first describe some of the basics of beamforming for receiving and transmitting. Next, we describe the three main categories of antenna array architectures, namely digital, analog, and hybrid architectures. Next, we describe the idea of antenna virtualization, which is the aggregation of multiple physical antennas in to a single functioning logical antenna element, and the concept of antenna ports for accounting for the different signals transmitted in the system. Finally, we describe how all these ingredients are put together to form a beam-based air interface.

3.3.1 Beamforming

Beamforming is a core physical layer technique at the heart of the NR air interface and can be performed during transmission and reception. Beamforming can improve system coverage performance and reliability by focusing the transmitted energy toward the intended receiver rather than spraying the energy indiscriminately in many directions. Beamforming can also increase system capacity by multiplexing multiple users on the same time-frequency resources, a process called "spatial multiplexing."

Figure 3.8 contains a basic diagram of beamforming for transmit (left) and receive (right). The transmit beamforming operation consists of splitting a signal into multiple identical copies, weighting those copies with a complex gain and phase weight (shown as a multiplier in Figure 3.8) and then sending the weighted signal copies to the respective antennas.

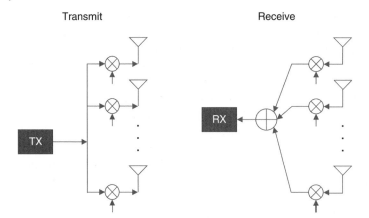

Figure 3.8 Single beam beamforming – general concept. Left: transmit. Right: receive.

The receive beamforming operation consists of taking the signals received from multiple antennas, weighting them with complex gain and phase weights and then summing the weighted signals to form a single output. Figure 3.8 shows what we call "single-beam" beamforming and is shown in a very generic context where the signals in the diagram can be digital signals at baseband, analog signals at RF, time-domain single carrier signals, or even a frequency-domain signal such as the signal to be transmitted on a single OFDM subcarrier for example. In some applications, the weighting coefficients are phase-only, meaning no gain adjustment is made; but generally speaking, the weighting coefficients can be varied in both gain and phase. For receive beamforming, the "pattern" is the response of the beamforming output as a function of the signal arrival angle and is determined by the weighting coefficients and the characteristics of the antennas. Correspondingly for transmit beamforming, the "pattern" is the response as a function of the departure angle and is also determined by the weighting coefficients and the characteristics of the antennas. Essentially the receive pattern indicates how sensitive the receive array is to signals arriving from different directions, whereas the transmit pattern indicates how the transmitted power is distributed in angle/space.

The "multi-beam" beamformer is an important extension of the single beam beamformer shown in Figure 3.8. A multi-beam beamformer can be realized in one of two ways: fully-connected (Figure 3.9) or sub-array-connected (Figure 3.10). In a fully-connected multi-beam beamformer, each input signal is beamformed across all elements in the array, whereas in the sub-array-connected multi-beam beamformer, each input signal is beamformed across a different (disjoint) set of elements from those used for the other input signals. A key feature of the sub-array multi-beam transmit beamformer is how there is no need for the summation devices behind every antenna element, which can be an important implementation consideration. Another point worth noting is that from a mathematical or signal processing perspective, a sub-array connected configuration can be realized within a fully-connected configuration by setting to zero the appropriate weighting coefficients in the fully connected configuration. However, in a real implementation, there are significant hardware related differences between a sub-array and fully-connected array, mostly in terms of the necessary hardware components.

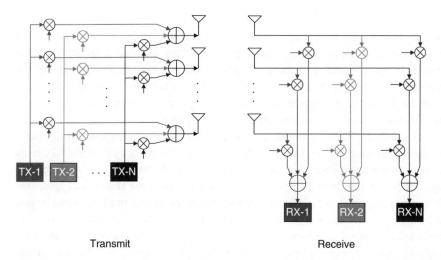

Transmit   Receive

Figure 3.9 Multi-beam beamforming antenna array – N beams in a fully connected configuration. Left: transmit. Right: receive.

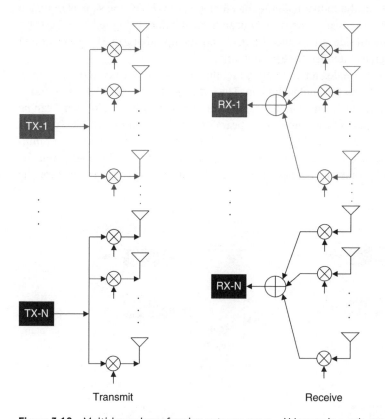

Transmit   Receive

Figure 3.10 Multi-beam beamforming antenna array – N beams in a sub-array connected configuration. Left: transmit. Right: receive.

3.3.2 Antenna Array Architectures

The antenna array architecture (sometimes called the beamforming architecture) refers to the configuration of the physical antenna elements of the antenna array and how the transceiver units are connected to those physical antenna elements. For classifying the antenna array architectures, there are three main categories, where a distinguishing feature is how the beamforming operation is controlled. These three main categories are (i) digital architectures, (ii) analog architectures, and (iii) hybrid architectures.

Figure 3.11 shows the digital antenna array architecture, where the beamforming operation is performed at baseband in the digital domain. The left diagram in Figure 3.11 is for transmitting a single data stream over a single beam, and the box labeled "single beam" can be replaced with the diagram of the single beam beamformer shown in Figure 3.8. The right diagram in Figure 3.11 is for simultaneously transmitting multiple data streams over multiple beams, and the box labeled "multi beam" can be replaced with the diagram of the multi-beam transmit beamformer shown in Figure 3.12. In a fully digital architecture, baseband signals are being beamformed in the digital domain, but various options are possible. For example, the baseband signals can be the entire wideband carrier, such as a single-carrier signal or a CDMA carrier, and the multiple signals are then mixed up to RF and transmitted out the individual antenna elements. In OFDM, the signals intended for each OFDM subcarrier can be separately beamformed before being fed into the IFFT operation at the transmit side, after which CP insertion and up-mixing to RF are performed separately on the signals intended for each antenna.

Figure 3.12 shows the analog antenna array architecture, where the beamforming operation is performed at RF in the analog domain. The left diagram in Figure 3.12 is for transmitting a single data stream over a single beam, and the box labeled "single beam" can be replaced with the diagram of the single beam beamformer shown in Figure 3.8. The right diagram in Figure 3.11 is for simultaneously transmitting multiple data streams over multiple beams, and the box labeled "multi beam" can be replaced with the diagram of the multi-beam transmit beamformer shown in Figure 3.9. In an analog array architecture, the

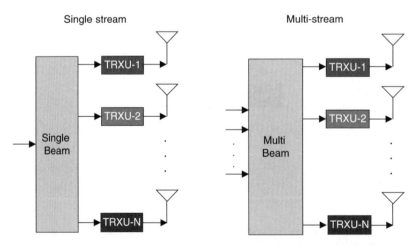

Figure 3.11 Digital antenna array architecture. Left: single beam. Right: multi-beam.

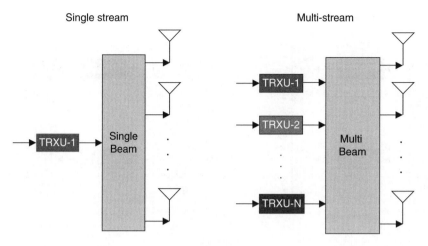

Figure 3.12 Analog antenna array architecture. Left: single beam. Right: multi-beam.

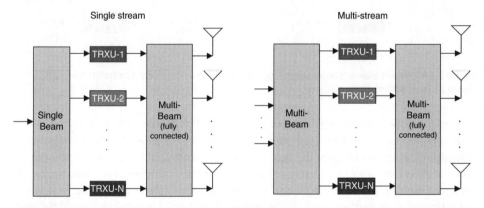

Figure 3.13 Hybrid antenna array architecture with fully-connected beamforming. Left: single beam. Right: multi-beam.

single or multiple data streams are first up-mixed to RF, and the beamforming is performed with analog phase-shifters that may or may not have gain-adjustment capability.

Figures 3.13 and 3.14 show the hybrid antenna array architecture, where there are two beamforming operations, one in the baseband digital domain, and a second in the RF analog domain. In a hybrid architecture, the term "precoding" is sometimes used to refer to the beamforming operation that is performed at baseband in the digital domain. One of the reasons behind the interest in hybrid arrays is the need to reduce the number of transceivers in the overall antenna array for various implementation considerations like cost or power consumption. For operating in the mmWave bands, current analog-to-digital and digital-to-analog converters have rather high power consumption requirements, so early mmWave implementations will typically be hybrid antenna arrays to provide a favorable tradeoff in performance, power consumption, and complexity.

These three architectures have various trade-offs and implementation considerations. The digital architectures have a high degree of flexibility given how the beamforming can

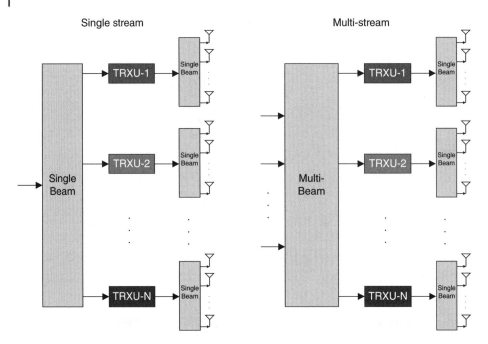

Figure 3.14 Hybrid antenna array architecture with sub-array-connected beamforming. Left: single beam. Right: multi-beam.

be performed with digital signal processing. With digital architectures combined with frequency–domain modulation schemes like OFDM, the beamforming can be adjusted in the frequency domain to adapt the beamforming response across the band to account for channel variations or even to perform user multiplexing in the frequency domain, where each user multiplexed in the frequency domain can have a different beam (i.e. beamforming weight vector). Performing multi-beam beamforming for multiplexing multiple users simultaneously in the spatial domain is relatively easy to manage in the digital domain with digital signal processing, whereas in the RF domain, each beam to be formed is created with actual physical hardware. Adjusting the beamforming characteristics across the frequency band is very difficult in the analog domain since analog RF weights are applied to the entire wideband RF signal.

3.3.3 Antenna Panels

Most cellular oriented antenna array implementations consist of one or more antenna "panels," where a panel is an arrangement of antenna elements having M rows, N columns, and P polarizations, denoted as the (M,N,P) configuration. Typically, the number of polarizations P is either 1 or 2, where P = 1 means all the elements in the panel array have the same polarization, and P = 2 means half the elements are one polarization, and the other half of the elements are a second polarization, typically orthogonal to the first polarization. Since exploiting polarization generally has significant performance benefits, the panel array is typically configured for P = 2, and the polarizations are often +45 and −45

Figure 3.15 A single antenna array panel of cross-polarized antennas (+45° and −45° elements) configured as (M,N,2).

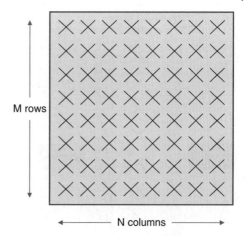

M rows

N columns

or 0 and +90. The elements in an antenna panel are often arranged in cross-pol pairs as shown in Figure 3.15. However, in some array implementations (e.g. those intended for mmWave operation), two panels may be used each with P = 1, but the second panel would have elements with a polarization orthogonal to the polarization of the elements in the first panel. Often the reason for having two panels each with a different polarization are because of implementation constraints where there is insufficient space contain the necessary hardware to support a cross-pol pair of antennas, which can happen in mmWave implementations due to the small wavelength at mmWave. For simplicity in this and subsequent discussions, we will assume antenna panels are cross-polarized panels consisting of cross-polarized element pairs (P = 2) as shown in Figure 3.15. Extensions to the dual single-pol panel configuration just described are relatively straightforward.

3.3.4 Antenna Virtualization

An important concept in operating large scale antenna arrays is the concept of antenna virtualization. Antenna virtualization involves combining multiple physical antennas into a single functioning logical antenna element that is then controlled by the physical layer of the air interface, where "controlled by" refers to the physical layer controlling the actual signals transmitted by and received from the logical antenna element. Typically, the virtualization operation can be represented as a simple beamforming operation of the type shown in Figures 3.8–3.10. Also, the term virtualization is typically intended to refer to cases where the combining operation is static, meaning that the combination of the multiple physical antennas is fixed for a long time and/or not adapted by the air interface. However, if the virtualization becomes adaptive and/or controlled by the air interface, then the term "virtualization" is typically not used, and this case is dealt with in the next subsection, where we will discuss the concept of beam-based operation. In this subsection, we will focus on static antenna virtualization, which basically forms static logical antenna elements that are controlled by the air interface with MIMO or beamforming transmission and reception techniques.

To illustrate the idea of antenna aggregation, consider a single column array of cross-pol elements as shown in Figure 3.16. Two transceivers are connected to the array, and all the

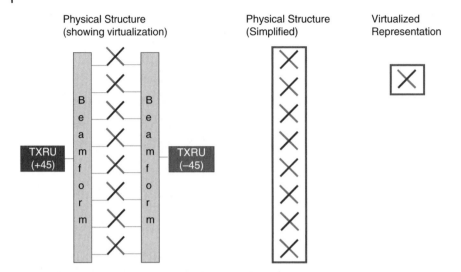

Figure 3.16 Single column antenna virtualization: single column aggregated into two ports.

co-pol elements within the column are combined via a beamformer and fed to a transceiver, one for each polarization. One purpose of such an arrangement is to provide control over the vertical pattern (beamwidth, pointing direction, gain, etc.) of the aggregated antennas that are driven by each transceiver. Figure 3.16 also shows a simplified physical structure showing only the physical elements. Then the far right of the figure shows a virtualized representation of the two-port array, where the elements are aggregated by the beamforming networks shown in the picture of the physical structure. In a typical legacy panel array, this single column structure is extended to multiple columns to form a linear arrangement in azimuth of virtualized ports as shown in Figure 3.17. The virtualization shown in Figures 3.16 and 3.17 essentially convert a two-dimensional array of physical antenna elements into a one-dimensional array of virtualized ports. The resulting structure has ports arranged in azimuth. Any beamforming that is done over those virtualized ports can adapt the overall array pattern only in the azimuth domain, whereas the vertical domain pattern is static and determined by the virtualization operation. Although the diagrams in Figure 3.16 imply that the virtualization operation is being performed in the RF domain, it should be noted that it is possible to perform the virtualization in the baseband digital domain, in which case each physical element would be driven by a separate transceiver.

The virtualization shown in Figures 3.16 and 3.17 can be extended to create a two-dimensional arrangement of virtualized elements as is shown in Figures 3.18 and 3.19. In Figure 3.16, the elements within a column can be subdivided into two or more sets, where the co-pol elements within each set are driven by a transceiver. In Figure 3.18, a column is divided into two sets and four transceivers drive the column. The idea in Figure 3.18 can then be extended to multiple columns as shown in Figure 3.19, where two rows of transceivers/virtual antenna elements are created. Such an arrangement (in contrast to the virtualization shown in Figures 3.16 and 3.17) can allow beamforming to adapt in both azimuth and elevation. An even further generalization is shown in Figure 3.20, where the virtualization operation occurs over both rows and columns.

Physical structure
(simplified)

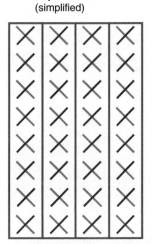

Virtualized
representation

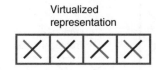

Figure 3.17 Multi column antenna virtualization: multiple columns aggregated into a single row of transceivers/virtual antennas ports.

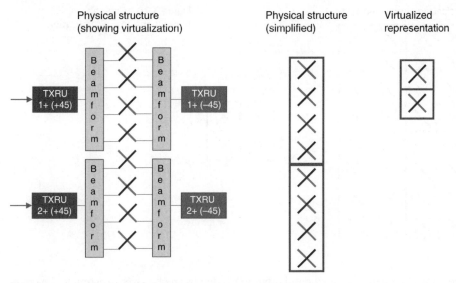

Figure 3.18 Single column antenna virtualization creating elevation ports: single column aggregated into four transceivers/virtual antennas.

3.3.5 Antenna Ports

In a beam-based air interface, the concept of an antenna port is the mechanism whereby the system (i.e. a transmitter or a receiver) can distinguish one transmitted signal in the system from another transmitted signal. In a broad sense, an antenna port is defined by its associated reference signal. The signals associated with an antenna port are transmitted over the same antenna configuration (e.g. physical antenna, logical antenna, or beam) and

Physical structure (simplified)

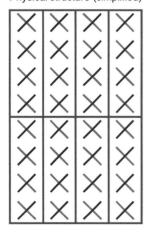

Virtualized
representation

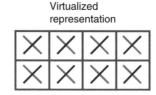

Figure 3.19 Multiple column antenna virtualization creating elevation ports: multiple columns aggregated into two rows of transceivers/virtual antennas.

Physical structure (simplified)

Figure 3.20 General two-dimensional virtualization.

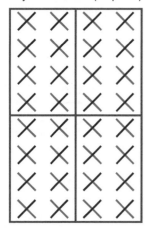

Virtualized
representation

have a specific reference signal configuration needed by the receiver to estimate the channel response or channel characteristics needed for processing the signals associated with that port. More specifically, an antenna port in 3GPP is defined such that the channel over which one symbol on the antenna port propagates can be inferred from the channel over which another symbol on the antenna port propagates. An antenna port is not necessarily associated with a specific antenna element (physical or logical) or beamformed signal, although in practice such associations are possible (and common) depending on the implementation choices made in the system design. The concept is basically a method of linking reference signals to the transmitted control or data information in a way that enables the receiver to effectively process the signals of that antenna port.

3.3.6 Beamforming for a Beam-Based Air Interface

Creating a beam-based interface with a large-scale antenna array then involves leveraging the previous ingredients, namely antenna virtualization and beamforming. Antenna virtualization operates on an array having a large number of physical elements and can be used to create a smaller more manageable number of static virtual antenna elements over which the system can operate. The virtualization operation creates logical antenna elements having the desired gain and directional response needed to enable the system to provide coverage over a service area. Next, MIMO and beamforming transmit and receive techniques are then applied over the logical antenna elements (or the physical elements directly if virtualization is not used). The system then defines antenna ports for transmitting the different types of signals in the system, such as control information or user data. In a beam-based air interface, all channels can be beamformed to cover the UEs located in the service area of the gNB.

It is worth mentioning that the above description is not meant to be restrictive since a large number of alternatives and variations are possible while still keeping within the confines of the NR specification. With this background in place, subsequent chapters will describe the details of how control and data are transmitted in the beam-based NR system leveraging the ingredients just described.

3.4 Frame Structure and Resource Allocation (Karri Ranta-aho, Nokia Bell Labs, Finland)

3.4.1 Resource Grid

The modulated multicarrier waveform creates a natural time/frequency grid of symbols and subcarriers. The symbols are grouped in time-domain to form slots, sub-frames, and radio frames that create the time-domain structure for mapping channels and signals for transmission, and the subcarriers are grouped together as PRBs to provide the frequency domain structure. This basic setup is the same as in LTE.

The main difference between LTE and NR in utilizing this basic resource grid comes from the property of scalable OFDM numerology introduced to NR. While LTE used 15 kHz sub-carrier spacing (SCS) and 14-symbols per ms, the NR numerology is scalable, providing a power-of-two scaling of the sub-carrier grid and number of symbols in a sub-frame. In both LTE and NR, a radio frame is defined to be 10 ms, and a sub-frame is 1 ms. LTE slot is somewhat unnecessary a concept due to historical reasons, as LTE scheduling takes place in a sub-frame level, whereas NR uses the term slot as the basic scheduling unit and thus the terms are not interchangeable between LTE and NR. The PRB in both is a set of 12 subcarriers, but the actual bandwidth of the PRB is a function of the subcarrier spacing. NR supports subcarrier spacings from 15 to 240 kHz as show in Table 3.2, although the subcarrier spacing of 240 kHz is only supported for SS/PBCH block and does not really have a PRB bandwidth as the SS/PBCH block is mapped directly to subcarriers.

NR slot is 14 symbols, with the exception of 60 kHz subcarrier spacing, where extended CP and 12-symbol slot is supported as well, making the 15 kHz NR slot equivalent to an LTE sub-frame of normal CP.

Table 3.2 OFDM numerologies.

	Subcarrier spacing (kHz)	PRB bandwidth (kHz)	Symbols/ sub-frame	symbols/ slot	Slot duration (ms)
LTE	15	180	14 (12)	7 (6)	0.5
NR	15	180	14	14	1
	30	360	28	14	0.5
	60	720	56 (48)	14 (12)	0.25
	120	1440	112	14	0.125
	240	N/A	224	14	0.0625

Figure 3.21 PRB/slot grid.

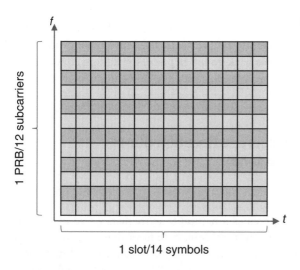

1 slot/14 symbols

When the subcarrier spacing is increased, the PRB bandwidth increases correspondingly, while the symbol duration shortens. As the NR slot structure is defined in terms of symbols, the basic PRB/slot square will not change as a function of the subcarrier spacing, but its physical dimensions in frequency and time scale. The PRB/slot grid is presented in Figure 3.21.

The NR subcarrier spacing scaling with steps of power-of-two was chosen to have the different numerologies generate a nested structure in both time and frequency domain. This is helpful if there is a need to operate different subcarrier spacings in the same cell at the same time. The nested time and frequency domain structures are depicted in Figures 3.22 and 3.23, respectively.

The minimum and maximum supported bandwidth is a function of the subcarrier spacing and frequency range. Subcarrier spacings of 15, 30, and 60 kHz are supported for data and control on frequency range 1, which was originally defined to span from 410 MHz to 6 GHz, but this was later extended to 7.125 GHz. Subcarrier spacings of 60 and 120 kHz are supported for data and control on frequency range 2, 24.25–52.6 GHz. Tables 3.3 and 3.4 show the maximum bandwidth part (BWP) size in PRBs and in MHz for a given nominal

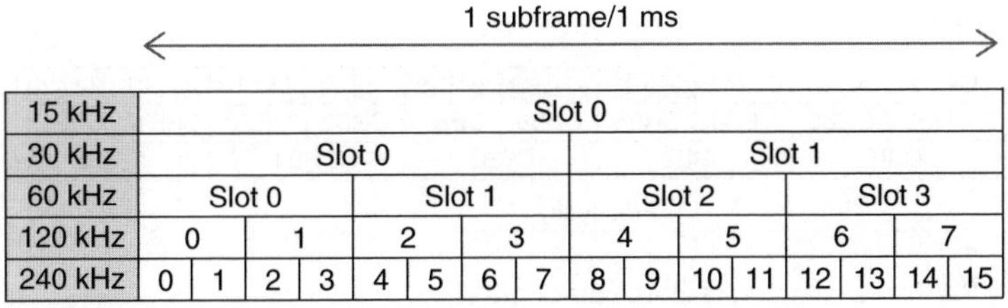

Figure 3.22 Nested time domain structure.

Figure 3.23 Nested frequency domain structure.

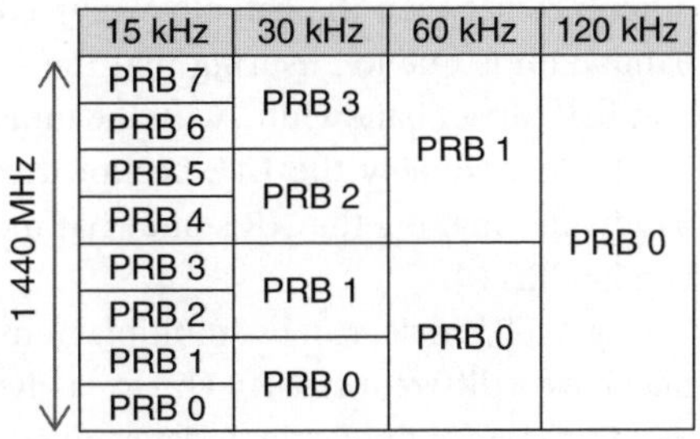

Table 3.3 Supported nominal carrier bandwidths and maximum BWP sizes for FR1.

SCS	5	10	15	20	25	30	40	50	60	80	90	100	MHz
15 kHz	25	52	79	106	133	160	216	270	—	—	—	—	PRB
	4.5	9.36	14.22	19.08	23.94	28.8	38.88	48.6	—	—	—	—	MHz
30 kHz	11	24	38	51	65	78	106	133	162	217	245	273	PRB
	3.96	8.64	13.68	18.36	23.4	28.08	38.16	47.88	58.32	78.12	88.2	98.28	MHz
60 kHz	—	11	18	24	31	38	51	65	79	107	121	135	PRB
	—	7.92	12.96	17.28	22.32	27.36	36.72	46.8	56.88	77.04	87.12	97.2	MHz

Table 3.4 Supported nominal carrier bandwidths and maximum BWP sizes for FR2.

SCS	50	100	200	400	MHz
30 kHz	66	132	264	—	PRB
	47.52	95.04	190.08	—	MHz
60 kHz	32	66	132	264	PRB
	46.08	95,04	190.08	380.16	MHz

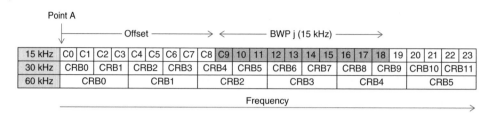

Figure 3.24 Defining a BWP on a common resource block grid.

carrier bandwidth. The maximum BWP size supported by the physical layer would be 275 PRBs, capping the maximum nominal carrier BW for each subcarrier spacing. The difference between the actually occupied maximum bandwidth and the nominal carrier bandwidth is due to ensuring that the transmit emissions are not leaking outside of the nominal carrier bandwidth, with the unused carrier edges acting as guard band, as defined in [28, 31]. Notably the LTE occupied carrier bandwidth is 90% of the nominal carrier bandwidth making the NR somewhat more spectrally efficient than LTE on most carrier bandwidths M.

Single cell/carrier can use multiple numerologies and as can be seen e.g. in Table 3.3 these may have a different size (in Hz) for a given nominal carrier bandwidth. Hence, the placement and size of the carrier is determined for each (used) numerology separately, providing the numerology's PRB grid. From UE perspective the different numerologies are represented by different BWPs that allow for separating the UE's current operating frequency bandwidth and OFDM numerology from other UEs in the same cell. From the cell's perspective there is a common resource block (CRB) grid relative to a common point in frequency called Point A. Each BWP of a given numerology is placed on the corresponding numerology's PRB grid, it's size in frequency is determined as number of PRBs, and its frequency location is determined relative to the Point A. The location of the Point A is provided by system information broadcast of the cell. The point A is just a logical point in frequency that carries no physical meaning other than that the locations of the BWPs are determined with a non-negative offset relative to the Point A. Specifically, the Point A is the frequency location of the first subcarrier of the CRB0 for all subcarrier spacings. An example in the Figure 3.24, the PRB0 of the BWP j corresponds to the CRB9 of the 15 kHz CRB grid.

3.4.2 Data Scheduling and HARQ

The basic principle of scheduling data in both uplink and downlink is very similar than in LTE. The PDCCH carries DCI, that can be used to schedule downlink and uplink data transmission. The DCI can be used for a few other purposes as well, for data scheduling it has four different formats, DCI format 0_0 and 0_1 to schedule uplink data on PUSCH, and format 1_0 and 1_1 to schedule downlink data on PDSCH. The reason for having two formats for each direction is to have a so-called fall-back format x_0 that can be used prior to the UE capabilities are known, or during a period of uncertainty of what the UE's configuration is. DCI formats x_1 are heavily link configuration dependent, so that they can exploit all the features configured for the UE, and the UE's configuration is naturally dependent on what the UE is capable of doing. Because of this, the formats x_1 may look quite different for different UEs in the same cell, or for the same UE when it operates under different cells. The UE is constantly searching for all the four formats (or just x_0 if it has not been provided with a dedicated configuration yet), and the UE identification is done with a C-RNTI, principally the same way as with LTE. In addition to providing information about the scheduled TB and the resources allocated for its transmission, the DCI also provides a vast array of additional information. The DCI fields related to data scheduling are outlined in high level in the Table 3.5 while omitting a number of other fields for brevity.

The primary scope for the DCI is to schedule a transport block (TB) for downlink or for uplink transmission, providing both the details of how the transport block is modulated and encoded, as well as on which resource blocks it is transmitted on. In addition, various timing related information is included. The LTE Hybrid Automatic Repeat Request (HARQ) design is providing a fixed DCI-to-data and data-to-HARQ-ACK timing. This is robust and convenient for frequency division duplex (FDD) operation with a fixed UE processing timeline, although it does require budgeting the maximum two-way propagation delay into the HARQ-ACK timing.

Figure 3.25 depicts the HARQ RTT loop. The DCI scheduling the PDSCH consists of both the time location of the PDSCH as well as the time location of the corresponding HARQ-ACK. In LTE the UE sees a nominally fixed delay from the end of the PDSCH to the beginning of the HARQ-ACK. The practical UE time budget needs to account for maximum timing advance (TA) that moves the HARQ-ACK location left. With NR, the specifications guarantee the minimum time before HARQ-ACK for the UE processing. For a larger timing advance the gNB needs to place the HARQ-ACK to a later slot to ensure that the UE processing time guarantee is kept. This way the HARQ-ACK timing does not need to budget for the maximum cell radius, and in smaller cells a shorter HARQ round trip time can be achieved. The next transmission (retransmission or new data) for a given HARQ process can only start after the UE has sent the HARQ-ACK corresponding to the previous transmission to that HARQ process.

3.4.3 Frequency Domain Resource Allocation and Bandwidth Part

In LTE, when the UE finds the sync channel of a cell it does not yet know what the cell's system bandwidth is. Only after reading the Master Information Block (MIB) on physical broadcast channel (PBCH) the UE knows what the overall operating bandwidth of the detected cell is. The LTE synchronization signals and the PBCH are confined within 1.4 MHz and have the same structure, no matter what the cell bandwidth is. This can be understood as switching from operating with an initial bandwidth used for the synchronization signal and PBCH to the actual operating bandwidth of the cell. The first LTE releases assumed that all UEs are capable of operating with all the possible system bandwidths, and it was not possible to have e.g. a 20 MHz cell with a 10 MHz UE. This caused some difficulties in a later stage when the support for low cost, narrower operating bandwidth MTC-type device categories was introduced to the specifications. The NR specifications covered this by introducing a concept of BWP that allows for the network to configure different UEs with different operating bandwidth within the overall carrier bandwidth. This is especially useful for support of UEs that are not able to support the full carrier bandwidth, while it

Table 3.5 Data scheduling related information fields of the data-scheduling DCI formats.

Field	Uplink DCI format		Downlink DCI format	
	0_0	0_1	1_0	1_1
Frequency domain resource assignment	Always present, size depending on BWP size and FD-RA configuration			
Time domain resource assignment	Always present, size depending on # of different TD-RA configured			
Carrier Indicator for cross-carrier scheduling	—	If configured	—	If configured
Bandwidth par Indicator for BWP switching	—	If configured	—	If configured
Frequency hopping flag	If configured	If configured	—	—
Modulation and coding scheme	Always 5 bits			Fields duplicated for 2 TBs with 5...8 layer MIMO
New Data Indicator	Always 1 bit			
Redundancy Version	Always 2 bits			
HARQ process ID	Always 4 bits			
Supplemental Uplink indicator		If configured	—	—
Precoding info and number of layers	—	Present with code-book based UL MIMO	—	—
Antenna ports	—	Always present	—	Always present
Code Blog Group Transmission info	—	Present if CBG configured	—	Present if CBG configured
Code blog group flushing out info	—	—	—	Present with CBG if configured
PTRS port to DMRS port association	—	Present with UL MIMO with PTRS	—	—
VRB-to-PRB mapping	—	—	Always present	If configured
PDSCH-to-HARQ-ACK timing indicator	—	—	Always present	Present if needed
PRB bundling size indicator	—	—	—	If configured
Rate matching indicator	—	—	—	If configured

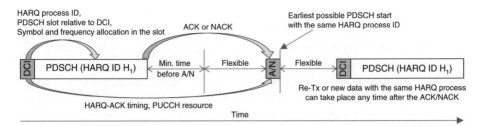

Figure 3.25 Flexible and dynamic HARQ-ACK feedback and retransmission timing.

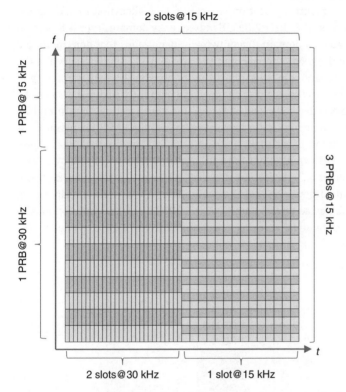

Figure 3.26 Multiplexing UEs with different subcarrier spacings in frequency and time in a cell.

can also be used switch the UE from one subcarrier spacing to another, or to narrow down its operating bandwidth for power saving purposes.

In frequency domain, all transmissions in both downlink and uplink apart from the SS/PBCH blocks are either scheduled or configured using the notion of the PRBs within the currently active BWP. This allows for the same structures and the same resource allocation mechanisms to work as-is with all the different subcarrier spacings, and the linkage to BWP allows for separating the UE's operating bandwidth from that of the cell bandwidth. In addition, NR supports carrier aggregation just like LTE, where each cell (active BWP of a cell) is configured and scheduled independently of other cells. An example of multiplexing UEs with different subcarrier spacings in frequency and time in a cell is presented in Figure 3.26.

The frequency location of the user data on PDSCH and PUSCH can be indicated with two different frequency domain resource allocation types creatively named Type 0 and Type 1. Type 0 is a bitmap-based resource allocation type, where the UE is provided a bitmap indicating which resource blocks are allocated for the PDSCH or PUSCH being scheduled. As the NR supports up to 275 PRBs and having 275 bits of frequency domain resource allocation on the scheduling message would have been too much of an overhead, for the purposes of Type 0 frequency domain resource allocation the PRBs are grouped into Resource Block Groups (RBGs), and the resource allocation is provided in the RBG level. The RBG size is a function of configuration and the active BWP size and ranges from 2 to 16 PRBs.

Type 1 frequency domain resource allocation is a contiguous allocation indicating the start and width of the resources allocated to the UE in frequency domain. Type 0 allocation is able to provide more frequency domain diversity for transmissions that don't require a large transmission bandwidth when comparing to the active BWP bandwidth, while the Type 1 gives a frequency-localized allocation that is more beneficial when attempting to exploit the frequency domain variations of the channel and map the scheduled data on the portion of the BWP that experiences the best channel. It is worth noting, that in uplink, non-contiguous frequency domain allocations can cause a very high PAPR to the transmitted signal, and for the transmitter to maintain the high peaks still within the linear region of the power amplifier, the UE would either have to implement a much beefier and less power-efficient power amplifier or, what is more practical, it has to take power back-off and reduce the average transmitted power so that the peaks are still within the linear region. Due to this property of the device power amplifier, the NR Rel-15 UEs only optionally support so-called "almost contiguous" resource allocation with the Type 0, while all UEs

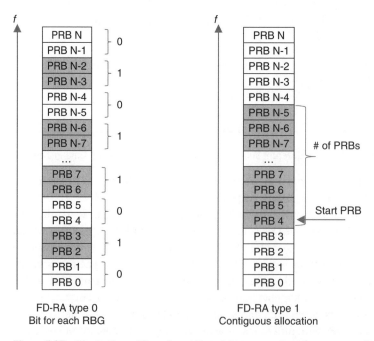

FD-RA type 0
Bit for each RBG

FD-RA type 1
Contiguous allocation

Figure 3.27 Illustration of Type 0 and Type 1 frequency domain resource allocation.

must support Type 0 resource allocation in downlink, and Type 1 resource allocation for both uplink and downlink (Figure 3.27). In addition, the UEs may optionally support the dynamic indication (as opposed to providing which resource allocation type is used with an radio resource control [RRC] configuration) of the resource allocation type being used with the scheduling message on PDCCH.

In addition to directly mapping the allocated (virtual) resource blocks to PRBs within the BWP, in downlink it is possible to employ interleaved mapping from the allocated (virtual) resource blocks to PRBs. Interleaved mapping redistributes the virtual resource blocks across the active BWP, allowing to achieve frequency diversity with Type 1 allocation. Notably interleaved mapping is only supported for downlink as the same difficulties related to non-contiguous resource allocation in uplink as discussed above for frequency domain resource allocation Type 0 cause the interleaved resource block mapping in uplink to be impractical.

3.4.4 Time Domain Resource Allocation

The time-domain resource allocation mechanism is very similar for both PDSCH and PUSCH, although there are some subtle differences. The network configures the UE with two tables, one for PDSCH and another for PUSCH, each consisting of up to 16 different time domain resource allocation configurations, or if no configuration is provided, specified default tables are used. Each entry in the table consists of

- A PDSCH/PUSCH mapping type, either Type A or Type B
- The slot number relative to the scheduling PDCCH
- The starting symbol number of the allocation in the indicated slot
- The number of symbols allocated within the indicated slot

When the UE is scheduled to receive a PDSCH or transmit a PUSCH, the scheduling message on the PDCCH points to one entry in the table determining the time-domain location and type of the scheduled data packet.

The main timing parameter is the slot number relative to the scheduling PDCCH, which in a typical setup would be 0 for PDSCH, unless there is a specific reason why the PDCCH and PDSCH should be separated to different slots, while for PUSCH it may be more typical that the PUSCH is scheduled to take place in a future slot, in FDD just because the UE needs some time to prepare for the PUSCH after it has decoded the PDCCH, and for time division duplex (TDD) in the more typical case that the DL control and UL data transmissions are separated to different slots due to the TDD constraint, although it is possible with some UE types and lower subcarrier spacings to schedule a PUSCH toward the end of the slot with a PDCCH in the beginning of the same slot.

The PDSCH/PUSCH mapping type determines the time-location of the first DMRS symbol and sets some constrains to location of the allocation within the indicated slot. For PDSCH, the mapping type A is designed for basic scheduling operation taking place at a slot granularity. The first DMRS symbol is on a fixed location after the nominal PDCCH region, either on the 3rd or 4th symbol. This location is common for all UEs in the cell and provided in the MIB, so that the PDSCH DMRS location is known already when receiving the System Information Blocks (SIBs). The PDSCH allocation may start as early as the first symbol of the slot, but no later than the indicated first DMRS symbol, i.e. it is possible for the PDSCH to start already before the PDCCH e.g. if the PDCCH is received on the 2nd symbol of the slot. The number of symbols in the allocation must be at least 3, and the allocation must end the latest at the last symbol of the slot. Such a violation in causality between the PDCCH scheduling the PDSCH happening only after the PDSCH start requires the UE to buffer the OFDM symbols until after it has received its PDCCH or determined that there was no PDCCH scheduling data to it in this slot.

PDSCH mapping type B is designed for shorter PDSCH allocations later in the slot. The first DMRS is always the first symbol of the PDSCH allocation, and the number of PDSCH symbols that can be allocated is either 2, 4, or 7, and the start of the allocation can be any symbol within the slot as long as the allocation does not span over the end of the slot. There is no real reason why the PDSCH allocations are limited to the 2, 4, or 7 symbols, but this restriction is more a matter of reducing the number of cases that one would need to test, and there are already talks to introduce additional valid durations to later 3GPP releases.

The PUSCH mapping type A is almost the same as for PDSCH. The location of the first DMRS is configured by the same parameter as with the PDSCH, but there is a small difference in the allocation flexibility. The PUSCH allocation must always start at the very beginning of the slot, and it must span at least 4 symbols. Similar difference to the PDSCH applies to the PUSCH mapping type B. The first DMRS symbol is the first symbol of the PUSCH allocation, but the duration can be freely selected from 1 to 14 symbols, again as long as the symbol boundary is not exceeded. The reason why the Type B allocation has no similar restrictions as with PDSCH was the need to multiplex PDCCH, DL/UL switching gap, PUCCH and SRS in many different combinations in a slot, so in the end no restrictions were defined for the PUSCH mapping type D allocation length. An example of resource allocation related information carried by the DCI is shown in Figure 3.28.

The main point of the PDSCH/PUSCH mapping type A is to have the DMRS symbols on the same locations for MU-MIMO purposes and to maintain commonality between the DL and UL for easier interference mitigation solutions when the received does not need to know if the interference being suppressed is uplink or downlink, while the mapping type B, especially in downlink can be used for scheduling a UE multiple times per slot, and it is subject to optional UE capabilities. In uplink mapping type B would also apply to bi-directional slots where the PDCCH takes place in the beginning of the slot, and PUSCH toward the end of the slot, and the large flexibility was deemed necessary to cover for also cases where the SRS and/or PUCCH are time-multiplexed after the end of the PUSCH.

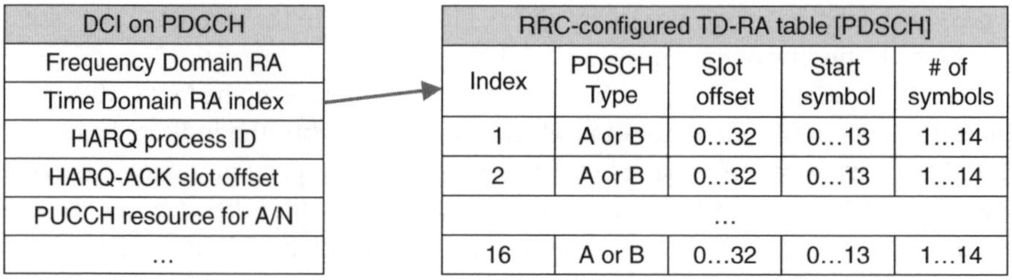

DCI on PDCCH		RRC-configured TD-RA table [PDSCH]				
Frequency Domain RA		Index	PDSCH Type	Slot offset	Start symbol	# of symbols
Time Domain RA index						
HARQ process ID		1	A or B	0…32	0…13	1…14
HARQ-ACK slot offset		2	A or B	0…32	0…13	1…14
PUCCH resource for A/N			…			
…		16	A or B	0…32	0…13	1…14

Figure 3.28 Among other things, the DCI carries resource allocation related information.

In addition to the scheduled PDSCH and PUSCH transmission, the network can configure semi-persistent scheduling (SPS) for downlink, as well as configured grant transmissions of two different types in the uplink, the two types this time titled as Type 1 and Type 2.

SPS for PDSCH is following the LTE SPS principle; the UE is RRC-configured with SPS parameters such as periodicity and PUCCH resource to use for HARQ-ACK feedback. The gNB then activates the SPS operation with a scheduling PDCCH scrambled with a specific CS-RNTI. The PDCCH scheduled the first packet normally, and the subsequent PDSCH transmissions follow the set periodicity and using the provided PUCCH resource for HARQ-ACK and the HARQ process ID is derived from the timing of the initial PDSCH transmission. The gNB can schedule retransmissions for the SPS PDSCH using CS-RNTI-scrambled PDCCH. The same mechanism is also used to deactivate the SPS operation.

Uplink Grant Free Transmission Type 2 operates exactly as the SPS in the downlink, but the Type 1 operation is a little bit different. With Type 1 configured grant there is no activation with PDCCH, but the configuration is fully provided over the RRC to the UE, and the UE can start to transmit at any time it gets data that is qualified for the grant free transmission on the time and frequency resources allocated for grant free operation. Still the retransmissions can be scheduled dynamically with CS-RNTI-scrambled PDCCH, as with Type 1. If a configured grant PUSCH overlaps with a scheduled PUSCH, the scheduled PUSCH takes priority. The logic behind this choice is that the gNB knows the configured grant transmission opportunities, and if it chooses to schedule data that would collide with those, then there must have been a good reason to do so.

3.5 Synchronization Signals and Broadcast Channels in NR Beam-Based System (Jorma Kaikkonen, Sami Hakola, Nokia Bell Labs, Finland)

In this section we cover the aspects related to the physical channel design of initial access and broadcast channels. The main focus of this section is on synchronization signal and channel design. Further aspects of related procedures related to the initial access are described in Section 4.1.

3.5.1 SS/PBCH Block

A core building block of the NR is a Synchronization Signal and PBCH block (SS/PBCH block). The SS/PBCH block is used e.g. for initial cell search, beam and cell measurements (RRM), radio link monitoring (RLM) and new beam identification in beam recovery procedure. The block consists of Primary synchronization signals and Secondary synchronization signals (PSS and SSS) and PBCH. Similar to LTE, these are located in concatenated symbols (Figure 3.29). PSS and SSS occupy each 1 symbol and 127 subcarriers and are centralized within the SS/PBCH block. The PBCH is allocated within 3 OFDM symbols and 240 subcarriers. In a PBCH symbol there are unused subcarriers in the middle of the frequency allocation where SSS is transmitted. The SS/PBCH block structure is illustrated in Figure 3.29.

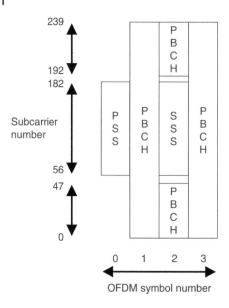

239

192

182

Subcarrier
number

56

47

0

0   1   2   3

OFDM symbol number

Figure 3.29 SS/PBCH block structure.

The SS/PBCH block transmission supports 15 and 30 kHz subcarrier spacing options in FR1 carrier frequency range, and 120 and 240 kHz subcarrier spacing options in FR2 carrier frequency range. To facilitate the initial cell selection of the UE, the default subcarrier spacing(s) is defined for each frequency band [28, 31]. For all the defined frequency bands (frequency band defines a lower frequency and an upper frequency where operating bandwidth can be occupied) at FR2, both 120 and 240 kHz subcarrier spacings are supported for SS/PBCH block. Typically, at FR1 a single subcarrier spacing option is defined per frequency band, but for carrier frequency range in operating bands n5, n41 and n66 both 15 and 30 kHz subcarrier spacings can be used (see Section 1.2.5 for spectrum related details). Naturally, all the SS/PBCH blocks of the cell have the same subcarrier spacing. In addition, after obtaining access, the UE can be explicitly configured with the used SS/PBCH block subcarrier spacing and, when configured, the UE can assume that the given subcarrier spacing is applied in the set of SS/PBCH blocks sharing the same center frequency.

The SS/PBCH blocks can be transmitted in certain time positions in a 5 ms half-frame, as depicted in Section 4.1. These locations are indexed, from 0 to L_{max}-1, and the index is carried in SS/PBCH block, to provide a slot level timing information. UE can determine the 2 LSB bits, for $L_{max} = 4$, or the 3 LSB bits, for $L_{max} > 4$, of an SS/PBCH block index per half frame from a one-to-one mapping with SS/PBCH block index and index of the DM-RS sequence transmitted in the PBCH. For $L_{max} = 64$, the UE determines the additional 3 MSB bits of the SS/PBCH block index per half frame from the PBCH payload bits.

3.5.2 Synchronization Signal Structure

In NR the number of the physical cell identities supported is 1008. The 1008 unique physical layer cell identities given by

$$N_{ID}^{cell} = 3N_{ID}^{(1)} + N_{ID}^{(2)}$$

where $N_{ID}^{(1)} \in \{0, 1, \dots, 335\}$ and $N_{ID}^{(2)} \in \{0, 1, 2\}$. The UE determines the cell identity from primary and secondary synchronization signals. When UE is performing initial cell search it searches first the PSS. PSS is located in a predefined synchronization signal raster in frequency domain. PSS is used for initial symbol boundary and coarse frequency synchronization to the NR cell.

Instead of using Zadoff-Chu sequences as in LTE, NR adopted a frequency domain based BPSK M-sequence. M-sequence was selected because of not having time offset – frequency offset ambiguity present with Zadoff-Chu sequences. Detection performance under initial frequency offset due to oscillator synchronization mismatch is improved and also UE complexity is reduced since the UE does not need to try with that many different PSS hypothesis in SSS detection as would be the case with Zadoff-Chu sequence as PSS. Correspondingly, joint PSS and SSS detection performance is improved.

Similar to LTE, there are three PSS sequences defined in NR. Definition for PSS is as follows:

$$d_{PSS}(n) = 1 - 2x(m)$$

$$m = (n + 43N_{ID}^{(2)}) \bmod 127$$

$$0 \le n < 127$$

where

$$x(i + 7) = (x(i + 4) + x(i)) \bmod 2$$

and

$$[x(6) \quad x(5) \quad x(4) \quad x(3) \quad x(2) \quad x(1) \quad x(0)] = [1 \quad 1 \quad 1 \quad 0 \quad 1 \quad 1 \quad 0].$$

Compared to LTE, 3 dB larger processing gain and higher frequency diversity can be achieved with two times longer sequence with the cost of increased UE complexity since bandwidth and sampling rate are doubled.

After the UE has detected PSS and acquired symbol timing and initial frequency synchronization, it detects SSS which carries the physical cell identity. Since the SSS is located on the same frequency location as PSS and one OFDM symbol apart, the UE may perform either non-coherent or coherent detection using channel estimates based on PSS.

SSS is a Gold sequence of length 127. There is one polynomial with 112 cyclic shifts and the other polynomial with 9 cyclic shifts forming together 1008 different PCIDs. The index of the detected PSS sequence (0, 1 or 2) is used in generation of nine cyclic shifts for the second polynomial. A Gold sequence $d_{SSS}(n)$ is defined as follows, where m_0 and m_1 are cyclic shifts for the first and second polynomial, respectively, $N_{ID}^{(1)}$ has values 0, 1, ..., 335 and $N_{ID}^{(2)}$ has values 0, 1, 2 corresponding to index carried by the PSS.

$$d_{SSS}(n) = [1 - 2x_0((n + m_0) \bmod 127)][1 - 2x_1((n + m_1) \bmod 127)]$$

$$m_0 = 15 \left\lfloor \frac{N_{ID}^{(1)}}{112} \right\rfloor + 5N_{ID}^{(2)}$$

$$m_1 = N_{ID}^{(1)} \bmod 112$$

$$0 \le n < 127$$

where

$$x_0(i+7) = (x_0(i+4) + x_0(i)) mod 2$$

$$x_1(i+7) = (x_1(i+1) + x_1(i)) mod 2$$

and

$$[x_0(6) \quad x_0(5) \quad x_0(4) \quad x_0(3) \quad x_0(2) \quad x_0(1) \quad x_0(0)] = [0 \quad 0 \quad 0 \quad 0 \quad 0 \quad 0 \quad 1]$$

$$[x_1(6) \quad x_1(5) \quad x_1(4) \quad x_1(3) \quad x_1(2) \quad x_1(1) \quad x_1(0)] = [0 \quad 0 \quad 0 \quad 0 \quad 0 \quad 0 \quad 1].$$

3.5.3 Broadcast Channels

3.5.3.1 PBCH

After detecting the PSS/SSS UE knows, in addition to the physical cell ID, the timing of the PBCH. The PBCH carrying the MIB, an integrated part of SS/PBCH block, is used for signaling the most essential system information related to access, frequency position and timing. The information is contained either in higher layer payload (i.e. MIB), as a part of the transport block payload, or in DMRS. It also includes the information whether the cell is barred or not, whether the UE is allowed to perform intra frequency reselection on this cell or not. Frequency domain position of SS/PBCH block in relation to common RB grid, timing info (e.g. MSBs of SFN) and MSBs of the SS/PBCH block position within a half-frame (index) are also carried in MIB and information for acquiring remaining minimum system information (RMSI), carried by SIB1, and whether it is present or not, as well as DMRS position for PDSCH and PUSCH.

As shown in Figure 3.29, PBCH is mapped to 3 OFDM symbols and 240 subcarriers. In the second PBCH symbol there are unused subcarriers in the middle of the frequency allocation where SSS is transmitted. The PBCH transmission is based on a single antenna port transmission using the same antenna port as the PSS and SSS within the SS/PBCH block. While frequency domain precoder cycling is precluded, the gNB may use time domain precoder cycling by changing the precoder from one PBCH transmission to another. The DM-RS of PBCH is mapped on every PBCH symbol to every 4th RE in every RB.

In the context of timing information, in addition to the SFN, the PBCH provides also the index of the time location of the given SS/PBCH block in the 5 ms pattern. At max three LSBs are carried by DMRS sequence and three MSBs (for FR2) are carried in PBCH transport block. The pattern and is discussed further in Section 4.1.1.

In order to enable the UE to monitor for the PDCCH scheduling the PDSCH carrying the SIB1, MIB provides the configuration needed for the PDCCH monitoring in *pdcch-ConfigSIB1*, together with the choice of numerology to be assumed for broadcast delivery (*subCarrierSpacingCommon*), 15 or 30 kHz for FR1 and 60 or 120 kHz for FR2. The Configuration of Common Control Resource Set (CORESET) configuration determines the number of symbols, $N_{symb}^{CORESET}$ and number of RBs, $N_{RB}^{CORESET}$ for the PDCCH scheduling the SIB1. The number of symbols can be selected among {1,2,3} symbols and the number of RBs among {24,48,96} RBs. Search space defining the time domain pattern of PDCCH monitoring is also provided by *pdcch-ConfigSIB1* and discussed in detail in Section 3.5.3.2.

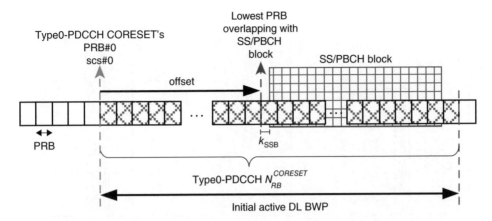

Figure 3.30 Illustration of the frequency location relation of Type0-PDCCH CORESET and SS/PBCH block.

The search space configuration determining the PDCCH monitoring occasions for SIB1 scheduling provided in MIB is called Type0-PDCCH Common search space (CSS). Correspondingly the CORESET determining the physical resources of the PDCCH scheduling the SIB1 is called as Type0-PDCCH CORESET.

As the Type0-PDCCH CORESET can be wider than the SS/PBCH block (>20 PRB), the UE needs to be provided also with information on the frequency offset between the first RB of the CORESET and the SS/PBCH block. Furthermore, the SS/PBCH block location is not fixed to the center of the carrier, different offset values are needed to enable the Type0-PDCCH CORESET to fall fully within the carrier. Also, as discussed in Section 4.1, in order to reduce the number of synchronization frequency raster locations, synchronization raster is sparser than the RF channel raster, and has different granularity. Thus, the SS/PBCH block may not be placed on the same common RB grid as data and control but the subcarriers will always be placed on same raster. To enable the UE to determine the common PRB raster of the cell used for data and control, an offset k_{SSB} (*ssb-SubcarrierOffset*) is provided to the UE indicating the number of tones from lowest subcarrier (#0) of the SS/PBCH block to closest subcarrier 0 of the common PRB. Figure 3.30 illustrates the frequency location relation of Type0-PDCCH CORESET and SS/PBCH block. The k_{SSB} definition is using 15 kHz subcarrier spacing at FR1 with value range of {0...23} to cover the case of 30 kHz subcarrier spacing and in FR2 definition uses the same subcarrier spacing as indicated for broadcast delivery (with value range of {0...11}). PBCH provides 4 bits of the *ssb-SubcarrierOffset* in the higher layer payload (MIB) and 1 additional MSB bit for FR1 is provided as part of the transport block payload. If the value indicated by *ssb-SubcarrierOffset* is larger than 23 (11) for FR1 (FR2), the *pdcch-ConfigSIB1* does not provide the Type0-PDCCH related configuration, but other assistance information for initial cell selection, as discussed further in Section 4.1.2.

3.5.3.2 SIB1

The SIB1 carries the RMSI required to obtain access to the cell, i.e. cell selection info required to determine the suitability of the cell based on quality, cell physical layer

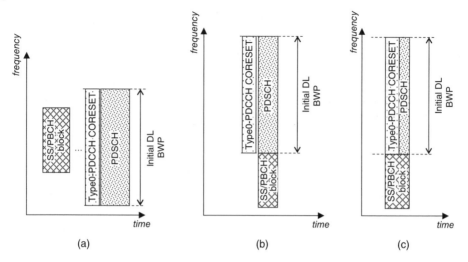

Figure 3.31 SS/PBCH block and Type0-PDCCH CORESET multiplexing patterns: (a) pattern 1, (b) pattern 2 and (c) pattern 3.

configuration, including the random access related configuration as well as indication of the actually transmitted SS/PBCH block(s).

To support different types of deployments, three different schemes for multiplexing the Type0-PDCCH (and PDSCH) with SS/PBCH block have been defined. The SS/PBCH block and Type0-PDCCH CORESET multiplexing patterns can be roughly divided into time domain multiplexing and frequency domain multiplexing as illustrated in Figure 3.31. While multiplexing pattern 1 was intended as more traditional scheduling of SIB1, so that SS/PBCH block and the PDCCH scheduling the SIB1 are time multiplexed, overlapping in frequency, patterns 2 and 3 are determined for scenarios where the available system bandwidth is wider and able to cover both SS/PBCH block and CORESET, and to minimize the beam sweeping overhead in time domain. Pattern 2 and pattern 3 are applicable only for FR2 deployments, while pattern 1 can be used in both frequency ranges.

For multiplexing pattern 1 the bandwidth requirement for system information delivery is always determined by $N_{RB}^{CORESET}$ of Type0-PDCCH CORESET as configurations where SS/PBCH block would be wider are precluded. The required bandwidth for different CORESET configurations, with multiplexing pattern 2 and 3 and different subcarrier spacings is shown in Table 3.6.

The multiplexing patterns define also the PDCCH monitoring occasions associated with each SS/PBCH block index i.e. when UE can expect the SIB1 to be scheduled in quasi-collocated manner with given SS/PBCH block index so that is all the transmitted signals can be assumed to be originating from the same transmission point (TRP). The UE, based on detecting and measuring the SS/PBCH blocks, can select the monitoring occasion(s) corresponding the strongest SS/PBCH block(s) for receiving the PDCCH scheduling. Of course, the content of SIB1 is cell specific, i.e. same for each SS/PBCH block index, so in principle the UE may choose to listen for all occasions until it has obtained SIB1.

Table 3.6 Frequency band allocation for SS/PBCH block and Type0-PDCCH CORESET multiplexing pattern 2 and 3.

Multiplexing pattern {SS/PBCHscs, SIB1scs} (kHz)	SIB1 scs (kHz)	$N_{RB}^{CORESET}$ (PRB)	Initial DL BWP (MHz)	SS/PBCH block scs (kHz)	SS/PBCH block BW (MHz)	Gap (PRB[a])	Tot BW (MHz)	Tot BW (PRB[a])
Pattern 2								
{120, 60}	60	48	34.56	60	28.8	2	64.8	90
{120, 60}	60	96	69.12	60	28.8	2	299.36	138
{240, 120}	120	24	34.56	120	57.6	2	295.04	66
Pattern 3								
{120, 120}	120	24	34.56	120	28.8	2	66.24	46
{120, 120}	120	48	69.12	120	28.8	2	100.8	70

a) Assuming subcarrier spacing of the SIB1.

For SS/PBCH block and CORESET multiplexing pattern 1 a configurable PDCCH monitoring occasion pattern is defined, so that for each SS/PBCH block index, two slot monitoring window is given. The configurable time pattern occurs every second radio frame (at SFN mod 2 = 0). The start of pattern the radio frame border can be adjusted by an offset O, e.g. {0, 2, 5, 7} ms for FR1 and {0, 2.5, 5, 7.5} ms for FR2, to give flexibility to the time multiplexing of the PDCCH monitoring pattern in relation to e.g. SS/PBCH block pattern. The pattern of the monitoring occasions can be further defined by step index, M, to determine the shift between the monitoring windows corresponding to different SS/PBCH block indexes. This parameter defines how much the slots in which the monitoring occasions corresponding to SS/PBCH block indexes overlap. Figure 3.32 illustrates the principle of determining the slots where the Type0-PDCCH monitoring occasions corresponding to SS/PBCH block index i. Note that PDCCH monitoring occasions corresponding to different SS/PBCH block indexes can fall to the same slots. In addition, the configuration gives options to determine the symbol indexes of the PDCCH monitoring occasion(s) in the slot. Depending on the configuration selected, there may be one or two PDCCH monitoring occasions within a slot. This allows further separation between the PDCCH monitoring occasions corresponding to different SS/PBCH block indexes, but the monitoring occasions can also be shared by different SS/PBCH block indexes. In Figures 3.33 and 3.34 show two exemplary illustrations for the pattern 1 configurations assuming different parametrization.

As these patterns only determine when the UE should monitor the scheduling PDCCH to acquire SIB1, in the case that monitoring occasions corresponding of two SS/PBCH block indexes fall to overlapping monitoring occasions, the network can select to which SS/PBCH block index (i.e. spatial direction) it will transmit the PDCCH and PDSCH, if at all. This is to say that the periodicity at which the network needs to send SIB1 is set at 160 ms, and the repetition transmission period within 160 ms, if any, is not fixed, i.e. can in principle vary.

For SS/PBCH block and CORESET multiplexing pattern 2 and 3, the PDCCH monitoring occasion pattern is fixed, and is defined in relation SS/PBCH block pattern (sharing the

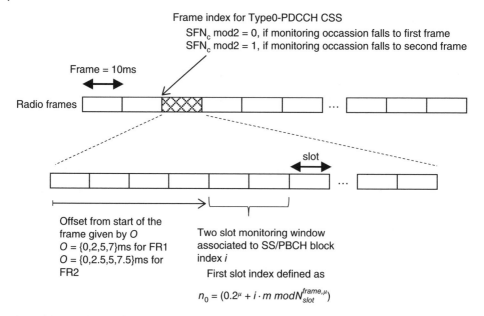

Figure 3.32 Determination of Type0-PDCCH monitoring slots for multiplexing pattern 1.

periodicity). The SS/PBCH block and CORESET multiplexing pattern 2 can be used when SS/PBCH block subcarrier spacing is 120 kHz and broadcast information subcarrier spacing is 60 kHz, or when SS/PBCH block subcarrier spacing is 240 kHz and broadcast subcarrier spacing is 120 kHz. With this multiplexing pattern the CORESET is always placed on RBs adjacent to the SS/PBCH blocks, on either side in frequency domain, but not in time. The PDCCH monitoring occasions corresponding to a given SS/PBCH block index are always placed in symbols preceding the SSB, either in same or previous slot. As in the SS/PBCH block pattern D and E there is slight difference in the placement of the SS/PBCH locations in two consecutive slots (see Figure 4.1 in Section 4.1.1.1) the PDCCH monitoring occasion definition varies per slot. When SS/PBCH block subcarrier spacing is 120 kHz and broadcast information subcarrier spacing is 60 kHz, the PDCCH monitoring occasions are in symbol indexes $\{0,1,6,7\}$ and that SS/PBCH block index corresponding to each monitoring occasion are $\{4k, 4k+1, 4k+2, 4k+3\}$, respectively, where $k = 0, 1,...,15$. When SS/PBCH block subcarrier spacing is 240 kHz and broadcast information subcarrier spacing is 120 kHz, the PDCCH monitoring occasions are in symbol indexes $\{0,1,2,3,0,1\}$ for the SS/PBCH block indexes $\{8k, 8k+1, 8k+2, 8k+3, 8k+6, 8k+7\}$ when SS/PBCH block falls in the same slot as the PDCCH monitoring occasion (based broadcast information subcarrier spacing) and in symbol indexes $\{12,13\}$ for the SS/PBCH block indexes $\{8k+4, 8k+5\}$, when the SS/PBCH block is in the next slot, where $k = 0, 1,...,7$. I.e. the monitoring occasions corresponding to SS/PBCH block indexes $8k+4$ and $8k+5$ are at the end of the slot preceding the corresponding SS/PBCH blocks. These are illustrated in Figure 3.35. In case of multiplexing pattern 2, the PDSCH delivering the SIB1 is assumed to be delivered on symbols corresponding to an SS/PBCH block. It is also good to note that for pattern 2 the number of symbols for PDCCH is restricted to one ($N_{symb}^{CORESET} = 1$)

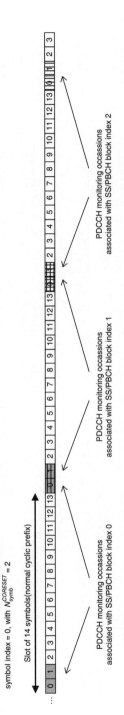

Figure 3.33 Illustration of determination of Type0-PDCCH monitoring occasions corresponding to SS/PBCH block indexes 0, 1 and 2, with $M = 1$ and SS first symbol index = 0, with $N_{symb}^{CORESET} = 2$.

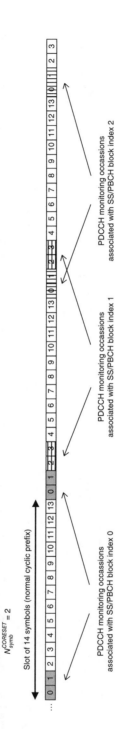

Figure 3.34 Illustration of determination of Type0-PDCCH monitoring occasions corresponding to SS/PBCH block indexes 0, 1 and 2, with $M = 1$ and SS first symbol index $= [0, N_{symb}^{CORESET}]$, with $N_{symb}^{CORESET} = 2$.

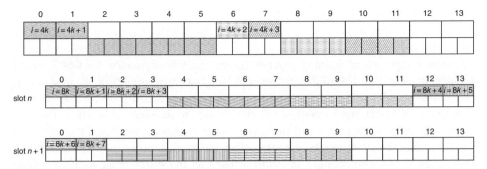

Figure 3.35 Illustration of multiplexing pattern 2 PDCCH monitoring occasions.

The SS/PBCH block and CORESET multiplexing pattern 3 is used when both the SS/PBCH block and broadcast are sent with same subcarrier spacing of 120 kHz. For pattern 3, PDCCH monitoring occasion corresponding to a given SS/PBCH block is located to symbols corresponding to SS/PBCH block symbols 0 and 1 and PDSCH is assumed to be multiplexed with symbols 2 and 3.

3.5.3.3 Delivery of Other Broadcast Information and Support of Beamforming

Major part of the SIB1 delivery design is related to the mechanisms to support beamforming. Correspondingly, also in delivery of other system information (OSI) and paging, mechanisms to support beam based delivery are supported. These follows the same principles as introduced for SIB1 in Section 3.5.3.2, i.e. defining time occasions where UE can monitor for control channel for scheduling the information, with an association to a certain SS/PBCH block index, giving the spatial association.

3.5.3.3.1 *Other System Information* In NR, OSI, including system information from SIB2 onwards, can be delivered through broadcast, at principle level in similar manner as e.g. in LTE, with additional support for using beams for broadcast. Alternatively, in NR, a mechanism has been introduced for delivering the OSI in "on-demand" manner. The "on-demand" procedure will trigger the network to initiate the broadcast of requested System Information (SI) message(s).

Like in LTE, the SI window is configured to determine where the UE can monitor for PDCCH scheduling the SI messages. For each entry in the SI messages list (*scheduling-InfoList*), at time window duration determined (*si-WindowLength*) occurs in certain radio-frames at configured periodicity (*si-Periodicity*). Then, to account the beam-based operation, monitoring pattern of time occasions associated with each SS/PBCH block is determined. The PDCCH monitoring occasions of the OSI is called as Type0A-PDCCH CSS. For OSI delivery, the monitoring pattern can be shared with SIB1, or defined separately.

Use of same PDCCH monitoring occasion definition as for SIB1, described in Section 3.5.3.2, is known as the default association, and determined by setting the search space ID to zero (*searchSpaceOtherSystemInformation* in SIB1). In this case the SI delivery follows the same (spatial) association assumptions with monitoring occasions with SIB1. In this case OSI also shares the CORESET configuration with SIB1. Depending on the size of the configured CORESET, the scheduling of OSI and SIB1 may need to be time multiplexed, i.e.

OSI is sent in different times than SIB1. Note that when default association is used, the SI window needs to be configured so that it overlaps with the PDCCH monitoring occasions defined for the Type0-PDCCH monitoring.

A second approach of defining separate search space configuration for OSI is called non-default association. The difference, in addition of having possibility of having different monitoring occasions, is that monitoring occasions are defined only for the actually transmitted SS/PBCH block indexes (provided by *ssb-PositionsInBurst* in SIB1). Furthermore, if the SI window length and the search space configuration allows, the SS/PBCH block specific occasions can be repeated. The monitoring occasions for the actually transmitted SS/PBCH blocks are determined based on the search space locations inside the SI window, so that each actually transmitted SS/PBCH block is associated to one valid PDCCH monitoring occasions (such that does not overlap with UL symbols) in consecutive manner. The UL symbols are determined by *tdd-UL-DL-ConfigurationCommon* in SIB1. The non-default association is illustrated in Figure 3.36.

In addition to these, the network can inform UE in SIB1 that "on-demand" procedure is to be used for acquisition of all or certain system information messages. This indicates that the SI messages are not broadcasted continuously. Network can configure SI message specific RACH resources, and the UE will use them to trigger the network to activate the broadcast of those messages. If SI message specific RACH resources are used, the UE will send a RACH using them and RAR will trigger the UE to start monitoring the configured occasions based on the given configuration, as described above. If network does not give specific RACH resources to certain or to any SI messages, the UE will use contention-based random access (CBRA) and then request the desired SI messages via RRC. Upon successful delivery of the request (L1 acknowledgment), the UE will start to monitor the PDCCH as described above, based on the configuration.

3.5.3.3.2 *Paging*
Paging delivery in NR follows the same basic principles as in LTE, accommodating changes needed to support a beam-based approach. Like in LTE, the paging occasion (PO) corresponding to set of UE IDs is determined based on configuration provided DRX cycle (T), giving the paging period, number of paging frames in period (N) and occasions (Ns) in paging frame (PF), and paging frame offset. The radio frame for a set of UE IDs thus occurs once per period T. It is good to note that in NR the paging occasion is not restricted within paging frame but is defined by the PDCCH monitoring occasions as described below. The paging monitoring occasions are called as Type2-PDCCH CSS.

Like for OSI delivery, there are two options how the monitoring occasions are determined for paging. If the default association-based approach is used, same PDCCH monitoring occasion definitions as for SIB1, described in Section 3.5.3.2, are used. In this case the paging delivery follows the same (spatial) association assumptions with monitoring occasions with SIB1 and possibly with OSI. In case of default association, if number of paging occasions in paging frame is two (Ns = 2), the paging occasion is either in first half-frame or second half frame based on the UE ID (This is only relevant/possible if multiplexing pattern 2 or 3 is used with 5 ms periodicity.). It is good to note though that the PDCCH monitoring occasions are independent of the paging configuration, thus the network needs to configure the paging frame and paging occasion consistently so that they correspond to the default association PDCCH monitoring occasions.

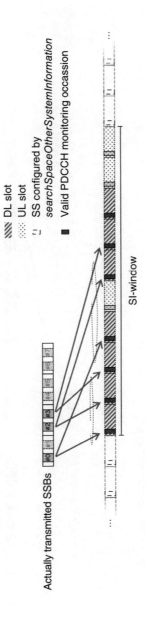

Figure 3.36 Illustration of non-default association for OSI delivery.

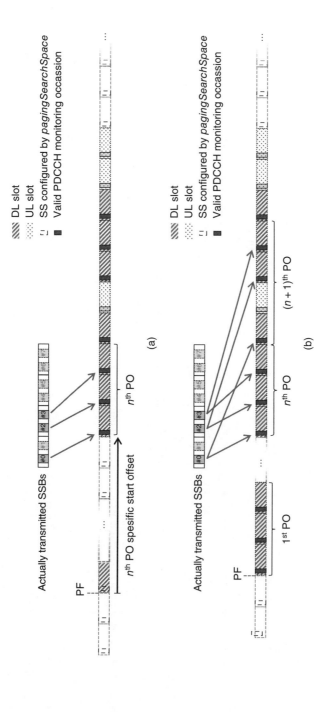

Figure 3.37 Illustration of PO definition with PO specific offset (a) and without (b) with non-default association.

Again, in the second option, separate search space configuration is given for paging. Like discussed for OSI, in case of non-default association, the monitoring occasions are defined only for the actually transmitted SS/PBCH block indexes and there is only one occasion for each SS/PBCH block. In this case the UE is provided an additional parameter determining the starting symbol of the PDCCH monitoring occasions in relation to the start of the paging frame (*firstPDCCH-MonitoringOccasionOfPO*). Then, in logical order, one valid PDCCH monitoring occasion (the search space configuration locations that do not overlap with UL symbols) is associated with one actually transmitted SS/PBCH block. So, each paging occasion covers as many valid PDCCH monitoring occasions as the is actually transmitted SS/PBCH blocks. If the paging occasion specific starting offset is not provided, the paging occasions are mapped in consecutive manner to valid PDCCH monitoring occasions. Figure 3.37 gives two illustrations for paging monitoring occasion determination.

In CONNECTED mode, when non-default association is used (i.e. paging search space ID $\neq 0$), UE does not follow the same paging occasion definition as in IDLE or INACTIVE but follows only the given search space configuration. In addition, if the UE has active transmission configuration indication (TCI) state for the CORESET corresponding the paging search space, when monitoring the PDCCH UE follows the TCI state (see TCI states description in Section 4.2). If not, in CONNECTED mode the UE follows the TCI-state determined by the initial access random access process.

3.6 Physical Random Access Channel (PRACH) (Emad Farag, Nokia Bell Labs, USA)

3.6.1 Introduction

The PRACH is used when the UE wants to communicate with the network and it doesn't have any uplink shared channel or control channel resources (PUSCH or PUCCH) to transmit on, and/or when the UE's uplink transmission is unsynchronized.[1] Furthermore, in NR, with beam-based operation, the PRACH is used to assist the UE in finding a beam to communicate on with the gNB during initial access, handover, and during beam failure recovery.

The PRACH consists of a preamble format made up of one or more preamble sequences transmitted in a time-frequency resource. The sequence used for PRACH should have good detection properties, with a low false alarm rate, and should be able to determine the round-trip propagation delay of the UE, to allow the network to synchronize the UE's UL transmitter through a time advance command. The Zadoff-Chu sequences, as described in Section 3.6.2, satisfy these properties. In Section 3.6.2, we discuss the construction and indexing of the preamble sequences. Section 3.6.3 describes the preamble formats available in NR and the characteristics of the various preamble formats. The UE transmits the PRACH preamble in a PRACH occasion as described in Section 3.6.4, which is a time-frequency resource that the network allocates for preamble transmission. Section 3.6.4 discusses the organization and configuration of the time-frequency resources used to transmit PRACH. Finally, Section 3.6.5 describes the PRACH baseband signal generation.

---

1 The UL transmission from the UE is unsynchronized when the UE's uplink signal at the gNB receiver is outside the CP window.

3.6.2 Preamble Sequence

The random access preamble sequence is based on the Zhadoff-Chu sequence. The Zhadoff-Chu sequence is given by:

$$x_u(n) = e^{-\frac{\pi n(n+1)u}{N_{ZC}}}$$

where,

- u is the root of the sequence with $u = 1, 2, \ldots, N_{ZC} - 1$.
- N_{ZC} is the length of the sequence and is denoted as L_{RA}. N_{ZC} is an odd prime number. There are two sequence length for the PRACH preamble; a long sequence with $L_{RA} = 839$ and a short sequence with $L_{RA} = 139$.

3.6.2.1 Useful Properties of Zhadoff-Chu Sequences

Now we look at some of the properties of the Zhadoff-Chu sequence.

1. The Zadoff-Chu sequence has a constant amplitude of unity.
2. The Zadoff-Chu sequence is periodic with period equal N_{ZC}. We can show that $x_u(n) = x_u(n + k \cdot N_{ZC})$, for any integer k.
3. Circular auto-correlation property:

    The circular auto-correlation of the sequence is given by:

$$R_{x_u(n),\, x_u(n)}(m) = \frac{1}{N_{ZC}} \sum_{n=0}^{N_{ZC}-1} x_u(n+m)x_u^*(n) = \begin{cases} 1 & m = 0 \\ 0 & \text{otherwise} \end{cases}$$

This leads to the circular cross-correlation of a sequence with a k-sample delayed version of itself having a single peak at sample k:

$$R_{x_u(n-k),x_u(n)}(m) = \frac{1}{N_{ZC}} \sum_{n=0}^{N_{ZC}-1} x_u(n-k+m)x_u^*(n) = \begin{cases} 1 & m = k \\ 0 & \text{otherwise} \end{cases}$$

This property is quite useful for the random access preamble, as it allows the network to determine the round trip delay, by correlating the received signal with a reference (non-delayed) version of that signal.

4. Circular cross-correlation of a Zadoff-Chu Sequence and a frequency shifted version of itself:

    Let the Zadoff-Chu sequence be $x_u(n)$ and frequency shifted sequence be $y_u(n,f) = e^{j2\pi \frac{fn}{N_{ZC}}} \cdot x_u(n)$.

    Where, f is the frequency shift relative to the sampling frequency of the Zadoff-Chu sequence. i.e. if the sampling frequency is F_s, the actual frequency shift is $f_{act} = fF_s$. For example, $f = 0.1$ when the frequency offset is equal to 0.1 of the sampling frequency. It can be shown that the cross-correlation function in this case is given by:

$$R_{y_u(n,f),x_u(n)}(m) = \frac{1}{N_{ZC}} \sum_{n=0}^{N_{ZC}-1} e^{j2\pi \frac{fn}{N_{ZC}}} \cdot x_u(n) \cdot x_u^*(n+m) = \frac{1}{N_{ZC}} x_u^*(m) \frac{1 - e^{j2\pi f}}{1 - e^{j2\pi f \frac{um+f}{N_{ZC}}}}$$

The amplitude of the correlation function is relevant in the preamble detector, as a large cross correlation amplitude at a sample different from the expected sample can lead to a false alarm.

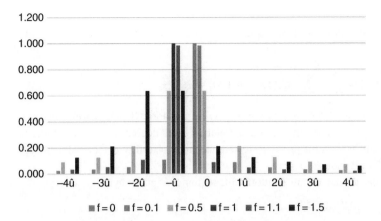

Figure 3.38 Cross-correlation of a Zadoff-Chu sequence having a frequency offset with a reference Zadoff-Chu Sequence. f is the relative frequency offset relative to the sampling frequency. ũ is the multiplicative inverse of the Zadoff-Chu sequence index u.

$$|R_{y_u(n,f),x_u(n)}(m)| = \frac{1}{N_{ZC}} \left| \frac{\sin(\pi f)}{\sin\left(\pi \frac{um+f}{N_{ZC}}\right)} \right|$$

Figure 3.38 shows a plot of the cross-correlation function for different frequency offset, f, values. The horizontal axis is a multiple of the multiplicative inverse of the Zadoff-Chu sequence index u modulo N_{ZC}, which is denoted as $\tilde{u}$. $\tilde{u}$ is found such that $\tilde{u} \cdot u = 1 + kN_{ZC}$, where k is an integer.

At $f = 0$, there is only one non-zero value for the auto-correlation function at sample point $m = 0$. All other auto-correlation values are zero. The auto-correlation function in this case, gives the expected delay position. As the frequency increases, additional peaks start to appear. At $f = 0.1$, additional peaks start to appear. The strongest are at $\pm\tilde{u}$, and these decrease at higher multiples of $\tilde{u}$. However, in this case, the main peak at $m = 0$ is much larger than the side peaks, and the probability of confusing the main peak with a side peak is quite small.

At $f = 0.5$, the peak at 0, and the peak at $-\tilde{u}$ are equal.[2] In this case, the receiver detects two peaks. If both peaks are within the delay range of the cell, the receiver will not be able to determine if the additional peak is from another UE or if it is a side loop of the main loop due to the larger frequency offset. As a result, restrictions will need to be placed on the allowed root sequences as described in Section 3.6.2.3.

At $f = 1$, the peak at $m = 0$, completely disappears and there is only one peak at $m = -\tilde{u}$, which corresponds to the first side loop. As f increases more to $f = 1.5$, now the magnitude of the second side loop at $m = -2\tilde{u}$ equals that of the first side loop at $m = -\tilde{u}$, suggesting that even more restrictions need to be placed on the allowed root sequences that can be used in a cell with higher frequency offset. More about this in Section 3.6.2.3.

---

2 Whether the peak appears at $-\tilde{u}$ or $+\tilde{u}$ depends on whether the frequency offset is positive or negative.

5. Circular cross-correlation between two sequences:
   The circular cross-correlation of two sequences with different roots, u_1 and u_2 is given by:

$$R_{x_{u1}(n),x_{u2}(n)}(m) = \frac{1}{N_{ZC}} \sum_{n=0}^{N_{ZC}-1} x_{u1}(n)x_{u2}^*(n+m) = \frac{1}{\sqrt{N_{ZC}}}$$

As discussed in Section 3.6.4, the preambles within a PRACH Occasion can span multiple root sequences. Having a cross-correlation that decreases as a function of the sequence length is a useful property to reduce the probability of false alarm from signatures on different root sequences.

6. It can be shown that the DF of the Zhadoff-Chu sequence is given by:

$$X_u(m) = x_u(\tilde{u}m)X_u(0)$$

Where $\tilde{u}$ is the multiplicative inverse of u modulo N_{ZC}.

Now let's discuss how to use these properties to construct the PRACH preamble sequences. We consider two cases, the first case is the low speed case, where the frequency offset is low. The second case is the high-speed case, where the frequency offset, due to Doppler, is high relative to the sampling frequency.

3.6.2.2 Unrestricted Preamble Sequences

Consider first the low speed case, i.e. the frequency offset of the received preamble signal is close to 0. The UE transmits a preamble based on the timing of the synchronization signals received by the UE. The preamble transmitted by the UE already includes the one-way delay from the gNB to the UE. As the preamble propagates to the base station, another one-way delay is added to the received signal. Thus, the received preamble at the gNB has a delay equal to the round-trip time between the UE and gNB. The received preamble signal at the gNB is circularly correlated with a reference preamble. The result of the correlation is a peak with a sample delay between 0 and the maximum round trip time.

Figure 3.39 shows an example of the circular correlation of a delayed received preamble signal with delay D and a reference preamble signal with zero delay, the result is a correlation peak at sample k given by:

$$k = L_{RA} \frac{D}{T_{seq}}$$

Where, L_{RA} is the length of the preamble sequence, T_{seq} is the duration of the preamble sequence in seconds and D is the delay in seconds. If the maximum round trip delay is D_{max}, we can find the corresponding k_{max}, with correlation peak k in the range $0 \leq k \leq k_{max}$. In this case, there are no correlation peaks, larger than k_{max}. Therefore, it is then possible to construct additional preamble sequences with a cyclic shift N_{CS} larger than or equal to $k_{max} + 1$ from the same root sequence according to the following equation:

$$x_{u,v}(n) = x_u((n + vN_{CS}) \bmod L_{RA})$$

These cyclically shifted sequences give correlation peeks in non-overlapping sample regions as shown in Figure 3.40. Hence, when the base station finds the correlation peak, it can determine, which sequence the UE has sent and the round-trip time of that UE. The range of v in the equation above is $0 \leq v \leq \left\lfloor \frac{N_{CS}}{L_{RA}} \right\rfloor - 1$.

Figure 3.39 Correlation of a delayed received preamble signal with a reference preamble signal.

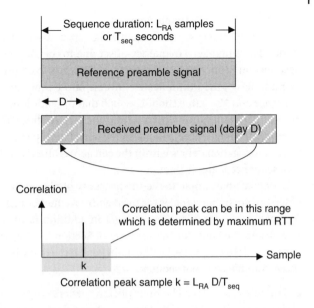

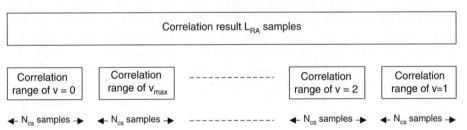

Figure 3.40 Correlation range of each cyclically shifted preamble sequence, where $v_{max} = \lfloor L_{RA}/N_{CS} \rfloor -1$.

The cyclic shift N_{cs} is determined by higher layer parameter *zeroCorrelationZoneConfig*. *zeroCorrelationZoneConfig* provides an index into a table, of 16 entries, that determines the N_{cs}. The network determines N_{cs} based on the maximum round trip time, including any multi-path delay spread.

Consider the following example on how to determine the N_{CS} value:

- Cell radius 2 km, i.e. RTT $= 2 \times 2000$ (m)/300 (m μs^{-1}) $+ 4.7 \,\mu s$ (propagation delay) $= 18 \,\mu s$.
- Preamble subcarrier spacing 1.25 kHz. This corresponds to $T_{seq} = 800 \,\mu s$.

$$L_{RA} = 839.$$

- Maximum sample delay $k_{max} = \left\lceil L_{RA} \dfrac{D_{max}}{T_{seq}} \right\rceil = 19.$
- N_{CS} is the smallest value in the "unrestricted set" of Table 6.3.3.1-5 of TS38.211 that is larger than or equal to $k_{max} + 1$. This value is $N_{CS} = 22$, which corresponds to *zeroCorrelationZoneConfig* of 4.

3.6.2.3 Restricted Preamble Sequences

Now consider the case of a cell with high speed users. The preamble received by the base station has a non-zero frequency offset due to the Doppler shift of the received signal. As described in Section 3.6.2.1 property 4, this leads to additional correlation peaks separated by multiples of the multiplicative inverse of the root sequence $\tilde{u}$. The selection of the root sequence and N_{CS} value should be such that the additional correlation peaks are distinct for all possible propagation delays within the cell. Additional preamble sequences with cyclic shifts can be introduced as long as the correlation peaks arising from the frequency offset for all propagation delays within the cell across all cyclically shifted preamble sequences of root sequence u are distinct.

Let's take an example to see the necessity of placing restrictions on root sequences used when the preamble signal is received with a frequency offset. The frequency is large enough for peaks to be significant at $m = \pm\tilde{u}$ in addition to $m = 0$. But no additional peaks are significant. We consider the scenario of Section 3.6.2.2, where maximum round trip time corresponds to $k_{\max} = 19$, i.e. the main peak has a delay between 0 and 19. Now we consider three examples for root sequence u:

- *Example 1*: $u = 710$. The multiplicative inverse of this root sequence is $\tilde{u} = 13$. The two side peaks for a UE with zero round trip time appear at $m = 826$ and $m = 13$. As the delay of the main peak ranges between $m = 0...19$, depending on the round-trip time of the UE in the cell, the delay of the negative side peak ranges between $m = 826...838$ and $m = 0...6$, and the delay of the positive side peak ranges between $m = 13...32$. This is shown in Figure 3.41. It is clear from that figure that the range of the negative side peak, the range of main peak and the range of the positive side peak overlap. This leads to ambiguity when trying to determine the round trip delay of the UE corresponding to the detected preamble.

  Accordingly, this leads to restriction in using this root sequence for UEs with frequency offset.

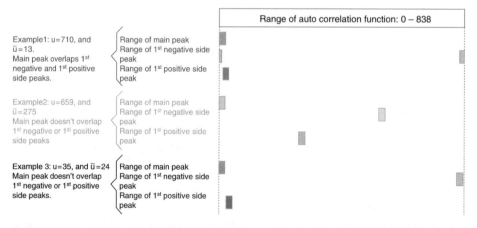

Figure 3.41 Examples showing the range of the main peak, the 1st negative and positive side peaks for different Zhadoff-Chu sequence index u. In these examples, $L_{RA} = 839$, subcarrier spacing is 1.25 kHz, and cell radius is 2 km.

- *Example 2*: $u = 659$. The multiplicative inverse of this root sequence is $\tilde{u} = 275$. The two side peaks for a UE with zero round trip time appear at $m = 564$ and $m = 275$. As the delay of the main peak ranges between $m = 0...19$, depending on the round-trip time of the UE in the cell, the delay of the negative side peak ranges between $m = 564...583$, and the delay of the positive side peak ranges between $m = 275...294$. This is shown in Figure 3.41. It is clear from that figure that the range of the negative side peak, the range of main peak and the range of the positive side peak don't overlap. Hence, the gNB receiver can determine the round trip delay of the UE corresponding to the detected preamble without ambiguity. However, in this case, there is restriction on the cyclic shifts that can be used. Additional sequences on this root sequence should avoid overlapping with negative peak, the main peak and the positive peak of other sequences on that root. This is shown in Figure 3.42.
- *Example 3*: $u=35$. The multiplicative inverse of this root sequence is $\tilde{u} = 24$. The two side peaks for a UE with zero round trip time appear at $m = 815$ and $m = 24$. As the delay of the main peak ranges between $m = 0...19$, depending on the round trip time of the UE in the cell, the delay of negative side peak ranges between $m = 815...834$, and the delay of the positive side peak ranges between $m = 24...43$. This is shown in Figure 3.41. It is clear from that figure that the range of the negative side peak, the range of main peak and the range of the positive side peak don't overlap. Hence, the gNB receiver can determine the round trip delay of the UE corresponding to the detected preamble without ambiguity. However, in this case, there is restriction on the cyclic shifts that can be used. Additional sequences on this root sequence should avoid overlapping with negative peak, the main peak and the positive peak of other sequences on that root. This is shown in Figure 3.42.

Preamble sequences where the main peak and the first negative side peak and the first positive side peek don't overlap within the delay range of the cell, are known as restricted set type A preamble sequences (Figure 3.42).

As the frequency offset increases more, additional side peaks become significant and the overlap of these peaks with the main peak and the first negative and positive side peaks should be avoided. Now consider that the peaks at $m = \pm 2\tilde{u}$ become significant. We reconsider example 2 above.

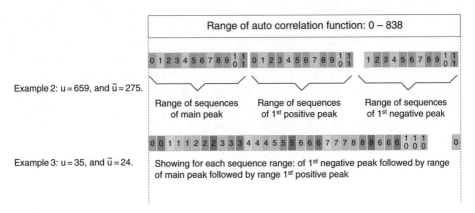

Figure 3.42 Examples of restricted set Type A preamble sequences for two different Zhadoff-Chu sequences.

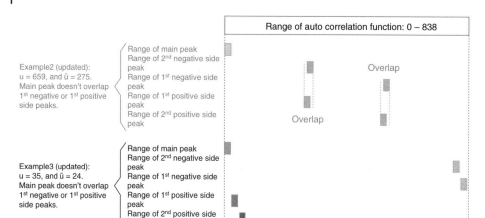

Figure 3.43 Examples showing the range of the main peak, the 1st and 2nd negative and positive side peaks for different Zhadoff-Chu sequence index u. In these examples, $L_{RA} = 839$, subcarrier spacing is 1.25 kHz, and cell radius is 2 km.

- *Example 2 (updated)*: Consider the additional peaks in example 2 above with $u = 659$ and a multiplicative inverse of $\tilde{u} = 275$. The two additional side peaks for a UE with zero round trip time appear at $m = 289$ and $m = 550$. The delay of the second negative side peak ranges between $m = 289...308$, and the delay of the second positive side peak ranges between $m = 550...569$. This is shown in Figure 3.43. It is clear from that figure that the range of the second negative side peak, and the range of first positive peak overlap. Similarly, the range of the second positive side peak and the range of the first negative side peak overlap. This leads to ambiguity when trying to determine the round trip delay of the UE corresponding to the detected preamble. Accordingly, this leads to restriction in using this root sequence for UEs with frequency offset large enough to cause the second side peak to be detectable.

- *Example 3 (updated)*: Consider the additional peaks in example 3 above with $u=35$ and a multiplicative inverse of $\tilde{u} = 24$. The two *additional* side peaks for a UE with zero round trip time appear at $m = 791$ and $m = 48$. The delay of the second negative side peak ranges between $m = 791...810$, and the delay of the second positive side peak ranges between $m = 48...67$. This is shown in Figure 3.43. It is clear from that figure that the range of the first and second negative side peaks, the range of main peak and the range of the first and second positive side peaks don't overlap. Hence, the gNB receiver can determine the round-trip delay of the UE corresponding to the detected preamble without ambiguity. However, in this case, there is restriction on the cyclic shifts that can be used, further restrictions should be placed on the additional sequences on this root sequence so as to avoid overlapping with first and second negative peaks, the main peak and the first and second positive peaks of other sequences on that root. This is shown in Figure 3.44.

Preamble sequences where the main peak and the first and second negative side peaks and the first and second positive sides peak don't overlap within the delay range of the cell, are known as restricted set type B preamble sequences.

Example 3: u = 35, and ũ = 24.

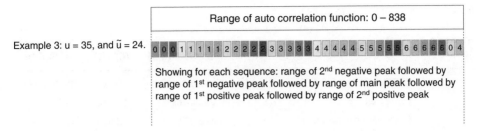

Figure 3.44 Examples of restricted set Type B preamble sequences for Zhadoff-Chu sequence with root sequence u = 35.

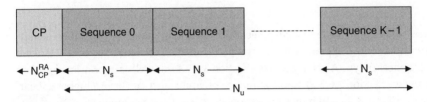

Figure 3.45 Preamble Format consisting of cyclic prefix followed by K sequence repetitions, with K ≥ 1.

3.6.3 Preamble Formats

As described in Section 6.3.2, PRACH preambles are based on Zadoff-Chu sequences with sequence length L_{RA}. NR supports two sequence length in release 15; $L_{RA} = 839$ which is known as the long sequence preamble, and $L_{RA} = 139$ which is known as the short sequence preamble. For each sequence length, there are several preamble formats defined. The preamble format, as in Figure 3.45, consists of a CP of length N_{CP}^{RA} samples, followed by K sequence repetitions each of length N_s samples, with a total sequence length of N_u samples and with K ≥ 1, at subcarrier spacing Δf^{RA}.

Different preamble formats are designed in NR to handle different scenarios.

- Preamble formats with smaller CP length are suitable for small cells, with less overhead. While preamble formats with larger CP are suitable for large cells with longer round-trip delays.
- Preamble formats with more sequence repetition provide greater coverage and can support a larger Maximum Coupling Loss, which could be useful for users in high propagation loss areas such as deep basements.
- Preamble formats with high subcarrier spacing are suitable for cells with high speed users which experience a high Doppler shift. While preamble formats with lower subcarrier spacing are suitable for larger cells.

When selecting the preamble format/subcarrier spacing of the preamble format, there is a compromise between coverage and resilience to Doppler shift.

In FR1, long sequence preamble formats are supported and short sequence preamble formats with subcarrier spacing 15 and 30 kHz are also supported. In FR2, only short sequence preamble formats with subcarrier spacing 60 and 120 kHz are supported.

Table 3.7 Long sequence preamble formats.

Preamble format	Use case	Δf^{RA} (kHz)	$N_u = K \cdot N_s \ (T_c)$	$N^{RA}_{CP} \ (T_c)$	Maximum cell radius (km)
0	LTE Refarming	1.25	$1 \cdot 24576\kappa$	3168κ	14.5
1	Large Cell	1.25	$2 \cdot 24576\kappa$	21024κ	99.7
2	Coverage Enhancement	1.25	$4 \cdot 24576\kappa$	4688κ	22.1
3	High speed users	5	$4 \cdot 6144\kappa$	3168κ	14.5

N_u and N^{RA}_{CP} are in units of $T_c = 0.5086$ ns the basic time unit of NR. $\kappa = 64$

Given their smaller subcarrier spacing (see Table 3.7), long sequence preamble formats are more susceptible to Doppler shift and frequency offset and hence support restricted preamble sequences of type A and type B (see Section 3.6.2.3), as well as unrestricted preamble sequences (see Section 3.6.2.2). Short sequence preamble formats have a higher subcarrier spacing (see Table 3.9), and hence only support unrestricted preamble sequences.

3.6.3.1 Long Sequence Preamble Formats
Long sequence preamble formats have a sequence length $L_{RA} = 839$, and two possible subcarrier spacings as shown in Table 3.7:

- $\Delta f^{RA} = 1.25$ kHz for preamble formats 0, 1 and 2.
- $\Delta f^{RA} = 5$ kHz for preamble format 3.

Let's take an example to see how to calculate the maximum cell radius of a preamble format. Consider preamble format 0, as shown in Figure 3.46, the CP length is $T_{CP} = 3164\kappa \ T_c = 103.125$ µs. The sequence length is $T_u = 24576\kappa \ T_c = 800$ µs. If we allocate one slot to the PRACH preamble, this leaves a guard period of $T_{GP} = 2976\kappa \ T_c = 96.875$ µs after the preamble till the end of the PRACH preamble allocation period. The guard period after the PRACH preamble is to ensure that the PRACH preamble transmission, which is not time-advanced at the UE, doesn't overlap with subsequent uplink transmissions, which are time aligned at the gNB receiver, and cause interference.

The cell should have a maximum round-trip time that is less than the guard period (T_{GP}). To preserve the circular convolution property, the maximum round-trip time + the multi-path delay spread (T_{DS}) should be less than the CP duration (T_{CP}). A cell with radius

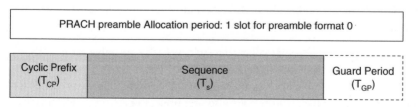

Figure 3.46 Preamble format 0.

R, has a maximum round-trip time of:

$$\mathrm{RTT}_{\mathrm{Max}} = \frac{2R}{c}$$

Where, c is the speed of light.

Therefore, the maximum cell radius can be determined by the following equation:

$$R = \frac{1}{2}c\,\min(T_{\mathrm{CP}} - T_{\mathrm{DS}}, T_{\mathrm{GP}})$$

Assuming a delay spread $T_{\mathrm{DS}} = 5\mu\mathrm{sec}$, we find that the maximum cell radius of preamble format 0 is 14.5 km. Similarly, the cell radius of other preamble formats is calculated and shown in Table 3.8.

3.6.3.2 Short Sequence Preamble Formats

Short sequence preamble formats have a sequence length $L_{\mathrm{RA}} = 139$. Each preamble format can have four possible subcarrier spacings $15 \cdot 2^{\mu}$ kHz as shown in Table 3.9. Where $\mu = 0$, 1, 2, and 3.

The maximum typical cell radius each short sequence preamble format supports depends on the preamble format and the subcarrier spacing as shown in Table 3.10.

Table 3.8 Long sequence preamble format delay spread.

	Format 0	Format 1	Format 2	Format 3
Cyclic prefix: T_{CP}	103.125 µs	687.375 µs	152.604 µs	103.125 µs
Sequence: T_{u}	800 µs	1600 µs	3200 µs	800 µs
Guard period: T_{GP}	96.875 µs	715.625 µs	147.396 µs	96.875 µs
Allocation size	1 slot	3 slots	3.5 slots	1 slot
Delay spread: T_{DS}	5 µs	16.7 µs	5 µs	5 µs
Max cell radius	14.5 km	100.2 km	22.1 km	14.5 km

Table 3.9 Short sequence preamble formats.

Preamble format	Δf^{RA} (kHz)	$N_u = K \cdot N_s\ (T_c)$	$N_{\mathrm{CP}}^{\mathrm{RA}}\ (T_c)$	Guard period (T_c)	Allocation size (symbols)
A1	$15 \cdot 2^{\mu}$	$2 \cdot 2048\kappa \cdot 2^{-\mu}$	$288\kappa \cdot 2^{-\mu}$	0	2
A2	$15 \cdot 2^{\mu}$	$4 \cdot 2048\kappa \cdot 2^{-\mu}$	$576\kappa \cdot 2^{-\mu}$	0	4
A3	$15 \cdot 2^{\mu}$	$6 \cdot 2048\kappa \cdot 2^{-\mu}$	$864\kappa \cdot 2^{-\mu}$	0	6
B1	$15 \cdot 2^{\mu}$	$2 \cdot 2048\kappa \cdot 2^{-\mu}$	$216\kappa \cdot 2^{-\mu}$	$72\kappa \cdot 2^{-\mu}$	2
B2	$15 \cdot 2^{\mu}$	$4 \cdot 2048\kappa \cdot 2^{-\mu}$	$360\kappa \cdot 2^{-\mu}$	$216\kappa \cdot 2^{-\mu}$	4
B3	$15 \cdot 2^{\mu}$	$6 \cdot 2048\kappa \cdot 2^{-\mu}$	$504\kappa \cdot 2^{-\mu}$	$360\kappa \cdot 2^{-\mu}$	6
B4	$15 \cdot 2^{\mu}$	$12 \cdot 2048\kappa \cdot 2^{-\mu}$	$936\kappa \cdot 2^{-\mu}$	$792\kappa \cdot 2^{-\mu}$	12
C0	$15 \cdot 2^{\mu}$	$1 \cdot 2048\kappa \cdot 2^{-\mu}$	$1240\kappa \cdot 2^{-\mu}$	$1096\kappa \cdot 2^{-\mu}$	2
C2	$15 \cdot 2^{\mu}$	$4 \cdot 2048\kappa \cdot 2^{-\mu}$	$2048\kappa \cdot 2^{-\mu}$	$2912\kappa \cdot 2^{-\mu}$	6

Table 3.10 Short sequence preamble format delay spread.

Preamble format	Delay spread	Cell radius (m)			
		15 kHz	30 kHz	60 kHz	120 kHz
A1	$\frac{2}{3}T_{CP}$	973	468	234	117
A2	T_{CP}	2109	1055	527	264
A3	T_{CP}	3515	1758	879	439
B1	$\frac{2}{3}T_{CP}$	352	176	88	44
B2	T_{CP}	1055	527	264	132
B3	T_{CP}	1758	879	439	220
B4	T_{CP}	3867	1934	967	483
C0	T_{CP}	5352	2676	1338	669
C2	T_{CP}	9297	4648	2324	1162

3.6.4 PRACH Occasion

Random Access preambles are transmitted in PRACH Occasions. A PRACH occasion (RO) is a time-frequency resource used to transmit the preamble. Each RO has 64 preambles. The first preamble, i.e. preamble index 0, corresponds to cyclic shift 0 of the logic root sequence index given by higher layer parameter *prach-RootSequenceIndex*. Subsequent preamble indices are numbered first in increasing order of cyclic shift, and then increasing order of logical root indices, until all 64 preamble indices are obtained, as shown in Figure 3.47. Assuming that the number of cyclic shifts per root sequence is m, the number of root sequence needed to get 64 preambles is $\left\lceil \frac{64}{m} \right\rceil$. Note that for restricted preamble sequences, the number of cyclic shifts supported per root sequence depends on the root sequence index. The number of consecutive logical root sequences used within a PRACH Occasion, starting with the logical root sequence given by higher layer parameter *rootSequenceIndex*, is such that the total number of cyclic shifts across these root sequences is at least 64.

The size of the PRACH occasion in the frequency domain, i.e. the number of PUSCH RBs the PRACH Occasion spans, depends on the preamble sequence length L_{RA}, the subcarrier spacing of the preamble, and the subcarrier spacing of the PUSCH.

Let's take an example of a long sequence preamble, i.e. $L_{RA} = 839$, with subcarrier spacing 5 kHz (i.e. preamble format 3). In this example, the number of RBs is calculated with respect to a PUSCH that has a subcarrier spacing of 15 kHz. The number of RBs is an integer value calculated such that all the preamble subcarriers can fit within the allocated RBs. Hence, the number of RBs is given by:

$$\left\lceil \frac{839 \times 5\,\text{kHz}}{12 \times 15\,\text{kHz}} \right\rceil = 24\,\text{RBs}$$

Within 24 PRBs based on 15 kHz subcarrier spacing, there are $24 \times 12 \times \frac{15}{5} = 864$ subcarriers with subcarrier spacing 5 kHz. The parameter $\bar{k}$ in Table 3.11 determines the starting subcarrier of the preamble sequence to position the preamble within the allocated PRBs. This is described in more detail in Section 3.6.5.

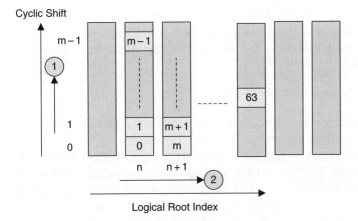

Where, n is given by higher layer
parameter *prach-RootSequenceIndex*

Figure 3.47 Enumeration of preamble index across cyclic shifts and logical root sequence index within PRACH Occasion.

Table 3.11 PRACH Occasion size in frequency domain.

L_{RA}	Preamble subcarrier spacing (kHz)	PUSCH subcarrier spacing (kHz)	Number of PUSCH RBs of RO, N_{RB}^{RA}	$\bar{k}$
839	1.25	15	6	7
839	1.25	30	3	1
839	1.25	60	2	133
839	5	15	24	12
839	5	30	12	10
839	5	60	6	7
139	15	15	12	2
139	15	30	6	2
139	15	60	3	2
139	30	15	24	2
139	30	30	12	2
139	30	60	6	2
139	60	60	12	2
139	60	120	6	2
139	120	60	24	2
139	120	120	12	2

Let's take another example, of a short sequence preamble, i.e. $L_{RA} = 139$, with subcarrier spacing 15 kHz. In this example, the number of PRBs is calculated with respect to a PUSCH that has a subcarrier spacing of 30 kHz. The number of RBs is an integer value calculated such that all the preamble subcarriers can fit within the allocated RBs. In this example, the number of RBs is given by:

$$\left\lceil \frac{139 \times 15 \text{ kHz}}{12 \times 30 \text{ kHz}} \right\rceil = 6 \text{ RBs}$$

Using the same principle, we can calculate the number of RBs for other preambles and PUSCH subcarrier spacings. This is given in Table 3.11.

The higher layer parameter *prach-ConfigurationIndex* determines the PRACH preamble format and the time domain resources of PRACH occasions, by pointing to a row in the random access configuration tables of TS 38.211 [34] (Tables 6.3.3.2-2/3/4). These tables are used for FR1 with paired spectrum/SUL, FR1 with unpaired spectrum, and FR2 with unpaired spectrum respectively. There are no paired spectrum frequency bands defined for FR2.

The columns of Tables 6.3.3.2-2/3/4 of TS 38.211 are shown in Figure 3.48. The "PRACH Configuration Index" is provided by higher layer parameter *prach-ConfigurationIndex*, which is a value between 0 and 255 and indicates which row to use in the corresponding table. The "Preamble Format" column determines which long sequence or short sequence preamble format to use as described in Section 3.6.3. The remaining columns describe the time domain structure of the PRACH Occasion.

For short sequence length preamble formats, two preamble formats are supported for some configurations, e.g. Ax/Bx, where $x \in \{1, 2, 3\}$. In this case, all the PRACH occasions, except the last one of each PRACH slot have preamble format Ax. The last PRACH occasion of the PRACH slot has format Bx. As described in Section 3.6.3.2, format Ax has no guard period at the end. The CP of the preamble format of the following PRACH occasion acts as the guard period. The last PRACH Occasion of the PRACH slot might not have a PRACH Occasion in the next symbol, hence that PRACH Occasion uses Format Bx which has guard period. PRACH configurations that use preamble format Ax only, can have a guard symbol following the last PRACH Occasion of the slot.

A PRACH configuration period spans several radio frames, it is always aligned with System Frame Number 0, i.e. $n_{SFN} = 0$. The PRACH configuration period can be {16, 8, 4, 2, 1} radio frames. The PRACH configuration period is given by column x of Figure 3.48. Within each PRACH configuration period one or more radio frames can have PRACH occasions, these are denoted by radio frames y of Figure 3.48 and are determined such that $n_{SFN} \bmod x = y$.

In FR1, each radio frame is divided into 10 subframes. In FR2, each radio frame is divided into 40–60 kHz slots. The column designated "Subframe number" or "Slot number" of

PRACH Configuration Index	Preamble Format	PRACH Configuration Period (x)	PRACH radio frames within period x (y)	"Subframe number" or 60 kHz "Slot number"	"Starting symbol" within PRACH slot of first RO	Number of PRACH slots within a subframe or 60 kHz slot	Number of time domain ROs within PRACH slot	PRACH duration in symbols

Figure 3.48 Columns of random access configuration tables ([Tables 6.3.3.2-2/3/4 of TS 38.211]).

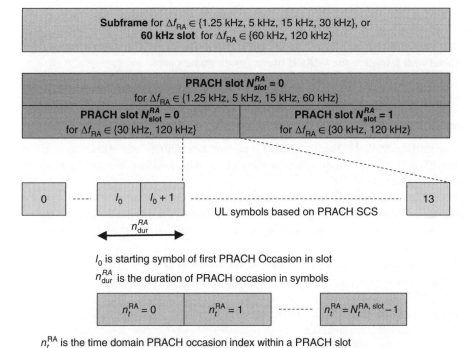

l_0 is starting symbol of first PRACH Occasion in slot

n_{dur}^{RA} is the duration of PRACH occasion in symbols

n_t^{RA} is the time domain PRACH occasion index within a PRACH slot

$N_t^{RA, slot}$ is the number of time domain PRACH occasions within a PRACH slot

Figure 3.49 Time domain organization of PRACH occasions within a subframe or 60 kHz slot.

Figure 3.48 determines the subframes or 60 kHz slots that have PRACH occasions in radio frame y within the PRACH configuration period.

In FR1, the supported subcarrier spacing for the short sequence preamble formats are 15 and 30 kHz as mentioned in Section 3.6.3. With 15 kHz, there can only be one PRACH slot within the subframe. With 30 kHz, there are two slots within a subframe. In FR1, the column designated "Number of PRACH slots within a subframe" determines the number of 30 kHz PRACH slots within a subframe. If the value is 2, both slots are used for PRACH. If the value is 1, only the second slot is used for PRACH.

In FR2, the supported subcarrier spacing for the short sequence preamble formats are 60 and 120 kHz as mentioned in Section 3.6.3. With 60 kHz, there can only be one PRACH slot within the 60 kHz slot. With 120 kHz, there are two slots within a 60 kHz slot. In FR2, the column designated "Number of PRACH slots within a 60 kHz slot" determines the number of 120 kHz PRACH slots within a 60 kHz slot. If the value is 2, both slots are used for PRACH. If the value is 1, only the second slot is used for PRACH.

The "starting symbol" determines the first symbol within a PRACH slot of the first PRACH Occasion. The "number of time-domain PRACH occasions within a PRACH slot" ($N_t^{RA,slot}$) determine the number of consecutive PRACH occasions within the PRACH slots with the first PRACH Occasion starting at "starting symbol," and each PRACH occasion has a duration "PRACH duration" (N_{dur}^{RA}) in symbols with respect to the numerology of the PRACH preamble.

For PRACH transmissions with $\Delta f_{RA} \in \{1.25 \text{ kHz}, 5 \text{ kHz}\}$, the starting symbol of the PRACH occasion is with respect to the 15 kHz numerology. For PRACH transmissions with $\Delta f_{RA} \in \{15 \text{ kHz}, 30 \text{ kHz}, 60 \text{ kHz}, 120 \text{ kHz}\}$, the starting symbol and duration of the PRACH occasion is with respect to the PRACH transmission numerology.

The starting symbol of the PRACH occasion is given by:

$$l = l_0 + n_t^{RA} N_{dur}^{RA} + 14 n_{slot}^{RA}$$

Where,

- l_0 is the starting symbol of the first PRACH occasion in a subframe or 60 kHz slot as given by the PRACH configuration tables by the column with heading "starting symbol" as shown in Figure 3.48.
- n_t^{RA} is the time domain PRACH transmission occasion index within a PRACH slot, the time domain PRACH transmission occasions are numbered sequentially starting from 0 to the "number of time-domain PRACH occasions within a slot" ($N_t^{RA,slot}$) minus one as shown in Figure 3.49.
- N_{dur}^{RA} is the duration of the PRACH occasion in symbols as given by the PRACH configuration tables by the column with heading "PRACH duration" as shown in Figure 3.48.
- n_{slot}^{RA} is the PRACH slot number within a subframe or 60 kHz slot.
  - For $\Delta f_{RA} \in \{1.25 \text{ kHz}, 5 \text{ kHz}, 15 \text{ kHz}, 60 \text{ kHz}\}$, $n_{slot}^{RA} = 0$.
  - For $\Delta f_{RA} \in \{30 \text{ kHz}, 120 \text{ kHz}\}$;
    - If the number of PRACH slots within a subframe or 60 kHz slot is 1 as given by the PRACH configuration tables by the corresponding column as shown in Figure 3.49, $n_{slot}^{RA} = 1$, i.e. the second slot of the subframe or 60 kHz slot is the PRACH slot.
    - If the number of PRACH slots within a subframe or 60 kHz slot is 2 as given by the PRACH configuration tables by the corresponding column as shown in Figure 3.49, $n_{slot}^{RA} \in \{0, 1\}$, i.e. both slots of the subframe or 60 kHz slot are PRACH slots.

PRACH occasions can be frequency division multiplexed in the frequency domain, the number of PRACH frequency resources multiplexed in the same time domain PRACH occasion is given by higher layer parameter *msg1-FDM = M*, the PRACH frequency resources

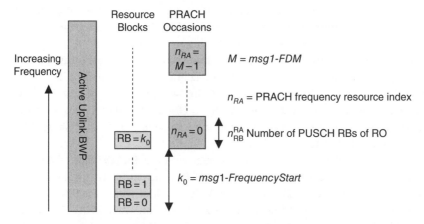

Figure 3.50 Frequency domain organization of PRACH occasions frequency division multiplexed in the same time domain PRACH occasion.

$n_{RA} \in \{0, 1, \ldots, M-1\}$ are multiplexed consecutively in increasing order of frequency within the initial or active uplink BWP starting with resource block given by higher layer parameter *msg1-FrequencyStart* for the first resource block of PRACH frequency resource $n_{RA} = 0$. Figure 3.50 shows the frequency domain organization of the PRACH occasions frequency division multiplexed in the same time domain PRACH occasion

3.6.5 PRACH Baseband Signal Generation

The generation of the time-continuous PRACH signal $s_l^{(p,\mu)}(t)$, where l is the PRACH symbol index, p is the antenna port and μ is the PRACH numerology, involves the following steps:

1. Determination of the complex amplitude of the PRACH subcarriers.
2. Mapping the PRACH sub-carriers within the frequency span of the PRACH Occasion.
3. Determination of the frequency of the first subcarrier of the first resource block of the PRACH occasion.
4. Prepending the PRACH signal with the CP.

Section 3.6.2 describes the generation of the preamble sequence, $x_{u,v}(n)$, this signal is converted to the frequency domain to map to the subcarriers used for PRACH transmission according to the following equation:

$$y_{u,v}(n) = \sum_{m=0}^{L_{RA}-1} x_{u,v}(m) \cdot e^{-j\frac{2\pi mn}{L_{RA}}}$$

To get the complex amplitude of the PRACH subcarriers, the frequency domain signal $y_{u,v}(n)$ is scaled by an amplitude scaling factor β_{PRACH},

$$a^{(p,RA)}(k) = \beta_{PRACH} \cdot y_{u,v}(k), \quad \text{where } k = 0, 1, \ldots, L_{RA} - 1$$

$a^{(p,RA)}(k)$ are then mapped to the PRACH subcarriers within the PRACH occasions. p is the antenna port of PRACH, which is antenna port 4000.

The number of PUSCH resource blocks (RBs) occupied by the PRACH occasion, N_{RB}^{RA}, is given by Table 3.11. The PUSCH subcarrier spacing and PRACH subcarrier spacing can be different. Within the PUSCH RBs allocated to the PRACH occasion, PUSCH subcarrier 0 (using the PUSCH numerology) is the first subcarrier in the first PRB of the PRACH occasion, this is aligned with PRACH subcarrier 0 (using the PRACH numerology). The first PRACH subcarrier of the PRACH preamble is determined by parameter $\bar{k}$ of Table 3.11.

In Figure 3.51 is an example of the mapping of the PRACH preamble subcarriers within a PRACH occasion. In this example, the PUSCH numerology is 15 kHz and the PRACH numerology is 1.25 kHz. The PRACH occasion has a frequency domain span of 6 PUSCH RBs (see Table 3.11), i.e. 72 PUSCH subcarriers numbered from 0 to 71. Subcarrier 0 of the PUSCH numerology is aligned with subcarrier 0 of the PRACH numerology. The first PRACH subcarrier of the PRACH preamble has an offset $\bar{k} = 7$.

Next, we determine the offset of first subcarrier of the PRACH occasion relative the start of the carrier corresponding to the PRACH transmission. Point A defined as a common reference point for the resource block grids. Figure 3.52 shows the components of the offset between the start of the carrier corresponding to PRACH and the first subcarrier of PRACH occasion n_{RA}. This offset consists of five components:

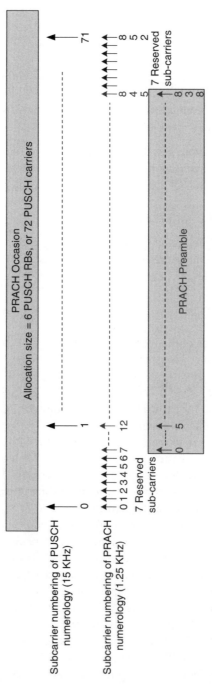

Figure 3.51 Example of allocation of PRACH subcarriers within the PRACH occasion. PUSCH numerology is 15 kHz, PRACH numerology is 15 kHz, PRACH numerology is 1.25 kHz, and L_{RA} = 839.

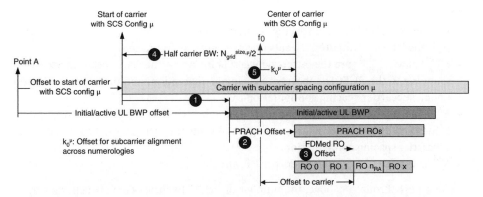

Figure 3.52 Determination of the first subcarrier of the first PUSCH RB of the PRACH occasion.

1. Offset within the carrier to the start of the BWP *i* corresponding to the PRACH transmission. This is given by (in number of PUSCH subcarriers):

$$(N_{\text{BWP},i}^{\text{start}} - N_{\text{grid}}^{\text{start},\mu}) \cdot N_{\text{sc}}^{\text{RB}}.$$

Where, $N_{\text{BWP},i}^{\text{start}}$ is the starting position of BWP *i* relative to point A in resource blocks with subcarrier spacing μ. $N_{\text{grid}}^{\text{start},\mu}$ is the starting position of the carrier relative to the point A in resource blocks with subcarrier spacing μ, which is given by higher layer parameter *SCS-SpecificCarrier*.

2. Offset within BWP *i* to the start of the first PRACH occasion. This is given by higher layer parameter *msg1-FrequencyStart* ($n_{\text{RA}}^{\text{start}}$) in number of RBs corresponding to the PUSCH subcarrier spacing configuration. This is represented by (in number of PUSCH subcarriers):

$$n_{\text{RA}}^{\text{start}} \cdot N_{\text{sc}}^{\text{RB}}$$

3. Offset between the first PRACH occasion and PRACH occasion n_{RA}. Each PRACH occasion has $N_{\text{RB}}^{\text{RA}}$ resource blocks corresponding to the PUSCH subcarrier spacing configuration. Therefore the number of PUSCH subcarriers between the first PRACH occasion, and PRACH occasion n_{RA} is given by:

$$n_{\text{RA}} \cdot N_{\text{RB}}^{\text{RA}} \cdot N_{\text{sc}}^{\text{RB}}$$

4. The offset between the start of the carrier with subcarrier spacing configuration μ and the middle of that carrier is (in number of PUSCH subcarriers):

$$\frac{N_{\text{grid}}^{\text{size},\mu}}{2} \cdot N_{\text{sc}}^{\text{RB}}$$

5. Offset for subcarrier alignment across numerologies:

$$k_0^{\mu} = \left(N_{\text{grid}}^{\text{start},\mu} + \frac{N_{\text{grid}}^{\text{size},\mu}}{2} \right) \cdot N_{\text{sc}}^{\text{RB}} - \left(N_{\text{grid}}^{\text{start},\mu_0} + \frac{N_{\text{grid}}^{\text{size},\mu_0}}{2} \right) \cdot N_{\text{sc}}^{\text{RB}} \cdot 2^{\mu_0 - \mu}$$

Where,

- μ is the PUSCH subcarrier spacing configuration.
- μ_0 is the largest configured UL subcarrier spacing configuration.
- $N_{\text{grid}}^{\text{start},\mu}$ and $N_{\text{grid}}^{\text{start},\mu_0}$ are the offsets between point A and the start of the carrier corresponding to the PUSCH subcarrier spacing configuration, and that with the largest uplink subcarrier spacing configuration.
- $N_{\text{grid}}^{\text{size},\mu}$ and $N_{\text{grid}}^{\text{size},\mu_0}$ are carrier bandwidth in number of RBs of the carrier corresponding to the PUSCH subcarrier spacing configuration, and the carrier with the largest uplink subcarrier spacing configuration.
- $N_{\text{sc}}^{\text{RB}}$ is the number of subcarriers per RB, which is 12.

Combining 1 through 5 above, the number of PUSCH subcarriers between the carrier frequency f_0 and the first subcarrier of PRACH occasion n_{RA} is k_1:

$$k_1 = k_0^\mu + (N_{\text{BWP},i}^{\text{start}} - N_{\text{grid}}^{\text{start},\mu}) \cdot N_{\text{sc}}^{\text{RB}} + n_{\text{RA}}^{\text{start}} \cdot N_{\text{sc}}^{\text{RB}} + n_{\text{RA}} \cdot N_{\text{RB}}^{\text{RA}} \cdot N_{\text{sc}}^{\text{RB}} - \frac{N_{\text{grid}}^{\text{size},\mu} \cdot N_{\text{sc}}^{\text{RB}}}{2}$$

Hence, the baseband PRACH signal can be represented as:

$$s_l^{(p,\mu)}(t) = \sum_{k=0}^{L_{\text{RA}}-1} a_k^{(p,\text{RA})} e^{j2\pi(k+Kk_1+\bar{k})\Delta f_{RA}(t - N_{\text{CP},l}^{\text{RA}} - t_{\text{start}}^{\text{RA}})}$$

Where, $t_{\text{start}}^{\text{RA}} \le t < t_{\text{start}}^{\text{RA}} + (N_u + N_{\text{CP},l}^{\text{RA}})T_c$

This signal is then upconverted to the carrier frequency according to the following equation:

$$Re\{s_l^{(p,\mu)}(t) \cdot e^{j2\pi f_0 \Delta f_{RA}(t - N_{\text{CP},l}^{\text{RA}} - t_{\text{start}}^{\text{RA}})}\}$$

Where,

- f_0 is the carrier frequency.
- Δf_{RA} is the subcarrier spacing of the PRACH transmission.
- K is the ratio between the PUSCH subcarrier spacing of uplink BWP corresponding to PRACH and the subcarrier spacing of PRACH, i.e. $K = \Delta f / \Delta f_{RA}$.
- k_1 is the offset between the carrier frequency f_0 and the first subcarrier of the PRACH occasion, as given earlier in this section.
- $t_{\text{start}}^{\text{RA}}$ is the start time of the PRACH occasion starting in symbol l.
- $N_{\text{CP},l}^{\text{RA}}$ is the CP length of time domain PRACH occasion starting in symbol l.
- N_u is the PRACH sequence duration in units of T_c.

The time domain organization of PRACH occasions and the location of the starting symbol of a PRACH occasion within a subframe or 60 kHz slot is described in Section 3.6.4 in conjunction with Figure 3.49. The length of the CP for a PRACH occasion stating in symbol l, $N_{\text{CP},l}^{\text{RA}}$, is given by:

$$N_{\text{CP},l}^{\text{RA}} = N_{\text{CP}}^{\text{RA}} + n \cdot 16\kappa$$

Where, n is the number of times the PRACH occasion overlaps with the start and middle of a subframe. This is to account for the fact that symbols at the start and middle of a subframe have a normal CP that is longer than the CP of the other symbols by $16\kappa \cdot T_c$. $N_{\text{CP}}^{\text{RA}}$ is given

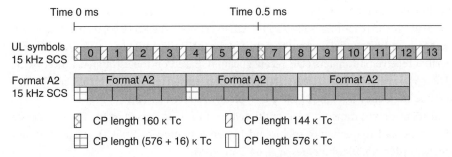

Figure 3.53 Example of CP length of preamble format A2 with subcarrier spacing of 15 kHz within a PRACH slot.

by Tables 3.7 and 3.9 for the long sequence preamble formats and short sequence preamble formats respectively.

Figure 3.53 shows an example of the CP length for the PRACH occasions of preamble format A2 with subcarrier spacing 15 kHz within a subframe. There are 3 PRACH occasions within a subframe, the first PRACH occasion spans symbols 0, 1, 2, and 3, which overlaps with the start of the subframe, and hence has a CP longer by 16κ. The second PRACH occasion spans symbols 4, 5, 6, and 7, which overlaps the middle of the subframe (time 0.5 ms), and hence has a CP longer by 16κ. The third PRACH occasion spans symbols 8, 9, 10, and 11, which doesn't overlap the start or middle of the subframe, hence the CP length is as given by Table 3.9.

3.7 CSI-RS (Stephen Grant, Ericsson, USA)

3.7.1 Overview

3.7.1.1 CSI-RS Use Cases

CSI-RS in NR are UE-specifically configured reference signals transmitted by the gNB used for the purposes of sounding the downlink radio channel. Such sounding provides various levels of knowledge of the radio channel characteristics ranging from limited knowledge on one extreme via reference signal received power (RSRP) estimates to detailed amplitude and phase estimates as a function of frequency, time, and space on the other extreme. Different use cases demand differing levels of knowledge of the radio channel characteristics. In NR, channel sounding with CSI-RS supports a more diverse set of use cases compared to LTE. With the exception of the first and last use cases in the following list, all other use cases in the below list are new to NR.

- Downlink CSI acquisition for link adaptation and codebook-based precoding for downlink MIMO (typically referred as "CSI-RS for CSI")
  - A CSI-RS resource is used by the UE to estimate the downlink (spatial) channel and provide feedback to the gNB on recommended MCS, recommended number of spatial layers (rank indication, or RI), and recommended precoder from a codebook (precoding matrix indication, or PMI).

- Downlink beam management (typically referred as "CSI-RS for BM")
  - A set of CSI-RS resources is used by the UE to estimate the RSRP corresponding to different downlink spatial domain transmit filters (gNB transmit beams) and/or different spatial receive parameters (UE receive beams). Based on the estimated RSRP(s), the UE may provide feedback to the gNB on a preferred pairing of transmit/receive beams.
- Fine time-frequency tracking at the UE (typically referred as TRS)
  - CSI-RS resources transmitted at multiple time locations is used by the UE to estimate fine time and frequency offsets to aid in demodulation of PDSCH, particularly for higher order modulation types, e.g., 64, 256-QAM.
- Radio resource management (RRM) measurements
  - To support mobility, a UE may be configured to perform RRM measurements on configured CSI-RS resources corresponding to the serving and one or more neighbor cells in addition to RRM measurements based on SS/PBCH block(s).
- Radio link monitoring (RLM) measurements
  - In NR, the default reference signal resource that the UE uses to perform RLM is CSI-RS, in contrast to LTE, where RLM is performed on cell-specific reference signals (CRSs). The CSI-RS resource(s) are used by the UE to estimate the BLER on a hypothetical PDCCH. If the estimated PDCCH BLER falls below a target threshold often enough and for long enough, the UE declares radio link failure and performs an RRC connection re-establishment procedure to restore the link.
- Beam failure detection (BFD)
  - Similar to RLM, CSI-RS is used to detect the event that the gNB transmit beam and UE receive beam become mis-aligned, resulting in a severe drop in received signal quality. While BFD and RLM procedures are similar, BFD does not involve L3 in the re-establishment of the link. This is handled by triggering new beam management measurements at L1/L2 to re-align transmit/receive beams.
- UL CSI acquisition for reciprocity-based (non-codebook based) uplink precoding
  - In NR, two UL MIMO modes are supported: codebook-based precoding and non-codebook-based precoding. In the latter, the UE acquires the downlink CSI through estimation of the radio channel based on a CSI-RS resource. The UE then adjusts its uplink spatial domain transmit filter (UE transmit beam) based on the assumption that the UL radio channel is reciprocal to the radio channel estimated in the downlink.
- PDSCH rate matching
  - As in LTE, zero-power channel-state information reference signals (ZP-CSI-RS) resources can be configured to inform the UE that certain resource elements are not available for PDSCH transmissions. The terminology ZP-CSI-RS is used to differentiate from the previous CSI-RS resource configurations which are referred to as non-zero-power channel-state information reference signals (NZP-CSI-RS). Due to the considerable flexibility of CSI-RS resource configuration in NR, PDSCH can be rate matched around virtually any combination of NR signals and channels, in addition to LTE signals and channels if a carrier is shared by NR and LTE.

3.7.1.2 Key Differences with LTE

In order to support the diverse set of use cases, the configuration of a CSI-RS resource in NR is much more flexible than in LTE with regards to the following characteristics:

- Time domain location
  - NR supports configuration of a CSI-RS resource starting at any of the 14 OFDM symbols within a slot, subject to multiplexing rules for various signals/channels described later in this chapter. In addition, for the larger number of antenna ports (24, 32) flexible starting location for each of the two pairs of OFDM symbol locations is allowed. In contrast, LTE supports CSI-RS configuration in up to only six fixed OFDM symbol locations of a subframe (see Figure 3.54). Such flexible configuration in NR, for example, allows a CSI-RS resource to be configured in the beginning of a slot, and then the corresponding CSI feedback at the end of the same slot, subject to UE processing capabilities. Such fast CSI feedback allows for greater CSI estimation accuracy, i.e. reduced aging of the CSI estimates.
- Frequency domain location
  - NR supports an arbitrary mapping to REs in the frequency domain, subject to a minimum number of contiguous REs, in most cases 2. In contrast, LTE supports only a limited set of RE locations according to Figure 3.54:
- Number of antenna ports
  - NR supports a wide range of antenna ports: 1, 2, 4, 8, 12, 16, 24, and 32 in contrast to LTE which supports only powers of 2. This allows for greater flexibility in antenna designs, for example, supporting different form factors for an array of cross polarized antenna elements.
- Code division multiplexing of antenna ports
  - While NR supports the same levels of code division multiplexing of ports as LTE (2, 4, and 8), it supports more flexible patterns for CDM grouping of ports across the time

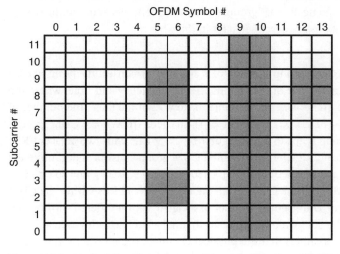

Figure 3.54 Limited time/frequency mapping locations for a CSI-RS resource in LTE. In contrast, NR allows for much greater flexibility in mapping locations.

and frequency domains commensurate with the enhanced flexibility of mapping to time/frequency resources.

- CSI-RS density
  - ○ NR supports density $\frac{1}{2}$, 1, and 3 REs/port/PRB in contrast to LTE which supports densities 1/3, $\frac{1}{2}$, and 1. For example, configuration of multiple single-port CSI-RS resources at different OFDM symbol locations all with density 3 (3 REs per PRB), allows for good time and frequency offset tracking performance to aid in PDSCH demodulation.
- CSI-RS bandwidth
  - ○ NR supports flexible CSI-RS bandwidth within a BWP. In contrast, CSI-RS in LTE occupies the entire bandwidth of a carrier.
- Time domain behavior
  - ○ NR supports an additional time domain CSI-RS behavior, semi-persistent transmission, on top of periodic and aperiodic transmission available in LTE. With semi-persistent CSI-RS transmission, medium access control elements (MAC-CE) are used to activate and deactivate a CSI-RS resource. While activated, the CSI-RS resource is transmitted with a configured periodicity; however, the MAC-CE activation/deactivation offers faster on/off control compared to an RRC re-configuration.

3.7.2 Physical Layer Design

3.7.2.1 Mapping to Physical Resources

The allowed time/frequency locations of a CSI-RS resource are defined in Table 3.12 reproduced from [34]. Eighteen base configurations are possible, where each base configuration is characterized by a number of antenna ports $X \in \{1, 2, 4, 8, 12, 16, 24, 32\}$, a density $\rho \in \{1, 3, 0.5\}$ REs/port/PRB, and a code-division-multiplexed (CDM) type with possible values FD-CDM2, CDM4 (FD2,TD2), CDM8 (FD2,TD4), and No CDM. The remaining columns in the table describe the mapping of CDM groups to time/frequency locations within a slot. Some exemplary configurations are described later in this section.

3.7.2.1.1 CDM Groups A CDM group consists of a number of consecutively numbered antenna ports mapped to a number of contiguous REs in frequency and time, where the number of ports per CDM group is L. For example in Row 12 of the table, the number of ports is $X = 16$, and CDM4(FD2,TD2) is used ($L = 4$), so there are $X/L = 4$ CDM groups. In this case each CDM group maps to a rectangular block of REs with two adjacent REs in frequency and two adjacent REs in time. The first CDM group consists of port numbers $p \in \{3000, 3001, \ldots, 3003\}$, the next CDM group $p \in \{3004, 3005, \ldots, 3007\}$, and so on up to port number 3015. All ports within a CDM group are mapped to all REs within the CDM group. Each port within a CDM group is further associated with a different orthogonal cover code (OCC) of length L, allowing the physical channel associated with each antenna port be distinguished. This assumes that the physical channel does not vary significantly over the REs within the CDM group. The fact that the REs within a CDM group are adjacent minimizes the variation for typical delay and Doppler spread values.

Table 3.12 CSI-RS locations within a slot.

Row	Ports X	Density ρ	cdm-Type	$(\bar{k}, \bar{l})$	CDM group index j	k'	l'
1	1	3	No CDM	$(k_0, l_0), (k_0+4, l_0), (k_0+8, l_0)$	0,0,0	0	0
2	1	1, 0.5	No CDM	(k_0, l_0),	0	0	0
3	2	1, 0.5	FD-CDM2	(k_0, l_0),	0	0, 1	0
4	4	1	FD-CDM2	$(k_0, l_0), (k_0+2, l_0)$	0,1	0, 1	0
5	4	1	FD-CDM2	$(k_0, l_0), (k_0, l_0+1)$	0,1	0, 1	0
6	8	1	FD-CDM2	$(k_0, l_0), (k_1, l_0), (k_2, l_0), (k_3, l_0)$	0,1,2,3	0, 1	0
7	8	1	FD-CDM2	$(k_0, l_0), (k_1, l_0), (k_0, l_0+1), (k_1, l_0+1)$	0,1,2,3	0, 1	0
8	8	1	CDM4 (FD2,TD2)	$(k_0, l_0), (k_1, l_0)$	0,1	0, 1	0, 1
9	12	1	FD-CDM2	$(k_0, l_0), (k_1, l_0), (k_2, l_0), (k_3, l_0), (k_4, l_0), (k_5, l_0)$	0,1,2,3,4,5	0, 1	0
10	12	1	CDM4 (FD2,TD2)	$(k_0, l_0), (k_1, l_0), (k_2, l_0)$	0,1,2	0, 1	0, 1
11	16	1, 0.5	FD-CDM2	$(k_0, l_0), (k_1, l_0), (k_2, l_0), (k_3, l_0), (k_0, l_0+1), (k_1, l_0+1), (k_2, l_0+1), (k_3, l_0+1)$	0,1,2,3, 4,5,6,7	0, 1	0
12	16	1, 0.5	CDM4 (FD2,TD2)	$(k_0, l_0), (k_1, l_0), (k_2, l_0), (k_3, l_0)$	0,1,2,3	0, 1	0, 1
13	24	1, 0.5	FD-CDM2	$(k_0, l_0), (k_1, l_0), (k_2, l_0), (k_0, l_0+1), (k_1, l_0+1), (k_2, l_0+1), (k_0, l_1), (k_1, l_1), (k_2, l_1), (k_0, l_1+1), (k_1, l_1+1), (k_2, l_1+1)$	0,1,2,3,4,5, 6,7,8,9,10,11	0, 1	0
14	24	1, 0.5	CDM4 (FD2,TD2)	$(k_0, l_0), (k_1, l_0), (k_2, l_0), (k_0, l_1), (k_1, l_1), (k_2, l_1)$	0,1,2,3,4,5	0, 1	0, 1
15	24	1, 0.5	CDM8 (FD2,TD4)	$(k_0, l_0), (k_1, l_0), (k_2, l_0)$	0,1,2	0, 1	0, 1, 2, 3
16	32	1, 0.5	FD-CDM2	$(k_0, l_0), (k_1, l_0), (k_2, l_0), (k_3, l_0), (k_0, l_0+1), (k_1, l_0+1), (k_2, l_0+1), (k_3, l_0+1), (k_0, l_1), (k_1, l_1), (k_2, l_1), (k_3, l_1), (k_0, l_1+1), (k_1, l_1+1), (k_2, l_1+1), (k_3, l_1+1)$	0,1,2,3, 4,5,6,7, 8,9,10,11, 12,13,14,15	0, 1	0
17	32	1, 0.5	CDM4 (FD2,TD2)	$(k_0, l_0), (k_1, l_0), (k_2, l_0), (k_3, l_0), (k_0, l_1), (k_1, l_1), (k_2, l_1), (k_3, l_1)$	0,1,2,3,4,5,6,7	0, 1	0, 1
18	32	1, 0.5	CDM8 (FD2,TD4)	$(k_0, l_0), (k_1, l_0), (k_2, l_0), (k_3, l_0)$	0,1,2,3	0,1	0, 1, 2, 3

Source: Table 7.4.1.5.3-1 in [34].

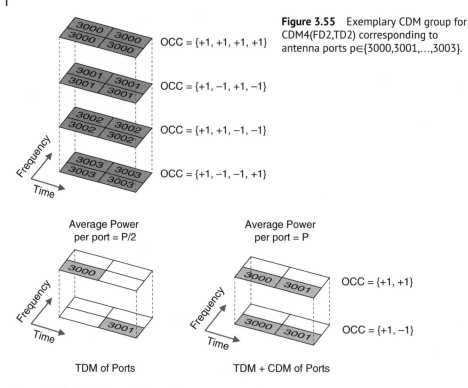

Figure 3.55 Exemplary CDM group for CDM4(FD2,TD2) corresponding to antenna ports $p \in \{3000,3001,\ldots,3003\}$.

Figure 3.56 Achieving full PA utilization through CDM.

Figure 3.55 shows the first CDM group in the above example illustrating the fact that all antenna ports of the CDM group are mapped to all REs of the CDM group. Further, a different OCC corresponds to each antenna port allowing the antenna ports to be distinguished at the UE during channel estimation.

The reason for multiplexing ports through CDM on each RE of a CDM group is to maximize the power amplifier (PA) utilization. For example (see Figure 3.56), consider two antenna ports transmitted in a purely time-division multiplex (TDM) manner in adjacent REs in time, with each port associated with a separate PA with output power P. For purely TDM transmission, the average transmit power per port is $P/2$ since each PA is utilized only half of the time over the two REs. If instead CDM is applied such that both ports are mapped to both REs using OCC $\{+1, +1\}$ for the first port and OCC $\{+1, -1\}$ for the second port, then the average transmit power per port is P instead of $P/2$ since both PA are fully utilized in both REs of the CDM group. Clearly a 3 dB boost in transmit power per port is beneficial for both coverage and channel estimation quality.

3.7.2.1.2 Example CSI-RS Configurations
Figure 3.57 shows exemplary 16 and 32-port CSI-RS resource configurations corresponding to Rows 12, 17, and 18 in Table 3.12. Such configurations are useful for antenna arrays with a large number of antenna elements in which codebook-based DL precoding is used. In such scenarios, the UE estimates the physical channels corresponding to a large number of antenna ports and feeds back recommendations on rank and precoding matrix to the gNB.

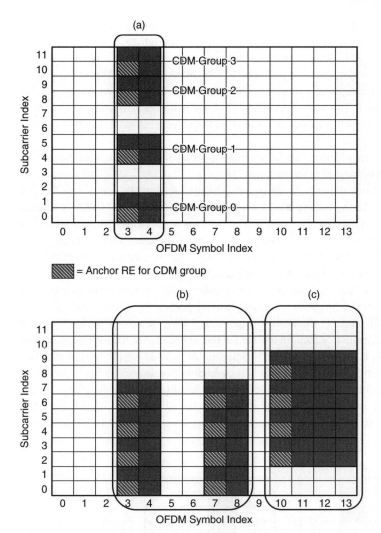

Figure 3.57 Exemplary CSI-RS resource configurations for various rows in Table 3.12: (a) Row12: 16 ports, CDM4(FD2,TD2); (b) Row17: 32 ports, CDM4(FD2,TD2); and (c) Row18: 32 ports, CDM8(FD2,TD4). All resources have density 1.

The 16-port CSI-RS example in Figure 3.57a corresponding to Row 12 serves as a good example to describe how the time/frequency mapping functions in Table 3.12. This particular resource configuration utilizes CDM4(FD2,TD2), hence there are $16/4 = 4$ CDM groups indexed by $j = 0, 1, 2, 3$ (6th column of the table). The quantity $(\bar{k}, \bar{l})$ in the 5th column gives the subcarrier index and OFDM symbol index, respectively, of an "anchor" RE for each CDM group. The anchor REs for each of the 4 CDM groups for this example are given by $(\bar{k}, \bar{l}) = (k_0, l_0), (k_1, l_0), (k_2, l_0), (k_3, l_0)$ where the subcarrier indices are given by $k_0 = 0, k_1 = 4$, $k_0 = 8$, and $k_0 = 10$ and the OFDM symbol index by $l_0 = 3$. The subcarrier index k' and the OFDM symbol index l' in the 7th and 8th columns are used to index the other REs in each CDM group for the purposes of mapping reference symbol sequence values to those REs.

In general, the OFDM symbol index of all anchor REs in a CSI-RS resource is a function of a single configured value for the starting OFDM symbol index of the CSI-RS resource. The only exceptions are Rows 13, 14, 16, and 17, where two starting symbol indices are configured which allows for non-contiguous pairs of OFDM symbols. Figure 3.57b corresponding to Row 17 shows an example of a non-contiguous 32-port CSI-RS resource with starting OFDM symbols 3 and 7. The starting position(s) of a CSI-RS resource may be located at any OFDM symbol index within a slot, provided that the CSI-RS resource is fully contained within one slot.

In general, the subcarrier indices of the anchor REs in a CSI-RS resource can be flexibly configured through a bitmap. For the examples in Figure 3.57 where each CDM group spans two REs in the frequency domain, there are six possible locations for an anchor RE. For example, in Figure 3.57a the bitmap $\mathbf{b} = [1\ 1\ 0\ 1\ 0\ 1]$ indicates that the 4 CDM groups are located at subcarrier indices 10, 8, 4, and 0. The bitmaps for Figure 3.57a,b are given by $\mathbf{b} = [0\ 0\ 1\ 1\ 1\ 1]$ and $\mathbf{b} = [0\ 1\ 1\ 1\ 1\ 0]$, respectively. Clearly, there is a significant degree of flexibility in the configuration of a CSI-RS resource both in the frequency and time domains.

Figure 3.58 shows several exemplary CSI-RS resource configurations corresponding to Rows 1, 3, and 4 in Table 3.12 corresponding to a smaller number of ports compared to the previous examples. Figure 3.58a shows an example of a 1-port configuration which has density of 3 REs/port/PRB. Since there is only one port, No CDM is used. A length-4 bitmap is used to configure the frequency domain location of the single anchor RE which can occupy subcarrier index 0, 1, 2, or 3. Such a 1-port configuration is used for the purposes of fine time/frequency tracking (TRS) to aid in PDSCH demodulation, especially for higher order modulations. CSI-RS for tracking is described in more detail later in this chapter.

Figure 3.58b,c show exemplary 2 and 4 port configurations corresponding to Rows 2 and 4. In both cases, FD-CDM2 is used. Again, bitmaps are used to configure the frequency

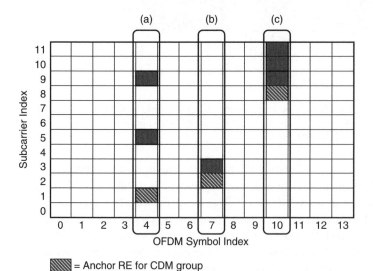

Figure 3.58 Exemplary CSI-RS configurations for various rows in Table 3.12: (a) Row1: 1 port, density 3, No CDM, (b) Row3: 2 ports, density 1, FD-CDM2; (c) Row4: 4 ports, density 1, FD-CDM2. 0-4.

domain location of the anchor RE. Note that the 4-port example in Figure 3.58c corresponding to Row 4 is a special case in which all 4 REs are contiguous in the frequency domain. Other rows in the table generally allow discontinuous pairs of REs in the frequency domain.

One use case for 2-port CSI-RS resources is beam management. In this case, a set of several 2-port resources is typically configured with each one at a different OFDM symbol location in a slot. For the case of gNB beam selection, the gNB may beamform each resource differently. Then the UE is configured to report L1-RSRP corresponding to the best, or top-N beams.

3.7.2.1.3 *CSI-RS Bandwidth*

The bandwidth of a CSI-RS resource is configurable such that it may occupy a part or whole of a BWP. Both the starting PRB and the number of PRBs of a CSI-RS resource within a BWP are configurable with a granularity of 4 PRBs under the constraint that the minimum CSI-RS bandwidth is 24 PRBs.

3.7.2.1.4 *Reduced Density CSI-RS*

With the exception of the 1-port CSI-RS resource in Figure 3.58a which has density 3 RE/port/PRB, all other CSI-RS resource configurations in the above examples have density 1 RE/port/PRB. As shown in Table 3.12, it is possible to configure a CSI-RS resource with a reduced density of $\frac{1}{2}$ RE/port/PRB. This is achieved using the same time/frequency pattern for a density 1 resource within a PRB, but leaving every 2nd PRB empty in the frequency domain. For density $\frac{1}{2}$, a CSI-RS resource can be configured to occupy either even or odd PRBs.

3.7.2.2 Antenna Port Mapping

Antenna pots for CSI-RS are numbered starting with 3000. As already discussed, an X port CSI-RS resource consists of a number of CDM groups where there are one or more CDM groups in the frequency domain and one or more CDM groups in the time domain. As illustrated in Figure 3.55, all antenna ports corresponding to a CDM group are mapped to all REs within that CDM group. For the case of multiple CDM groups, the antenna port numbers are mapped to CDM groups in the order of increasing frequency first and then increasing time second. Figure 3.59 illustrates the port mapping for the 32-port example in Figure 3.57c.

3.7.2.3 Sequence Generation and Mapping

The reference symbol in each RE of a CSI-RS resource is a QPSK value given by

$$ r_{l,n_{\text{s,f}}}(m) = \frac{1}{\sqrt{2}}(1 - 2 \cdot c(2m)) + j\frac{1}{\sqrt{2}}(1 - 2 \cdot c(2m+1)) $$

where $c(i)$ is a pseudo-random sequence defined by a length-31 Gold sequence. The pseudo-random sequence generator is initialized with a different seed value for each OFDM symbol of a slot indexed by $l \in \{0, 1, \ldots, 13\}$:

$$ c_{\text{init}} = (2^{10}(N_{\text{symb}}^{\text{slot}} n_{\text{s,f}}^{\mu} + l + 1)(2n_{\text{ID}} + 1) + n_{\text{ID}}) \bmod 2^{31} $$

The seed value is also a function of the slot number within a radio frame denoted by $n_{\text{s,f}}$ as well as a UE-specifically configured 10-bit scrambling ID n_{ID} that is configured per CSI-RS resource.

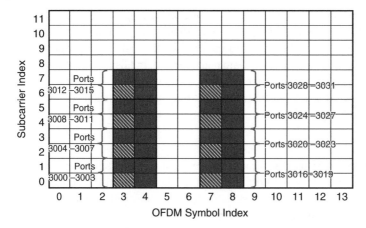

Figure 3.59 Illustration of antenna port mapping – frequency first, time second.

With this randomization, a different pseudo-random sequence is mapped to the REs of each OFDM symbol of a CSI-RS resource. For a given symbol l of a CSI-RS resource in slot number $n_{s,f}$, consecutive values of the reference signal sequence $r_{l,n_{s,f}}(m)$ are mapped to the REs across the frequency domain of the CSI-RS resource. In other words, the sequence is not sub-sampled in any way. Further, the mapping is such that a different pair of consecutive sequence values is associated with each PRB of the CSI-RS resource in the order of increasing PRB index. For a given PRB, the same pair of values is mapped to the one or more CDM groups across the frequency domain within that PRB. Such repetition across CDM groups in the frequency domain, while similar to the mapping used in LTE, can be sub-optimal in terms of minimizing PAPR in cases where a CSI-RS resource is precoded with a non-diagonal precoding matrix.

3.7.2.4 Time Domain Behavior

A CSI-RS resource can be configured to have one of three different time domain behaviors:

- Periodic
- Semi-persistent
- Aperiodic

For the case of periodic and semi-persistent CSI-RS, a periodicity is semi-statically configured such that the resource is transmitted once every N slots where the allowed configurable values are

$$N \in \{4, 5, 8, 10, 16, 20, 32, 40, 64, 80, 160, 320, 640\}$$

In addition, an offset O is configured where $O \in \{0, 1, \ldots, N-1\}$ measured in number of slots. The reference point for the slot offset is with respect to the first slot (slot 0) of radio frame 0.

While the periodicity and offset are configured in the same way for both periodic and semi-persistent CSI-RS, the difference in the two types lies in what assumptions the UE may make about the presence of the CSI-RS resource. For the case of periodic CSI-RS, once the resource is configured to the UE, the UE may assume that it is present in the slots governed by the configured periodicity and offset values. In contrast, for semi-persistent CSI-RS, the UE may not assume that the CSI-RS is present until it receives an explicit MAC-CE activation message. Once the CSI-RS resource is activated, the UE may assume that the CSI-RS is present in the configured slots up until the UE receives an explicit MAC-CE deactivation message. Note that while the UE may generally assume the presence of CSI-RS according to the configured periodicity and offset values, this assumption is subject to a slot being classified as a downlink by either semi-static and/or dynamic slot format indication (see [34] for details).

For the case of aperiodic CSI-RS, a periodicity value is not configured. For this type of resource, the UE may only assume the presence of a configured resource once it receives a PDCCH with DCI that explicitly triggers the CSI-RS resource. The UE assumes that the resource is present in the same or later slot than the one containing the PDCCH depending on a configured slot offset. The slot offset for aperiodic CSI-RS is a semi-statically configured value between 0 and 6 slots, where 0 refers to the same slot in which the DCI trigger is received.

Strictly speaking, the time domain behavior (periodic, semi-persistent, aperiodic) is configured not at CSI-RS resource level, but at the level of a CSI-RS resource set which contains one or more resources. Furthermore, for the case of aperiodic CSI-RS, the slot offset is also configured at resource set level. Hence, CSI-RS resources are always configured/activated/triggered on a per-set basis, even if a set only contains a single resource. CSI-RS resource sets are described in more detail later in this chapter.

3.7.2.5 Multiplexing with Other Signals
As described previously, configuration of CSI-RS is extremely flexible in both the frequency and time domain. However, there are specified rules about multiplexing of CSI-RS with some other signals/channels that are described in [34]. Such multiplexing rules are

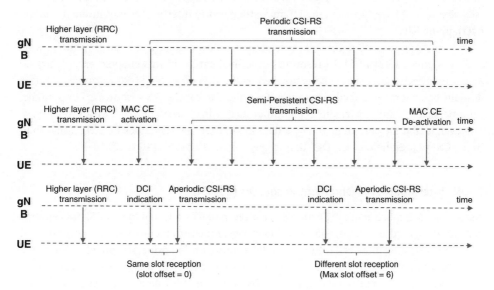

Figure 3.60 Periodic, semi-persistent, aperiodic CSI-RS transmission.

introduced to ease processing at the UE. The specified rules state that the UE shall not expect that PRBs of a configured CSI-RS resource overlap PRBs of any of the following signals/channels:

- A PDCCH search space set within a CORESET
  - Note: For the special case of a CSI-RS resource set configured with the parameter repetition set to "on" (described in more detail later in this chapter), the UE shall not expect a PDCCH search space set within a CORSET to be configured in the same OFDM symbols as any of the CSI-RS(s) in the set
- A scheduled PDSCH and its associated DMRS
- A broadcast PDSCH carrying SIB1
- A SS/PBCH block

3.7.3   Zero Power CSI-RS

When a UE is scheduled for PDSCH reception, it may assume that the gNB avoids mapping PDSCH and PDSCH DMRS to all REs that overlap those of the CSI-RS resource(s) that are configured to that UE. However, CSI-RS resources in different time/frequency locations of a slot may be configured to other UE(s). Hence, the scheduled UE needs a mechanism to be informed about the REs that are used for CSI-RS resources of other UE(s). The use of ZP-CSI-RS resources offers such a mechanism.

If a UE is configured with a ZP-CSI-RS resource, it may assume that the gNB avoids mapping PDSCH and PDSCH DMRS to all REs that overlap those of the configured ZP CSI-RS resource(s). It is important to note that this is the only assumption that the UE shall make about REs occupied by a ZP CSI-RS resource. In [34], it is stated very clearly that "The UE performs the same measurement/reception on channels/signals except PDSCH regardless of whether they collide with ZP CSI-RS or not." For example, if a ZP-CSI-RS and a CSI-IM (see next section) overlap, then the UE may still perform interference measurement on the overlapping REs.

The configuration of a ZP-CSI-RS resource follows exactly the same principles as non-zero-power (NZP) CSI-RS resources described earlier in this chapter, except that no actual reference signal sequence is mapped to the REs of a ZP-CSI-RS resource. All time domain behaviors described previously are valid for ZP-CSI-RS resources, i.e. periodic, semi-persistent, and aperiodic. One difference with aperiodic ZP-CSI-RS resources, however, is that they are triggered by a field in DL DCI that schedules PDSCH rather than in the CSI request field of UL DCI that triggers an aperiodic NZP-CSI-RS.

3.7.4   Interference Measurement Resources (CSI-IM)

As described in more detail in Section 4.3, a pattern of REs referred to as a CSI-IM resource can be configured for the purposes of interference measurement to enable accurate CSI reporting reflecting inter-cell interference. The configuration principles are identical to the configuration of an NZP-CSI-RS resources described earlier in this chapter, except that a restricted number of patterns is supported. Two different 4-RE patterns are supported: Pattern 0 is a 2×2 pattern consisting of 2 contiguous subcarriers and two contiguous OFDM

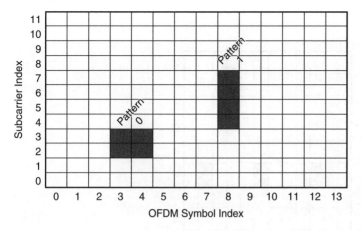

Figure 3.61 Exemplary configurations of CSI-IM Pattern 0 and Pattern 1.

symbols. Pattern 1 is a 4×1 pattern consisting of 4 contiguous subcarriers in a single OFDM symbol. These patterns are the same as in Rows 5 and 4, respectively, in Table 3.12, and have the same flexibility in where within a slot they may be configured. Two exemplary locations are illustrated in Figure 3.61.

3.7.5 CSI-RS Resource Sets

CSI-RS resources are configured within a CSI-RS resource set, where the set can include either a single CSI-RS resource or multiple resources. The same is true for CSI-IM and ZP-CSI-RS resources. This configuration mechanism simplifies the activation (for semi-persistent CSI-RS) and triggering (for aperiodic CSI-RS) since multiple resources can be activated/triggered simultaneously. There are at least two important use cases in which a CSI-RS resource set is configured for a specific purpose, and these are described here with examples.

3.7.5.1 CSI-RS for Tracking
For the purposes of fine time/frequency tracking to aid in demodulation of PDSCH, particularly for higher order modulations, the UE expects to be configured with a set of resources configured in a specific way. To distinguish that the CSI-RS resources are to be used for time/frequency tracking, a higher layer parameter called *trs_Info* is configured within the set. The collection of CSI-RS resources in the set used for tracking is commonly referred to as tracking reference signal (TRS). The CSI-RS set contains either 2 or 4 periodic CSI-RS resources with periodicity $2^{\mu} X_p$ slots where $X_p = 10, 20, 40$, or 80 and where μ is related to the SCS, i.e. $\mu = 0, 1, 2, 3, 4$ for 15, 30, 60, 120, 240 kHz, respectively. The slot offsets for the 2 or 4 CSI-RS resources are configured such that the first pair of resources are transmitted in one slot, and the 2nd pair (if configured) are transmitted in the next (adjacent) slot. All four resources are single port with density 3. Figure 3.62 shows an exemplary configuration of one period and one PRB of CSI-RS for tracking. While this diagram shows that the CSI-RSs are located at OFDM symbols {4,8}, other configurations are possible, e.g. symbols {5,9}, {6,10} for FR1, and even more possibilities for FR2 (see [34]). Note, however, that for the

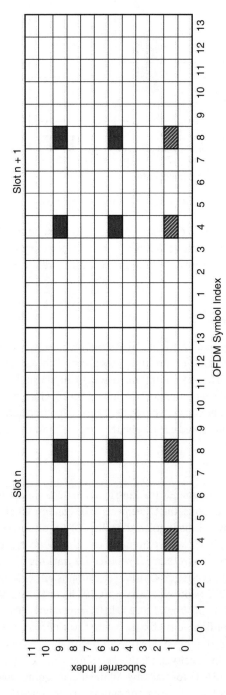

Figure 3.62 Exemplary configuration (one period) of CSI-RS for tracking.

case of FR2, the 2nd pair of CSI-RS resources in the second slot is not configured to avoid fixing the transmit beam for two slots. The minimum bandwidth of each CSI-RS resource is either 52 PRBs or equal to the number of PRBs of the active BWP, whichever is smaller.

Aperiodic TRS, i.e. a set of aperiodic CSI-RS for tracking, may be optionally configured; however, periodic TRS must always be configured. If aperiodic TRS is configured, then the time and frequency domain configurations (except for the periodicity) must match those of the periodic TRS. Further, the UE may assume that the aperiodic TRS resources are quasi-co-located with the periodic TRS resources. One example use case for aperiodic TRS is for the scenario where a UE wakes up from C-DRX and monitors PDCCH. Triggering an aperiodic TRS is beneficial to re-establish fine sync in preparation for data reception.

The time separation of the resources in each slot allows for the estimation of frequency offset through measurement of the phase variation across each pair of resources on each occupied subcarrier. Similarly, the sparse pattern in frequency allows for the estimation of time offset through measurement of the phase variation across the REs in the frequency domain within a resource. Averaging over both frequency and time reduces the noise in the time and frequency offset estimates.

3.7.5.2 CSI-RS for L1-RSRP Measurement

Another important use case for a set of CSI-RS resources is beam management, discussed in more detail in Section 4.2. For beam management purposes, the UE is configured to measure and potentially report L1-RSRP on each CSI-RS resource in the set. More details on the configuration of CSI measurement and reporting, including L1-RSRP, are included in Section 4.3. To distinguish that the CSI-RS resources are to be used for L1-RSRP measurement, a higher layer parameter called *repetition* is configured within the set. If this parameter is configured, then the CSI-RS resources in the set are restricted to be either all 1-port or all-2 port resources, configured according to either Row 1, 2, 3 of Table 3.12.

The value of *repetition* parameter can be "on" or "off," which affects what assumptions the UE may make about the downlink transmit spatial filtering configuration that is applied at the gNB to each CSI-RS resource in the set. In other words, the parameter controls what the UE may assume about gNB beamforming of the CSI-RS resources.

For *repetition* = "off," the UE may not assume that downlink transmit spatial filtering configuration is the same amongst the CSI-RS resources in the set. In other words, the gNB is free to apply different beamforming to each CSI-RS resource in the set. Typically the UE maintains the same receive spatial filterning configuration during these measurements. The use case for this configuration is for the UE to measure L1-RSRP corresponding to multiple gNB transmit beam candidates and report the best candidate to the network. The beam selection is in the form of an index of the CSI-RS resource in the set with the maximum measured L1-RSRP value as well as the measured L1-RSRP value itself. Alternatively, the UE may also be configured to report the top-N beams where N is up to 4. This procedure is illustrated in Figure 3.63.

With this procedure, the gNB may use the measurement reports corresponding to the candidate beams as a basis for later beamforming of other DL signals/channels, e.g., PDSCH, PDCCH, or other CSI-RS(s). When it does so, the gNB can refer back to the CSI-RS resource that the UE reported by signaling a "pointer" to that CSI-RS resource. The pointer is in the form of a TCI state which contains a quasi-colocation (QCL) configuration and the ID of the CSI-RS resource. TCI states are described in more detail in Section 4.2. Typically, the UE

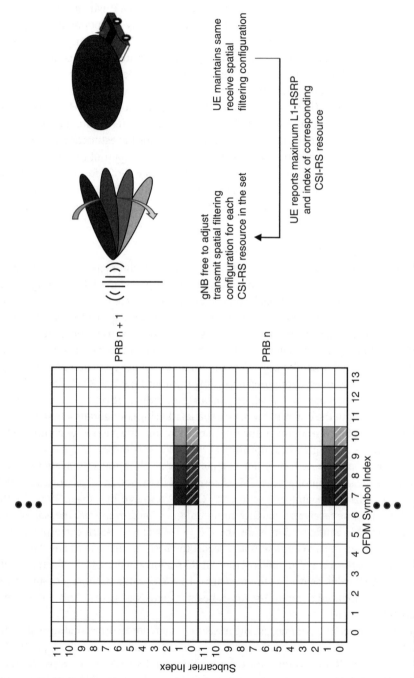

Figure 3.63 UE measurement and reporting on a set of CSI-RS resources with repetition = "off."

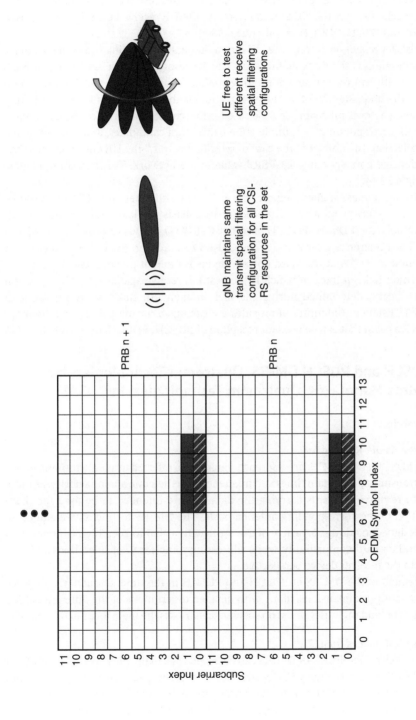

UE free to test
different receive
spatial filtering
configurations

gNB maintains same
transmit spatial filtering
configuration for all CSI-
RS resources in the set

PRB n + 1

PRB n

Figure 3.64 UE measurement on a set of CSI-RS resources with repetition = "on."

keeps track of the UE receive spatial filter configuration that is used during the measurement on a particular CSI-RS resource. When the gNB indicates a TCI state in the future, it provides a mechanism for the UE to recall (lookup) the UE receive beam that can be used for reception of PDSCH, PDCCH, or other CSI-RS(s).

In contrast, for *repetition* = "on," the UE may assume that downlink transmit spatial filtering configuration is the same amongst the CSI-RS resources in the set. Hence, the UE is free to test different receive spatial filtering configurations for each CSI-RS resource in the set with the understanding that the gNB maintains the same transmit spatial filtering configuration, i.e. transmit beam. The use case for this configuration is for the UE to measure L1-RSRP corresponding to multiple UE receive beam candidates, and thus refine the UE beam selection. In contrast to the case of *repetition* = "off," the UE does not report the CSI-RS index and corresponding L1-RSRP value to the network. This procedure is illustrated in Figure 3.64.

Even without a measurement report, the gNB can still refer to a CSI-RS resource measured in this beam refinement procedure as a basis for later transmission of DL signals/channels, e.g. PDSCH, PDCCH, or other CSI-RS(s). In this case the indicated TCI state contains a reference to any one of the CSI-RS resources in the CSI-RS resource set with repetition = "on" that was used during the measurement procedure. It is up to UE implementation to keep track of which is the best UE receive spatial filter configuration (UE receive beam) determined during the measurement. In this way, when the gNB indicates a TCI state in the future, it provides a mechanism for the UE to recall (lookup) the UE receive beam that can be used for reception of PDSCH, PDCCH, or other CSI-RS(s).

3.8 PDSCH and PUSCH DM-RS, Qualcomm Technologies, Inc. (Alexandros Manolakos, Qualcomm Technologies, Inc, USA)

3.8.1 Overview

3.8.1.1 What Is DM-RS Used for?

DM-RS for PDSCH and PUSCH, are reference signals, also referred to as data training pilots, which are transmitted at certain times and frequencies on the same antenna port(s) as those used for the resource elements carrying the data and are intended to be used for channel estimation of the corresponding antenna ports. The DM-RS for the data channels of NR were designed to address a variety of use cases with a goal to keep high configuration flexibility and forward-compatibility, while addressing complexity considerations and constraints from the receiver device perspective.

The placement of the DM-RS of PDSCH and PUSCH in the time, frequency, and space (i.e. antenna port(s)) dimensions, along with detailed descriptions of the different configurability options, and their usefulness on specific scenarios is the main focus of this chapter.

3.8.1.2 Key Differences from LTE

Several of the key principles of NR DM-RS can be traced back in the PUSCH DM-RS, PDSCH UE-RS and CRS used in the early LTE specifications, including sequence designs (for both CP-OFDM and DFT-S-OFDM waveform), placement on the time-frequency grid, mapping to physical resources and PRB bundling. Yet, it will be instructive to summarize below a few key differences with a short description for each one.

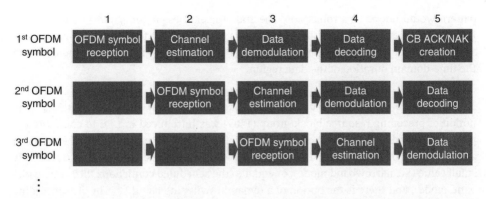

Figure 3.65 Example of "pipeline processing" enabled by a front-loaded DM-RS pattern.

3.8.1.2.1 Symmetric Mapping to Physical Resources for both DL and UL DM-RS. In NR UL, in contrast to LTE UL, both CP-OFDM and DFT-S-OFDM based waveforms are supported. Commonalities between the PHY layer designs of CP-OFDM and DFT-S-OFDM-based waveforms were considered beneficial, at least from specification simplicity and receiver implementation standpoint. Furthermore, the DM-RS patterns for NR PDSCH and PUSCH are symmetric in time, frequency, and space, in contrast to LTE PDSCH and PUSCH, in an attempt to reduce implementation complexity at the transmitter side and receiver side, while facilitating flexible duplex operation and dynamic TDD scenarios.

3.8.1.2.2 Scalable DM-RS Patterns across Numerologies The DM-RS patterns were designed to be applicable for all supported numerologies (i.e. for different subcarrier spacings and CP configurations). Such design choice allowed for a scalable design depending on the carrier frequency and deployment scenario, and common transmit and receive blocks (e.g. the channel estimation implementation used in one numerology can be reused for another numerology).

3.8.1.2.3 Front-Load DM-RS Configuration for Low-Latency Applications with Floating Placement Spanning One or Two Consecutive Symbols Enabling low-latency applications for both DL and UL has been one of the main considerations on the forward-compatibility design of NR DM-RS of the data channels. Placing training pilots toward the beginning of the data region allows the UE to acquire a channel estimate as soon as the first few symbols arrive, without having to buffer the full package, and kick-off the data demodulation and decoding soon after that. Such a DM-RS placement, which is referred to as front-load DM-RS, facilitates what is also known as "pipeline processing" (Figure 3.65) of the main processing blocks at a receiver: OFDM symbol reception, channel estimation, data demodulation, data decoding, ACK/NACK creation without having to receive and store all the packages. Note that several other aspects in NR were needed to be tailored to facilitate such processing, e.g. frequency-first mapping of the encoded data on the time-frequency resources or single-codeword for up to rank 4 PDSCH or PUSCH transmissions.

3.8.1.2.4 Configurable and Scalable Time-Domain Locations of the DM-RS Symbols Due to the time-varying nature of wireless channels, employing only a front-load DM-RS pattern would not be as a robust and spectral efficient design compared to a distributed DM-RS placement. For this reason, NR supports configuring up to three additional DM-RS symbols based on a combination of semi-static configuration and dynamic indication of the time-domain duration of the data region.

3.8.1.2.5 Narrowband and Wideband PRB Bundling with an Option of Dynamic Switching The notion of precoding resource block group (PRG) was introduced in LTE to configure the precoding granularity in the frequency-domain for the UE to perform channel estimation across a group of consecutive PRBs. In NR, a PRG can be chosen to be either a relatively small value (i.e. narrowband mode), or equal to the scheduled contiguous PRBs (i.e. wideband mode), and there is the option of a dynamic switching using 1 bit in the scheduling DCI to switch between different values.

3.8.2 Physical Layer Design

3.8.2.1 Mapping to Physical Resources

In a broadband MIMO-OFDM system, the data DM-RS pilot placement in time, frequency and space (i.e. transmit antennas, or antenna ports) is a key part of the design of an efficient MIMO communication system. On the one hand, DM-RS pilot symbols need to be spaced close enough in time and frequency to capture the time and frequency variations of the multipath channel environment. On the other hand, if DM-RS pilots are placed too close, either in frequency or in time, it may result to higher overhead than necessary, which is not counter-balanced by a better channel estimation, resulting to lower overall data channel throughput. Therefore, having in mind a variety of deployments and scenarios, the DM-RS of both DL and UL data channels can be mapped in a variety of different ways on the time-frequency-space physical resources.

3.8.2.1.1 Frequency-Domain Mapping Two DM-RS configuration types have been introduced, referred to as DM-RS Type 1 and DM-RS Type 2, each one supporting a single-symbol and a double-symbol configuration, with some key properties summarized in Table 3.13.

By observing Figures 3.66 and 3.67, it is noted that both DM-RS types use a similar approach to multiplex pilots of different antenna ports in the code domain. Specifically, for the single-symbol DM-RS, two antenna ports are orthogonalized on the same resource elements using a length-2 OCC, whereas for the double symbol DM-RS, 4 antenna ports are orthogonalized using a length-4 OCC.

A key difference between the two DM-RS types is the density of pilots dedicated to each antenna port, and as a result, the maximum number of antenna ports that each DM-RS configuration type can support. Specifically, DM-RS Type 1 is denser in frequency domain compared to DM-RS Type 2, rendering it more resilient to larger frequency-domain channel variations, whereas Type 2 DM-RS can pack more orthogonal antenna ports per PRB, by reducing the density by 33% compared to DM-RS Type 1.

Note that the configuration of the DM-RS Type is provided through higher-layer signaling independently for each PDSCH and PUSCH, each mapping Type (A or B) and each BWP independently. In other words, the same UE can be scheduled with a different DM-RS Type

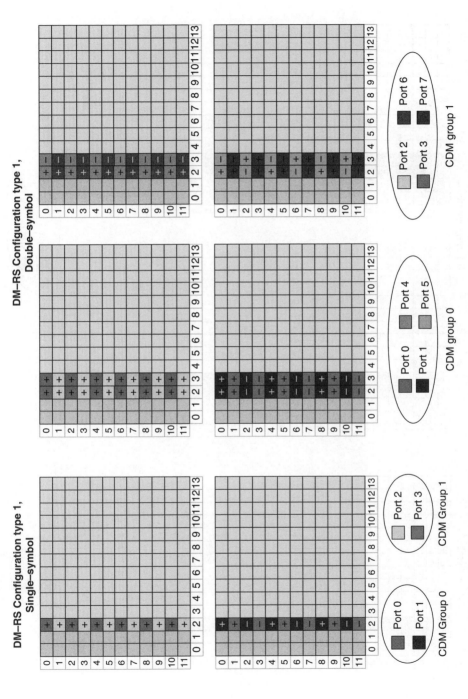

Figure 3.66 Front-loaded DM-RS Configuration Type 1.

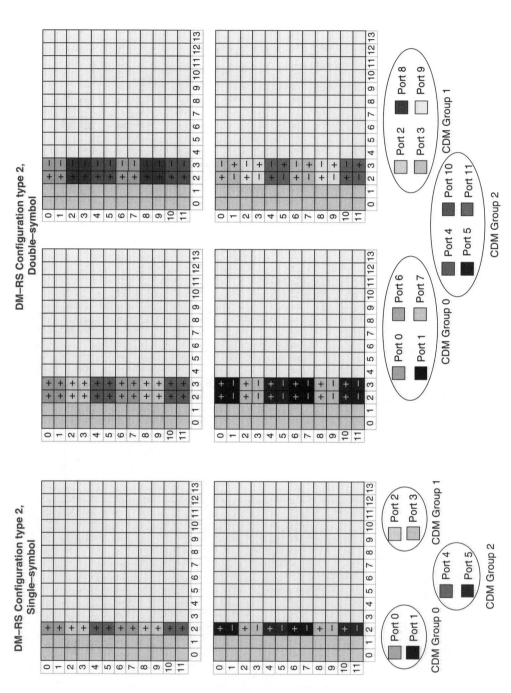

Figure 3.67 Front-loaded DM-RS configuration Type 2.

Table 3.13 Main properties of the DM-RS Type 1 and 2.

DM-RS configuration type and length	Maximum number of antenna ports	Multiplexing type	Number of REs per port
Single-symbol Type 1	4 ports	2 (2) antenna ports are mapped in even (odd) numbered subcarriers in the frequency domain, orthogonalized using a length-2 OCC	6
Double-symbol Type 1	8 ports	4 (4) antenna ports are mapped in even (odd) numbered subcarriers in the frequency domain over 2 consecutive OFDM symbols, orthogonalized using a length-4 OCC	6
Single-symbol Type 2	6 ports	2 (2) antenna ports are mapped in consecutive subcarriers in the frequency domain, orthogonalized using a length-2 OCC.	4
Double-symbol Type 2	12 ports	4 (4) antenna ports are mapped in consecutive numbered subcarriers in the frequency domain over 2 consecutive OFDM symbols, orthogonalized using a length-4 OCC	4

if it is scheduled with PDSCH mapping Type A compared to being scheduled with a PDSCH mapping Type B, or between PDSCH and PUSCH, and for different BWPs.

With regards to configuration of single-symbol and double-symbol DM-RS, the UE is higher-layer configured whether

- the DM-RS is single-symbol only, or
- it can be up to two consecutive symbols (referred to as double-symbol DM-RS).

Furthermore, if the higher-layer parameter *maxLength* is equal to "len2," whether single-symbol or double-symbol DM-RS shall be used in any specific PDSCH or PUSCH is determined using the scheduling DCI.

3.8.2.1.2 *Time-Domain Mapping* Across time domain, the DM-RS patterns for NR Rel-15 PDSCH and PUSCH can be decomposed to:

- the DM-RS pattern used for the front-load DM-RS (See Figures 3.66, 3.67), and
- a set of additional DM-RS symbols distributed inside the scheduled data channel duration which are also single-symbols, or double-symbols, following the length of the front-load DM-RS. That is, if the front-load DM-RS is single-symbol, then also the additional DM-RS shall be single-symbol.

NR Rel-15 supports up to three additional single-symbol DM-RS when the front-load DM-RS is single-symbol, and up to one additional double-symbol DM-RS when the front-load DM-RS is double-symbol. Inside the scheduled time-domain allocation of a PDSCH or PUSCH, the UE may expect up to 4 DM-RS symbols. Note however that, during the duration of a slot, a maximum of 7 DM-RS symbols can be received if a UE is capable of receiving multiple back-to-back PDSCHs, each one with a duration of 2 OFDM symbols.

The location of the DM-RS across time in a PDSCH allocation is dependent on both higher-layer configuration and dynamic (DCI-based) signaling according to the following principles:

- *dmrs-TypeA-Position*: The location of the front-load DM-RS for PDSCH mapping type A. Specifically, for PDSCH mapping type A, the front-load DM-RS appears on the 3rd (*dmrs-TypeA-Position* = 'pos2') of 4th OFDM symbol (*dmrs-TypeA-Position* = 'pos3') of a slot for PDSCH mapping type A. Note that for PDSCH mapping type B, the first symbol of the scheduled allocation would carry the DM-RS independently of this parameter unless there is a collision with control resources (details on these special cases are described later in this subsection).
- *maxLength:* The higher-layer parameter "*maxLength*" determines the maximum number of OFDM symbols for DL front-loaded DM-RS. "len1" corresponds to value 1 (single-symbol DM-RS) and "len2" corresponds to value 2 (double-symbol DM-RS). Note that if the field is absent in the RRC configuration, the UE applies value "len1," or in other words single-symbol DM-RS are configured. Also it should be emphasized that if this field is set to "len2," it does not mean that always double-symbol DM-RS is scheduled; the final determination is based on the signaling in the scheduling DCI.
- *dmrs-AdditionalPosition:* Maximum number of additional DM-RS inside the time-domain PDSCH allocation. Note that if the field is absent in the RRC configuration, the UE assumes that up to two additional DM-RS are configured (i.e. *dmrs-AdditionalPosition* = "pos2"). Note also that if *maxLength* equals to "len2," the *dmrs-AdditionalPosition* cannot be set to more than "pos1," or in other words, when double-symbol DM-RS is used, there can be up to one more double-symbol DM-RS (total 4 DM-RS symbols inside the PDSCH allocation).

When it comes to PUSCH, there are a lot of similarities to the PDSCH DM-RS, especially for the case of PUSCH without frequency domain hopping. To be more precise, for PUSCH without frequency hopping, **dmrs-TypeA-Position, maxLength, dmrs-AdditionalPosition** configure the location of the front-loaded DM-RS of PUSCH, the maximum number of OFDM symbols for the front-loaded DM-RS, the maximum number of additional DM-RS inside the time-domain PUSCH allocation, respectively, in the same way as the corresponding higher-layer parameters are configured for PDSCH (Figure 3.68).

Based on these higher-layer configured parameters and the actual time-domain allocation of PDSCH, the location of the front-loaded and additional DM-RS symbol is given by $\bar{l}$ shown in Table 3.14, where the definition of l_d changes depending on the mapping type (See Figures 3.69–3.71):

- for PDSCH mapping type A, l_d is the duration is between the first OFDM symbol of the slot and the last OFDM symbol of the scheduled PDSCH resources in the slot
- for PDSCH mapping type B, l_d is the duration is the number of OFDM symbols of the scheduled PDSCH resources

The definition of l_d for different mapping Types is depicted in Figure 3.68.

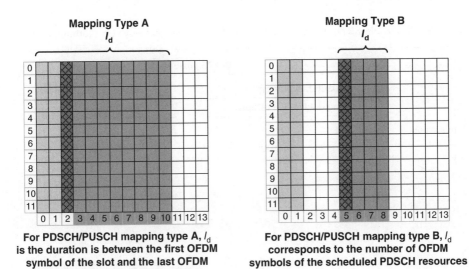

For PDSCH/PUSCH mapping type A, l_d is the duration is between the first OFDM symbol of the slot and the last OFDM symbol of the scheduled PDSCH resources in the slot

For PDSCH/PUSCH mapping type B, l_d corresponds to the number of OFDM symbols of the scheduled PDSCH resources

⬙ DM-RS ▦ Data (PDSCH or PUSCH)

Figure 3.68 Definition of l_d for different mapping Types. For mapping Type A, l_d is defined with respect to the beginning of the slot, whereas for Type B with respect to the first scheduled data symbol.

When PUSCH with intra-slot frequency domain hopping is scheduled, different DM-RS pattern is supported compared to the case of PUSCH without hopping, since at each PUSCH hop, the channel estimation is performed independently using the DM-RS symbols appearing inside the corresponding hop. Table 3.15 and Figure 3.72 provide the details with all the DM-RS patterns supported depending on the duration of each hop (Figure 3.73).

In addition to the DM-RS patterns shown above, there are some special cases that needed a separate treatment (Figures 3.74–3.76):

- For PDSCH Type B, if the PDSCH duration l_d is 2, 4, or 7 OFDM symbols for normal CP or 2, 4, 6 OFDM symbols for extended CP, and the PDSCH allocation collides with resources reserved for a search space set associated with a CORESET, the symbol location of the DM-RS shall be incremented such that the first DM-RS symbol occurs immediately after the CORESET and
  - if the PDSCH duration l_d is 2 symbols, the UE is not expected to receive a DM-RS symbol beyond the second symbol, whereas if the PDSCH duration l_d is 4 symbols, the UE is not expected to receive a DM-RS symbol beyond the third symbol
  - If the PDSCH duration l_d is 7 symbols for normal CP or 6 symbols for extended CP,
    - the UE is not expected to receive the first DM-RS beyond the fourth symbol, and
    - if one additional single-symbol DM-RS is configured, the UE only expects the additional DM-RS to be transmitted on the 5th or 6th symbol when the front-loaded DM-RS symbol is in the 1st or 2nd symbol, respectively, of the PDSCH duration, otherwise the UE should expect that the additional DM-RS is not transmitted.

Table 3.14 PDSCH DM-RS positions for single-symbol (top) and double-symbol (bottom) DM-RS.

	DM-RS positions $\bar{l}$					
	PDSCH mapping type A			PDSCH mapping type B		
	dmrs-AdditionalPosition			dmrs-AdditionalPosition		
l_d in symbols	0	1	2	0	1	2
<4				—	—	
4	l_0	l_0		—	—	
5	l_0	l_0		—	—	
6	l_0	l_0		l_0	l_0	
7	l_0	l_0		l_0	l_0	
8	l_0	l_0		—	—	
9	l_0	l_0		—	—	
10	l_0	$l_0, 8$		—	—	
11	l_0	$l_0, 8$		—	—	
12	l_0	$l_0, 8$		—	—	
13	l_0	$l_0, 10$		—	—	
14	l_0	$l_0, 10$		—	—	

	DM-RS positions $\bar{l}$							
	PDSCH mapping type A				PDSCH mapping type B			
	dmrs-AdditionalPosition				dmrs-AdditionalPosition			
l_d in symbols	0	1	2	3	0	1	2	3
2	—	—	—	—	l_0	l_0		
3	l_0	l_0	l_0	l_0	—	—		
4	l_0	l_0	l_0	l_0	l_0	l_0		
5	l_0	l_0	l_0	l_0	—	—		
6	l_0	l_0	l_0	l_0	l_0	$l_0, 4$		
7	l_0	l_0	l_0	l_0	l_0	$l_0, 4$		
8	l_0	$l_0, 7$	$l_0, 7$	$l_0, 7$	—	—		
9	l_0	$l_0, 7$	$l_0, 7$	$l_0, 7$	—	—		
10	l_0	$l_0, 9$	$l_0, 6, 9$	$l_0, 6, 9$	—	—		
11	l_0	$l_0, 9$	$l_0, 6, 9$	$l_0, 6, 9$	—	—		
12	l_0	$l_0, 9$	$l_0, 6, 9$	$l_0, 5, 8, 11$	—	—		
13	l_0	l_0, l_1	$l_0, 7, 11$	$l_0, 5, 8, 11$	—	—		
14	l_0	l_0, l_1	$l_0, 7, 11$	$l_0, 5, 8, 11$	—	—		

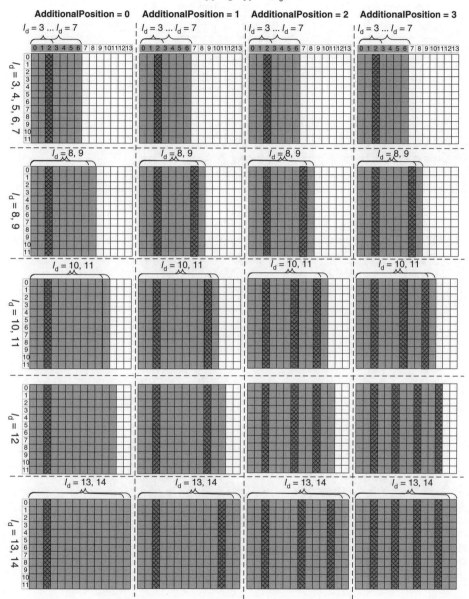

Figure 3.69 DM-RS patterns for mapping Type A with front-load DM-RS single-symbol on the 3rd symbol of the slot.

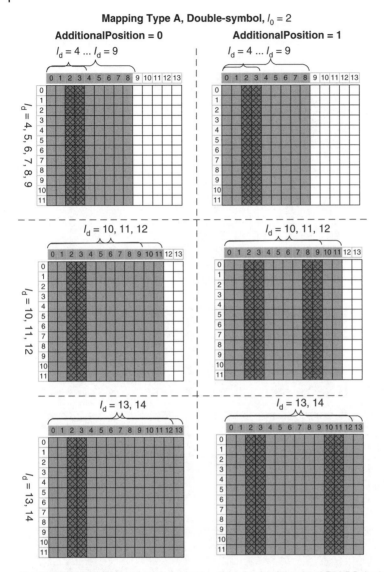

Figure 3.70 DM-RS patterns for mapping Type A with front-load DM-RS double-symbol on the 3rd and 4th symbol of the slot.

- When LTE and NR Rel-15 co-exist on the same carrier, the NR UE can, if operating on 15 kHz subcarrier spacing, be informed about the position of the CRS using the RRC parameter *lte-CRS-ToMatchAround*. However, as can be seen in Figure 3.77, when an additional DM-RS symbol is configured to the UE, it collides with CRS, leading to degraded performance. Therefore, for PDSCH Type A, when all the following conditions are satisfied, the location of additional DM-RS appears on the 13th symbol of the slot:

Figure 3.71 DM-RS patterns for PDSCH mapping Type B.

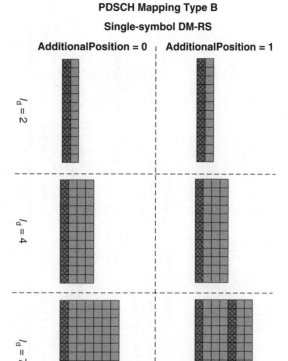

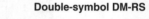

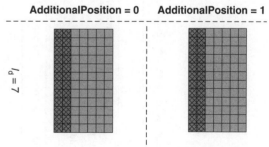

- o the higher-layer parameter *lte-CRS-ToMatchAround* is configured and any PDSCH DM-RS symbol coincides with any symbol containing LTE CRSs as indicated by *lte-CRS-ToMatchAround*,
- o with the higher-layer parameters *dmrs-AdditionalPosition* is equal to "pos1,"
- o $l_0 = 3$, and
- o the UE has indicated it is capable of this additional DM-RS pattern.

Table 3.15 PUSCH DM-RS positions within a slot for single-symbol DM-RS and intra-slot frequency hopping enabled.

l_d in symbols	DM-RS positions $\bar{l}$											
	PUSCH mapping type A								PUSCH mapping type B			
	$l_0 = 2$				$l_0 = 3$				$l_0 = 0$			
	dmrs-AdditionalPosition				dmrs-AdditionalPosition				dmrs-AdditionalPosition			
	0		1		0		1		0		1	
	1st hop	2nd hop	1st hop	2nd hop	1st hop	2nd hop	1st hop	2nd hop	1st hop	2nd hop	1st hop	2nd hop
≤3	—	—	—	—	—	—	—	—	0	0	0	0
4	2	0	2	0	3	0	3	0	0	0	0	0
5, 6	2	0	2	0, 4	3	0	3	0, 4	0	0	0, 4	0, 4
7	2	0	2, 6	0, 4	3	0	3	0, 4	0	0	0, 4	0, 4

PUSCH mapping type A DM-RS with intra-slot frequency hopping enabled

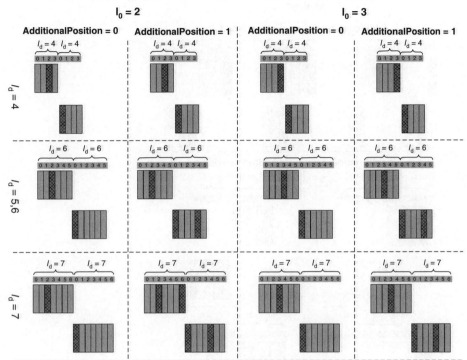

Figure 3.72 PUSCH mapping Type A DM-RS with intra-slot frequency hopping enabled with front-load DM-RS single-symbol on the 3rd symbol of the slot.

Note that there exist additional DM-RS patterns which would result in collision between the NR DM-RS and CRS, e.g. the PDSCH with duration of 13 of 14 symbols with three additional DM-RS symbols, but for these cases DM-RS shift was not introduced in NR Rel-15.

3.8.2.2 Default DM-RS Pattern for PDSCH and PUSCH

As it is described above, the DM-RS pattern used in a PDSCH, or PUSCH depends on a multiple higher-layer configured parameters and DCI indication. However, there exist several cases that a "default" DM-RS pattern needs to be employed as described in detail below (Figure 3.78).

Specifically, when receiving PDSCH scheduled by DCI format 1_0 (in which case no DCI bits are used for antenna port signaling), or receiving PDSCH before dedicated higher-layer configuration of any of the parameters *dmrs-AdditionalPosition*, *maxLength*, and *dmrs-Type*, the UE shall make the following assumptions regarding the DM-RS pattern:

- single-symbol front-load DM-RS of DM-RS configuration type 1 on DM-RS port 1000 is transmitted.
- the remaining orthogonal antenna ports are not associated with transmission of PDSCH to another UE and in addition (in other words SU-MIMO operation).

PUSCH mapping type B DM-RS with intra-slot frequency hopping enabled

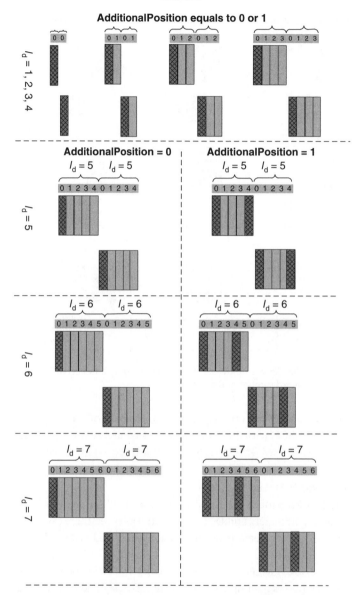

Figure 3.73 PUSCH mapping Type B DM-RS with intra-slot frequency hopping enabled.

- For PDSCH with mapping type A,
  - the UE shall assume *dmrs-AdditionalPosition* = "pos2" and up to two additional single-symbol DM-RS present in a slot according to the PDSCH duration indicated in the DCI
- For PDSCH with mapping type B,
  - with allocation duration of seven symbols for normal CP (or six symbols for ECP), the UE shall assume one additional single-symbol DM-RS present in the 5th or 6th symbol

2-symbol and 4-symbol PDSCH Mapping Type B

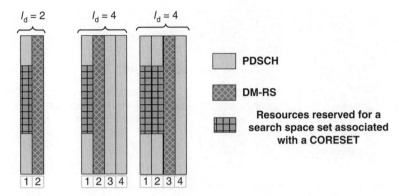

Figure 3.74 Front-load DM-RS location in case of collision of single-symbol DM-RS with CORESET for PDSCH mapping Type B of length 2 or 4.

7-symbol PDSCH Mapping Type B with single-symbol DM-RS

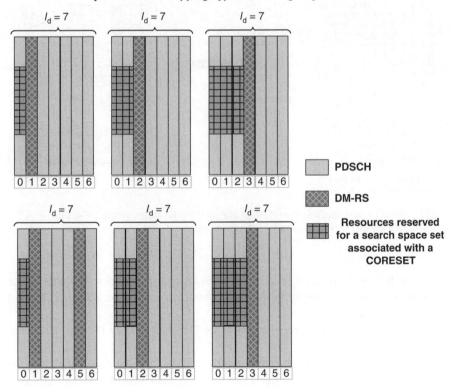

Figure 3.75 Front-load DM-RS location in case of collision of single-symbol DM-RS with CORESET for PDSCH mapping Type B of length 7.

7-symbol PDSCH Mapping Type B with double-symbol DM-RS

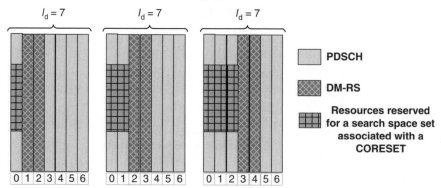

Figure 3.76 Front-load DM-RS location in case of collision of double-symbol DM-RS with CORESET for PDSCH mapping Type B of length 7.

PDSCH Mapping Type A – Collision of DM-RS with CRS rate matching pattern

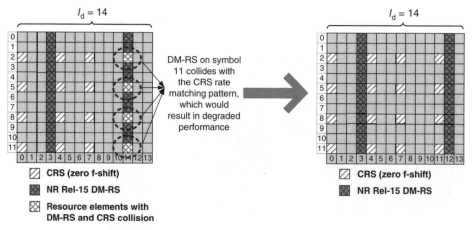

Figure 3.77 NR-LTE Coexistence scenario: When one additional single-symbol DM-RS is configured, it appears on the 13th symbol of the slot to avoid collision with the CRS rate matching resources.

when the front-loaded DM-RS symbol is in the 1st or 2nd symbol respectively of the PDSCH allocation duration, otherwise the UE shall assume that the additional DM-RS symbol is not present

o with allocation duration of four symbols, the UE shall assume that no additional DM-RS are present

o with allocation duration of two symbols with mapping type B, the UE shall assume that no additional DM-RS are present, and the UE shall assume that the PDSCH is present in the symbol carrying DM-RS

• For all cases expect PDSCH with mapping type B with 2 symbol duration, PDSCH is not present in the symbol carrying DM-RS.

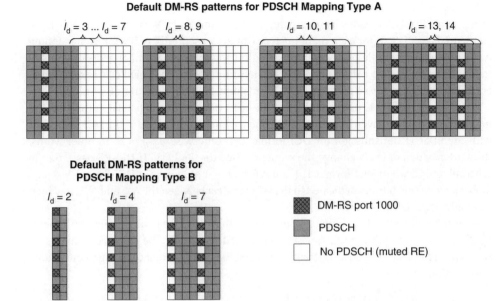

Figure 3.78 Default DM-RS patterns for PDSCH.

Turning our attention to UL, for PUSCH not scheduled by PDCCH format 0_1 with CRC scrambled by C-RNTI, CS-RNTI, or MCS-C-RNTI, the UE will transmit PUSCH with a DM-RS according to the following configurations:

- DM-RS Type 1 with DM-RS port 0
- If transform precoding is disabled, on the OFDM symbols carrying DM-RS, the resource elements are not used for any PUSCH transmission except for the case of PUSCH with allocation duration of one or two OFDM symbols.
- If transform precoding is enabled, PUSCH is never mapped on resource elements on the OFDM symbols carrying DM-RS.
- If frequency hopping is disabled, the UE will transmit DM-RS with up to two additional DM-RS according to PUSCH duration, whereas if frequency hopping is enabled, up to one additional DM-RS shall be transmitted per hop.

3.8.2.3 Sequence Generation and Scrambling

3.8.2.3.1 Sequence Generation for CP-OFDM Waveform The reference symbol in each resource element carrying DM-RS is a QPSK symbol given by

$$r(m) = \frac{1}{\sqrt{2}}(1 - 2 \cdot c(2m)) + j\frac{1}{\sqrt{2}}(1 - 2 \cdot c(2m + 1))$$

where $c(i)$ is a pseudo-random sequence defined by a length-31 Gold sequence (the same to the one used in LTE and other physical layer channels in NR). The pseudo-random sequence generator is initialized with a different seed value for each OFDM symbol of a slot indexed by $l \in \{0, 1, \ldots, 13\}$:

$$c_{\text{init}} = (2^{17}(N_{\text{symb}}^{\text{slot}} n_{\text{s,f}}^{\mu} + l + 1)(2N_{\text{ID}}^{n_{SCID}} + 1) + 2N_{\text{ID}}^{n_{SCID}} + n_{\text{SCID}})\text{mod}2^{31}$$

The seed value is also a function of the slot number within a radio frame denoted by $n_{s,f}^\mu$ as well as up to two UE-specifically configured 16-bit scrambling IDs, N_{ID}^0 and N_{ID}^1, which are configured per PDSCH for each mapping type (up to two scrambling IDs for each mapping type). The above formula provides randomization across time at symbol and slot levels to randomize interference within one PDSCH allocation, across multiple PDSCH scheduling occasions, and even across different PDSCH mapping types and different BWPs.

Whether N_{ID}^0 or N_{ID}^1 is used to initialize the seed value c_{init}, it depends on the quantity $n_{SCID} \in \{0, 1\}$ which is given by the DM-RS sequence initialization field in the DCI associated with the PDSCH transmission for DCI format 1_1, otherwise $n_{SCID} = 0$. In other words, the gNB may dynamically change the sequence used for the PDSCH DM-RS based on the scheduling DCI when DCI format 1_1 is employed. For scenarios of DCI format 1_0 (e.g. fallback mode), no bits are dedicated in the DCI to select between the N_{ID}^0 and N_{ID}^1, in which case only N_{ID}^0 is used.

After the sequence $c(i)$ is initialized using c_{init}, the sequence $r(m)$ is scaled by a factor β_{PDSCH}^{DMRS} to conform with the transmission power and mapped to the resource elements $(k, l)_{p,\mu}$ within the CRBs which are allocated for PDSCH transmission according to the following equation:

$$a_{k,l}^{(p,\mu)} = \beta_{PDSCH}^{DMRS} w_f(k') w_t(l') r(2n + k')$$

$$k = \begin{cases} 4n + 2k' + \Delta & \text{Configuration type 1} \\ 6n + k' + \Delta & \text{Configuration type 2} \end{cases}$$

$$k' = 0, 1$$

$$l = \bar{l} + l'$$

$$n = 0, 1, \ldots$$

where $w_f(k')$, $w_t(l')$ and Δ are given by Tables 3.16 and 3.17.

The parameter Δ controls the mapping a specific port on a specific subcarrier offset within the PRB. For example, in DM-RS Type 1 single-symbol, there are 2 CDM groups with 4 ports, where ports 1000, 1001 consist the first CDM group (occupying subcarriers

Table 3.16 Parameters for PDSCH DM-RS configuration type 2.

p	CDM group λ	Δ	$w_f(k')$		$w_t(l')$	
			$k' = 0$	$k' = 1$	$l' = 0$	$l' = 1$
1000	0	0	+1	+1	+1	+1
1001	0	0	+1	−1	+1	+1
1002	1	1	+1	+1	+1	+1
1003	1	1	+1	−1	+1	+1
1004	0	0	+1	+1	+1	1
1005	0	0	+1	−1	+1	−1
1006	1	1	+1	+1	+1	−1
1007	1	1	+1	−1	+1	−1

Table 3.17 Parameters for PDSCH DM-RS configuration type 2.

p	CDM group λ	Δ	$w_f(k')$		$w_t(l')$	
			$k' = 0$	$k' = 1$	$l' = 0$	$l' = 1$
1000	0	0	+1	+1	+1	+1
1001	0	0	+1	−1	+1	+1
1002	1	2	+1	+1	+1	+1
1003	1	2	+1	−1	+1	+1
1004	2	4	+1	+1	+1	+1
1005	2	4	+1	−1	+1	+1
1006	0	0	+1	+1	+1	−1
1007	0	0	+1	−1	+1	−1
1008	1	2	+1	+1	+1	−1
1009	1	2	+1	−1	+1	−1
1010	2	4	+1	+1	+1	−1
1011	2	4	+1	−1	+1	−1

0,2,4,6,8,10 of each PRB) whereas ports 1002, 1003 appear on the second CDM group (occupying subcarriers 1,3,5,7,9,11 of each PRB). The parameter $w_f(k')$ is used to perform the OCC in the frequency domain (FD-OCC), either [34] or [1,−1], the parameter $w_t(l')$ is used to perform the OCC in the time domain (TD-OCC). For example, for port 1000 we observe that $w_f(k')$ is always 1, whereas for port 1001, $w_f(0) = 1$ and $w_f(1) = -1$, resulting in being orthogonal in the code domain with port 1000.

The reference point for k is subcarrier 0 in CRB 0, unless the corresponding PDCCH is associated with CORESET 0 and Type0-PDCCH CSS is addressed to SI-RNTI, in which case the reference point for k is subcarrier 0 of the lowest-numbered resource block in CORESET 0. In other words, whenever the UE is aware of the location of the CRB with respect to the PDSCH allocation, the sequence starts from the first subcarrier of that resource block (which may even in a distance of $275*8 - 1 = 2199$ resource blocks away from the start of current component carrier), and uses only the sequence values mapped inside the resource blocks of the data allocation, as shown in Figure 3.79.

Using the above formulas, it can be observed that the same sequence is mapped on the frequency domain for different port indices as illustrated in Figure 3.80 for the DM-RS Type 1 with single-symbol DM-RS.

3.8.2.3.2 Updates in NR Rel-16 for CP-OFDM Low-PAPR DM-RS Design
The sequence mapping supported in NR Rel-15 may result in higher PAPR in the DM-RS symbols from the data symbols under some antenna port combinations, due to two main reasons.

- First, the reference signal sequence to resource mapping results in a repetitive symbol structure in the frequency domain, and
- second, multiple CDM groups can be transmitted through the same power amplifier (PA).

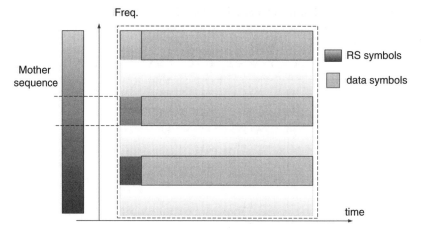

Freq.

Mother
sequence

RS symbols

data symbols

time

Figure 3.79 The sequence of the PDSCH DM-RS is initialized starting from CRB0, unless the corresponding PDCCH is associated with CORESET 0 and Type0-PDCCH common search space and is addressed to SI-RNTI.

DMRS antenna ports

Subcarrier Index in PRB	0	1	2	3
11	0	0	r(5)	−r(5)
10	r(5)	−r(5)	0	0
9	0	0	r(4)	r(4)
8	r(4)	r(4)	0	0
7	0	0	r(3)	−r(3)
6	r(3)	−r(3)	0	0
5	0	0	r(2)	r(2)
4	r(2)	r(2)	0	0
3	0	0	r(1)	−r(1)
2	r(1)	−r(1)	0	0
1	0	0	r(0)	r(0)
0	r(0)	r(0)	0	0

Figure 3.80 An example of sequence to subcarrier mapping for DMRS type 1, within one OFDM symbol.

Due to these reasons, for NR Rel-16, an updated sequence initialization was done to solve the high PAPR problem for the CP-OFDM waveform. To quantify the problem, note that a statistical estimate of the PAPR of a waveform can be determined from the CCDF of the power-to-average power ratio. Typically, the PAPR estimate is evaluated at the 99.99% percentile of this CCDF. To illustrate the PAPR issue on NR Rel-15 sequences with a specific example, the CCDFs of the power-to-average power ratios when mapping two layers using NR MIMO precoders. The CCDFs of the PAPR are shown in Figure 3.81 for an allocation of 133 PRBs and DM-RS ports {0,2}. Based on this Figure, we observe that 2–3 dB higher PAPR occurs for the DM-RS symbols compared to the data symbols.

For this reason, in NR Rel-16, the sequence mapping was updated in the following way. Since the reason of the increase in PAPR is the fact that the same frequency domain sequence is used for ports of different CDM groups, for NR Rel-16, a different sequence is employed for ports of different CDM group. Specifically, for both DM-RS type 1 and 2, the

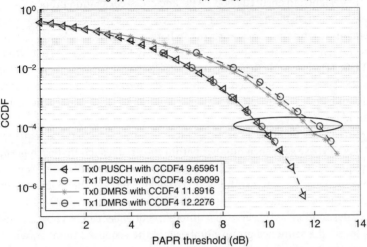

Figure 3.81 DM-RS type 1 PAPR from the CCDF of the power-to-average power ratio.

c_{init} for CDM group λ is used for NR Rel-16 DM-RS sequence generation is

$$c_{init}(\lambda) = \left(2^{17}(N_{symb}^{slot} n_{s,f}^{\mu} + l + 1)(2N_{ID}^{n'_{SCID}(\lambda)} + 1) \right.$$
$$\left. + 2N_{ID}^{n'_{SCID}(\lambda)} + n'_{SCID}(\lambda) + 2^{17} \left\lfloor \frac{\lambda}{2} \right\rfloor \right) mod2^{31}$$

where, $n'_{SCID}(\lambda = 0) = n_{SCID}, n'_{SCID}(\lambda = 1) = 1 - n_{SCID}, n'_{SCID}(\lambda = 2) = n_{SCID}$ and n_{SCID} is provided by the scheduling DCI.

As an illustration on how the above formula would resolve the PAPR issue, consider the DM-RS Type 1 which has 2 CDM groups. For the first CDM group, the sequence is initialized as follows:

$$c_{init}(0) = (2^{17}(N_{symb}^{slot} n_{s,f}^{\mu} + l + 1)(2N_{ID}^{n_{SCID}} + 1) + 2N_{ID}^{n_{SCID}} + n_{SCID})mod2^{31}$$

whereas, for the second CDM group, the sequence is initialized with:

$$c_{init}(1) = (2^{17}(N_{symb}^{slot} n_{s,f}^{\mu} + l + 1)(2N_{ID}^{1-n_{SCID}} + 1) + 2N_{ID}^{1-n_{SCID}} + 1 - n_{SCID})mod2^{31}$$

which means that the c_{init} of the first CDM group depends on $N_{ID}^{n_{SCID}}$, whereas c_{init} of the second CDM group depends on $N_{ID}^{1-n_{SCID}}$.

3.8.2.3.3 Sequence Generation for DFT-S-OFDM

When transform precoding is enabled (DFT-S-OFDM waveform), the sequence design should be such that the PAPR of the OFDM symbols carrying DM-RS is similar to that of the OFDM symbols carrying data, otherwise a higher back-off of the Transmit Power would be needed from the maximum transmission power of the UE. Similar to LTE, when allocation of PUSCH is larger than a threshold, the sequences used are the well-known Zadoff-Chu sequences: Complex-valued sequences which, when applied to radio signals, give rise to signals of constant amplitude, whereby cyclically shifted versions of the sequence imposed on a signal result in zero correlation with one another at the receiver. For small PUSCH allocation, specific Computer Generated Sequences (CGSs) are designed to ensured good PAPR and cross-correlation properties.

To be more precise, when transform precoding is enabled, the reference-signal sequence $r(n)$ shall be generated according to $r(n) = \bar{r}_{u,v}(n)$, where $n = 0, 1, \ldots, M_{ZC} - 1$, with $M_{ZC} = N_{PUSCH}/2$ is the length of the sequence and N_{PUSCH} is the number of subcarriers in the scheduled PUSCH allocation. Recall that for PUSCH with transform precoding enabled, only DM-RS Type 1 is supported which has six resource elements per PRB per DM-RS symbol, and therefore, in an allocation containing N_{PUSCH} subcarriers, the sequence length would be $\frac{N_{PUSCH}}{2}$.

The sequences $\bar{r}_{u,v}(n)$ are divided into 30 groups, where $u = \{0, 1, \ldots, 29\}$ is the sequence group number and v is the base sequence number within the group, such that each group contains one base sequence ($v = 0$) of each with length M_{ZC}, when the allocation is up to 5 PRBs, and two base sequences ($v = 0, 1$) otherwise. For details on how the sequences are generated according to the $\{u, v\}$ parameters see 38.211 [34].

Based on the above description, we observe that different sequences can be chosen based on two parameters $\{u, v\}$, the sequence group number and the sequence number, which can higher-layer configured to the following values. The sequence group u is given by the following formula

$$u = (f_{gh} + n_{ID}^{RS}) mod 30,$$

where n_{ID}^{RS} is either higher-layer configured for each UE or is equal to the N_{ID}^{cell}. The value f_{gh} along with the sequence number v are as follows:

if group hopping is enabled and sequence hopping is disabled	$f_{gh} = v = 0$ $f_{gh} = \left(\sum_{m=0}^{7} 2^m c(8(N_{symb}^{slot} n_{s,f}^{\mu} + l) + m) \right) mod 30$ $v = 0$ where the pseudo-random sequence $c(i)$ is initialized with $c_{init} = \left\lfloor \frac{n_{ID}^{RS}}{30} \right\rfloor$ at the beginning of each radio frame.
if sequence hopping is enabled and group hopping is disabled	$f_{gh} = 0$ $v = \begin{cases} c(N_{symb}^{slot} n_{s,f}^{\mu} + l) & \text{if } M_{ZC} \geq 6N_{sc}^{RB} \\ 0 & \text{otherwise} \end{cases}$ where the pseudo-random sequence $c(i)$ is initialized with $c_{init} = n_{ID}^{RS}$ at the beginning of each radio frame.

In other words, the sequence used for the DM-RS transmission, depending on whether sequence/group hopping is enabled, could depend on the OFDM symbol number, resulting to a different sequence on different DM-RS symbols of the same PUSCH allocation, or across PUSCH allocations.

3.8.2.3.4 Updates in NR Rel-16 for DFT-S-OFDM Waveforms
For cell-edge UEs, the PAPR of the transmit waveform is an important factor to determine the maximum transmit power. In NR Rel-15, $\pi/2$ BPSK modulation with frequency domain spectrum shaping (FDSS) is supported for the PUSCH channel with transform precoding due to its lower PAPR property than other modulation orders. However, since the DM-RS sequence for PUSCH with

DFT-s-OFDM waveform is a Zadoff-Chu sequence, the OFDM symbols carrying DM-RS have a higher PAPR than the OFDM symbols carrying PUSCH with $\pi/2$ BPSK modulation as shown in Figures 3.82 and 3.83 where the PAPR of the $\pi/2$ BPSK modulated random PUSCH data with the PAPR of NR ZC sequences is compared.

Specifically, the ZC sequences plotted in Figure 3.82 are the 60 NR ZC sequences with length 96. The PAPR of both filtered and un-filtered ZC sequences are plotted. As can be

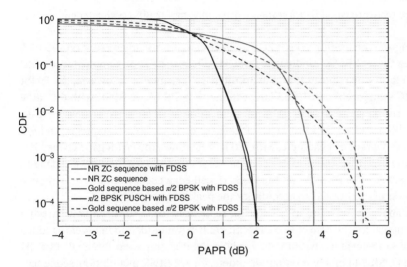

Figure 3.82 PAPR comparison among $\pi/2$ BPSK modulated random PUSCH data, $\pi/2$ BPSK based DM-RS sequences and NR ZC sequences; FDSS corresponds to a time-domain response of [0.28, 1, 0.28]; Number of Allocated Tones = 96.

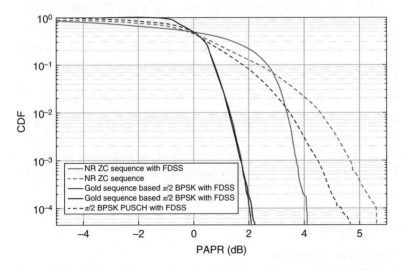

Figure 3.83 PAPR comparison among $\pi/2$ BPSK modulated random PUSCH data, $\pi/2$ BPSK based DM-RS sequences and NR ZC sequences; FDSS corresponds to a time-domain response of [0.28, 1, 0.28]; Number of Allocated Tones = 180.

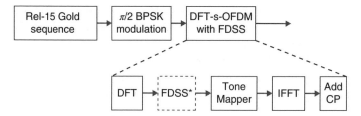

Figure 3.84 Gold sequence based DMRS sequence generation for π/2 BPSK with filtering.

seen from the Figure, the un-filtered NR ZC sequences has 3.2 dB larger PAPR (at the 10^{-4} CDF point) than the π/2 BPSK modulated PUSCH. When the FDSS is applied to the ZC DM-RS, the PAPR may be reduced from the un-filtered ZC sequences. However, the PAPR of the filtered ZC sequence is still 1.6 dB larger than the PAPR of π/2 BPSK with the same FDSS (which corresponds to [0.28,1,0.28] in the time domain). Moreover, as we increase the number of allocated RBs for PUSCH, the PAPR gap between NR ZC sequence and π/2 BPSK may gets even larger. For example, with 180 DM-RS tones, the PAPR gap between NR ZC sequences and π/2 BPSK (both with FDSS) increases to 2 dB (see Figure 3.83). Due to the PAPR gap between the DM-RS and the PUSCH, cell edge users would have to reduce the transmit power of the PUSCH to account for the peak PA power limit.

Based on the above motivation, in NR Rel-16, sequences for OFDM symbols carrying DM-RS were designed for the case of PUSCH with π/2 BPSK modulation in order to reduce the PAPR to the same level as that of data symbols. As it is employed for small PUSCH allocations with QPSK and higher modulation orders, for π/2 BPSK modulation, sequences with length less than 30 (i.e. length 6, 12, 18, and 24 which correspond to PUSCH allocation of 1, 2, 3, 4 PRBs respectively), are CGSs. For sequences with length 30 or larger, the DM-RS for ππ/2 BPSK modulation for PUSCH is generated based on Gold-sequence followed by π/2 BPSK modulation, followed by transform precoding resulting in a DM-RS Type 1 pattern. The diagram for the sequence generation is shown in Figure 3.84.

- Step 1) Generate a set of gold sequences of desired length. The NR pseudo-random sequence generator is reused.
- Step 2) π/2 BPSK modulation.
- Step 3) Modulate the π/2 BPSK sequence by DFT-s-OFDM with a proper FDSS.

As shown in the diagram above, Gold sequences are modulated using π/2 BPSK and DFT-s-OFDM with a FDSS filter. Note that, since both the DM-RS and data are generated using π/2 BPSK and DFT-s-OFDM, their PAPR are the same. In particular, the new π/2 BPSK based DM-RS have 1.6~2 dB less PAPR compared with the ZC based DM-RS in NR Rel-15.

3.8.3 Procedures and Signaling

3.8.3.1 Physical Resource Block Bundling
The notion of PRG was introduced in LTE to configure the precoding granularity in the frequency-domain and facilitate the channel estimation at the UE by allowing it to perform channel estimation across a group of consecutive PRBs. Performing a joint estimation

across a group of PRBs can in general improve the performance of the channel estimation algorithm compared to a per-PRB channel estimation procedure. On the other hand, enforcing the same transmit precoder across a large number of contiguous PRBs may result to lower frequency-domain transmit precoding gains in scenarios of wireless channels of high frequency selectivity. Therefore, multiple configurations for the PRB bundling are supported to ensure the most appropriate value is used depending on the channel characteristics, or the deployment scenario. Specifically, in NR Rel-15, two different modes of operation are supported:

- *Narrowband PRG*: PRG can be set to either 2 or 4 PRBs, to primarily address the following use-cases: high beamforming gains when frequency selective transmit precoding is employed, efficient MU-MIMO UE pairing, and transparent PRG-based transmit precoding cycling when channel state information (CSI) at the transmitter is unavailable.
- *Wideband PRG*: PRG can be set the value "wideband," in which case the UE may assume that it shall be scheduled with a set of contiguous PRBs in which the same precoding is applied in all the scheduled PRBs. Such mode of operation could be useful when the transmitter uses a wideband closed-loop transmit precoding (e.g. when UE has fed-back a wideband PMI), or when a transmit precoding scheme with small precoder variations in the frequency domain is employed (e.g., small Delay CDD).

3.8.3.1.1 Narrowband Physical Resource Block Group (PRG) A smaller PRG size allows for the serving base station to perform a more frequency selective precoding compared to a higher PRG size. However, a smaller PRG size may lead to a worse channel estimation performance at the UE compared to a larger PRG value. As an illustration, in Figure 3.85, a gNB with 16 antennas transmits toward a UE with four receive antennas over a CDL-B channel with

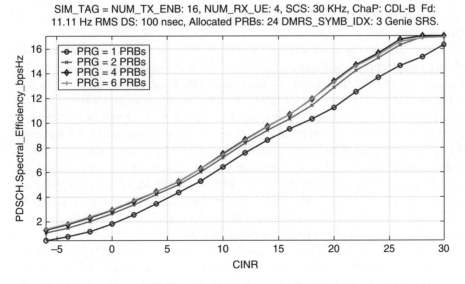

Figure 3.85 Comparison of PDSCH spectral efficiency with link and rank adaptation with front-load DM-RS and different PRG values.

100 ns R.M.S. delay spread. We observe that a PRG of 1 PRB results to spectral efficiency losses compared to a larger PRG value, even if a genie SVD-based transmit precoding is applied on the downlink, especially due to the DM-RS channel estimation losses. On the other hand, increasing the PRG value results to diminishing gains in channel estimation quality with increasing losses due to worse transmit precoding frequency selectivity.

It should also be noted that an increasing PRG value, e.g. PRG = 8, does not necessarily imply that the receiver would perform a joint processing of DM-RS over 8 PRBs, since this entails to a more sophisticated procedure with higher complexity and memory requirements at the receiver, without noticeable gains in performance. Based on the above considerations, NR Rel-15 supports setting the PRG to either 2 PRBs or 4 PRBs when it comes to the narrowband PRG options.

3.8.3.1.2 *Wideband Physical Resource Block Group (PRG)* When the PRG is configured to be equal to the "wideband" value, the UE could perform a wideband channel estimation over the contiguous scheduled PRBs, and the network could either use a constant precoder over all the scheduled PRBs, or it can use any other per-tone precoding design (or differently stated, per-tone frequency selective precoding design) over such contiguous allocations. Such a configuration could potentially have a variety of benefits compared to a narrowband PRG:

- First, more flexibility is given to the network for precoding designs and MU-MIMO pairing of the users. For example, the network can schedule UEs on the same allocation and signal to them that the PRG equals to the allocation they have been scheduled. In this scenario the network is not required to keep the same precoder across consecutive subcarriers. The precoding matrices can be optimized even further compared to the case that PRG equals a small set of PRBs. Note that the network still has the option to design a constant precoder if CSI at the transmitter is low quality, or if it finds it unnecessary to design per-tone precoders.
- Second, the UEs can perform wideband channel estimation across all the scheduled PRBs which could potentially lead to energy savings, or better throughput performance in some scenarios. Specifically, the UE can take advantage of the frequency continuity of the precoded channel and perform a FFT-based channel estimation if the allocation is large enough, such that it is justified to choose such an algorithm over a narrowband channel estimation method. Such gains could be more prominent in scenarios of low geometries. For example, in Figure 3.86, we compare the spectral efficiency gains in a TDL-C 300 ns wireless channel with a 2-symbol front-load DM-RS achieved from a FFT-based wideband Channel Estimation procedure, and a MMSE-based Channel Estimation procedure with PRG equals to 4 PRBs for the scenario of constant open-loop precoder. It is observed that at a low spectral efficiency, e.g. 0.5 bps Hz^{-1} (cell edge throughput), 1 dB of gain is expected by using a wideband PRG over PRG equal to 4 PRBs.
- Third, if the network indeed chooses to employ a per-tone continuous precoding design, and signal to the UE that a wideband PRG is used, the effective channel may potentially have a significantly compressed delay spread, which can result to a better channel estimation at the UE.

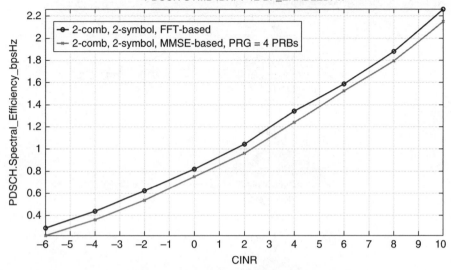

SIM_TAG = NUM_TX_ENB: 8, NUM_RX_UE: 4, SCS: 30 KHz, ChaP: TDL-C, Fd: 11 Hz (2.973Kmh), RMS DS: 300 nsec, MIMO_NUM_LAYER: 1, Allocated PRBs: 168, PDSCH SYMB IDX: 1-12 BF_ENABLED: 1.

Figure 3.86 Comparison of PRG = "Wideband" and PRG = 4 at low geometries.

3.8.3.1.3 Configuration Details of PRB Bundling The value of PRB bundling applied for a specific PDSCH can be set to the allowable options as follows:

- For a PDSCH scheduled by PDCCH with DCI format 1_0, PRB bundling is always equal to 2. Recall that PDCCH with DCI format 1_0 is mainly used for robust communication with the UE, or fall-back operation if RRC connection is lost, and therefore specifying one value was deemed sufficient.
- For PDSCH scheduled by PDCCH with DCI format 1_1 with CRC scrambled by C-RNTI, MCS-C-RNTI, or CS-RNTI, PRB bundling can be set either semi-statically, through RRC configuration, or dynamically, for UEs supporting the dynamic PRB bundling feature.

When PRB bundling is configured semi-statically, PRG is indicated by one single value in the RRC parameter *bundleSize*. When PRB bundling can be dynamically signaled, one bit in the DCI format 1_1 is dedicated for the determination of the PRG value. In this scenario, the following procedure is used:

- The UE is configured with two sets of values.
  - o The first set S_1 can take either one value from {2,4,'wideband'}, or two values {2, 'wideband'} or {4, 'wideband'}
  - o The second set S_2 takes only one value from the {2,4, 'wideband'}.
- If the PRB bundling size S_2 indicator is set to
  - o "0," PRG equals to the value in the 2nd set.
  - o "1" and the first set contains one value, PRG equals to the value of the 1st set
  - o "1" and the first set contains two values, PRG equals to 'wideband' only if the scheduled PRBs are contiguous and larger than half of the active BWP size $\frac{N_{BWP}^{size}}{2}$, otherwise PRG equals to 2 or 4 respectively (Figure 3.87).

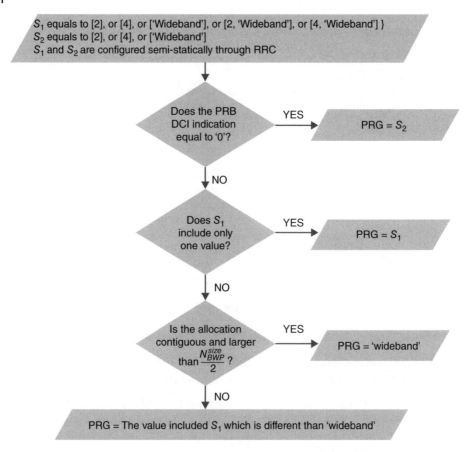

Figure 3.87 Procedure for determining the PRB bundling value in DCI format 1_1 with dynamic PRG signaling enabled.

To understand the above procedure, it would be instructive to explain the motivation behind such a procedure and provide a few examples. Using the dynamic PRG signaling, it should be possible to dynamically switch from a narrowband value to a wideband, and vice versa, or switch between the two narrowband values. It should also be noted that gains of the wideband PRG values are expected when the bandwidth is large enough and there is one contiguous allocation. If the allocation is distributed across the BWP bandwidth, the UE cannot perform an FFT-based procedure since there are PRBs in the bandwidth of the BWP which do not contain any DM-RS. Furthermore, even if the allocation is contiguous, if the size is small, e.g. 10 PRBs, then performing a wideband-based procedure may not result to significant gains over using a narrowband value (e.g., PRG = 4 PRBs).

Using the above procedure, the following dynamic switches between the three different PRG values are possible:

- Consider the case that $S_1 = [2,\,'wideband']$ and $S_2 = [4]$. The gNB could use the PRB bundling dynamic indication bit to switch between
    - PRG = 2 and PRG = 4 if the allocation is non-contiguous or smaller than half the BWP size,
    - PRG = 'wideband' and PRG = 4 if the allocation is contiguous and larger than half of the BWP size.
- Consider the case that $S_1 = [2]$, $S_2 = [4]$, or vice versa. The gNB could use the PRB bundling dynamic indication bit to switch between PRG = 2 and PRG = 4.

3.8.3.2 DM-RS to PDSCH and PUSCH EPRE Ratio

The PDSCH/PUSCH to DM-RS Energy Per Resource Element (EPRE) ratio refers to the power ratio of one DM-RS layer over the corresponding data layer from UE perspective. For DM-RS type 1, this ratio can take the value 0 or −3 dB, whereas for DM-RS type 2, also the value −4.77 dB can be configured, depending the number of DM-RS CDM groups without data on the OFDM symbols carrying DM-RS using the following formula:

$$\frac{Data\ EPRE}{DMRS\ EPRE}(dB) = -10 \cdot \log_{10}(Number\ of\ CDM\ groups\ without\ data)$$

In the NR Rel-15 specification, the following table summarizes all the possible options (Table 3.18):

To understand the above formula, consider the example of DM-RS type 2 where the only allocated DM-RS on the front-load DM-RS symbol is the CDM group 0 as shown in the Figure 3.88. There are three difference values that the DM-RS to PDSCH EPRE ratio could take:

- CDM group 1 and 2 do not carry any PDSCH, in which case, the gNB is expected to transfer the energy of those resource elements to the resource elements carrying DM-RS (pilot power boosting). In that case the resource elements carrying DM-RS (CDM group 0) will have three times more power than a resource element carrying PDSCH on any other OFDM symbol, in which case the ratio of PDSCH EPRE to DMRS EPRE is $10 \cdot \log_{10}\left(\frac{1}{3}\right) = -4.77\ dB$. Note that in this scenario all three CDM groups do not carry any data.
- Only CDM group 1 carries PDSCH, in which case the energy of the resource elements of the CDM group 2 is transferred to the DM-RS in CDM group 0, resulting in two times

Table 3.18 The ratio of PDSCH EPRE to DM-RS EPRE [38.214].

Number of DM-RS CDM groups without data	DM-RS configuration type 1 (dB)	DM-RS configuration type 2 (dB)
1	0	0
2	−3	−3
3	—	−4.77

Single symbol front-load DM-RS configuration Type 2

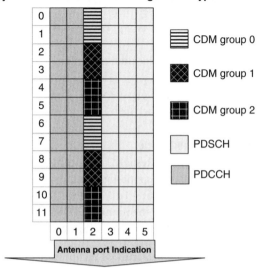

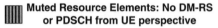

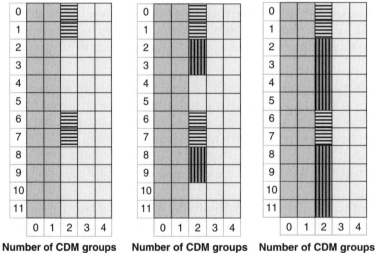

Figure 3.88 Example of front-load DM-RS with configuration type 2 with DM-RS on CDM group 0 assigned to the UE.

more power than a resource element carrying PDSCH on any other OFDM symbol, in which case the ratio of PDSCH EPRE to DM-RS EPRE is $10 \cdot \log_{10}\left(\frac{1}{2}\right) = -3 \, dB$. Note that in this scenario two CDM groups do not carry any data.

- Both CDM group 1 and 2 carry PDSCH, in which case no energy is transferred to the DM-RS of CDM group 0, resulting in a ratio of PDSCH EPRE to DM-RS EPRE of $10 \cdot$

$\log_{10}\left(\frac{1}{1}\right) = 0 \, dB$. Note that in this scenario only one CDM group does not carry any data, the one that carries DM-RS.

It should be clarified that the same PDSCH to DM-RS EPRE ratio is applied to all the DM-RS symbols of the PDSCH time-domain allocation.

3.8.3.3 Antenna Port DCI Signaling

In DCI formats 0_1 and 1_1 the gNB signals to the UE which DM-RS antenna ports are being used for the corresponding PUSCH and PDSCH respectively.

Starting from the DM-RS of PDSCH, the field of the antenna port indication in DCI format 1_1 is 4, or 5 or 6 bits depending on the DM-RS type and the higher-layer configured parameter *maxLength* as shown in Table 3.19.

Based on the value of the antenna port DCI field, the following transmission properties are signaled:

- PDSCH DM-RS port indices scheduled to the UE
- PDSCH rank is implicitly determined by the number of DM-RS port signaled. For example, if ports {0,2} are signaled, then the PDSCH rank is 2.
- Number of CDM groups without PDSCH on the OFDM symbols carrying DM-RS (which in turn selects one of the specified DM-RS to PDSCH EPRE ratios as described in Section 3.8.2.3.2)
- Whether the UE may assume it is SU-MIMO transmission, or it cannot make this assumption, in which case the PDSCH is scheduled in an SU-MIMO or MU-MIMO mode.

As an example on how to interpret the Antenna port tables, consider the antenna port indication table for DM-RS Type 1 with single-symbol DM-RS shown in Table 3.20 (all tables can be found in 38.212 [35] Section 7.3.1.1.2).

In this case, the UE is indicated in the DCI a 4-bit number ranging from 0 to 11, such that one of the rows shown in Table 3.20 is selected. For example, a value of 0 means that the DM-RS port 0 is scheduled with 1 CDM group without data. This means that all the resource elements of the second CDM group on the OFDM symbol with DM-RS will also contain data (See also Figure in Section 3.5.2.3.2). However, if value 3 is indicated, DM-RS port 0 is scheduled with 2 CDM groups without data, which refers to the case that PDSCH and DM-RS are TDMed on different OFDM symbols.

Table 3.19 Size of antenna port DCI field for DCO format 1_1.

DM-RS type	maxLength	Number of bits in the DCI for antenna port indication
1	1	4
1	2	5
2	1	5
2	2	6

Table 3.20 Antenna port(s) signaling for PDSCH and DM-RS Type 1 with maxLength equals to 1.

One Codeword:Codeword 0 enabled,Codeword 1 disabled		
Value	Number of DMRS CDM group(s) without data	DMRS port(s)
0	1	0
1	1	1
2	1	0,1
3	2	0
4	2	1
5	2	2
6	2	3
7	2	0,1
8	2	2,3
9	2	0-2
10	2	0-3
11	2	0,2
12–15	Reserved	Reserved

Even though in the row with value 0 and 3 the same DM-RS port is indicated to the UE, the PDSCH rate matching and the DM-RS to PDSCH EPRE ratio are different since these rows serve a different scheduling purpose. For example, a UE may be scheduled with the row with value 0 when no additional power on the DM-RS is required to perform a good channel estimation (e.g. high SINR regime) and scheduling data on more resource elements would result to higher spectral efficiency. On the other hand, if the scheduler considers that the UE is in a low SINR regime, it may use the row with value 3 to schedule rank 1 transmission with power boosted DM-RS (DM-RS to PDSCH EPRE ratio is 3 dB as described in Section 3.5.2.3.2) such that channel estimation is improved.

When it comes to MU-MIMO scheduling, rows with value 3 and 4 are useful to schedule 2 different UEs on a different CDM group, such that the scheduler could signal to the first UE the row with value 3 and to the second UE the row with value 4. Similarly, for MU-MIMO scheduling with 2 UEs each one with rank 2, one UE would be scheduled with the row with value 7 and the second UE with the row with value 8.

Finally, for certain of the rows, the UE may assume that it is an SU-MIMO scheduling. Specifically, for row with value 11, the UE is scheduled with DM-RS port 0 and 2, each one appearing on a different CDM group. In this case, the UE may assume that the remaining ports (port 1 which is CDMed with port 0 and port 3 which is CDMed with port 2) are not being scheduled to some other UE on the same PRBs and OFDM symbols.

Turning over to DM-RS of PUSCH, the size of the field used for antenna port indication in the DCI format 0_1 depends on the following aspects: the DM-RS Type, the maxLength, whether transform precoder is enabled or disabled. Specifically, if transform precoding is

Table 3.21 Size of Antenna Port DCI field for DCO format 0_1.

DM-RS type	max Length	Transform precoding	Number of bits in the DCI for antenna port indication
1	1	Enabled	2
1	2	Enabled	4
1	1	Disabled	3
1	2	Disabled	4
2	1	Disabled	4
2	2	Disabled	5

enabled, only rank 1 can be transmitted with DM-RS Type 1, whereas if transform precoding is disabled, higher rank can be transmitted with either DM-RS Type 1 or 2 (Table 3.21).

3.8.3.4 Quasi-Colocation Considerations for DM-RS of PDSCH

Knowledge of a priori information related to the long-term statistics of the PDSCH channel, e.g. delay spread, doppler spread, average delay, and doppler shift, can be used at the receiver to construct an appropriate channel estimation block tailored for the specific channel characteristics. Such long-term statistics of the wireless channel are measured at the receiver using reference signals (e.g. CSI-RS for tracking in NR, or CRS in LTE) which can be configured as QCL source for a PDSCH.

In NR Rel-15, in the absence of CSI-RS configuration, and unless otherwise configured, the UE may assume PDSCH DM-RS and SS/PBCH block to be quasi co-located with respect to Doppler shift, Doppler spread, average delay, delay spread, and, spatial Rx parameters (if applicable), and therefore the UE may use the quasi co-located SS/PBCH block to acquire an estimate of the long-term statistics of the channel. Using as a QCL reference a CSI-RS for tracking is considered a more appropriate reference signal for the estimation of such quantities compared to an SS/PBCH block, since the former signal was specifically designed for this reason (e.g. larger bandwidth than an SS/PBCH block).

Several scheduling restrictions were also specified in NR which can be exploited by the UE to simplify the implementation of the DM-RS channel and noise estimation block. Specifically,

- A UE may assume that the PDSCH DM-RS within the same CDM group are quasi co-located with respect to Doppler shift, Doppler spread, average delay, delay spread, and spatial Rx. It may also assume that DM-RS ports associated with a PDSCH are all QCL with QCL Type A, Type D (when applicable) and average gain. In other words, using the above assumptions, the UE may perform a joint estimation of DM-RS ports which are CDMed using the same long-term statistics, and it is not required to measure, or use, different long-term statistics for different DM-RS ports of the same PDSCH.
- When it comes to MU-MIMO scheduling, a UE may assume that any co-scheduled UE would have the same transmission properties in their PDSCH with respect to the following parameters:

o The actual number of front-loaded DM-RS symbol(s)
o The actual number of additional DM-RS
o The DM-RS symbol location
o DM-RS configuration type
o Same precoding in other DM-RS ports of the same CDM group
o Same resource allocation in other DM-RS ports of the same CDM group on the same PRG grid
- Lastly, since any CDM group without data on an OFDM symbol carrying DM-RS can be used by the UE for noise estimation, the UE may assume that these CDM groups would not overlap with any configured CSI-RS resource(s) for that UE, otherwise the noise estimation would be distorted by the reception of the CSI-RS signals.

3.9 Phrase- Tracking RS (Youngsoo Yuk, Nokia Bell Labs, Korea)

3.9.1 Phase Noise and its Modeling

3.9.1.1 Phase Noise in mm-Wave Frequency and its Impact to OFDM System

The introduction of the 5G NR system in higher carrier bands (mmWave) provides the opportunity of utilizing a large amount of transmission bandwidth of several GHz, resulting in very high data rate and low latency services. However, the operation in high frequency bands brings new challenges for the design and performance of the analog front-end. One of the highest difficulties to overpass in analog design is the effect of a noisy oscillator, and it grows quadratically with the carrier frequency, the phase noise (PN) becoming critical. Thus, keeping the effect of phase noise below a certain level is one of the key objectives for mmWave communications.

Phase noise (PN), caused by oscillator implementation technology destroys the orthogonality of subcarriers in OFDM. In the following we consider an OFDM transmission with N subcarriers, carrier spacing Δf, and a vector $X_k = [X_0,...,X_{N-1}]$ to be transmitted within an OFDM symbol. The digital baseband signal at time instance $t = n \cdot Ts$ where $n = \{0, ..., N-1\}$ is given by

$$x_n = \frac{1}{\sqrt{N}} \sum_{k=0}^{N-1} X_k e^{j2\pi kn/N},$$ (3.1)

Also, we assume that the channel impulse response is $\mathbf{h} = [h_0, h_1, ..., h_{L-1}]^T$, and it does not change during one OFDM symbol duration. Phase noise is inherently present in an oscillator, and its effect is equivalent to a random phase modulation of the carrier, resulting in the modulation signal modeled as $s(t) = e^{j2\pi f_c t + \phi(t)}$, where $\phi(t)$ is the random phase-noise process. The modulation signals are used for up-/down-conversion to/from the carrier frequency f_c.

With such assumptions, we can model the signal after demodulation as:

$$\mathbf{r} = \text{diag}(e^{j\Phi})(\mathbf{x} \odot \mathbf{h})$$

$$\mathbf{R} = \mathbf{J} \odot (\mathbf{XH})$$

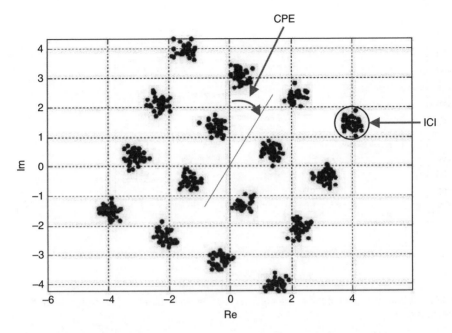

Figure 3.89 Impact of local oscillator frequency offset on received constellation.

$$R_k = X_k H_k CPE \underbrace{J_0}_{} + ICI \underbrace{\sum_{l=0, l \neq k}^{N-1} X_l H_l J_{k-l}}_{} + \eta_k, \tag{3.2}$$

where $J_i = \frac{1}{N} \sum_{n=0}^{N-1} e^{j\phi_n} e^{-j2\pi ni/N}$

The DC coefficient J_0 acts on all subcarriers as a common factor, and it can be approximated for small phase nose as $J_0 \approx e^{j\bar{\phi}}$, and we can call it common phase error (CPE). Also, remaining terms in the Eq. (3.2) is called as ICI [36].

Figure 3.89 is illustrating the impact from CPE and ICI on the transmitted constellation of a 64 QAM. In OFDM based systems, CPE causes a constant rotation in the angle of the modulation constellation, and ICI is causing scattering of the constellation points. CPE is constant in one OFDM symbol because it is an average phase rotation in one OFDM symbol, while ICI is observed as phase variation within one OFDM symbol. This limits the maximum received SNR especially at higher carrier frequencies. The PN impact may be handled by using large subcarrier spacing and PN estimation and compensation at the receiver.

3.9.1.2 Principles of Oscillator Design and Practical Phase Noise Modeling

Fabrication material of RFIC is one key factor to determine the phase noise model with respect to the commercialization cost, size, and low power consumption. While there are many different fabrication methods, the most common fabrication materials are CMOS, GaAs, SiGe, and GaN. A summary of the phase noise levels achieved by different state of the art fabrication methods and materials is given in Figure 3.90 [37].

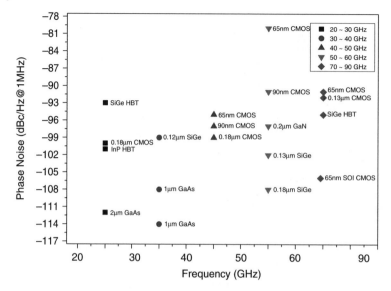

Figure 3.90 A brief summary of the phase noise level achieved by different fabrication methods and materials [37].

While GaAs-based devices can provide a lower phase noise level, it is still expensive and power-consuming. The CMOS-based devices are available at lower cost and have less power consumption. Taking the cost and power constraint at the UE side into consideration, it appears reasonable to assume CMOS-based design for the UE. For the gNB, depending on the gNB class, architecture, etc. GaAs may be considered, as the performance gains may outweigh the power consumption/cost.

The PN impact is dependent on the utilized oscillator performance, modeled by oscillator power spectral density (PSD). Figure of merit (FOM) is a parameter characterizing the component performance and it is typically announced by the component manufacturer. The oscillator performance is also dependent on the amount of consumed power. Here, it can be estimated that less than ~50 mW total oscillator power consumption would be feasible at least regarding the user equipment. In other words, oscillator models with good performance but with high power consumption are not found feasible for a user device.

As an important factor to determine design parameters such as subcarrier spacing and modulation schemes relative to EVM target, phase noise models for NR system have been studied in 3GPP [37, 38]. Figure 3.91 shows the basic structure of oscillator with phase-locked loop (PLL). The PSD of phase noise is mostly influenced by three components: the reference clock, the voltage-controlled oscillator (VCO), and the loop filter. Figure 3.92a illustrates contribution from different components to the total PSD. Total PSD of a PLL oscillator consists of the different sub-components on the oscillator circuit, namely reference clock, PLL loop components, and the VCO. The PSD of the phase noise for a certain frequency range can be interpreted as CPE and/or ICI in an OFDM system with respect to subcarrier spacing, as illustrated in Figure 3.92b. The level of CPE can be estimated with the accumulation of PSD in lower frequency region relative to

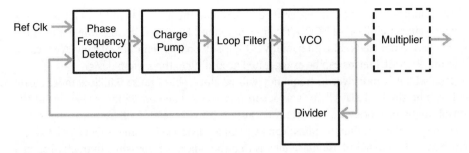

Figure 3.91 Basic structure of a phase-locked loop (PLL) circuit.

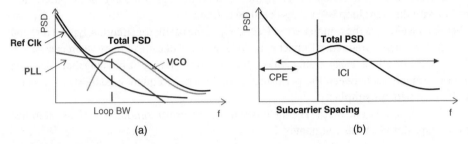

Figure 3.92 Illustration of PSD contribution: (a) by oscillator components (b) by phase noise components to OFDM receiver.

subcarrier spacing, while ICI terms are contributed from the relatively large frequency offset components around subcarrier spacing.

In general, for the purpose of performance evaluation of the receiver, two types of phase noise models, multiple pole-zero model and PLL-based model, are considered. The former models the PSD of phase noise as a combination of poles and zeros. For example, the well-known PN PSD formula utilized in 802.11ad is as follows.

$$S_{Tot}(f) = PSD_0 \left[\frac{1 + (f/f_z)^2}{1 + (f/f_p)^2} \right].$$

The pole-zero model is hard to reflect the practical design parameters into the model. PLL-based model can provide better approximation to the real model by considering practical implication to the model as shown in Figure 3.91, the total PSD of a PLL oscillator consists of the different sub-components on the oscillator circuit, namely reference clock, PLL loop components and VCO. VCO part can be typically understood to further consist of $1/f^2$ and $1/f^3$ sub-components. Typically, the reference clock and PLL components dominate the PSD inside the closed loop bandwidth of the PLL, and the VCO is the main impacting sub-component outside the loop bandwidth. When considering relatively small subcarrier spacing values (smaller than the PLL loop bandwidth), the significance of the terms dominating the PSD inside the loop bandwidth range will increase also in terms of ICI. Thus, neglecting these components from the oscillator modeling may result in too optimistic model, where especially in the case of large loop bandwidths it may be understood to contain some of the basic PLL oscillator sub-components: PSD_0 may be interpreted to represent the PLL floor level, exponent 2 in the formula represents to slope of VCO

$1/f^2$ sub-component, f_p can be understood to estimate the loop BW and f_z is representing the lower saturation level of VCO $1/f^2$ sub-component. However, in this simple model the impact of reference clock, PLL jitter, and VCO $1/f^3$ sub-components are missing. As a result, either of CPE or ICI terms can be exaggerated according to the system design.

A PSD which is narrower in frequency will produce phase noise which is more correlated in time than a PSD which is wider in frequency. The correlation properties of the utilized oscillator model are important because a fully correlated phase noise, regardless how strong, will mean that the phase can be completely tracked from symbol block to symbol block. Uncorrelated phase noise means that the phase is impossible to track in such a manner. However, if the phase noise has low magnitude it will not affect the performance that severely. In order to enable the evaluation of this effect, an oscillator model parameterization with different loop BW values could be defined.

Let us consider f_c is the output carrier frequency of the oscillator circuit. A typical method for adjusting carrier frequency impact to a model is that a decade of increase in carrier frequency increases the corresponding PSD with 20 dBc Hz^{-1}. It may be possible that also some other dependencies between the performance and carrier frequency exist. These impacts are ignored here for simplicity.

Let $S_{Ref}(f)$, $S_{PLL}(f)$, and $S_{VCO}(f)$ be the phase noise spectrum components of the reference clock, loop, and VCO sub-components.

- $S_{VCO}(f)$ consists further of $1/f_2$ and $1/f_3$ components: $S_{VCO}(f) = S_{VCO\_v2}(f) + S_{VCO\_v3}(f)$

The total oscillator PSD $S_{Tot}(f)$ can be constructed by summing these sub-components together including the transfer function of the utilized loop filter [39]. Typically, 3rd or 4th order filters are used as a loop filter. A quick and simplified estimation of the closed loop performance of the PLL can be obtained by assuming that the loop bandwidth equals approximately to the frequency where the reference clock + PLL and the free running VCO phase noise are equal and by utilizing a simple step function to model the loop filter:

$$S_{Tot}(f) = \begin{cases} S_{Ref}(f) + S_{PLL}(f), & \text{when } f \leq loop\ BW \\ S_{VCO}(f), & \text{when } f > loop\ BW \end{cases} \tag{3.3}$$

The listed sub-components can be modeled separately utilizing the following common formula:

$$S_{Ref/\ PLL/VCO\_v2/VCO\_v3} = PSD0 \cdot \left[\frac{1 + (f/f_z)^k}{1 + f^k} \right] \tag{3.4}$$

where

$$PSD0 = FOM + 20 \log f_c - 10 \log \left(\frac{P}{1\ mW} \right) \tag{3.5}$$

and where FOM is the FOM, $f_c f_c$ is the carrier frequency and P is the consumed power. Considering the expectation for the phase noise level achievable with reasonable cost and power consumption as presented above, in this example the following parameters are suggested for the phase noise model at the UE (CMOS-based) and gNB (GaAs-based) side,

Table 3.22 Parameters for phase noise models.

	Model 1, UE, Loop BW = 187 kHz				Model 2, BS, Loop BW = 112 kHz			
	REF clk	**PLL**	**VCO V2**	**VCO V3**	**REF clk**	**PLL**	**VCO V2**	**VCO V3**
FOM	−215	−240	−175	−130	−240	−245	−187	−130
f_z	Inf	1.00E+04	50.30E+06	Inf	Inf	1.00E+04	8.00E+06	Inf
P (mW)	10	20	20		10	20	50	
k	2	1	2	3	2	1	2	3

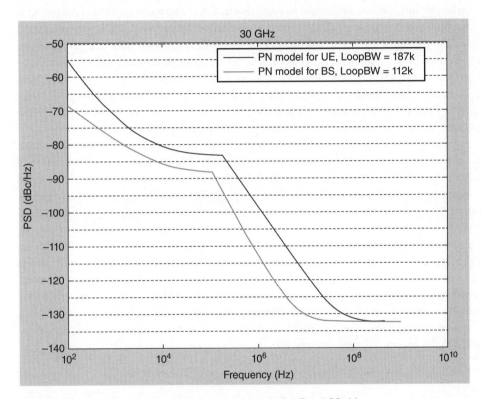

Figure 3.93 PSD of proposed phase noise model at both UE and BS side.

respectively (see Table 3.22). The PSD of the described phase noise models at both UE and gNB for 30 GHz are depicted in Figure 3.93.

Due to limitations on power consumption and cost, the phase noise in UE is higher than it in gNB, and in most case, only the UE phase noise is necessary to be considered. That is, in downlink, phase noise in UE receiver to be compensated while in uplink, the phase noise at UE transmitter is the dominant factor.

In the following sections, we discuss on the principles of phase noise compensation.

3.9.2 Principle of Phase Noise Compensation

The compensation of phase noise in mmWave system is one of the key missions for achieving the desired EVM requirement for higher modulation.

The influence of oscillator phase noise to the system performance can be analyzed in two aspects. The phase noise is due to local oscillator design, so it is constant regardless of transmission channel or transmission scheme. In other words, when the SINR of the received signal is relatively low, thermal noise and interference are much larger than the phase noise at the receiver, and phase noise compensation is not required. However, when SINR of the received signal is high, phase noise is relatively higher than thermal noise and interference, and phase noise becomes dominant to determine received EVM at the receiver.

Another aspect is how much sensitive a transmission scheme (e.g. MCS, symbol-length) from phase noise is. Different scheduled MCS determines the immunity to phase rotation. For example, low modulation order such as QPSK has enough immunity to the phase rotation (e.g. up to $\pm 45°$) thanks to large separation of each constellation point, and no serious performance degradation from phase noise is observed. However, for higher modulations such as 16 QAM, 64 QAM or 256 QAM, modulation constellations are sensitive to the phase rotation, and phase noise may cause severe performance degradation at the receiver.

In fact, these two aspects do not work independently, because the modulation order used for transmission is determined by the SNR regime, and hence higher-order modulation is scheduled in high SNR region. Figure 3.94 illustrates the explanation of the phase noise enhancement when higher order modulation is applied with high SINR of the received signal. In high SNR region, when phase noise becomes dominant compared to the required EVM level at the receiver, phase noise compensation is required.

Figure 3.95 shows the simulation results of spectral efficiency performance with/without PN compensation when fc = 30 GHz and subcarrier spacing = 120 kHz are used. When MCS is relatively small (16 QAM and R = 1/2), no performance degradation is observed, however, as the MCS increases (64 QAM R = 5/6 or 256 QAM R = 3/4), it is observed that the phase noise degrades the spectral efficiency severely.

In order to overcome the poor EVM due to phase noise, two different approaches can be taken at the transmitter and receiver side.

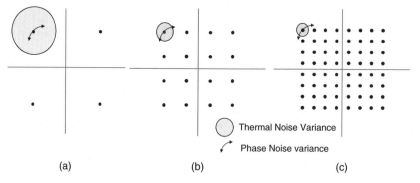

(a)   (b)   (c)

Figure 3.94 Illustration of Thermal noise and phase noise impact for various modulation schemes. (a) QPSK. (b) 16 QAM. (c) 64 QAM.

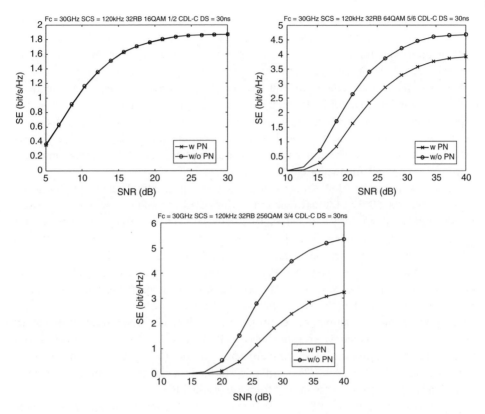

Figure 3.95 Spectral Efficiency Performance comparison with/without phase noise with different MCS.

The first approach is to design the transmission signal (i.e. numerology) to be robust to phase noise. 3GPP NR introduces larger subcarrier spacing than LTE (i.e. 15 kHz), up to 120 kHz, which implies the short OFDM symbol reducing the maximum phase rotation in an OFDM symbol with a given phase noise. However, there is a limitation to increase subcarrier spacing with certain conditions. System bandwidth is one condition to limit the subcarrier spacing, and small bandwidth with larger subcarrier spacing is not efficient to support higher throughput. Propagation channel (i.e. delay spread) also limits the maximum subcarrier spacing, since a channel with long delay requires longer CP to degrade the spectral efficiency due to longer CP overhead as well as higher resolution in frequency channel estimation for frequency selectivity. Thus, such design strategy cannot be solely applied for overcoming phase noise impact.

The receiver side technique compensating the phase noise impact should be taken into account together with employing a large subcarrier spacing at the transmitter. Without specification support, a blind detection algorithm can be applied with the derivation of phase noise by averaging detected constellation points in time, however, this method requires a number of samples in time to be measured also while introducing additional receiver complexity. As an alternative solution, a dedicated reference signal for estimation of phase rotation can be considered. The mission of the new reference signal is to provide the reference for estimating the time-varying channel caused by phase variation. Though DM-RS can also be used for estimating channel variations in time by repeating the DM-RS symbols up to four times, due to its higher density in frequency domain, its penalty to spectral efficiency is much larger than its gain combating phase noise.

Thanks to the shorter delay spread of the mmWave channel, the frequency selectivity of the transmission channel is not high compared to the channel variation in time due to phase noise, a new reference signal for phase noise compensation, which is sparse in frequency-domain but dense in time-domain, has been justified to be introduced for 5G NR. The so-called PT-RS is similar to the DM-RS, emphasizing on the estimation of the time-varying channel due to phase noise.

The basic structure of PT-RS in NR is illustrated in Figure 3.96 PT-RS cannot be solely used without DM-RS. The estimated channel with DM-RS should be a reference for comparing phase rotation from PT-RS. PT-RS REs are placed in fixed subcarrier locations with low density in frequency but dense in time. Assuming small channel variation between OFDM symbols, the receiver may compare estimated channels from PT-RS REs in different OFDM symbols. This happens at every subcarrier location with PT-RS used to improve the performance of phase estimation. The receiver may compensate the phase rotation of the all subcarriers based on the estimated phase rotation.

Figure 3.97 shows the simulation results of spectral efficiency when different frequency PT-RS densities are applied. The results show that PT-RS based CPE compensation provides clear performance gain over the spectral efficiency without CPE compensation. However, with high MCS (256 QAM R = 3/4), a certain performance degradation over the ideal compensation is observed due to the impact from the residual ICI.

As discussed above, the PT-RS in NR system only provides the functionality to estimate CPE, and there is a limitation to support higher modulation order (e.g. 256 QAM) which is severely influenced by ICI. Figure 3.98 shows the concept of phase noise estimation for CPE only and CPE + ICI. With severe variation of phase in an OFDM symbol, CPE only estimation is hard to reflect all phase noise impact to the OFDM symbols, and residual phase noise after CPE compensation can impact to the performance fn the communication system (Figure 3.99).

Figure 3.100 shows the comparison of the simulation results of the spectral efficiency with/without ICI compensation with the scenario of higher MCS (64 QAM R = 5/6, 256

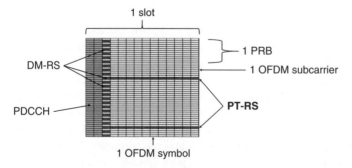

Figure 3.96 Basic structure of PT-RS in NR.

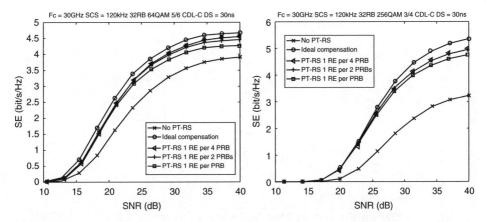

Figure 3.97 Evaluation results for spectral efficiency with CPE compensation with different PT-RS frequency densities.

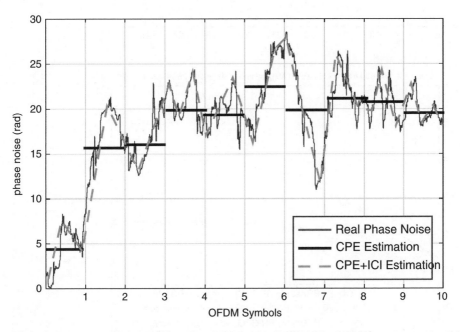

Figure 3.98 Illustration of phase noise estimation for CPE and ICI with a given phase noise (10 OFDM symbols).

QAM R = 3/4) and high mobility (500 kmph) [41]. In the simulation results, by adopting ICI compensation, 6–25% of gain were obtained in terms of spectral efficiency.

ICI is caused by the interference from the adjacent subcarriers due to drifting of frequency offset, and it is restricting the transmission with higher-order modulation (e.g. 256 QAM) requiring higher EVM requirement. ICI can be estimated by measuring interference among the adjacent subcarriers, and PT-RS with a set of consecutive subcarriers in consecutive symbols is required to estimate ICI [36, 42]. for further improvement for higher modulation,

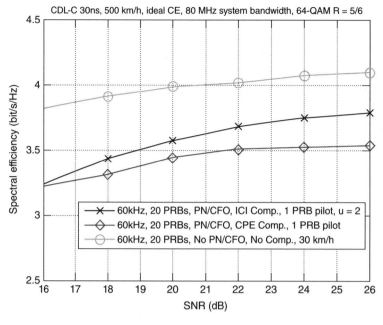

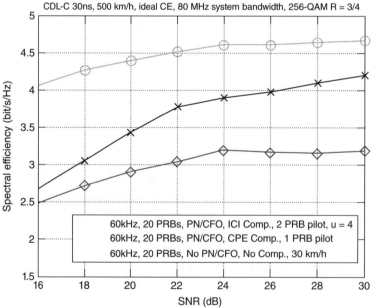

Figure 3.99 Comparison of the simulation results of the spectral efficiency with/without ICI compensation [40] (u = 2 or 4 indicates the number of frequency component of estimated ICI).

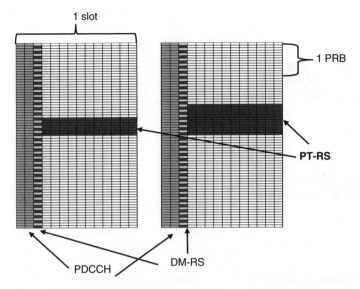

Figure 3.100 Example of block PT-RS structures with different densities.

256 QAM or more, a new PT-RS pattern (e.g. localized or block PT-RS pattern) is required to be supported. The example of block PT-RS structure is shown in Figure 3.100.

The NR consists of different PT-RS designs for CP-OFDM and DFT-s-OFDM. In the following sections we provide the details on the physical layer design and the related procedure for each PT-RS option.

3.9.3 NR PT-RS Structure and Procedures

3.9.3.1 PT-RS Design for Downlink

In downlink, a single port PT-RS is used for a UE associated to single TRP, and the port is associated with one DM-RS port as a phase reference. Though different oscillators can be used for different ports, as we have discussed in the previous section, gNB oscillators have good phase noise property, and just a single antenna port is enough regardless of the number of scheduled DM-RS ports. Multiple PT-RS ports can be considered when multiple TRPs are communicating with one UE. Due to high probability of blockage in mmWave channel, each TRP should deliver one PT-RS port to be used if one of the transmission paths from multiple TRPs are blocked.

Also, in order to reduce the PT-RS overhead, various time and frequency domain patterns can be used according to the UE/network capability, scheduling information such as DM-RS ports, bandwidth, MCS, and so on.

3.9.3.1.1 *PT-RS Port Association with Scheduled DM-RS Ports.* Only a PT-RS port is configured regardless of the number of the scheduled DM-RS ports. When scheduled with a single codeword, the PT-RS port is associated with the DM-RS port having the lowest index among the scheduled DM-RS ports. If two codewords are scheduled, the PT-RS port is associated with the DM-RS port with the lowest index among the scheduled DM-RS ports used for the

Figure 3.101 Illustration of PT-RS port mapping to DM-RS port when one or two codewords are scheduled. (a) One codeword. (b) Two codewords.

Table 3.23 The parameter k_{ref}^{RE}.

	k_{ref}^{RE}							
	DM-RS Configuration type 1				DM-RS Configuration type 2			
	resourceElementOffset				resourceElementOffset			
DM-RS antenna port p	00	01	10	11	00	01	10	11
1000	0	2	6	8	0	1	6	7
1001	2	4	8	10	1	6	7	0
1002	1	3	7	9	2	3	8	9
1003	3	5	9	11	3	8	9	2
1004	—	—	—	—	4	5	10	11
1005	—	—	—	—	5	10	11	4

codeword with higher MCS than the other codeword, Figure 3.101. Once the PT-RS port is associated with the DM-RS port, the PT-RS is transmitted with the same subcarrier as the associated DM-RS, and the same sequence with the DM-RS is repeated for the PT-RS in a slot.

3.9.3.1.2 The Frequency Domain Patterns. The PT-RS RE location in a PRB is determined by the Eq. (3.6).

$$k = k_{ref}^{RE} + (iK_{PT\text{-}RS} + k_{ref}^{RB})N_{sc}^{RB}$$

$$k_{ref}^{RB} = \begin{cases} n_{RNTI} \bmod K_{PT\text{-}RS} & \text{if } N_{RB} \bmod K_{PT\text{-}RS} = 0 \\ n_{RNTI} \bmod (N_{RB} \bmod K_{PT\text{-}RS}) & \text{otherwise} \end{cases} \quad (3.6)$$

According to the DM-RS port index to be associated with, the PT-RS allocation starts from the different RE (k_{ref}^{RE}) to distribute the interferences from different ports. Also, different mapping rules can be applied by higher layer parameter *resourceElementOffset* to differentiate among TRPs. Table 3.23 provides the offset according to the associated DM-RS port.

The frequency density of PT-RS ($K_{PT\text{-}RS}$) can be either 2 or 4, which indicate PT-RS is present every 2nd PRBs or every 4th PRBs respectively. The density is related to the scheduled bandwidth. Because the number of PT-RS REs is more related to the scheduled bandwidth, to avoid a deficiency in the number of allocated PT-RS REs, high density $K_{PT\text{-}RS} = 2$ may be used when scheduled bandwidth is smaller than a threshold. The UE may indicate to the network a set of preferred bandwidth thresholds as a UE capability, and the bandwidth threshold for each frequency density is indicated to the UE by the higher layer parameter *frequencyDensity*. If this frequency density is not signaled, default values (i.e. $K_{PT\text{-}RS} = 2$) is used. The exact PRB position for the PT-RS transmission is determined by the PRB offset ($k_{ref}^{RB} = \{0,..,K_{PT\text{-}RS} - 1\}$), which can be differentiated by RNTI of the scheduled PDSCH/PUSCH.

3.9.3.1.3 PT-RS Time Domain Patterns

Three different time densities of $L_{PT-RS} = \{1, 2, 4\}$ are supported, which means PT-RS is present in every OFDM symbol, in every 2nd OFDM symbol, or in every 4th OFDM symbol. The density is determined by the phase noise level, subcarrier spacing and the MCS to be scheduled. If the oscillator's phase noise is high with a certain subcarrier spacing or if the MCS is higher than a threshold requiring higher EVM, the time domain density may be high (e.g. $L_{PT\text{-}RS} = 1$), hence PT-RS is transmitted in every symbol. The UE may indicate to the network a set of preferred MCS thresholds as UE capability, and the MCS threshold for each time density is indicated to the UE by the higher layer parameter *timeDensity*. If it is not signaled, default value $L_{PT\text{-}RS} = 1$ is used.

As phase noise compensation is critical for the proper decoding/reception of the PDSCH, the PT-RS symbol is present from the first symbol in the PDSCH allocation and then the PT-RS is present at every $L_{PT\text{-}RS}$ symbols. As we can see, there are quite a few reference signals to be transmitted and quite a set of possible configurations in terms of frequency densities, time periodicities, etc. There is the possibilty that a collision/overlap, of the DM-RS and PT-RS can happen. In such a case the PT-RS transmission is skipped. If DM-RS is located before the new PT-RS symbol position, the DM-RS position is used as new reference position of PT-RS. Figure 3.102 shows the examples of the PT-RS time domain allocation for different time density and the number of DM-RS symbols when different time densities $L_{PT\text{-}RS} = 2$, and 4 and different number of DM-RS symbols are configured.

3.9.3.1.4 Power Allocation for PT-RS

Because a single port PT-RS is used regardless of the number of DL layers and the REs of other antenna ports overlapped with PT-RS should remain empty, PT-RS RE can be transmitted with power boosting by utilizing the power from the empty REs. Such power boosting improves the performance of the phase estimation by with the increase of SINR of PT-RS reception. The level of power boosting as a form of the ratio of PT-RS EPRE to PDSCH EPRE per layer per RE is determined by the number of PDSCH layers scheduled and the higher-layer parameter *epre-Ratio*, where the *epre-Ratio* is a parameter reflecting different gNB transmitter implementation. If all the antenna ports delivering PDSCH can be coherently combined with all the other ports, *epre-Ratio* = 0 may be configured and PT-RS can be power boosted with *10log10(the number of PDSCH layers)* [dB]. If all the transmission ports are not fully coherently combined, *epre-Ratio* = 1 may be configured and PT-RS uses the same EPRE with PDSCH REs.

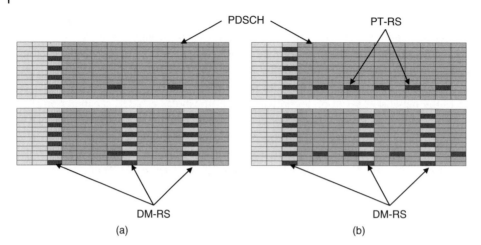

Figure 3.102 PT-RS time domain allocation for different density when configured with different number of DM-RS symbols. (a) $L_{PT-RS} = 4$. (b) $L_{PT-RS} = 2$.

3.9.3.2 PT-RS Design for Uplink CP-OFDM

For uplink CP-OFDM, the same PT-RS design is used as the downlink design with small modification about the terminologies. (e.g. UL DM-RS ports index starts from 0 while DL DM-RS ports are indexed from 1000). The same procedure for determining PT-RS time and frequency domain density is applied as for downlink PT-RS, and the higher layer parameters are configured separately for downlink and uplink.

3.9.3.2.1 PT-RS Port Association with Scheduled DM-RS Ports
The key difference in uplink transmission compared to the downlink PT-RS is the number of PT-RS ports. For uplink, one or two PT-RS port(s) can be scheduled based on the UE capability. Two PT-RS ports can be used for non/partial-coherent UL transmission, with two local oscillators in a UE. This would be the case of a UE equipped with two or more transmission panels for UL. If a UE has reported the capability of supporting full-coherent UL transmission, only a single PT-RS port is used.

For non-codebook based UL transmission, the actual number of transmitted UL PT-RS port(s) is determined based on SRS resource indicators (SRIs). A UE is configured with the PT-RS port index for each configured SRS resource by the higher layer parameter *ptrs-PortIndex* configured by *SRS-Config*. If the PT-RS port index associated with different SRIs are the same, the corresponding UL DM-RS ports are associated to the one UL PT-RS port.

For partial-coherent and non-coherent codebook-based UL transmission, the actual number of UL PT-RS port(s) is determined based on TPMI and/or TRI in DCI format 0_1. And, a DM-RS port transmitted with the antenna ports 1000 and 1002 may associate with PT-RS port 0, and a DM-RS port transmitted with the antenna ports 1001 and 1003 may associate with PT-RS port 1.

From a signaling perspective, 2 bit DCI parameter *PTRS-DMRS association* in DCI format 0_1 are used to indicate the DM-RS port to be associated a PTRS port. When a single PT-RS port is used, DCI indicates one of four antenna ports, and the DM-RS port mapped to the

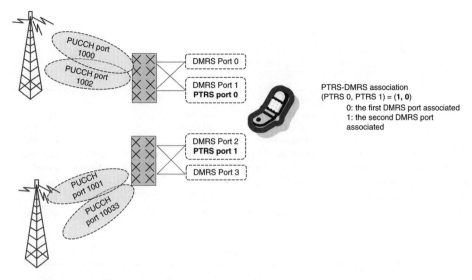

Figure 3.103 Illustration of PT-RS ports association with DM-RSs for a UE equipped two transmission panels (when rank 4 and PTRS-DMRS association = (1,0)).

indicated antenna port is associated to PTRS port. When two PT-RS ports are used, MSB and LSB of DCI field indicate one of antenna ports for PT-RS port 0/1. For example, as shown in Figure 3.103, if *PTRS-DMRS association* = (1,0), PT-RS port 0 is associated with a DM-RS port mapped to PUSCH port 1002 and PT-RS port 1 is associated with a DM-RS port mapped to PUSCH port 1001. If MSB = 0, PT-RS port 0 is associated with a DM-RS port mapped to PUSCH port 1000.

3.9.3.2.2 *Power Allocation for PT-RS* When the UE is scheduled with $Q_p = \{1, 2\}$ PT-RS port(s) in uplink, the PT-RS power is power boosted according to the number of PUSCH layers. The power boosting value is different for different transmission schemes, which transmission codebook (full/partial or non-coherent) is used. Also, the higher layer parameter *UL-PTRS-power* = {00,01,...} can indicate different options to be used for power boosting. Table 3.24 shows the power boosting value for each condition.

3.9.3.3 PT-RS Design for Uplink DFT-s-OFDM

Regarding to the design of PT-RS for DFT-s-OFDM, pre-DFT and post-DFT insertion methods were discussed in standardization. The former method is to insert PT-RS as time-domain signal like PUSCH data processing, while the latter is to add PT-RS in frequency domain similar to PT-RS in CP-OFDM. Due to the PAPR increase from post-DFT insertion, even if it provides the common design for both CP-OFDM and DFT-s-OFDM, the pre-DFT insertion method was finally adopted.

In Figure 3.104 we show the example of PT-RS pattern for DFT-s-OFDM in time domain. The PT-RS for DFT-s-OFDM consists of $N_{group}^{PT\text{-}RS}$ groups ($N_{group}^{PT\text{-}RS} = \{2, 4, 8\}$) included within a DFT-s-OFDM symbols, where a PT-RS group is a unit of allocation composed of N_{samp}^{group} consecutive DFT-s-OFDM samples ($N_{samp}^{group} = \{2, 4\}$) before DFT precoding. Table 3.25 shows the PT-RS symbol mapping with different values of the number of groups and the number

Table 3.24 PT-RS power boosting value in dB.

| | The number of PUSCH layers | | | | | | | |
| | 1 | 2 | | 3 | | 4 | | |
UL-PTRS-power	All cases	Full coherent	Partial and non-coherent and non-codebook based	Full coherent	Partial and non-coherent and non-codebook based	Full coherent	Partial coherent	Non-coherent and non-codebook based
00	0	3	$3Q_p-3$	4.77	$3Q_p-3$	6	$3Q_p$	$3Q_p-3$
01	0	3	3	4.77	4.77	6	6	6
10				Reserved				
11				Reserved				

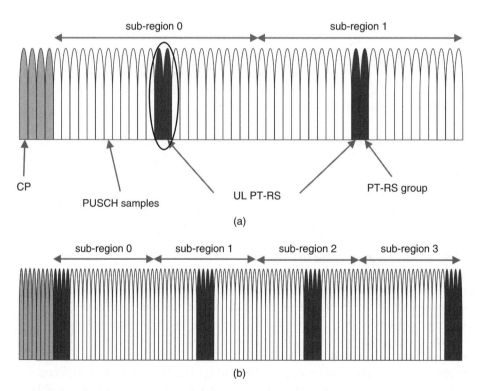

Figure 3.104 Example of PT-RS design for DFT-s-OFDM ($N_{group}^{PT-RS} = 2, N_{samp}^{group} = 2$). (a) $N_{group}^{PT-RS} = 2, N_{samp}^{group} = 2$. (b) $N_{group}^{PT-RS} = 4, N_{samp}^{group} = 4$.

Table 3.25 PT-RS symbol mapping.

Number of PT-RS groups $N_{group}^{PT\text{-}RS}$	Number of samples per PT-RS group N_{samp}^{group}	Index m of PT-RS samples in OFDM symbol l prior to transform precoding
2	2	$s\lfloor M_{sc}^{PUSCH}/4\rfloor + k - 1$ where $s = 1, 3$ and $k = 0, 1$
2	4	$sM_{sc}^{PUSCH} + k$ where $\begin{cases} s = 0 \text{ and } \quad k = 0, 1, 2, 3 \\ s = 1 \text{ and } k = -4, -3, -2, -1 \end{cases}$
4	2	$\lfloor sM_{sc}^{PUSCH}/8\rfloor + k - 1$ where $s = 1, 3, 5, 7$ and $k = 0, 1$
4	4	$sM_{sc}^{PUSCH}/4 + n + k$ where $\begin{cases} s = 0 \text{ and } \quad k = 0, 1, 2, 3 \qquad\qquad n = 0 \\ s = 1, 2 \text{ and } \quad k = -2, -1, 0, 1 \quad n = \lfloor M_{sc}^{PUSCH}/8\rfloor \\ s = 4 \text{ and } k = -4, -3, -2, -1 \qquad n = 0 \end{cases}$
8	4	$\lfloor sM_{sc}^{PUSCH}/8\rfloor + n + k$ where $\begin{cases} s = 0 \qquad\quad \text{ and } \quad k = 0, 1, 2, 3 \qquad\qquad n = 0 \\ s = 1, 2, 3, 4, 5, 6 \text{ and } \quad k = -2, -1, 0, 1 \quad n = \lfloor M_{sc}^{PUSCH}/16\rfloor \\ s = 8 \qquad\quad \text{ and } k = -4, -3, -2, -1 \qquad n = 0 \end{cases}$

Table 3.26 PT-RS group pattern as a function of scheduled bandwidth.

Scheduled bandwidth	Number of PT-RS groups	Number of samples per PT-RS group
$N_{RB0} \le N_{RB} < N_{RB1}$	2	2
$N_{RB1} \le N_{RB} < N_{RB2}$	2	4
$N_{RB2} \le N_{RB} < N_{RB3}$	4	2
$N_{RB3} \le N_{RB} < N_{RB4}$	4	4
$N_{RB4} \le N_{RB}$	8	4

of samples in a group. Different patterns are used for different scheduled bandwidths. The phase rotation for each sample can be estimated by, for example, linear interpolation from the estimated phase rotation in each PT-RS group.

When $N_{samp}^{group} = 2$, one DFT-s-OFDM symbol is divided into the same number of sub-regions. One PT-RS port group is mapped to each sub-region in the center of the sub-region. When $N_{samp}^{group} = 4$, the first group and the last group are mapped to the start and the end of the symbol, and the remaining groups are mapped to the center of sub-regions as the same way when used for $N_{samp}^{group} = 2$.

The UE can be configured with the higher layer parameter *timeDensity* set to $L_{PT\text{-}RS} = 2$, otherwise default values $L_{PT\text{-}RS} = 1$ is used.

The higher layer parameter *sampleDensity* indicates the sample density thresholds $\{N_{RB,i}\}$ ($i = 0, \ldots 4$). In Table 3.26 we show the mapping of PT-RS parameter for each scheduled bandwidth.

Table 3.27 PT-RS scaling factor (β′) when transform coding enabled.

Scheduled modulation	PT-RS scaling factor (β')
π/2-BPSK	1
QPSK	1
16 QAM	$3/\sqrt{5}$
64 QAM	$7/\sqrt{21}$
256 QAM	$15/\sqrt{85}$

3.9.3.3.1 PT-RS Sequence and Port Multiplexing For PT-RS sequence, the pseudo-random sequence randomized by the higher-layer parameter *nPUSCH-Identity* is applied. The pseudo-random sequence is modulated by π/2 BPSK modulation for reducing the PAPR of the signal, and it is mapped to each PT-RS sample. An OCC is applied to each PT-RS group, and up to $N_{\text{samp}}^{\text{group}} = \{2,4\}$ ports can be multiplexed by different OCC.

For different modulation orders, the PT-RS sequence is scaled with the amplitude of the outer-most constellation point, as shown in Table 3.27. This provides the effect of PT-RS power boosting without an increase of PAPR when higher modulation is used. Because higher-order modulation such as 64 QAM and 256 QAM requires high EVM requirement, this enables obtaining high accuracy in phase estimation with the increase of modulation order.

3.10 SRS (Stephen Grant, Ericsson, USA)

3.10.1 Overview

3.10.1.1 SRS Use Cases

SRS in NR are UE-specifically configured reference signals transmitted by the UE used for the purposes of the sounding the uplink radio channel. Like for CSI-RS, such sounding provides various levels of knowledge of the radio channel characteristics. On one extreme, the SRS can be used at the gNB simply to obtain signal strength measurements, e.g. for the purposes of UL beam management. On the other extreme, SRS can be used at the gNB to obtain detailed amplitude and phase estimates as a function of frequency, time, and space. In NR, channel sounding with SRS supports a more diverse set of use cases compared to LTE.

- *Downlink CSI acquisition for reciprocity-based gNB transmit beamforming (downlink MIMO)*
  - o One or more SRS resources are used by the gNB to acquire detailed CSI for the downlink (spatial) channel based on an assumption that the uplink and downlink channels are reciprocal, e.g. in TDD deployments. Using the channel estimates obtained from SRS reception, the gNB computes (non-codebook based) precoding weights for DL MIMO transmissions.

o For the case when the number of transmit radio chains at the UE is less than the number of receive chains, SRS transmissions from the UE can be switched in a TDM manner between subsets of UE antennas in order for the gNB to progressively build up channel estimates for the full spatial channel between all gNB transmit antennas and all UE receive antennas.

o Similarly, for the case when the number of carriers supported in the UL is less than the number of carriers supported in the DL, SRS transmissions from the UE can be switched in a TDM manner between subsets of carriers in order for the gNB to progressively build up channel estimates for all downlink carriers.

- *Uplink CSI acquisition for link adaptation and codebook/non-codebook based precoding for uplink MIMO*

o In NR, two UL MIMO modes are supported: codebook-based precoding and non-codebook-based precoding. The former/latter is applicable in scenarios in which reciprocity cannot/can be assumed between downlink and uplink.

o For codebook-based precoding, the gNB may configure the UE to transmit SRS for purposes of acquiring UL CSI. Based on the CSI estimated at the gNB, the gNB indicates a suitable MCS, a number of spatial layers (rank indication), and a precoding matrix (PMI indication) to the UE for subsequent UL MIMO transmissions.

o For non-codebook based transmission, the precoding weights are instead determined at the UE based on CSI-RS reception under the assumption of reciprocity. The UE then transmits one or more precoded SRS resources, e.g. corresponding to different antenna panels. In turn, the gNB then indicates a suitable modulation and code scheme corresponding to a suitable subset of the SRS resources (SRI indication) to the UE to enable subsequent UL MIMO transmissions.

- *Uplink beam management*

o Uplink beam management based on a set of SRS resources transmitted by the UE is the analogue to downlink beam management based on a set of CSI-RS resources transmitted by the gNB. If the UE maintains a fixed spatial domain transmit filter over the set of SRS resources, it provides the gNB with an opportunity to adjust its spatial domain receive filter (receive beam) to optimize reception. If the UE adjusts its spatial domain transmit filter over the set of SRS resources, it provides the gNB with an opportunity to select a preferred UE transmit "beam." This can be indicated to the UE as a "spatial relation" for future SRS or PUSCH transmissions. For more on beam management, see Section 4.2.

3.10.1.2 Key Differences with LTE

In order to support the diverse set of use cases for NR, SRS configuration has been enhanced in order to increase flexibility compared to LTE. The following enhancements have been added:

- Time domain location and SRS resource size

o In LTE, SRS is restricted to the last symbol of a slot, and for the case of TDD also within the UpPTS. In contrast, in NR an SRS resource can be configured as either one, two, or four adjacent OFDM symbols. Moreover, the SRS resource can be located anywhere within the last six OFDM symbols of a slot.

- Intra-slot hopping
    - In LTE, only inter-slot hopping is supported since SRS is only allowed to occupy a single OFDM symbol within a slot. In contrast, in NR since the size of an SRS resource is expanded to be either one, two, or four OFDM symbols, it offers the opportunity to complete a frequency hopping cycle all within one slot.
- Intra-slot repetition
    - In NR, all ports of an SRS resource are sounded in each OFDM symbol of the resource. If frequency hopping is not configured, this means the same subcarriers are occupied across all OFDM symbols within the resource. Such repetition in a multi-symbol SRS resource is beneficial for enhancing coverage if needed.
    - If frequency hopping is configured, flexible 2x or 4x repetition may be configured in addition, such that repetition occurs over two or four OFDM symbols within a slot before the next hop.
- Frequency domain location
    - In NR, additional bandwidth configuration flexibility has been added such that an arbitrary portion of a BWP can be sounded, with or without frequency hopping within the sounded portion.
- Time domain behavior
    - NR supports an additional time domain SRS behavior, semi-persistent transmission, on top of periodic and aperiodic transmission available in LTE. With semi-persistent SRS transmission, MAC-CE are used to activate and deactivate a set of one or more SRS resources. While activated, an SRS resource is transmitted with a configured periodicity and slot offset. MAC-CE activation/deactivation offers faster on/off control compared to periodic SRS resources which are configured purely by RRC.

3.10.2 Physical Layer Design

3.10.2.1 Mapping to Physical Resources

The time/frequency mapping of an SRS resource is defined by the following characteristics

- Time duration $N_{\text{symb}}^{\text{SRS}}$
    - The time duration of an SRS resource can be one, two, or four consecutive OFDM symbols within a slot, in contrast to LTE which allows only a single OFDM symbol per slot.
- Starting symbol location l_0
    - The starting symbol of an SRS resource can be located anywhere within the last six OFDM symbols of a slot provided the resource does not cross the end-of-slot boundary (see Figure 3.105).
- Repetition factor R
    - For an SRS resource configured with frequency hopping, repetition allows the same set of subcarriers to be sounded in R consecutive OFDM symbols before the next hop occurs. The allowed values of R are {1, 2, 4} where $R \leq N_{\text{symb}}^{\text{SRS}}$.
- Transmission comb spacing K_{TC} and comb offset $\bar{k}_{\text{TC}}$
    - An SRS resource always occupies resource elements (REs) of a frequency domain comb structure, where the comb spacing is either two or four REs like in LTE. Such

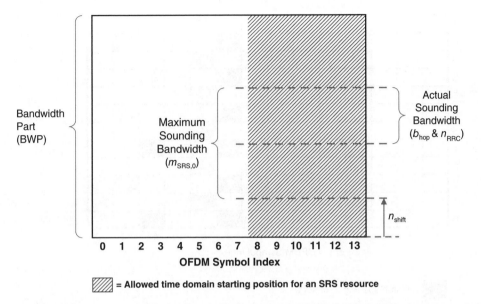

= Allowed time domain starting position for an SRS resource

Figure 3.105 Time and frequency domain location of an SRS resource. In the time domain, an SRS resource can be located within the last 6 symbols of a slot. The bandwidth configuration parameters C_{SRS}, B_{SRS}, b_{hop}, n_{RRC}, and n_{shift} can be adjusted such that an SRS resource can occupy part or a whole of a BWP.

a structure allows frequency domain multiplexing of different SRS resources of the same or different users on different combs, where the different combs are offset from each other by an integer number of REs. The comb offset is defined with respect to a PRB boundary, and can take values in the range $\{0, 1, ..., K_{TC} - 1\}$ REs. Thus, for comb spacing $K_{TC} = 2$, there are two different combs available for multiplexing if needed, and for comb spacing $K_{TC} = 4$, there are four different available combs. Figure 3.106 shows a few examples for the case of a single symbol SRS resource.

- Periodicity and slot offset for the case of periodic/semi-persistent SRS described in more detail below
- Sounding bandwidth within a BWP described in more detail below

3.10.2.1.1 *SRS Time Domain Behavior* An SRS resource can be configured to have one of three different time domain behaviors:

- Periodic
- Semi-persistent
- Aperiodic

For the case of periodic and semi-persistent SRS, a periodicity is semi-statically configured such that the resource is transmitted once every N slots where the allowed configurable values are

$$N \in \{1, 2, 4, 5, 8, 10, 16, 20, 32, 40, 64, 80, 160, 320, 640, 1280, 2560\}$$

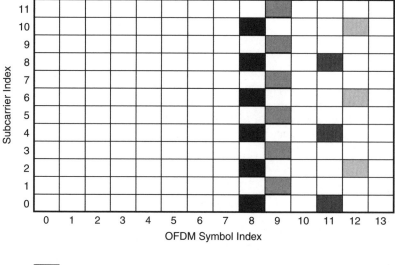

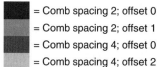 = Comb spacing 2; offset 0
= Comb spacing 2; offset 1
= Comb spacing 4; offset 0
= Comb spacing 4; offset 2

Figure 3.106 Transmission comb examples for a single-symbol SRS resource. A comb is defined by the comb spacing K_{TC} and the comb offset $\bar{k}_{TC}$.

In addition, an offset O is configured where $O \in \{0, 1, ..., N-1\}$ measured in number of slots. The reference point for the slot offset is with respect to the first slot (slot 0) of radio frame 0.

While the periodicity and offset are configured in the same way for both periodic and semi-persistent SRS, the difference in the two types lies in when the UE begins to transmit the SRS resource. For the case of periodic SRS, once the UE receives an SRS configuration by RRC, the UE begins to transmit in the slots governed by the configured periodicity and offset values. In contrast, for semi-persistent SRS, the UE does not begin to transmit SRS until it receives an explicit MAC-CE activation message. Once the SRS resource is activated, the UE begins to transmit the SRS in the configured slots up until the UE receives an explicit MAC-CE deactivation message. Note that while the UE transmits SRS according to the configured periodicity and offset values, the transmission is subject to a slot being classified as an uplink by either semi-static and/or dynamic slot format indication (see [9] for details).

For the case of aperiodic SRS, a periodicity value is not configured. For this type of resource, the UE only transmits a configured resource once it receives a PDCCH with DCI that explicitly triggers the SRS resource. The UE transmits the resource in a later slot than the one containing the PDCCH depending on a configured slot offset. The slot offset for aperiodic SRS is a semi-statically configured value between 1 and 32 slots, where 1 refers to the next slot after the one in which the DCI trigger is received.

Strictly speaking, SRS resources are always configured to be within a set of resources of the same time domain behavior (periodic, semi-persistent, aperiodic). For the case of

Table 3.28 Example SRS bandwidth configuration for the case of $C_{SRS} = 13$.

C_{SRS}	$B_{SRS} = 0$		$B_{SRS} = 1$		$B_{SRS} = 2$		$B_{SRS} = 3$	
	$m_{SRS, 0}$	N_0	$m_{SRS, 1}$	N_1	$m_{SRS, 2}$	N_2	$m_{SRS, 3}$	N_3
13	48	1	24	2	12	2	4	3

Source: Table 6.4.1.4.3-1 in [34].

aperiodic SRS, the slot offset is configured at resource set level. Hence, SRS resources are always configured/activated/triggered on a per-set basis, even if a set only contains a single resource. SRS resource sets are described in more detail later in Section 3.10.3.

3.10.2.1.2 SRS Bandwidth Configuration The bandwidth (BW) configuration of an SRS resource is controlled by the RRC parameters C_{SRS}, n_{shift}, B_{SRS}, b_{hop}, and n_{RRC}. Together these parameters define which portion of a BWP is sounded by an SRS resource. The parameter $C_{SRS} \in \{0, 1, ..., 63\}$ selects a bandwidth configuration for the SRS resource corresponding to a particular row of the length-64 Table 6.4.1.4.3-1 in [22]. For explanation purposes, the example of $C_{SRS} = 13$ is considered here and is shown in Table 3.28.

The value of $m_{SRS, 0}$ in the first column of Table 3.28 determines the maximum bandwidth within the BWP that can be sounded by the configured SRS resource. For the example of $C_{SRS} = 13$, the maximum bandwidth is 48 PRBs. If it is desired to sound the full BWP, then a value of C_{SRS} should be chosen such that $m_{SRS, 0}$ is close as possible to the size of the BWP. If only partial sounding of the BWP is desired, then C_{SRS} can be chosen with values of $m_{SRS, 0}$ that are less than the BWP size.

For partial BWP sounding, the position of the maximum sounding bandwidth within the BWP is determined by the parameter n_{shift} which determines the index of the first PRB of the maximum sounding bandwidth (see Figure 3.105). In the examples shown later in this chapter, a BWP of size 106 PRBs is considered (e.g., 40 MHz BWP using 30 kHz SCS) and $n_{shift} = 24$. This means that the maximum sounding bandwidth starts at PRB 24 and extends to PRB $24 + 48 = 72$. The granularity of the n_{shift} parameter is a single PRB, while the granularity of the SRS transmission bandwidth is 4 PRBs since all values of $m_{SRS, b}$ in the bandwidth configuration table are multiples of 4. Here b indexes columns in the table, i.e. $b \in \{0, 1, 2, 3\}$.

Within the maximum sounding bandwidth, the bandwidth and the bandwidth position over which SRS is *actually* transmitted are controlled by the parameters b_{hop} and n_{RRC}, respectively (see Figure 3.105). The parameter b_{hop} is configured as either 0, 1, 2, or 3, which determines the actual sounding bandwidth as $m_{SRS,b}$ for $b = b_{hop}$.[3] For the case of $b_{hop} = 0$ the actual sounding bandwidth and the maximum sounding bandwidth are the same, and the parameter $b_{hop} = 0$ has no effect. From Table 3.28, if $b_{hop} = 1$, for example, the sounding bandwidth is $m_{SRS, 1} = 24$ PRBs out of the maximum 48 PRBs. The position of the actual sounding bandwidth within the maximum is determined by the parameter n_{RRC} which has a configurable range of 0 ... 67. Despite this wide range, there are only

---

3 Note that in the case when frequency hopping is disabled, the actual sounding bandwidth is given by $m_{SRS, b}$ where $b = \min(b_{hop}, B_{SRS})$.

$\prod_{b=0}^{b_{\text{hop}}} N_b$, non-overlapping (orthogonal) positions of the actual sounding bandwidth where the values of N_b are shown in Table 3.28. For example, if $b_{\text{hop}} = 1$, there are $N_1 \cdot N_2 = 2$ possible non-overlapping positions of the 24 PRBs within 48. In this way, the parameter n_{RRC} can be used for frequency domain multiplexing of SRS resources from the same or different users. Furthermore, if frequency hopping is enabled (see discussion below), the parameter n_{RRC} can also be used to select a particular hopping pattern amongst a number of orthogonal patterns to support additional multiplexing within the same sounding bandwidth.

3.10.2.1.3 SRS Frequency Hopping

The final parameter of the SRS bandwidth configuration is B_{SRS} which controls whether all or a subset of the PRBs in the actual sounding bandwidth are used either (i) by hopping over a number of smaller BW allocations in different OFDM symbols, or (ii) by a fixed (non-hopped) BW allocation. For frequency hopped SRS, the frequency domain starting position of each hop varies over time according to a pre-defined hopping pattern, whereas for non-hopped SRS the frequency domain starting position is fixed over time. It is useful to recognize that non-hopped SRS is just a special case of SRS hopping in which the number of hops in a hopping cycle is only one.

The parameter B_{SRS} is configured as either 0, 1, 2, or 3, which determines the number of PRBs within the actual sounding bandwidth that are sounded in each hop of a hopping cycle which is given by $m_{\text{SRS}, b}$ for $b = B_{\text{SRS}}$. In other words, B_{SRS} is the SRS bandwidth *per hop*. The relationship between B_{SRS} and b_{hop} determines whether frequency hopping is enabled or disabled.

Hopping is enabled if $b_{\text{hop}} < B_{\text{SRS}}$. For example, from Table 3.28, if $B_{\text{SRS}} = 2$, the bandwidth per hop is 12 PRBs. If $b_{\text{hop}} = 1$, the bandwidth over which hopping occurs is equal to $m_{\text{SRS}, 1} = 24$ PRBs (half of the maximum 48), and a hopping cycle completes in $N_2 = 2$ hops. If $b_{\text{hop}} = 0$, the bandwidth over which hopping occurs is equal to $m_{\text{SRS}, 0} = 48$ PRBs (equal to the full 48), and a hopping cycle completes in $N_1 \cdot N_2 = 4$ hops. In general, a hopping cycle completes in $N = \prod_{b=b_{\text{hop}}+1}^{B_{\text{SRS}}} N_b$ hops.

In contrast, hopping is disabled if $b_{\text{hop}} \geq B_{\text{SRS}}$. Note that it makes no difference to the actual sounding bandwidth if b_{hop} is greater than B_{SRS} or if b_{hop} is equal to B_{SRS}. In both cases the actual sounding bandwidth is given by $m_{\text{SRS}, b}$ where $b = \min(b_{\text{hop}}, B_{\text{SRS}})$. In other words, the actual sounding bandwidth and the per-hop sounding bandwidth are the same since there is only one hop. Using the example in Table 3.28, if $b_{\text{hop}} = B_{\text{SRS}} = 2$, the actual sounding bandwidth is 12 PRB out of 48 PRBs, and the same 12 PRBs are sounded in all OFDM symbols of the SRS resource. As described previously, which 12 out of the 48 PRBs is controlled by the parameter n_{RRC}.

3.10.2.1.4 Example SRS Configurations

In this section, some example SRS configurations are illustrated considering intra-slot hopping, inter-slot hopping, and repetition. Both aperiodic and periodic/semi-persistent SRS resources are considered. For all examples shown here, the following parameter values are assumed:

- Size of BWP is 106 PRBs
- The same SRS bandwidth configuration as shown in Table 3.28 is used, i.e. $C_{\text{SRS}} = 13$
  - This results in a maximum sounding bandwidth of $m_{\text{SRS}, 0} = 48$ PRBs within the 106 PRB BWP
  - The shift parameter is configured as $n_{\text{shift}} = 24$ such that the maximum sounding bandwidth is located starting at PRB index 24

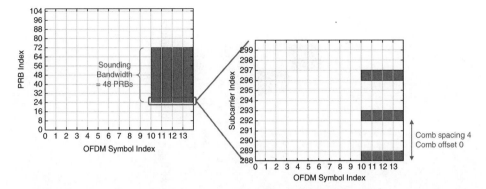

Figure 3.107 Aperiodic SRS transmission of 4-symbol resource with $b_{hop} = B_{SRS} = 0$ (no hopping). The diagram on the right shows a zoomed-in view of the first PRB of the SRS resource showing a comb spacing of four subcarriers.

- The transmission comb spacing is configured as $K_{TC} = 4$ and the comb offset is configured as $\bar{k}_{TC} = 0$
  o All ports are mapped to the same comb (see discussion in Section 3.10.2.2)
- The starting symbol of the SRS resource is configured such that the last OFDM symbol(s) within the slot are used

Figure 3.107 shows the transmission of a 4-symbol aperiodic SRS resource with $b_{hop} = B_{SRS} = 0$, i.e. hopping is disabled, and the sounding bandwidth is 48 PRBs. The illustration on the right-hand side shows a zoomed in view of the first PRB of the SRS resource showing that the comb spacing is 4 subcarriers, and the comb offset is 0, i.e. aligned with the PRB boundary.

Figure 3.108 shows three different examples of an aperiodic SRS transmission with hopping enabled and where the hopping cycle completes within one slot (intra-slot hopping). In all cases $b_{hop} = 0$ such that the sounding bandwidth is 48 PRBs. In (a), a 2-symbol resource is transmitted with $B_{SRS} = 1$ (24 PRBs per hop). In (b), a 4-symbol resource is transmitted with $B_{SRS} = 2$ (12 PRBs per hop). In (c), a 4-symbol resource is transmitted with $B_{SRS} = 1$ (24 PRBs per hop), but 2X repetition is configured ($R = 2$). The order of repetition and hopping is always repetition first, then hopping.

The next two examples consider periodic SRS transmission in which the period of the SRS resource is configured as two slots. While transmission in every other slot is quite frequent, it is useful for the purposes of illustration. Figure 3.109 shows the transmission of a 2-symbol periodic SRS resource with $b_{hop} = 0$ and $B_{SRS} = 2$ (12 PRBs per hop). This requires 4 hops to sound the 48 PRBs. In (a), no repetition is configured, thus since the resource is 2 symbols, the hopping cycle extends across 2 periods of the SRS resource. In other words, both intra-slot hopping and inter-slot hopping occur. In (b), 2X repetition is configured ($R = 2$), such that repetition occurs intra-slot and hopping occurs inter-slot. With 2X repetition, the hopping cycle is twice as long as without repetition.

Figure 3.110 shows the transmission of a single-symbol periodic SRS resource with $B_{SRS} = 2$ (12 PRBs per hop) and no repetition ($R = 1$). Since four hops are required to complete a cycle, and the SRS resource is only a single symbol, the hopping cycle extends over four periods of the SRS resource. Furthermore, hopping occurs only inter-slot. In (a), $b_{hop} = 0$ meaning all 48 PRBs are sounded in four hops. In (b) and (c), $b_{hop} = 1$ meaning

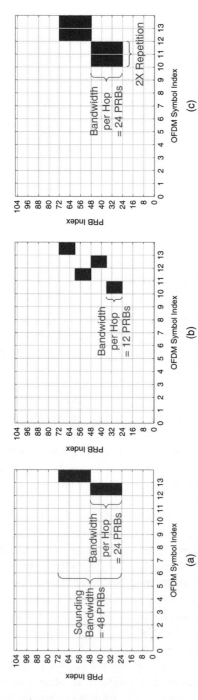

Figure 3.108 Aperiodic SRS transmission with intra-slot frequency hopping of (a) 2-symbol resource with $b_{hop} = 0$ and $B_{SRS} = 2$; (b) 4-symbol resource with $b_{hop} = 0$ and $B_{SRS} = 1$; and (c) 4-symbol resource with $b_{hop} = 0$ and $B_{SRS} = 1$ and 2X repetition (R = 2).

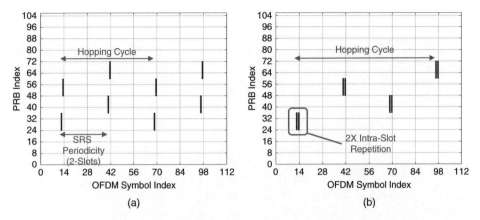

Figure 3.109 Periodic SRS transmission with intra + inter-slot frequency hopping of a 2-symbol resource with $b_{hop} = 0$ and $B_{SRS} = 2$ for (a) no repetition (R = 1); and (b) 2X repetition (R = 2).

that the actual sounding bandwidth is only 24 PRBs (half of the maximum 48). Which half of the PRBs that is sounded is controlled by the parameter n_{RRC}. In (b), n_{RRC} is set such that the first half is sounded, and in (c) n_{RRC} is set such that the second half is sounded. Note that in (b) and (c) where only half of the maximum bandwidth is sounded by setting $b_{hop} = 1$, the hopping pattern inherently repeats similar to the behavior when the repetition parameter R is configured to support intra-slot repetition.

3.10.2.2 Antenna Port Mapping

Antenna ports for SRS are numbered starting with 1000. The number of antenna ports N_{ap} for an SRS resource may be configured as 1, 2, or 4. Generally, the multiple ports are all multiplexed onto the same comb through assignment of equally spaced cyclic shifts to the antenna ports, where all ports use the same base sequence. See Section 3.10.2.3.1 for more details on cyclically shifted base sequences. There is one exception, and that is for the case of a 4-port SRS resource when the cyclic shift d_0 (corresponds to the first antenna port) is RRC configured in the 2nd half of the value range $0 \ldots D_{max} - 1$. In this case, the two pairs of ports {1000,1002} and {1001,1003} are multiplexed onto two different evenly spaced combs.

In NR, all antenna ports are sounded in every OFDM symbol of the SRS resource. This is true regardless of whether or not frequency hopping is configured. This mapping is particularly beneficial in the case where hopping is disabled, in which case all ports are mapped to the same set of subcarriers in every OFDM symbol of the resource. Such repetition allows for enhanced SRS coverage since more double the energy can be captured compared to sounding on just one OFDM symbol.

Figure 3.111 shows two examples of the mapping of port numbers to combs for the case of a 4-port, 2-symbol SRS resource. On the left, all 4 ports are mapped to the same comb in each OFDM symbol, where the comb spacing is 2. On the right, different port pairs are mapped to two different combs, each with spacing 4. In both cases, the same ports are repeated on each OFDM symbol of the resource.

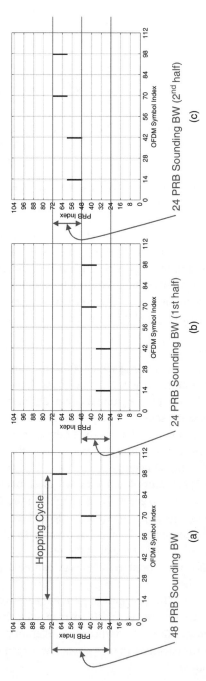

Figure 3.110 Periodic SRS transmission with inter-slot frequency hopping of a 1-symbol resource with (a) $b_{hop} = 0$ and $B_{SRS} = 2$; (b) $b_{hop} = 1$ and $B_{SRS} = 2$ with $n_{RRC} = 0$; and (c) $b_{hop} = 1$ and $B_{SRS} = 2$ with $n_{RRC} = 6$.

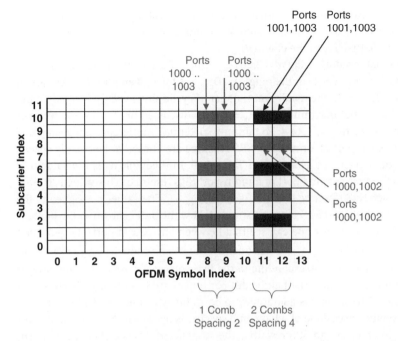

Figure 3.111 Mapping of port numbers to combs for two examples of a 4-port, 2 symbol SRS resource.

3.10.2.3 Sequence Generation and Mapping

The reference signal sequence corresponding to the pth SRS antenna port is given by

$$r_l^p(n) = e^{j\alpha n}\bar{r}_{u,v}(n), \quad n = 0, 1, \dots, M_b - 1$$

where n is a subcarrier index and M_b is the length of the sequence. The sequence $r_l^p(n)$ is mapped to the lth OFDM symbol of the SRS resource. The number of REs occupied by SRS in each OFDM symbol (same for all symbols) depends on the number of RBs per hop (determined by B_{SRS}) and the comb spacing K_{TC}. Hence,

$$M_b = m_{\mathrm{SRS},b} \cdot (N_{\mathrm{sc}}^{\mathrm{RB}}/K_{\mathrm{TC}})$$

for $b = B_{\mathrm{SRS}}$ where $N_{\mathrm{sc}}^{\mathrm{RB}} = 12$ is the number of subcarriers per RB. For the SRS configuration examples discussed previously with 12/24/48 RBs per hop and comb spacing 4, the sequence length is thus 36/72/144.

3.10.2.3.1 Cyclic Shifts of Base Sequences

The term $e^{j\alpha n}$ is a phase rotation applied to the SRS base sequence $\bar{r}_{u,v}(n)$ in the frequency domain for the pth SRS antenna port. SRS base sequences are described in more detail below. In the time domain, the phase rotation translates to a cyclic shift of the time domain version of the base sequence. The same base sequence is used for all antenna ports of the SRS resource; however, a different cyclic shift is applied per port. Since the reference signal sequences for multiple antenna ports are multiplexed onto the same comb, the cyclic shifts are selected to ensure mutual orthogonality among all antenna ports. This allows the uplink channel to be estimated for each port individually without interference from other ports. Strictly speaking, the zero interference assumption is met as long as the spacing among the multiple cyclic shifts is greater than the maximum multipath delay in the channel.

Orthogonality is achieved by using $\alpha = 2\pi d / D_{max}$ where D_{max} is the maximum number of cyclic shifts and d is an integer in the range $0...D_{max} - 1$ for a given port. For the case of comb spacing two ($K_{TC} = 2$), the maximum number of cyclic shifts is $D_{max} = 8$. For comb spacing two ($K_{TC} = 4$), the maximum is $D_{max} = 12$. As discussed previously, the number of SRS antenna ports is either 1, 2, or 4. To minimize potential interference between ports, the values of d for each port of a multiport SRS resource are chosen to be equally spaced across the range of $0...D_{max} - 1$ as shown in Table 3.29. In this table d_0 is the value of d corresponding to the first antenna port of a multi-port SRS resource. This value is configured by RRC for each SRS resource in the range $0...D_{max} - 1$. The values d assigned to the remaining ports of a multiport SRS resource simply follow the order shown in the table.

By observing the table closely one can see that between any pair of ports, $|\alpha_1 - \alpha_2|$ can only take the values $2\pi \cdot \left\{ 0, \frac{1}{4}, \frac{1}{2}, \frac{3}{4} \right\}$. Consequently, the inner product $\sum_{n=0}^{N_{sc}^{RB} - 1} e^{j(\alpha_1 - \alpha_2)n}$ of the phase rotation sequences corresponding to the two ports is always 0 over any integer multiple of 12 subcarriers. Since the bandwidth per hop of an SRS resource is always an integer multiple of 4 PRBs, there is guaranteed to be an integer multiple of 12 subcarriers in every OFDM symbol regardless of what comb spacing is used (2 or 4). In summary, for any pair of ports p_1 and p_2 of an SRS resource, the sequence $r_l^{p_1}(n)$ is orthogonal to $r_l^{p_2}(n)$ since the base sequence $\bar{r}_{u,v}(n)$ is the same for all ports.

Another observation from the table is that only a subset of the maximum number of cyclic shifts is used for an SRS resource. This means that if another SRS resource, either from the same or different user is configured, the two SRS resources can share the same comb as long as different values of the first cyclic shift d_0 are configured for the two resources. If the two SRS resources share the same base sequence and the same bandwidth, they will be orthogonal to each other. The maximum multiplexing capacity in the cyclic shift domain is shown in Table 3.30. It is important to note that when multiplexing SRS resources on the same comb, one must be careful to account for the delay spread of the channel so as to ensure that the cyclic shift spacing does not become too small relative to the delay spread.

Table 3.29 Cyclic shifts assigned to multiple SRS antenna ports. d_0 is the RRC configured cyclic shift for the first antenna port of the SRS resource.

| | | 1 antenna port | | 2 antenna ports | | 4 antenna ports | |
| | | $N_{ap} = 1$ | | $N_{ap} = 2$ | | $N_{ap} = 4$ | |
Comb spacing K_{TC}	Maximum # cyclic shifts D_{max}	Used cyclic shifts	Values of d	Used cyclic shifts	Values of d	Used cyclic shifts	Values of d
2	8	1 of 8	d_0	2 of 8	$(d_0 + \{0, 4\})$ mod 8	4 of 8	$(d_0 + \{0, 2, 4, 6\})$ mod 8
4	12	1 of 12	d_0	2 of 12	$(d_0 + \{0, 6\})$ mod 12	4 of 12	$(d_0 + \{0, 3, 6, 9\})$ mod 12

Table 3.30 Maximum number of SRS resources that may be multiplexed on the same comb using different cyclic shifts of the same base sequence.

Comb Spacing	1 antenna port	2 antenna ports	4 antenna ports
K_{TC}	$N_{ap} = 1$	$N_{ap} = 2$	$N_{ap} = 4$
2	8	4	2
4	12	6	3

If the spacing is less than the delay spread, there will be a loss in orthogonality between the SRS resources. In this case, it is better to FDM multiplex the resources by configuring them with different comb offset values $\bar{k}_{TC}$, i.e. the resources occupy different combs.

3.10.2.3.2 SRS Base Sequences Like in LTE, multiple base sequences $\bar{r}_{u,v}(n)$ of flexible length are available. The multiple sequences for a given length are organized into 30 different sequence groups where $u \in \{0, 1, ..., 29\}$ indexes a sequence group, and $v \in \{0, 1\}$ indexes the sequences within a group (at most 2). The number of sequences per group depends on the sequence length. For sequences of length longer than 72, there are two sequences per group; otherwise there is only one and $v = 0$. As discussed previously, cyclic shifts of the *same* base sequence are orthogonal. While *different* base sequences (i.e. within or between different groups) are not completely orthogonal, they still have low cross-correlation, meaning that interference due to SRS transmissions within/between cells with different base sequences can be maintained at a low level.

For the case when neither group nor sequence hopping is configured (see next section for discussion on group/sequence hopping), the same base sequence is used in each OFDM symbol of the SRS resource. In this case, the group and sequence indices are given by

$$u = n_{ID}^{SRS} \bmod 30$$

$$v = 0$$

where n_{ID}^{SRS} is a 10-bit sequence ID that is configured on a per-UE and per-SRS resource basis. This is in contrast to LTE where the sequence ID is cell-specific.

The introduction of UE-specific sequence IDs in NR enables the same ID to be assigned to one group of UEs and/or cells, and different IDs to different groups of UEs/cells. For example, to emulate LTE functionality, the same ID can be configured for all UEs in a given cell, but different IDs configured to UEs in different cells. With this approach, each different cell (within a cluster of 30 cells) is assigned a different base sequence, and inter-cell interference can be maintained at a level as low as possible. Another approach is to assign different IDs to different groups of UEs within the same cell, e.g., UEs grouped on a per-beam basis. Yet another is to form groups of co-scheduled UEs in a MU-MIMO scenario such that clean channel estimates are obtained for UEs within a group multiplexed through different cyclic shifts of the same base sequence while simultaneously maintaining low inter-group interference.

As in LTE, two types of base sequences are defined. For both types, the joint design goals are the following:

1. Maintain low PAPR and cubic metric (CM) at the UE to enable lower cost power amplifier designs
2. Ensure low cross-correlation between different base sequences for the reasons described above
3. For each sequence length allow for the formation of 30 sequence groups, each with one or two sequences, i.e. up to 60 different sequences.

3.10.2.3.3 Z-C Base Sequences Like in LTE, for each base sequence length greater than a certain value, a set of Zadoff-Chu (Z-C) sequences mapped to the 30 different groups are defined which satisfy all three design goals. Z-C sequences have the attractive property that they have a constant amplitude in the frequency domain ensuring SRS reference signal sequences have low cross correlation. Another attractive property is that the IFFT of a Z-C sequence is also a Z-C sequence, meaning constant amplitude in the time domain, ensuring the low PAPR/CM design goal is satisfied.

To satisfy the 3rd design goal, the number of Z-C sequences of a given length should be maximized. Z-C sequences have the property that the number of Z-C sequences of length M is equal to one less than the number of integers that are relative prime to M. Hence, to maximize the number of Z-C sequences, M should be a prime number. However, there is also a constraint that the SRS sequence length is always an integer multiple of 12. To satisfy this constraint, a Z-C sequence of prime length is cyclically extended to a length equal to an integer multiple of 12.

Based on the above, the base Z-C sequences used for SRS are defined as

$$\bar{r}_{u,v}(n) = x_q(n \bmod N_{ZC}), \quad n = 0, 1, \ldots, M_b - 1$$

where

$$x_q(i) = e^{-j\frac{\pi q i(i+1)}{N_{ZC}}}$$

Here, N_{ZC} is the largest prime number such that $N_{ZC} < M_b$ and M_b is the length of the SRS reference signal sequence (integer multiple of 12). Furthermore, the Z-C sequence root q is a function of the group and sequence indices u and v such that all groups contain unique Z-C roots. The shortest Z-C sequence defined in NR is length 36, since for lengths 12 and 24 there are not enough roots to occupy the 30 different groups. Furthermore, only for Z-C sequences of length 72 and longer are there two sequences per group, since for lengths 36, 48, and 60 there are fewer than 60 different roots.

3.10.2.3.4 Computer Generated Base Sequences (CGS) For base sequences of 12 and 24 for SRS, rather than Z-C sequences, a set of 30 different QPSK sequences are defined as

$$\bar{r}_{u,v}(n) = e^{j\phi_u(n)\pi/4}, \quad n = 0, 1, \ldots, M_b - 1$$

where $v = 0$ (1 sequence per group). The sequences $\phi_u(n) \in \{\pm 1, \pm 3\}$ are found by computer search for $u \in \{0, 1, \ldots, 29\}$ such that the mutual cross-correlation between the 30 different base sequences is as low as possible. The phase sequences for length 12 and 24 are tabulated in [34].

3.10.2.3.5 Group and Sequence Hopping As described above, when neither group nor sequence hopping is configured, the group index u is semi-statically configured as a function of the RRC configured SRS sequence ID $n_{\text{ID}}^{\text{SRS}}$, and the sequence index v is zero. Furthermore, the same base sequence is used in every OFDM symbol of the SRS resource.

To facilitate interference randomization between cells and/or UEs with different sequence IDs, and hence different base sequences, two different hopping approaches are specified, where only one or the other is configured. In the first approach, called group hopping, the sequence index $v = 0$ and the group index u is pseudo-randomly selected in every OFDM symbol of the SRS resource and in every slot occupied by the resource. The pseudo-random sequence $c(i)$ governing the group hopping is initialized as $c_{\text{init}} = n_{\text{ID}}^{\text{SRS}}$ at the beginning of each radio frame.

In the second approach, called sequence hopping, the group index $u = n_{\text{ID}}^{\text{SRS}} \bmod 30$ is the same as when hopping is not used; however, the sequence index v is pseudo-randomly selected between 0 and 1 in every OFDM symbol of the SRS resource and in every slot occupied by the resource. Again, the pseudo-random sequence $c(i)$ governing the sequence hopping is initialized by $c_{\text{init}} = n_{\text{ID}}^{\text{SRS}}$ at the beginning of each radio frame.

3.10.2.4 Multiplexing with Other UL Signals

While configuration of SRS is quite flexible in both the frequency and time domain, there are specified rules described in [22] about multiplexing of SRS with other uplink channels (PUSCH, PUCCH) and with other SRS resources.

o Multiplexing with PUSCH
   o Only TDM multiplexing of SRS and PUSCH is allowed, and SRS may only be transmitted after PUSCH and its corresponding DMRS
o Multiplexing with PUCCH
   o Generally, PUCCH takes priority over SRS; hence SRS is (partially) dropped in the OFDM symbols that overlap with PUCCH. The exception is if aperiodic SRS overlaps OFDM symbol(s) of PUCCH carrying only periodic/semi-persistent CSI or L1-RSRP reports in which case PUCCH is dropped.
o Multiplexing with other SRS
   o Generally, SRS can be TDM, FDM, and CDM multiplexed with other SRS resources. However, several special cases limit transmission of FDM/CDM'd SRS resources in the same OFDM symbol.
   o In case one or more SRS resources overlap in the same OFDM symbol, the priority order for SRS transmission is aperiodic SRS > semi-persistent SRS > periodic SRS. SRS resource(s) of lower priority are (partially) dropped in the OFDM symbols that overlap with SRS resources of higher priority.
   o For antenna switching, SRS resources from different sets my not be transmitted simultaneously, since the different sets correspond to different antennas which are switched in TDM fashion.
   o For UL beam management, if multiple SRS sets are configured, at most one SRS resource from each set may be transmitted in the same OFDM symbol.

3.10.3 SRS Resource Sets

SRS are always configured within an SRS resource set consisting of one or more SRS resources. This configuration mechanism simplifies the activation (for semi-persistent SRS) and DCI triggering (for aperiodic SRS) since multiple resources can be activated/triggered simultaneously. Several use cases have been identified for SRS, and thus the RRC configuration of an SRS resource set contains a parameter called "*usage*." Depending on the value of *usage*, SRS set(s) will have different configurations appropriate for the indicated use case, e.g., the number of allowed sets, the number of allowed resources per set, etc. The valid values of this parameter are *antennaSwitching, codebook, nonCodebook,* and *beamManagement*.

3.10.3.1 SRS for Downlink CSI Acquisition for Reciprocity-Based Operation

To support reciprocity-based operation in TDD deployments, i.e. where the gNB acquires detailed CSI for the downlink (spatial) channel based on estimates of the UL (spatial) channel, the gNB configures one or more SRS resource sets with *usage = antennaSwitching*. This value setting is a bit of a misnomer, since it does not imply that the UE always performs antenna switching.

How the SRS resource set(s) are actually configured depends on UE capability which is expressed in terms of the number of simultaneously usable Tx and Rx chains (antennas). For example, some UEs are capable of transmitting simultaneously in the uplink on as many antennas as used for reception in the downlink. In this case, the gNB configures the UE with SRS resource set(s) containing an SRS resource for which the number of SRS ports is equal to the number of UE antenna ports. The gNB is then able to acquire downlink CSI based on estimates of the full UL spatial channel based on these SRS resource(s). This is referred to as full channel sounding. In this case, the UE indicates capability from the following set of values: 1T1R, 2T2R, and 4T4R where "T" represents transmit chains (Tx antennas) and "R" indicates receive chains (Rx antennas).

For UEs not capable of full channel sounding, then partial sounding is performed through antenna switching. For example, if the UE has four receive antennas, but it is only able to transmit on two of them simultaneously in the uplink, then it indicates "2T4R" during capability exchange. This means that in order for the gNB to acquire CSI for the full spatial channel, the UE first transmits a two-port SRS resource on one pair of UE antenna ports during one time instance and then switches to the second pair of antenna ports in a later time instance where it transmits a second 2-port SRS resource. Other valid values that may be indicated by the UE are "1T2R," and "1T4R" for supporting sounding one antenna at a time out of 2 or 4, respectively. Figure 3.112 illustrates the example of 2T4R antenna switching and the associated SRS resource configuration. Two 2-port SRS resources are configured within a single SRS resource set. In OFDM symbols 9 and 10, UE antenna ports 0 and 1 are sounded. In OFDM symbols 12 and 13, UE antenna ports 1 and 2 are sounded. The gap in between the two SRS resources (symbol 11) allows time for the UE to physically switch the antennas.

Analogous to the spatial domain (antennas), asymmetrical channel sounding scenarios also occur in the frequency domain (carriers). Often, a UE is often able to aggregate more carriers in the downlink than the uplink. For example, a UE may be able to receive on two

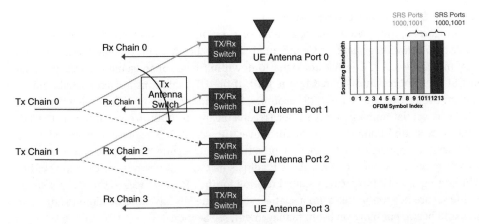

Figure 3.112 2T4R antenna switching and associated SRS resource configuration. Two symbol SRS resources are used, for example, where the first SRS resource (green) is for sounding UE antenna ports 0,1, and the second one (red) for UE antenna ports 2, 3. The gap in between the two resources allows the UE time to switch antennas.

aggregated carriers in the downlink but transmit on only a single carrier in the uplink. Like in LTE, NR supports carrier-based SRS switching to handle such scenarios. In this case, the UE transmits SRS on the primary carrier during one time instance, and then it switches (re-tunes) to the 2nd carrier, referred to as a "PUSCH-less SCell," That is, the UE configured with multiple serving cells in the UL; however, one or more are only used for the purposes of SRS transmission for UL channel sounding. These carrier(s) are used neither for UL data nor UL control transmissions.

3.10.3.2 SRS for Uplink CSI Acquisition

To support CSI acquisition for the two UL MIMO modes in NR (codebook and non-codebook-based precoding), the gNB configures the UE with a single set of SRS resources with *usage = codebook* or *nonCodebook* containing up to 2 or 4 SRS resources, respectively. The former/latter is applicable in scenarios in which reciprocity cannot/can be assumed between downlink and uplink.

For codebook-based precoding, the CSI estimated at the gNB based on the transmitted SRS resource(s) is used to indicate a suitable MCS, a number of spatial layers (rank indication), and a precoding matrix (PMI indication) to the UE for subsequent UL MIMO transmissions. In addition, the scheduling DCI contains an SRI that indicates which of the up to two SRS resources in the SRS resource set upon which PUSCH transmission should be based. This mechanism supports UEs with two antenna panels (a UE capability). In this way SRI effectively indicates which panel should be used for PUSCH transmission.

For non-codebook based transmission, the precoding weights are instead determined at the UE based on reception of an associated CSI-RS resource using the assumption of DL-UL reciprocity. The UE transmits one or more precoded SRS resources, e.g., corresponding to different antenna panels. The CSI estimated by the gNB based on the precoded SRS resource(s) is used to indicate a suitable MCS for subsequent UL MIMO transmission. In addition the SRI field is used to indicate one or more of the up to four SRS resources in the SRS resource set to support multi-panel operation.

3.10.3.3 SRS for Uplink Beam Management

To support uplink beam management, one or more sets of SRS resources with *usage = beamManagement* are configured. Uplink beam management based on SRS transmitted by the UE is the analogue to downlink beam management based on a set of CSI-RS resources transmitted by the gNB. If the UE maintains a fixed spatial domain transmit filter over the set of SRS resources, it provides the gNB with an opportunity to adjust its spatial domain receive filter (receive beam) to optimize reception. If the UE adjusts its spatial domain transmit filter over the set of SRS resources, it provides the gNB with an opportunity to select a preferred UE transmit "beam." While somewhat loosely specified, whether or not the UE adjusts its transmit beam is controlled by either configuring or not configuring a spatial relation for the SRS resources. If the SRS resources in the set are not configured with a spatial relation, the UE has more freedom to adjust its transmit beam. For more on beam management, see Section 4.2.

3.11 Power Control (Mihai Enescu, Nokia Bell Labs, Finland)

As in the case of LTE, the NR specifies power control mechanisms associated to the transmission of UL signals and channels. In other words, there are power control adjustments for the transmission of PRACH, PUCCH, PUSCH, and SRS. The details of PRACH and SRS power control are detailed in the PRACH and SRS chapters, in the following we focus more on the power control for PUSCH.

The fundamental difference between LTE and NR is the use of beams in both downlink and uplink for FR2 operation at least. In LTE, the uplink transmission was performed in an isotropic way, that is energy was radiated by the UE in all directions. Perhaps a good way of looking into the differences of the LTE and NR is by comparing the two formulas for power control for PUSCH.

Let us first introduce the LTE power control formula for PUSCH:

$$P_{PUSCH,c} = \min$$

$$\times \begin{cases} P_{CMAX,c}(i), \\ P_{0\_PUSCH,c}(i) + 10\log_{10}(M_{PUSCH,c}(i)) + \alpha_c(j) \cdot PL_c + \Delta_{TF,c}(i) + f_c(i) \end{cases} [dBm] \quad (3.7)$$

The formula in (3.7) is constructed based on the *open-loop* power control elements:

- the eNB received target power P_{0_PUSCH}
- the maximum allowed power used by the UE for UL transmission P_{CMAX},
- the amount of scheduled resources/PRBs M_{PUSCH},
- the pathloss computed by the UE PL which is also adjusted by a so-called fractional factor α configured by the network (where $0 \leq \alpha \leq 1$) with the value 1 disabling the fractional power control,
- the MCS used for the PUSCH transmission Δ_{TF}, and the *closed-loop* power
- adjustment command f_c.

The NR formula is constructed according to the same logic as above, however, it takes into account the beam domain.

$$P_{PUSCH,b,f,c}(i,j,q_d,l) = \min$$
$$\begin{cases} P_{CMAX,f,c}(i), \\ P_{0\_PUSCH,b,f,c}(j) + 10\log_{10}(2^{\mu} \cdot M_{RB,b,f,c}^{PUSCH}(i)) \end{cases}$$
$$+ \alpha_{b,f,c}(j) \cdot PL_{b,f,c}(q_d) + \Delta_{TF,b,f,c}(i) + f_{b,f,c}(i,l) \} \, [dBm] \qquad (3.8)$$

Looking at the LTE and NR formulas, we note the larger amount of indexing in the NR formula. Indeed, this is perhaps a NR particularity in general, the NR standard is constructed in a more modular way compared to LTE. While in LTE the PUSCH power control was applied per cell c in a subframe i, in NR the power control has higher resolution, being applied in addition to the LTE dimensions, also to the active UL BWP b, for a carrier f, using a parameter set configuration j while the PUSCH power control adjustment state can have an index l. One similarity to LTE is that the power control is applied to the PUSCH bandwidth expressed as the number of resource blocks while in addition, in NR it is scaled with the configured subcarrier spacing μ as multiple subcarrier spacings are supported in 5G. The parameter j allows the possibility of creating multiple power control options depending on the UL channel which is transmitted: scheduled or grant-free PUSCH, random access Message 3; or allowing multiple pairs of PUSCH open-loop power control options associated with SRI, as we will describe later in this section.

The essence of the NR beam-based power control is reflected in the configuration and computation of the pathloss component PL. As the formula states, PL is calculated based on an indicated reference signal index q_d, for an active DL BWP of a serving cell c. In the absence of a reference signal index, the UE would utilize an RS resource from the SS/PBCH block that the UE is using for MIB reception. On the other hand, several reference signal indices may be provided, and the respective set may contain both SS/block and CSI-RS resource index. In other words, the PUSCH pathloss reference may map to a SS/PBCH block index or to CSI-RS. This is illustrated in Figure 3.113 where the UE is provided with a set of possible reference signals used for PL calculation, and it is also indicated a specific reference signal index q_d, which is applied in that particular cell/subframe selection. Hence, for PUSCH power control, the UE would perform measurements for a set of multiple beams, being indicated the transmission beam for which the PUSCH transmission is performed

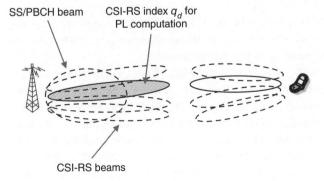

Figure 3.113 Configuration of SS/PBCH and CSI-RS and indication of reference signal index q_d for pathloss computation.

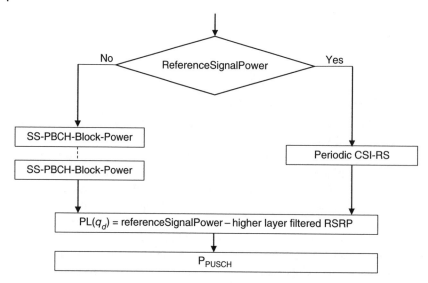

Figure 3.114 Pathloss Computation for PUSCH power control.

and the subsequent PL measurement which needs to be applied. The pathloss computation is the only parameter of the power control formula which is beam dependent. Note that the reference signals which are configured for PL computation are updated based on higher layer indication, hence SS/PBCH block indices or CSI-RS may be added or removed from the set of reference signals used in this procedure.

The pathloss PL is based on the difference between the higher layer indicated *referenceSignalPower* and the higher layer filtered RSRP computed on the indicated reference signal index:

$$PL_{f,c}(q_d) = referenceSignalPower - \text{higher layer filtered RSRP}$$

The above computation rules is based on the configuration of periodic CSI-RS used for the computation of *referenceSignalPower*, in the absence of CSI-RS the UE using the SS/PBCH-block power, a parameter provided by the higher layers, as shown in the diagram from Figure 3.114

The RSRP *referenceSignalPower* is defined [4, 38.215] as the linear average over the power contributions (in [W]) of the resource elements that carry either secondary synchronization signals or CSI-RS occasions, depending on the configuration.

There is yet another important component of the PUSCH power control formula and this is based on the configuration of SRI (SRS Resource Indicator which can be seen in a non-formal way as *an UL beam indicator*). Similar to LTE, the PUSCH power control formula contains an adjustment term $f_{b,f,c}(i,l)$. If a single parameter configuration j is provided, the UE would use close loop power control. When SRI configuration is available, hence the UE is provided with SRS beam indication information, the power control adjustment is possible per SRI/beam, hence the mapping of closed loop processed is done to each state of the SRI. Explained more freely, the UE would make an association between the DL reference signals used for the PL computation and the SRI, as there is an assumption of beam correspondence between the two. When the network indicates an SRI in the DCI, the UE would automatically use the corresponding computed PL.

Readers Take

- NR power control retains similar LTE elements of open loop and closed loop power control.
- The NR power control has higher resolution, being applied in addition to the LTE dimensions of cell c and subframe i, also to the active UL BWP b, for a carrier f,
- Multiple configurations j are allowed in order to configure power for scheduled or grant-free PUSCH, or for random access Message 3.
- Multiple closed-loop parameter sets controlled by parameter l are allowing the association of SRI to gNB TPC commands.
- A reference signal index q_d allows for the selection of the reference signal used for PL computation from a set of configured SS/PBCH blocks or CSI-RS.

3.12 DL and UL Transmission Framework (Mihai Enescu, Nokia, Karri Ranta-aho, Nokia Bell Labs, Finland)

The information originating from the higher layers is carried in downlink and uplink physical layer based on resource elements. A number of resource elements are grouped to form downlink and uplink physical channels. For downlink, these channels are PDSCH, PDCCH and PBCH. In uplink, the physical channels are defined as PUSCH, PUCCH and PRACH. That is, the physical layer channelization relies on the time/frequency resource element grid introduced in Section 3.4.

3.12.1 Downlink Transmission Schemes for PDSCH

In downlink, a single transmission scheme is defined for the PDSCH, and it is used for all PDSCH transmissions. This is based on the use of DM-RS as reference signals, while various CSI feedback options are enabling different levels of closed loop transmission (described in Section 4.3), still under a single transmission scheme strategy. Up to 8 DM-RS ports can be used for Single user MIMO (SU MIMO) while up to 4 DM-RS ports can be used per UE for Multi user MIMO (MU MIMO). This means that in SU MIMO the maximum transmission rank may be 8 while in MU MIMO the maximum transmission rank may be 4. For MU MIMO, the maximum number of orthogonal DM-RS ports is set to 12. This leads to different MU MIMO multiplexing possibilities, ranging from 12 UEs each having a single layer, up to 3 UEs each having four layers.

One may wonder why open loop transmission is not supported in 5G NR, after all, transmit diversity is a very good technique for providing coverage in LTE. Such techniques have been considered in the early standardization discussions, but it has been concluded that for example precoding cycling can be supported in a standard transparent way, in which different precoders can be applied on DM-RS in frequency domain and utilizing a low amount of bundled PRBs. Hence while there is no specification support, it is still a possibility for such transparent transmission.

3.12.2 Downlink Transmit Processing

3.12.2.1 PHY Processing for PDSCH

The 5G NR transport channel processing is very similar to what we have in LTE. For DL transmission the Downlink Shared Channel (DL-SCH) transport channel is processed for mapping to the PDSCH physical channel. In Figure 3.115 the main building blocks of the DL transport channel processing chain are illustrated. CRC attachment functionality is applied to one or two transport blocks which are sent by the Medium Access Control (MAC) layer. This is followed by the channel coding, which is done based on LDPC, a main difference from LTE. The number of coded bits are adjusted to the scheduled resources during the rate matching procedure. These bits are then further scrambled and modulated, possibilities being QPSK, 16 QAM, 64 QAM and 256 QAM with 1024 QAM a potential enhancement in the future. The modulated symbols are mapped to the physical resources, with a first stage of layer mapping which maps the up to two codewords to different layers (up to 8 in NR), followed by precoding, resource block mapping and antenna port mapping. The codeword to layer mapping is slightly different than in LTE. While in LTE the two codewords could be mapped to the eight layers when the number of layers was larger than two, in 5G NR this split is possible only when the number of layers is larger than four. In other words, a single codeword is mapped to up to four layers, while two codewords are used when the

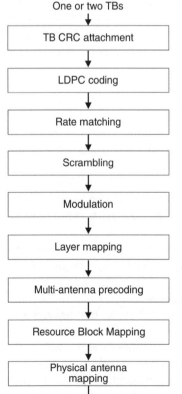

Figure 3.115 Transport channels processing in downlink.

number of layers is larger than four, the mapping being done in a similar way as when a single codeword is transmitted.

3.12.2.2 PHY Processing for PDCCH

The main purpose of the PDCCH is to schedule the transmission of PDSCH in downlink and the transmission of PUSCH in uplink, while it is also used for some other purposes too. The DCI is the entity of the PDCCH which provides in a dynamic way the indication of control information to the UE, in other words, the PDCCH is read by the UE before it knows where to find and demodulate the PDSCH and how to decode it, or how to encode the PUSCH and where and how to transmit it. There are two important aspects of PDCCH, the resources used to deliver the control information to the UE, which are based on the so called CORESETs, and the information it contains, which is referred as DCI formats.

As it was the case of LTE, to find the control information, the UE monitors a set of monitoring occasions in one or more CORESETs. This is done according to a search space configuration. Similar to LTE, CORESETs can be found in one of the first three OFDM symbols. A CORESET is formed by a set of Resource Element Groups (REGs), which are groups of resource elements carrying control information, these being further mapped into logical Control Channel Elements (CCEs) carrying 108 bits modulated by QPSK and coded based on polar coding. Each CORESET has an associated CCE-to-REG mapping which can be interleaved or non-interleaved.

While there are quite a few interesting details of the CORESETs design, we would like to focus next on CORESET types and the information delivered by CORESETs. Multiple CORESETs can be configured in a UE using higher layer signaling (RRC), however, the first control information obtained by the UE is in the form of CORESET#0 and the CORESET#0 configuration is delivered by MIB received on PBCH.

The control information is grouped in DCIs, where several formats having various payload sizes and embedding different functionalities have been designed for DL and UL, these are summarized in Table 3.31.

Components of DCI formats are summarized in Table 3.32.

Table 3.31 DCI formats in 5G NR Release 15.

DCI format number	DCI usage
0_0	scheduling PUSCH (fallback format)
0_1	scheduling PUSCH
1_0	scheduling PDSCH (fallback format)
1_1	scheduling PDSCH
2_0	Slot format indicator
2_1	Preemption indicator
2_2	TPC command for PUSCH/PUCCH power control
2_3	TPC commands for SRS power control

Table 3.32 Information carrier by the data-scheduling DCI formats.

Field	Uplink scheduling DCI format		Downlink scheduling DCI format	
	0_0	0_1	1_0	1_1
Freq domain resource assignment	Always present, size depending on BWP size and FD-RA configuration			
Time domain resource assignment	Always present, size depending on number of different TD-RA possibilities configured			
Carrier Indicator for cross-CC scheduling	—	If configured	—	If configured
Bandwidth par Indicator for BWP switch	—	If configured	—	If configured
Frequency hopping flag	If configured	If configured	—	—
Modulation and coding scheme		Always 5 bits		Fields duplicated for 2 TBs (>4 layer MIMO Tx)
New Data Indicator		Always 1 bit		
Redundancy Version		Always 2 bits		
HARQ process ID	Always 4 bits			
DL assignment index	—	Always present	Always present	Always present
TPC command for PUSCH	Always present	Always present	—	—
TPC command for PUCCH	—	—	Always present	Always present
Supplemental Uplink indicator	—	If configured	—	—
SRS resource indicator	—	If configured	—	—
Precoding info and number of layers	—	Present with code-book based UL MIMO	—	—
Antenna ports	—	Always present	—	Always present

Field	Col 1	Col 2	Col 3
SRS request	—	Always present	Always present
CSI request	—	If configured	—
Code Blog Group Transmission info	—	Present if CBG configured	Present if CBG configured
Code Blog Group flusing out info	—	—	Present with CBG if configured
PTRS port to DMRS port association	—	Present with UL MIMO with PTRS	—
Beta offset indicator for HARQ-ACK and CSI multiplexing on PUSCH	—	If configured	—
DMRS sequence initialization	—	1 bit with CP-OFDM	—
UL SCH indicator to request CSI without UP data	—	Always present	Always present
VRB-to-PRB mapping	Always present	—	If configured
PUCCH resource indicator	Always present	—	Always present
PDSCH-to-HARQ-ACK timing indicator	Always present	—	Present if needed
PRB bundling size indicator	—	—	If configured
Rate matching indicator	—	—	If configured
ZP CSI-RS trigger	—	—	If configured
Transmission configuration indicator	—	—	If configured

3.12.3 Uplink Transmission Schemes for PUSCH

In uplink the situation is a bit different from downlink in the sense that two transmission schemes are possible in NR, codebook and non-codebook-based transmission. The UE knows which type of transmission to use based on indication from the network which, based on a higher layer parameter "*TxConfig*" can select one of the two uplink transmission schemes. In both methods the gNB has various degrees of involvement in controlling how the UL transmission is performed. Yet, the two techniques are complementing each other as we will describe next. Up to four layers are possible to be transmitted in uplink and the use of transform precoding implies the transmission of a single layer.

3.12.3.1 Codebook Based UL Transmission

In codebook-based transmission the gNB has a tight control on the choice of precoding done by the UE. Codebook based operation would also fit better in cases where channel reciprocity is not sustained. In Figure 3.116 we summarize the procedures performed by gNB and UE in this situation. The UE would transmit several SRS resources to the gNB (in Figure 3.116 the second SRS is identified as best SRS) which estimates the UL channel based on these resources. Based on these measurements, the gNB is able to compute the best SRS, which is identified by the SRI (SRS Resource indication) and also the rank and precoder (TPMI, in UL the PMI is denoted by Transmit PMI to differentiate from the DL ones) which should be used by the UE for the transmission. This precoder is selected from a predefined UL codebook over the SRS ports in the selected SRS resource by the SRI. At the end of all these processing stages, the gNB has: the SRI, TPMI and rank, which can be used by the UE for the UL transmission. If a single SRS resource is configured, no SRI needs to be indicated to the UE, the TPMI is used to indicate the preferred precoder over the SRS ports in the configured single SRS resource. The TPMIs are supposed to be wideband. While frequency selective TPMIs might provide further performance improvement, the cost of signaling such information would be tremendously high as the baseline wideband information would be factored by the amount of sub-bands. The gNB indicates to the

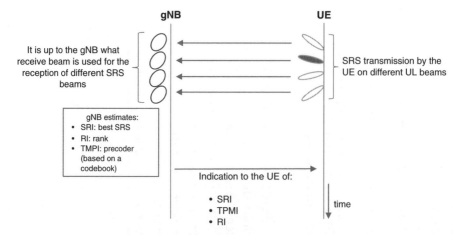

Figure 3.116 Diagram for codebook based UL transmission procedure.

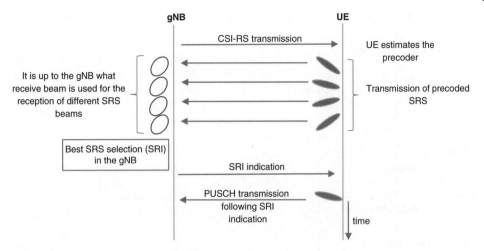

Figure 3.117 Diagram for non-codebook based UL transmission procedure.

UE the SRI, TPMI and hence the UE has now the necessary information for performing codebook-based UL transmission.

Under the codebook-based operation, three forms of transmission are possible in UL:

- *Full coherence*: where all the ports can be transmitted coherently.
- *Partial coherence*: where port pairs can be transmitted coherently. This can happen in the situation when the UE has calibrated antenna ports within a panel but also have multiple panels not calibrated to each other. The coherent transmission happens inside a panel where an SRS resource is configured as such.
- *Non-coherent*: where no port pairs can be transmitted coherently. In this case the transmit chains of the UE operates in a more independent fashion and there is no phase adjustment between them.

3.12.3.2 Non-Codebook Based UL Transmission

Non-codebook-based UL transmission fits more channel reciprocity and in some sense it is "lighter" with respect to the operation chain. In Figure 3.117 we summarize the procedures performed by gNB and UE in this situation. The UE computes the precoder based on the estimation of the CSI-RS. One or multiple SRS are precoded and transmitted in UL while the gNB estimates the best received SRS and its corresponding SRI which is further indicated to the UE. In other words, the UE knows the best beam which can be further used for UL transmission following such SRI indication.

3.12.4 Uplink Transmit Processings

3.12.4.1 PHY Processing for PUSCH

The uplink transport channel (UL-SCH) processing is similar with the downlink, however with a few differences. A single codeword is mapped to up to four layers and when transform precoding is used (hence when operating with single carrier DFT-S-OFDM waveform), the transmission is restricted to single layer only. As we can see, the uplink operation is to some

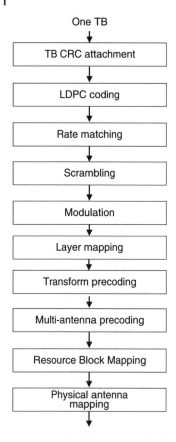

One TB

TB CRC attachment

LDPC coding

Rate matching

Scrambling

Modulation

Layer mapping

Transform precoding

Multi-antenna precoding

Resource Block Mapping

Physical antenna mapping

Figure 3.118 Transport channels processing in uplink.

extent a scaled down version of the downlink, the main reason being the complexity which naturally is different when we compare the gNB with the UE processing capability. For coverage reasons, and also to achieve low PAPR transmission, the single carrier DFT-based operation is also supported in UL, in addition to OFDM. The building blocks of the processing chain are illustrated in Figure 3.118.

3.12.5 Bandwidth Adaptation

3.12.5.1 Overview

As discussed in Section 3.4, the way the different carrier bandwidths in LTE (1.4, 3, 5, 10, 15, and 20 MHz) are supported with a common 1.4 MHz PSS, SSS, and PBCH structure can be seen as the base on which the NR bandwidth flexibility was defined. In LTE, the UE doing initial cell search is searching for the LTE's equivalent of NR's SS/PBCH block. Both the LTE and NR have the same three building blocks, PSS, SSS, and PBCH used for cell search and acquisition. In LTE these are confined within a 1.4 MHz bandwidth, no matter what the LTE cell's bandwidth is. Only after reading the MIB on the PBCH the UE knows what the actual cell bandwidth is. The one downside of this way of operating the system is that if the UE is not able to support the carrier bandwidth of the cell, then the UE cannot operate in that cell. Thus, all the mentioned bandwidths were mandatory (subject to a given frequency band) for all the UEs from the very beginning of LTE. This caused some difficulties later down the road when low cost devices with narrower supported RF bandwidths were introduced and was also somewhat rigid in other respects. One could say that it also prevented introduction of wider bandwidth carriers, but there were other design related obstacles that made carrier aggregation anyway a more natural path for wider bandwidths than specifying wider carrier bandwidth.

In NR the following concepts contribute to the bandwidth flexibility and bandwidth adaptation

- UE specific and adaptable UE specific BWP
- No link between the actual operating bandwidth of the uplink and the link
- Ability to configure the frequency location of the downlink and uplink control channels
- No fixed link between the frequency location of the SS/PBCH block and the other channels.

All in all, the NR bandwidth adaptation can be used at least for

- Supporting UEs with different bandwidth capabilities in a wide bandwidth cell
- Reducing the UE power consumption
- Improved spectrum flexibility over LTE, e.g. for avoiding harmful interference to other systems with better granularity than specified carrier bandwidths offer.

3.12.5.2 Support for Narrow-Band UE in a Wide-Band Cell

In NR, the initial operation is very similar to the LTE. As discussed in Section 3.5, the SS/PBCH block has a structure that is not tied to the cell's operating bandwidth, although the operating band does define the subcarrier spacing the SS/PBCH block is using, and thus also its occupied carrier bandwidth. The UE, after reading the MIB from the PBCH receives information on how to read the PDCCH scheduling SIB1, but it does not yet know the actual bandwidth with which the cell operates, nor does the network yet know what bandwidths the UE supports.

The MIB provides configuration for the PDCCH to schedule SIB1. The frequency location and bandwidth of this PDCCH determines the initial bandwidth the UE and the gNB communicate with. With this the UE is able to provide the network with its capabilities, including what bandwidths the UE is able to support on any given band it can operate with. After this step the rest is easy; the UE is provided with a BWP to operate on, or it keeps on operating with the initial BWP, and as discussed in Section 3.4, one UE's BWP is independent of what some other UE's BWP is. All the downlink operation is confined within the UE's active BWP, and the same goes for the uplink. This allows for accommodating UEs with different bandwidth capabilities in a cell that is supporting a wider carrier bandwidth.

3.12.5.3 Saving Battery with Bandwidth Adaptation

As it happens, the bandwidth adaptation can be used for other purposes as well as just supporting UEs with different capabilities. One important reason for being able to adapt the UEs operating bandwidth is to reduce UE's power consumption. This can be achieved by changing the active BWP to a different, narrower one, when the UE is not in need of the full bandwidth data services. This allows for the UE to reduce the ADC sampling rate,

Table 3.33 RNTI fields used in 5G.

RNTI	Value (hexa-decimal)	Usage	Description
N/A	0000		
RA-RNTI	0001–FFEF	Random Access Response	If RA-RNTI is configured, the UE needs to monitor Type-1 PDCCH common search space with CRC scrambled by RA-RNTI in the configured set, in response to a PRACH transmission (described in TS 38.213).
Temporary C-RNTI	0001–FFEF	Contention Resolution (when no valid C-RNTI is available)	
Temporary C-RNTI	0001–FFEF	Msg3 transmission	This is used for identifying the Random Access Response (RAR), known also as message 3 (msg3)
C-RNTI	0001–FFEF	Dynamically scheduled unicast transmission	If CS-RNTI is configured, the UEs receives a DCI format indicating that the UE shall receive PDSCH/PUSCH or CSI-RS in the set of symbols of the slot. (described in TS 38.213)
C-RNTI	0001–FFEF	Dynamically scheduled unicast transmission	If C-RNTI is configured, it provides to each UE a unique identifier for
C-RNTI	0001–FFEF	Triggering of PDCCH ordered random access	dynamically scheduled unicast transmission.
CS-RNTI	0001–FFEF	Configured scheduled unicast transmission (activation, reactivation, and retransmission)	If CS-RNTI is configured, the UEs receives a DCI format indicating that the UE shall receive PDSCH/PUSCH or CSI-RS in the set of symbols of the slot (described in TS 38.213)
CS-RNTI	0001–FFEF	Configured scheduled unicast transmission (deactivation)	
INT-RNTI (Interruption RNTI)	0001–FFEF	used to indicate pre-emption in DL. It is configured in the *DownlinkPreemption* IE	If INT-RNTI is configured, the UE needs to monitor Type-3 PDCCH common search space for DCI format 2_1 with CRC scrambled by INT-RNTI in the configured set of the serving cells. If a UE detects a DCI format 2_1 for a serving cell from the configured set of serving cells, the UE may assume that no transmission to the UE is present in PRBs and in symbols, from a set of PRBs and a set of symbols of the last monitoring period, that are indicated by the DCI format 2_1. (described in TS 38.213)
SFI-RNTI (Slot Format Indication RNTI)	0001–FFEF	used to indicate slot format. It is configured in the *SlotFormatIndicator* IE	If SFI-RNTI is configured, the UE needs to monitor Type-3 PDCCH common search space for DCI format 2_0 with CRC scrambled by SFI-RNTI, which indicates a slot format for each slot in a number of slots for each DL BWP or each UL BWP of a serving cell (described in TS 38.213)

Table 3.33 (Continued)

RNTI	Value (hexa-decimal)	Usage	Description
SP-CSI-RNTI (Semi-Persistent CSI RNTI)	0001–FFEF	used to activate semi-persistent CSI reporting on PUSCH. It is configured in the *CSI-ReportConfig* IE	Fr a set of symbols of a slot that are indicated by higher layer parameters or DCI format 2_0, the UE can only perform semi-persistent CSI report in the set of symbols when a DCI format with CRC scrambled by SP-CSI-RNTI is detected (described in TS 38.213).
TPC-CS-RNTI (Transmit Power Control-Configured Scheduling-RNTI)	0001–FFEF	used to control the uplink power of configured scheduling. This RNTI is only specified in TS 38.321. There is no definition in other specifications.	
TPC-PUCCH-RNTI	0001–FFEF	PUCCH power control	If TPC-PUCCH-RNTI is configured, the UE needs to monitor Type-x PDCCH [common] search space for DCI format 1_0, 1_1, or 2_2 with CRC scrambled by TPC-PUCCH-RNTI (As described in TS 38.213)
TPC-PUSCH-RNTI	0001–FFEF	PUSCH power control	If TPC-PUSCH-RNTI is configured, the UE needs to monitor Type-x PDCCH [common] search space for DCI format 0_0, 0_1, or 2_2 with CRC scrambled by TPC-PUSCH-RNTI (described in TS 38.213).
TPC-SRS-RNTI	0001–FFEF	SRS trigger and power control	If TPC-SRS-RNTI is configured, the UE needs to monitor Type-x PDCCH [common] search space for DCI format 2_3 with CRC scrambled by TPC-SRS-RNTI (described in TS 38.213).
Reserved	FFF0–FFFD		
P-RNTI		Paging and System Information change notification	If P-RNTI is configured, the UE needs to monitor Type-2 PDCCH common search space on a primary cell (described in TS 38.213)
SI-RNTI	FFFF	Broadcast of System Information	If SI-RNTI is configured, the UE needs to monitor Type-0 PDCCH common search space on a primary cell (described in TS 38.213).

which translates to saved energy while still maintaining an active connection even if the scheduling bandwidth may be restricted from what the full carrier could support. 3GPP Release 16 is investigating further power saving techniques, but the ability to switch the UE to a narrower bandwidth is the main Release 15 energy saving feature for connection that is fully active, as the UE is still constantly schedulable. DRX operation or RRC inactive state can only be applied when the UE is not continuously exchanging data with the network. For example, a UE in an 100 MHz NR carrier could be configured to a 10 MHz BWP, reducing its maximum data rate to 10% of what could be achieved. This may feel like a big cut in the data rate, but this is still a typical LTE carrier bandwidth that the UE can be served if the needed data rates don't require the high data rates supported by the full carrier bandwidth.

3.12.5.4 Spectrum Flexibility

Another problem with the LTE's somewhat rigid cell bandwidth regime was problematic when the operator's available bandwidth was not fitting to one of the supported LTE carrier bandwidths. The number of different carrier bandwidths all need to have specific OBE requirements to protect the spectrum outside the carrier. In addition, there are in-band emission requirements to ensure that the different UEs' transmissions stay within the scheduled transmit allocation after the RF impairments, or that the UE receiving a particular scheduling allocation is not troubled with another UE receiving its transmission on a neighboring part of the carrier with a very different power level.

Configuring a cell for a given nominal carrier bandwidth provides out-of-band guarantees for both harmful interference from neighboring part of the spectrum as well as guarantees the neighbor from interference generated by the cell. However, it is entirely possible to e.g. have a very sensitive or very interfering neighbor, that requires more guard band. This can be achieved by leaving part of the carrier closest to the troublesome neighbor unused. In LTE this was not possible in the downlink as the PDCCH was always full-band, and in uplink it was somewhat cumbersome as the PUCCH was hopping from carrier edge to carrier edge. It was possible to avoid the cell edge, but then you had to have the same margin on both sides of the uplink carrier. In NR the PDCCH and PUCCH, as well as the data channels are confined to whichever part of the carrier the configuration tells them to be, and the BWP concept can be used for this purpose as well.

The ability to flexibly place the PDCCH and the PUCCH anywhere within the carrier also provides the needed flexibility to protect the important control channels from the potentially harmful interference of a problematic neighbor, and still leaves the possibility to schedule the HARQ-protected data channels on the full bandwidth, if that is seen as a good way to operate in the given spectrum band, be that problematic neighbor the operators own system using some other radio technology, or some other system altogether.

3.12.6 Radio Network Temporary Identifiers (RNTI)

Specific radio channels are identified by the means of RNTIs (Radio Network Temporary Identifiers). The 5G RNTI is based on 16 bits and are summarized as in Table 3.33.

4

Main Radio Interface Related System Procedures

Jorma Kaikkonen[1], Sami Hakola[1], Emad Farag[2], Mihai Enescu[1], Claes Tidestav[3], Juha Karjalainen[1], Timo Koskela[1], Sebastian Faxér[3], Dawid Koziol[4], and Helka-Liina Määttänen[5]

[1] Nokia Bell Labs, Finland
[2] Nokia Bell Labs, USA
[3] Ericsson, Sweden
[4] Nokia Bell Labs, Poland
[5] Ericsson, Finland

4.1 Initial Access (Jorma Kaikkonen, Sami Hakola, Nokia Bell Labs, Finland, Emad Farag, Nokia Bell Labs, USA)

4.1.1 Cell Search

In cell search procedure the User Equipment (UE) acquires the time and frequency synchronization to a cell, and determines the physical layer cell ID. The UE does this by searching for primary synchronization signal (PSS) and secondary synchronization signal (SSS) and decoding Physical Broadcast Channel (PBCH) carried by synchronization signal (SS)/PBCH block, described in Section 3.5.

Similar to other radio access systems, the UE looks for the SSs in predefined frequency locations. For initial cell selection purposes, the valid frequency locations for the SS/PBCH blocks are determined by synchronization raster given in [28, 31]. UE uses these locations to look for SS/PBCH blocks and system acquisition when it has not been provided any other information regarding frequency location of the SS/PBCH blocks. For both frequency ranges, FR1 and FR2, a global synchronization raster is defined via global synchronization channel number (GSCN), which corresponds to a given frequency position of the SS/PBCH block. The mapping between GSCN and the actual frequency position depends on the frequency range. For FR1, to accommodate different Radio Frequency (RF) channel raster spacings, GSCN maps to absolute frequency positions in a clustered manner (i.e. locations at offset of {50, 150, 250 kHz} at every 1.2 MHz), while for FR2 the frequency locations are evenly spaced, every 1.44 MHz. For each frequency band a sub-set of GSCN locations are valid. The GSCN location points to the center frequency of the SS/PBCH block. Like in earlier radio access technologies (RF) channel raster is also determined. For new radio (NR) the channel raster is dependent of the frequency range, so that for bands where co-existence with LTE is required, 100 kHz raster is used, while for other bands it depends of the applied

5G New Radio: A Beam-based Air Interface, First Edition. Edited by Mihai Enescu.
© 2020 John Wiley & Sons Ltd. Published 2020 by John Wiley & Sons Ltd.

sub-carrier spacing (SCS) (e.g. {15, 30 kHz}). Corresponding to GSCN, a new radio absolute radio frequency channel number (NR-ARFCN) is defined in 38.101 [28, 31], mapping to RF reference frequency for different frequency ranges. While ARFCN is used to indicate different frequency locations in NR, channel raster itself is not signaled to the UE. It however determines the valid placements for the NR carriers.

As can be seen, in NR the synchronization raster is sparser than the channel raster, alleviating the cell search complexity at the UE. To enable this the SS/PBCH block is not required to be placed to the center of a carrier, nor to the same common resource block (RB) grid as data, but the sub-carrier locations are aligned between the SS/PBCH block and the common RB grid. The necessary information to determine the common RB grid is provided by Master Information Block (MIB) and SIB1, as noted in Section 3.5. Additionally, as discussed in Section 3.5, the default subcarrier spacing(s) for a given frequency band are given in [28, 31].

4.1.1.1 SS/PBCH Block Time Pattern

As a disruptive functionality compared to earlier cellular generations, NR supports beamforming for the broadcast signaling like SSs, PBCH, system information delivery including both SIB1 and other system information and paging. The use of beamformed signals for initial access is due to the need of scaling the system across the carrier frequencies, and, as explained in previous chapters, due to the nature of mandatory beamforming in higher carriers. For the SS/PBCH block carrying the SSs and PBCH, the beamforming is supported by having multiple candidate SS/PBCH blocks and corresponding time-domain locations within a half-frame (5 ms). During the half-frame, different SS/PBCH blocks may be transmitted in different spatial directions (i.e. using different beams) spanning the coverage area of a cell. The time-domain candidate locations of the SS/PBCH blocks, i.e. mapping pattern into the slots within a half-frame is determined by the SCS.

The SS/PBCH block time patterns are named cases A–E and shown in Figure 4.1. All the other SCSoptions than 30 kHz have a single mapping pattern defined. Typically, two SS/PBCH candidate locations are mapped within a slot, except for the 240 kHz SCS option where four locations are mapped across two slots. For 30 kHz SCS option, two different patterns are defined, Case B and Case C. Case B is typically used in bands where co-existence with LTE is required. With Case A and Case C it would be possible to support a bi-directional slot with both DL and UL control (at a start and end of the slot) by multiplexing with higher SCS (i.e. 30 or 60 kHz), e.g. for low latency services. Correspondingly, the SS/PBCH block pattern of 120 kHz (Case D) and 240 kHz (Case E) would allow multiplexing both DL and UL control with 60 and 120 kHz numerology to the slots carrying/overlapping with SS/BPCH block transmission. Slot format.

The maximum number of SS/PBCH block positions within a 5 ms half-frame, L_{max}, depends also on the carrier frequency range as illustrated in Table 4.1.

The SS/PBCH blocks are indexed in an ascending order in time within a half frame from 0 to $L_{\text{max}} - 1$. A UE determines the 2 LSB bits, for $L_{\text{max}} = 4$, or the 3 LSB bits, for $L_{\text{max}} > 4$, of an SS/PBCH block index per half frame from a one-to-one mapping with and index of the DeModulation Reference Signals (DM-RS) sequence transmitted in the PBCH. For $L_{\text{max}} = 64$, the UE determines the 3 MSB bits of the SS/PBCH block index per half frame from the PBCH payload bits.

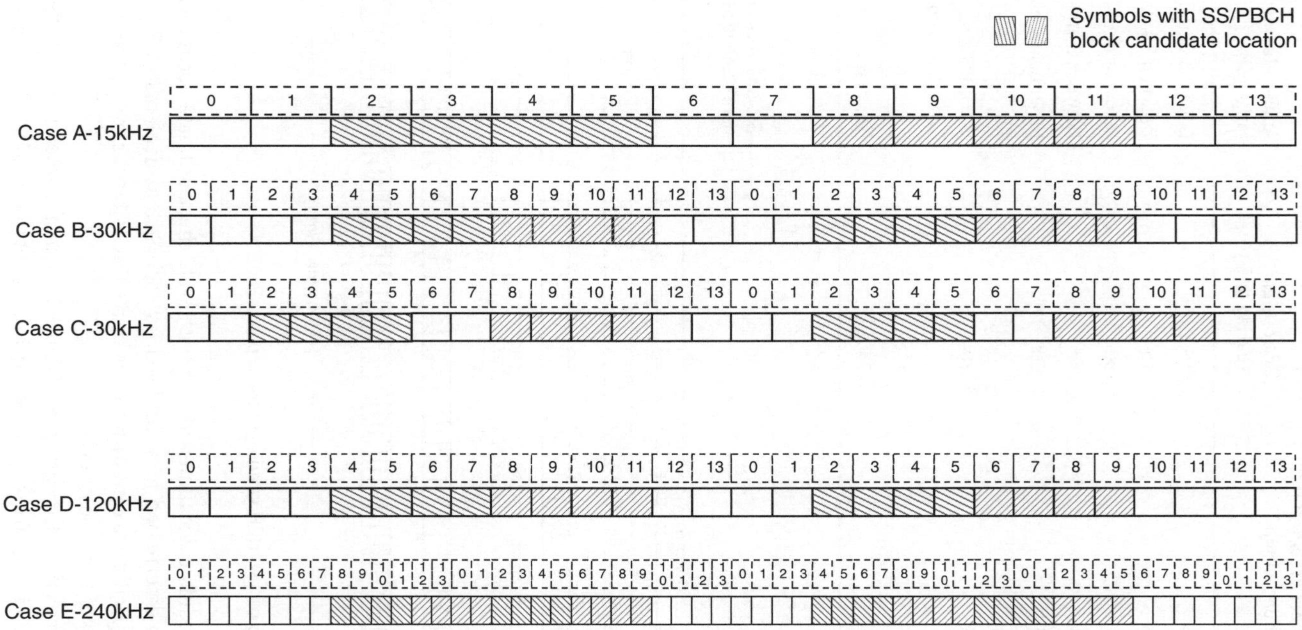

Figure 4.1 Mapping of SS/PBCH candidate locations to symbols in slot(s).

Table 4.1 Maximum number of candidate SS/PBCH block positions (L_{max}) per carrier frequency range.

Pattern case and number of SS/PBCH block candidate locations within a 5 ms half-frame	Subcarrier spacing (kHz)	Applicable frequency range (GHz)
Case A and $L_{max} = 4$	15	Below or equal to 3
Case A and $L_{max} = 8$	15	Between 3 and 6
Case B and $L_{max} = 4$	30	Below or equal to 3
Case B and $L_{max} = 8$	30	Between 3 and 6
Case C and $L_{max} = 4$	30	FDD: Below or equal to 3 TDD: Below or equal to 2.4
Case C and $L_{max} = 8$	30	FDD: Between 3 and 6 TDD: Between 2.4 and 6
Case D and $L_{max} = 64$	120	Above 6
Case E and $L_{max} = 64$	240	Above 6

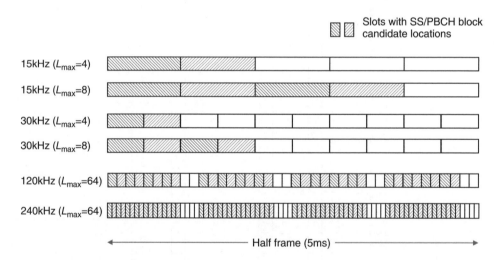

Figure 4.2 Mapping of SS/PBCH candidate locations to slot.

Similarly, to candidate locations in symbol level, the candidate locations within a half-frame are also determined in a slot level for each SCS as depicted in Figure 4.2. Up to the frequency specific maximum number of SS/PBCH blocks, L_{max}, the network can choose freely the used time locations.

For initial cell selection, the UE may assume that half frames with SS/PBCH blocks occur with a periodicity of 20 ms. After acquiring access to the cell, the UE can obtain the information on periodicity of the half-frames where SS/PBCH blocks are transmitted per serving cell (*ssb-periodicityServingCell*). In addition to the periodicity in serving cell, the UE can be provided with the Synchronization Signal/Physical Broadcast Channel block

Measurement Time Configuration (SMTC), based on which UE can expect to be able to perform measurements on the frequency layer. If the UE is not configured the periodicity, the UE assumes 5 ms periodicity. The UE may also assume that the periodicity is same for all SS/PBCH blocks in the cell and that only SS/PBCH blocks transmitted with same SS/PBCH block index (i.e. same location with the half-frame) on the same center frequency as quasi co-located.

For a serving cell without transmission of SS/PBCH blocks, a UE acquires time and frequency synchronization with the serving cell based on SS/PBCH blocks on the PCell, or on the PSCell, of the cell group for the serving cell.

4.1.1.2 Initial Cell Selection Related Assistance Information

In NR two types of cells/deployments were considered. Standalone cells for which UE may acquire access directly need to broadcast corresponding minimum system information, that is SIB1, in relation to at least one of the SS/PBCH blocks transmitted by the cell. For non-standalone deployments, the cells do not broadcast this information. For example, in case of multi-radio access technology (RAT) dual connectivity where the NR access is added as an additional RAT in addition to LTE RAT. For SS/PBCH blocks that do not carry minimum system information, some information elements (IEs) (related to providing the information to access SIB1) can be re-used. As discussed in Section 3.4, k_{SSB} can be used to indicate the lack of SIB1 related information and then *pdcch-ConfigSIB1* can be re-interpreted as an assistance information regarding where a SS/PBCH block with corresponding SIB1 can be (or cannot) be found. Based on the k_{SSB} value, *pdcch-ConfigSIB1* can indicate the offset to the next synchronization raster location (GSCN) where SS/PBCH block with information to obtain SIB1 can be found. Alternatively, *pdcch-ConfigSIB1* can indicate a range (down to 0) where UE can expect not to find any SS/PBCH block with SIB1.

It is good to note that as the amount of "always-on" type of signals is reduced in NR, down to SS/PBCH blocks, the periodicity of these can be sparse, power sensing type of initial cell search mechanism are not necessarily feasible, and UE may need to check each synchronization location raster individually. In certain cases the frequency bands can be rather wide, and number of bands to be considered maybe high, resulting that the number of synchronization raster locations the UE needs to check can be rather large, and as a consequence, the initial cell selection may be quite time and energy consuming.

4.1.2 Random Access

4.1.2.1 Introduction

The physical random access channel (PRACH) is described in Section 3.6. In this section we look at the higher layer and physical layer procedures for random access using PRACH.

Section 4.1.2.2 describes the higher layer random access procedures including contention-based random access (CBRA) and contention-free random access as well as Physical Dedicated Control Channel (PDCCH) order. Section 4.1.2.3 looks at the use cases and triggers of the random access procedure. In Section 4.1.2.4, we look at the physical layer random access procedures such beam management during initial access, beam failure recovery (BFR) and power control.

4.1.2.2 Higher Layer Random Access Procedures

There are two types of random access procedures; CBRA procedure, and contention free random access (CFRA) procedure. In the CBRA procedure, multiple users randomly select a preamble from a pool of preambles, a contention resolution phase determines which user, if any, had its data successfully received by the network. Contention resolution resolves contention between two or more users selecting the same preamble. In the CFRA procedure, a preamble is uniquely pre-allocated to the user, hence there is no possibility of preamble collision between users and no need for contention resolution.

4.1.2.2.1 *Contention-Based Random Access* CBRA is a four-step procedure as shown in Figure 4.3. In the first step, the UE randomly selects a preamble in the PRACH Occasion associated with the reference signal of the beam the UE wants to transmit on. This preamble is detected by the gNB receiver. The detection of the preamble involves determination of the presence of a preamble index in a PRACH Occasion, and estimation of the round trip delay of the detected preamble.

If the serving cell of the random access procedure is configured with supplementary uplink (SUL) carrier and the carrier used for random access is not explicitly signaled, the UE selects the carrier to use for the random access procedure based on the Reference Signal Received Power (RSRP) of the downlink reference signal. If that RSRP is less than higher layer parameter *rsrp-Threshold-SUL*, the UE selects the SUL carrier for the random access procedure, else the UE selects the normal uplink (NUL) carrier. Once the random access procedure starts, all uplink transmissions remain on the same carrier.

Random access preambles can be configured into two groups: group A and group B. If random access preambles group B is configured, the UE selects group B, if either of these conditions is true:

- The size of Msg3 exceeds the higher layer parameter *ra-Msg3SizeGroupA*, and the path loss is less than a value determined by the UE's maximum power and higher layer configuration parameters.

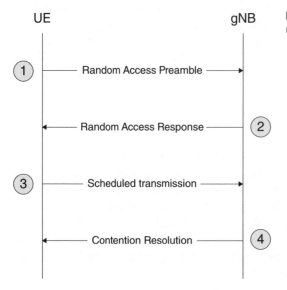

Figure 4.3 Four-step contention-based random access procedure 38.300 [10].

- The random access procedure is initiated for a logical channel and the Common Control Channel (CCCH) service data unit (SDU) size plus Medium Access Control (MAC) subheader is greater than *ra-Msg3SizeGroupA*.

In the second step, the gNB acknowledges the reception of the preamble by sending a random access response (RAR) on a PDSCH channel. The RAR is scheduled by a Downlink Control Information (DCI) with a CRC scrambled by the Random Access Radio Network Temporary Identifier (RA-RNTI) on the corresponding PDCCH. The PDCCH of the RAR is transmitted in Type1-PDCCH Common Search Space (CSS). The RA-RNTI is a function of the time and frequency of the PRACH occasion the preamble is detected on according to the following equation 38.321 [43]:

$$RA - RNTI = 1 + s_{id} + 14 \times t_{id} + 14 \times 80 \times f_{id} + 14 \times 80 \times 8 \times ul\_carrier_{id}$$

where,

- s_{id} is the symbol ID of the first OFDM symbol of the PRACH Occasion, where $0 \leq s_{id} < 14$.
- t_{id} is the slot ID of the first slot of the PRACH Occasion, the range of t_{id} depends on the SCS. The maximum range is $0 \leq t_{id} < 80$, with a SCS of 120 KHz.
- f_{id} is the frequency domain index of the PRACH Occasion, where $0 \leq f_{id} < 8$.
- $ul\_carrier_{id}$ is the uplink carrier ID of the uplink carrier of the PRACH occasion. 0 corresponds to the NUL carrier, and 1 corresponds to the supplemental uplink carrier.

The UE attempts to receive the RAR during the RAR window as shown in Figure 4.4. The RAR window starts at the first symbol of the earliest CORESET (control resource set) of Type1-PDCCH CSS that starts at least one symbol after the PRACH occasion corresponding to the PRACH transmission. The RAR window size is configured in number of slots and is less than 10 ms. In case of CBRA, the DMRS port of the PDSCH channel carrying the RAR and the DMRS port of the corresponding PDCCH channel are quasi-co-located with the reference signal used for association and transmission of the corresponding preamble.

The RAR MAC protocol data unit (PDU) transmitted in the PDSCH channel, consists of one or more subPDUs and optional padding. There are three types of RAR subPDUs

- A subPDU signaling the Backoff indicator. This consists of a MAC subheader for Backoff indicator. This subPDU should be at the beginning of the RAR MAC PDU if it is transmitted. Figure 4.5 shows the MAC subheader for Backoff Indicator of size 1 byte. The type field "T" is set to 0. The Backoff Indicator field "BI" is a four-bit field that signals the overload condition in the cell. "E" is the extension field, this is a one-bit flag indicating if this subPDU is the last subPDU in the MAC PDU or not. A value of "1" indicates that there are more subPDUs in the MAC PDU. A value of "0" indicates that this is the last subPDU of the MAC PDU. "R" is a reserved bit set to "0".

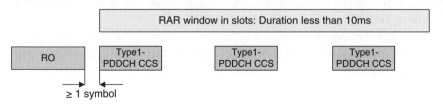

Figure 4.4 Random access response (RAR) window.

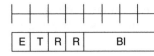

Figure 4.5 MAC subheader for backoff indicator.

T = 0 for MAC subheader
carrying the Backoff Indicator

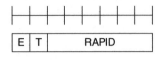

Figure 4.6 MAC subheader for RAPID.

T = 1 for MAC subheader
carrying the RAPID

Figure 4.7 MAC subPDU with subheader for RAPID and MAC RAR.

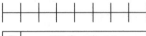

Figure 4.8 MAC random access response (RAR).

- A subPDU signaling the Random Access Preamble Identifier (RAPID) only. This subPDU is used to acknowledge System Information (SI) request. It consists of a MAC subheader for the RAPID. Figure 4.6 shows the MAC subheader for RAPID of size 1 byte. The type field "T" is set to 1. The "RAPID" field is a six-bit field that signals the preamble identifier corresponding to this MAC subPDU. The field "E" is as explained earlier in this section.
- A subPDU signaling the RAPID with a MAC RAR. This subPDU consists of a MAC subheader for RAPID followed by a MAC RAR as shown in Figure 4.7. The MAC subheader for the RAPID is as shown in Figure 4.6. The MAC RAR is seven bytes and is shown in Figure 4.8. Hence, the total size of this subPDU is eight bytes. The MAC RAR consists of the following fields:
  o "R" is a reserved bit set to 0.
  o Time advance command that is 12 bits. For a time advance value T_A, the time adjustment N_{TA} is given by:

$$N_{TA} = T_A \cdot 16 \cdot \kappa \cdot \frac{\Delta f_{ref}}{\Delta f_{sc}} \cdot T_c$$

Table 4.2 UL grant fields in random access response (RAR).

RAR grant field	Number of bits	Comment
Frequency hopping flag	1	If this flag is "0" PUSCH is transmitted without frequency hopping, else frequency hopping is used.
PUSCH frequency resource allocation	14	The frequency domain location (starting PRB) and size of the PUSCH transmission scheduled by the RAR UL grant, as well as the frequency hop offset if frequency hopping is configured.
PUSCH time resource allocation	4	Provides the slot, starting symbol, length and PUSCH mapping type of the PUSCH transmission scheduled by the RAR UL grant through an index to a default table, or a configured time-domain allocation list.
MCS	4	Use the first 16 entries of the MCS table. The selection of the MCS table depends on whether or not transform precoding is enabled for Msg3.
TPC command for PUSCH	3	Allowed power step size $\{-6, -4, -2, 0, 2, 4, 6, 8\}$ dB. -6 dB corresponds to b000, values are mapped consecutively till 8 dB which corresponds to b111.
CSI request	1	This field is reserved

where, $\kappa = 64$, Δf_{sc} is the SCS of the first uplink transmission after the RAR, $\Delta f_{ref} = 15 \, KHz$ is a reference SCS and $T_c = \frac{1}{(4096 \times 480 \, KHz)} = 0.5086$ ns is the basic time unit for NR. The range of T_A is 0 … 3846. The maximum time adjustment delay of the Tracking Area (TA) command when the SCS of the first uplink transmission after the RAR is 15 KHz corresponds to a cell with radius approximately 300 km (i.e. round-trip time of 2 ms).

o Uplink grant that is 27 bits. The fields of the uplink grant are given by Table 4.2. The uplink grant schedules the first PUSCH transmission after the RAR.

o Temporary C-RNTI (TC-RNTI) that is 16 bits. The TC-RNTI is used when scheduling Msg3, it is also used when scheduling Msg4 if the UE doesn't have a C-RNTI.

At the start of the random access procedure, the UE sets the *PREAMBLE_TRANSMISSION_COUNTER* to 1. If the UE doesn't receive the RAR during the RAR window with a RAPID that matches the transmitted preamble, it increments the *PREAMBLE_TRANSMISSION_COUNTER* by 1 and retransmits the PRACH preamble after a back off period. After several PRACH preamble retransmissions attempts, when the *PREAMBLE_TRANSMISSION_COUNTER* exceeds higher layer parameter *preambleTransMax*, without receiving a RAR with a RAPID that matches the transmitted preamble, the Random Access Channel (RACH) procedure fails.

When the UE successfully decodes a RAR with a RAPID that matches the transmitted preamble and that includes a MAC RAR, the UE performs the following:

– It processes the received TA command.
– It processes the UL grant in the MAC RAR, and in response transmits a PUSCH (a.k.a. Msg3). The power setting of the PUSCH transmission is determined as described in Section 4.1.2.4.4.

In the third step of the CBRA procedure, the UE transmits Msg3 on the Uplink Shared Channel (UL-SCH) on a PUSCH in response to the uplink grant of the RAR. Msg3 contains C-RNTI MAC CE if the UE triggering the CBRA procedure is in CONNECTED Mode with a C-RNTI, or it contains the CCCH SDU with the Contention Resolution Identity MAC CE. The trigger of the random access procedure, as described in Section 4.1.2.3, determines the content of Msg3. For example;

- During initial access, Msg3 contains *RRCSetupRequest*.
- During Radio Resource Control (RRC) re-establishment, Msg3 contains *RRCReestablishmentRequest*.
- To transition from the RRC_INACTIVE state to the RRC_CONNECTED state, Msg3 contains *RRCResumeRequest* or *RRCResumeRequest1*.
- To get system information, Msg3 contains *RRCSystemInfoRequest*.
- During contention-based BFR Msg3 contains C-RNTI MAC CE.

After the UE transmits Msg3 it starts the *ra-ContentionResolutionTimer* and monitors the PDCCH channel. The UE can receive a PDCCH with a CRC scrambled by the TC-RNTI requesting a retransmission of Msg3, if the gNB didn't successful decode Msg3. When the UE retransmits Msg3, it restarts the *ra-ContentionResolutionTimer* after the Msg3 retransmission.

In the fourth step of the CBRA procedure, the gNB transmits a message to the UE.

- If Msg3 contained the C-RNTI MAC CE, the gNB transmits a PDCCH with CRC scrambled by the C-RNTI. Upon reception of this PDCCH, the UE stops *ra-ContentionResolutionTimer* and considers the CBRA procedure successful.
- If Msg3 contained the CCCH SDU, the gNB transmits a PDCCH scheduling a PDSCH with CRC scrambled by the TC-RNTI, indicated to the UE in the RAR. The corresponding PDSCH echoes back the contention resolution identity received in Msg3. Upon reception of the PDCCH scheduling a PDSCH and with CRC scrambled by the TC-RNTI, the UE decodes the corresponding PDSCH. If the contention resolution identity received in the PDSCH matches that transmitted in Msg3, the UE stops *ra-ContentionResolutionTimer* and considers the CBRA procedure successful and sends uplink Hybrid Automatic Repeat Request (HARQ) acknowledgement to the gNB to stop any further retransmissions of this message. Else when the contention resolution identity received in the PDSCH doesn't match that transmitted in Msg3, the UE stops *ra-ContentionResolutionTimer* and considers the Contention Resolution unsuccessful. If the CBRA procedure is successful and it was not triggered by an SI request, the UE promotes the TC-RNTI to become the C-RNTI.

If the *ra-ContentionResolutionTimer* expires before the UE successfully receives the downlink message transmitted by the gNB, the UE considers Contention Resolution unsuccessful.

If Contention Resolution is unsuccessful, the UE increments the *PREAMBLE_TRANSMISSION_COUNTER* by 1 and restarts the random access procedure, with preamble selection, after a back off period. After several PRACH preamble retransmissions attempts, when the *PREAMBLE_TRANSMISSION_COUNTER* exceeds higher layer parameter *preambleTransMax*, without successful completing the random access procedure, the RACH procedure fails.

4.1.2.2.2 Contention Free Random Access In CFRA, shown in Figure 4.9, the preamble assignment is predetermined by the network. The preamble assignment determines:

– The preamble index.
– The mask index according to Table 4.4, to determine the PRACH Occasion(s) to use for preamble transmission within the set of PRACH occasions associated with the SS/PBCH Block or the channel-state information reference signals (CSI-RS) used for preamble transmission.

The UE transmits on a preassigned preamble index and PRACH occasion(s) associated with a reference signal (SS/PBCH block or CSI-RS) that exceeds a higher layer configured threshold *rsrp-ThresholdSSB* or *rsrp-ThresholdCSI-RS*.

As the preamble is preassigned by the network, there is no contention resolution phase. The random access procedure is considered successful after reception of a RAR with a RAPID that matches the transmitted preamble.

4.1.2.2.3 PDDCH Order The network can trigger the UE to initiate a random access procedure through a PDCCH order. The PDCCH order is trigger by DCI Format 1_0. The PDCCH order can trigger,

– A CFRA procedure, in this case the preamble index is provided by *ra_PreambleIndex* in PDDCH order and is not equal to b000000.
– A CBRA procedure, in this case the preamble index provided by *ra_PreambleIndex* in the PDCCH order is equal to b000000.

Table 4.3 shows the structure of the DCI of the PDCCH order. The contents of each field are explained in Table 4.3. The PDCCH order has DCI with CRC scrambled by C-RNTI.

For a PDCCH order triggered with a non-zero "Random Access Preamble Index," the "PRACH Mask Index" indicates which of the PRACH occasions associated with "SS/PBCH Index" are used for the preamble transmission triggered by the PDCCH order as shown in Table 4.4. The PRACH occasions are mapped consecutively for each SS/PBCH block index and reset every cycle of consecutive PRACH occasions per SS/PBCH block.

Figure 4.9 Contention-free random access procedure [38.300].

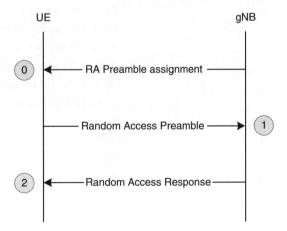

Table 4.3 Fields of DCI of PDCCH order.

PDCCH order field	Size	Description
Identifier for DCI formats	1 bit	Value should be "1" indicating DCI Format 1_0
Frequency domain resource assignment	$\lceil \log_2(N_{RB}^{DL,BWP} \cdot (N_{RB}^{DL,BWP} + 1)/2) \rceil$	$N_{RB}^{DL,BWP}$ is the size of DL BWP. This field is all ones indicating that this is a PDCCH order.
Random Access Preamble index	6 bits	*ra_PreambleIndex*. If this is all 0 a CBRA procedure is triggered, else the *ra_PreambleIndex* is used for a CFRA procedure.
UL/SUL indicator	1 bit	If SUL is configured, and ra_PreambleIndex is not all 0, this field indicates UL carrier, else it is reserved. When indicating UL carrier: "0" is for normal uplink and "1" is for SUL.
SS/PBCH Index	6 bits	If *ra_PreamblexIndex* is not all zeros, the SS/PBCH block index used to determine the associated RO to transmit the preamble on, else reserved.
PRACH Mask Index	4 bits	If *ra_PreamblexIndex* is not all zeros, this parameter is used to determine the RO to transmit the preamble on, else reserved.
Reserved bits	10 bits	

Table 4.4 PRACH mask index values [38.321].

PRACH mask index	Allowed PRACH occasion(s) of SSB
0	All
1	PRACH Occasion Index 1
2	PRACH Occasion Index 2
3	PRACH Occasion Index 3
4	PRACH Occasion Index 4
5	PRACH Occasion Index 5
6	PRACH Occasion Index 6
7	PRACH Occasion Index 7
8	PRACH Occasion Index 8
9	All even PRACH Occasions
10	All odd PRACH Occasions
11	Reserved
12	Reserved
13	Reserved
14	Reserved
15	Reserved

Figure 4.10 shows an example with three SS/PBCH Blocks transmitted. Each SS/PBCH Block is associated with eight consecutive PRACH occasions, i.e. ssb-perRACH-Occasion is 1/8. In the PDCCH order, the SS/PBCH index signaled is 1 and the PRACH Mask Index signaled is 2. This determines the PRACH occasion to use for the PDCCH order preamble transmission as the second PRACH occasion of the set of PRACH occasions associated with SSB1.

When a PDCCH order triggers a CFRA procedure, the PRACH preamble is quasi-collocated with the same DL RS (SS/PBCH Block or CSI-RS) as the DMRS port of the PDCCH, as shown in Figure 4.11. The path loss and preamble transmission power calculation are based on that DL RS, rather than the SS/PBCH block index provided in the PDCCH order.

When a PDDCH order triggers a CFRA procedure;

- The DMRS port of the PDCCH of the RAR,
- The DMRS port of the PDSCH of the RAR,
- And the DMRS port of the PDCCH order

are quasi co-located with the same DL RS (SS/PBCH Block or CSI-RS), as shown in Figure 4.11, with respect to Doppler shift, Doppler spread, average delay, delay spread, and spatial RX parameters.

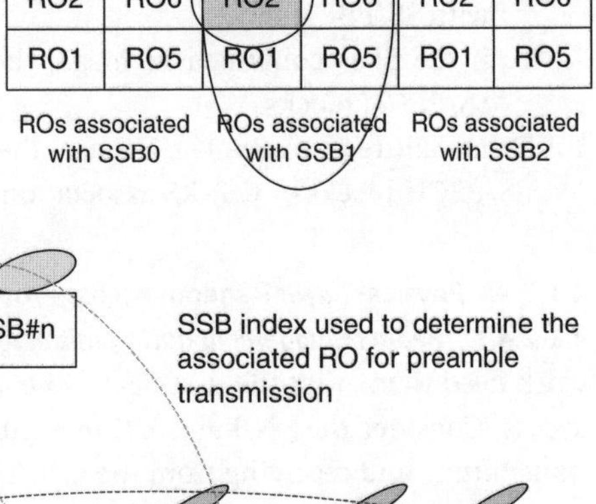

Figure 4.10 Example of the indication of the SS/PBCH index and PRACH Mask index for PDCCH order.

Number of Transmitted SSBs: 3
SS/PBCH Index: 1
PRACH Mask Index: 2
Preamble transmitted in RO2 associated with SSB1

RO4	RO8	RO4	RO8	RO4	RO8
RO3	RO7	RO3	RO7	RO3	RO7
RO2	RO6	RO2	RO6	RO2	RO6
RO1	RO5	RO1	RO5	RO1	RO5

ROs associated with SSB0   ROs associated with SSB1   ROs associated with SSB2

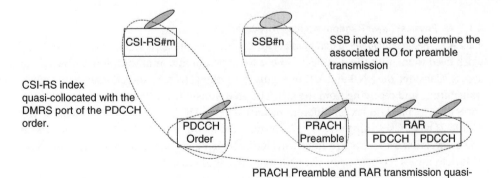

CSI-RS index quasi-collocated with the DMRS port of the PDCCH order.

SSB index used to determine the associated RO for preamble transmission

PRACH Preamble and RAR transmission quasi-collocated with DMRS port of PDCCH order

Figure 4.11 Quasi-collocation for preamble triggered by PDCCH order and corresponding RAR.

4.1.2.3 Random Access Use Cases

In NR, the RACH procedure is triggered by one of the following events:

1. Initial System Access from RRC_IDLE State. The UE uses the CBRA procedure based on the SS/PBCH blocks.
2. RRC Connection Re-establishment after radio link failure (RLF). The UE uses the CBRA procedure based on the SS/PBCH blocks.
3. Handover. This uses the contention based or CFRA procedures, and it can be based on SS/PBCH blocks or CSI-RS.
4. UL/DL data arrival when the UE is in RRC_CONNECTED state, with non-synchronized UL.
   a. In case of UL data arrival with non-synchronized UL, the UE uses the CBRA procedure based on the SS/PBCH blocks.
   b. In case of DL data arrival with non-synchronized UL, the network uses a PDCCH order with contention based or CFRA procedure based on the SS/PBCH blocks.
5. UL data arrival when the UE is in RRC_CONNECTED state, with no resource allocated on PUCCH for the UE to send a scheduling request. The UE uses the CBRA procedure based on the SS/PBCH blocks.
6. Scheduling request failure. The UE sends a scheduling request in response to UL data arrival but fails to receive an UL grant from the network. The UE uses the CBRA procedure based on the SS/PBCH blocks.
7. Transition from RRC_INACTIVE state to RRC_CONNECTED state. The UE uses the CBRA procedure based on the SS/PBCH blocks.
8. Establishing time alignment when adding SCell. the network uses a PDCCH order with CBRA procedure based on the SS/PBCH blocks.
9. Request of Other SI (on demand SI).
   a. If the SI is contained in Msg1, the UE uses a dedicated preamble based on association with SS/PBCH block.
   b. If the SI is contained in Msg3, the UE uses the CBRA procedure based on the SS/PBCH blocks.
10. Beam failure recovery. The UE uses the contention based or CFRA procedure based on SS/PBCH blocks or CSI-RS association.

4.1.2.4 Physical Layer Random Access Procedures

4.1.2.4.1 Beam Management during Initial Access In a beam-based system the user and network need to identify the best beam to use for subsequent communication during initial access. Consider the gNB and UE of Figure 4.12, the gNB uses directional beams when transmitting and receiving from the cell. The gNB transmits SS/PBCH blocks, with different SS/PBCH block index on different beams. When a UE is powered on, it listens to the SS/PBCH blocks, scanning across its Rx beams, and identifies a SS/PBCH block index with power level that exceeds a threshold provided by higher layer parameter *rsrp-ThresholdSSB*. This determines the beam-pair the gNB and the UE use to communicate with each other. Let's assume that the gNB and the UE have beam-correspondence, i.e. the beam used for reception determines the beam used for transmission, and vice versa. The UE transmits the preamble based on the selected SS/PBCH block index on a beam determined by the Rx

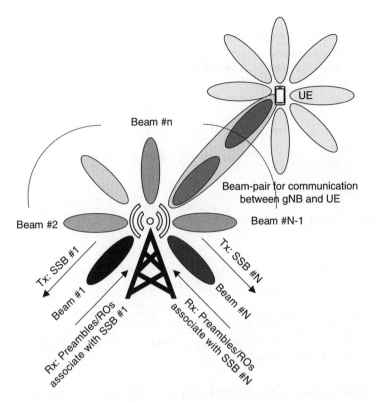

Figure 4.12 Beam-based transmission in gNB with N beams. SS/PBCH Block (SSB) #n is transmitted on beam #n. Preambles/PRACH Occasions (ROs) associated with SS/PBCH block index #n are received on the same beam.

beam used to receive the SS/PBCH block. The gNB receives the preamble and determines the best beam to communicate with UE. Subsequent random access procedure messages, i.e. the RAR, Msg3 and the contention resolution message are transmitted and received using the same beam-pair as shown in Figure 4.13.

When the UE transmits the preamble to the gNB, it conveys the selected SS/PBCH block index to the gNB, so that subsequent transmissions from the gNB to that UE use the same beam corresponding to the selected SS/PBCH block. This is conveyed by the preamble index and the PRACH occasion used to transmit the preamble. The network configures the association between the SS/PBCH Block indices and the PRACH occasions and preamble indices by broadcast signaling. Through this association, the network is informed about the selected SS/PBCH block index, subsequent transmissions to the UE use the corresponding beam.

There are two higher layer parameters that define the association pattern between the SS/PBCH blocks and the PRACH preambles/ROs:

– The number of transmitted SS/PBCH blocks N_{Tx}^{SSB} through higher layer parameter *ssb-PositionsInBurst*.
– The number of SS/PBCH blocks per RO N through higher layer parameter *ssb-perRACH-OccasionAndCB-PreamblesPerSSB* in *RACH-ConfigCommon* IE, or *ssb-perRACH-*

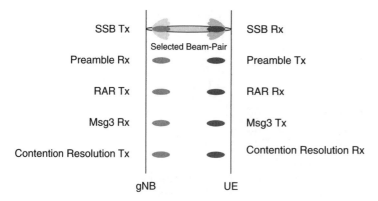

Figure 4.13 Beam operation during RACH procedure, assuming beam correspondence at the gNB and at the UE.

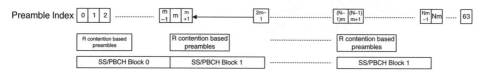

Figure 4.14 Association of preambles within PRACH occasions to contention-based preambles of SS/PBCH Blocks associated to that PRACH occasion.

Occasion in *CFRA* or *BeamFailureRecoveryConfig* IEs. Where, $N \in \left\{ \frac{1}{8}, \frac{1}{4}, \frac{1}{2}, 1, 2, 4, 8, 16 \right\}$. A value less than 1, indicates that the Synchronization signal/Physical Broadcast Channel Block (SSB) is associated with $1/N$ ROs.

As described in Section 3.6.4, each PRACH occasion has 64 preambles. Of the 64 preambles $N_{\text{preamble}}^{\text{total}}$ preambles, as provided by higher layer parameter *totalNumberOfRA-Preambles* and starting with preamble index 0, are available for association with SS/PBCH blocks, as depicted in Figure 4.14.

The association is done cyclically over the PRACH preambles and the valid PRACH occasion. The association period is X times the PRACH configuration period (defined in Section 3.6.4), and doesn't exceed 160 ms, with $X \in \{1, 2, 4, 8, 16\}$. For example, if the PRACH configuration is 40 ms, the maximum value of X is 4, i.e. in this case X can be 1, 2, or 4.

The mapping order of SS/PBCH blocks that are indicated to the UE as transmitted, to preambles and ROs, is in the following order:

– First, in increasing order of preamble index within an RO.
– Second, in increasing order of frequency resource index within a time-domain PRACH occasion.
– Third, in increasing order of RO within a PRACH slot.
– Fourth, in increasing order of PRACH slot.

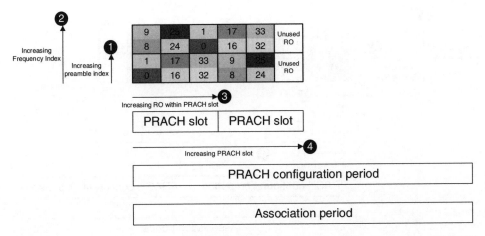

Figure 4.15 Example of mapping SS/PBCH Blocks to ROs. In this example, there are two SS/PBCH blocks mapped to one RO. There are two frequency division multiplexed ROs. There are three time-domain ROs per PRACH slots and two PRACH slots in a PRACH configuration period.

The duration of the association period is minimum period such that within the association period, each SS/PBCH block is associated with at least one RO. If after an integer number of cycles of mapping N_{Tx}^{SSB} SS/PBCH blocks to ROs there are left over ROs that are not enough to map all the N_{Tx}^{SSB} SS/PBCH blocks, these ROs are not used for PRACH transmission.

In unpaired spectrum, a PRACH occasion as determined by the *prach-ConfigurationIndex* (see Section 3.6.4) is considered valid if it is within the common semi-statically configured UL symbols (configured by higher layer parameter *TDD-UL-DLConfigurationCommon*) or if it doesn't overlap with or precede a common semi-statically configured DL symbol (configured by higher layer parameter *TDD-UL-DLConfigurationCommon*) or SS/PBCH block in a slot, and is at least N_{gap} symbols after the last DL symbol or SS/PBCH block. Where, $N_{gap} = 0$ for the long sequence preamble formats, and $N_{gap} = 2$ for the short sequence preamble formats. In paired spectrum all PRACH occasions are considered valid.

Figure 4.15 shows one example of how SS/PBCH blocks are associated with ROs. In this example:

- The number of SS/PBCH blocks indicated to the UE as being transmitted is 10, i.e. $N_{Tx}^{SSB} = 10$. The transmitted SS/PBCH blocks have indices {0, 1, 8, 9, 16,17,24,25,32,33}.
- The number of SS/PBCH blocks mapped to one RO is 2. i.e. $N = 2$.
- The number of frequency division multiplexed ROs is 2.
- The number of time-domain ROs in one PRACH slot is 3.
- The number of PRACH slots in one PRACH configuration period is 2.

In this example, we assume that all ROs are valid. Within one PRACH configuration period, there are $2 \times 2 \times 3 \times 2 = 24$ sets of preamble-indices/ROs that SS/PBCH blocks can

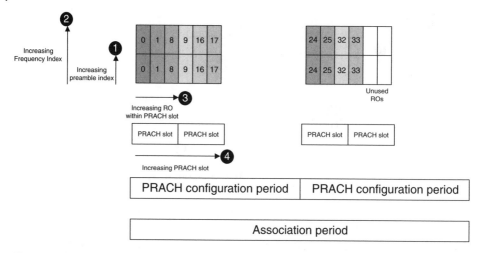

Figure 4.16 Example of mapping SS/PBCH Blocks to ROs. In this example, one SS/PBCH block is mapped to two ROs, i.e. N = 1/2. There are two frequency division multiplexed ROs. There are three time-domain ROs per PRACH slots and two PRACH slots in a PRACH configuration period.

be mapped. Hence, each SS/PBCH blocks is mapped twice, and the last two ROs are left unused. The mapping order is as described earlier in this section and shown in Figure 4.15. In this example the association period is equal to the PRACH configuration period.

Figure 4.16 shows another example of how SS/PBCH blocks are associated with ROs. In this example:

- The number of SS/PBCH blocks indicated to the UE as being transmitted is 10, i.e. $N_{Tx}^{SSB} = 10$. The transmitted SS/PBCH blocks have indices {0, 1, 8, 9, 16,17,24,25,32,33}.
- Each SS/PBCH block is mapped to two ROs. i.e. $N = \frac{1}{2}$.
- The number of frequency division multiplexed ROs is 2.
- The number of time-domain ROs in one PRACH slot is 3.
- The number of PRACH slots in one PRACH configuration period is 2.

In this example, we assume that all ROs are valid. Within one PRACH configuration period, there are $\frac{1}{2} \times 2 \times 3 \times 2 = 6$ sets of preamble-indices/ROs that SS/PBCH blocks can be mapped to. Hence, one PRACH configuration period is not enough to map all SS/PBCH blocks, two PRACH configuration periods can map all SS/PBCH blocks, with the last four ROs in the second PRACH configuration period left unused. The mapping order is as described earlier in this section and shown in Figure 4.16. In this example the association period is equal to two PRACH configuration periods.

The association pattern should be long enough such that every SS/PBCH block is associated at least once with a valid PRACH occasion. As the number of valid PRACH occasions in a RACH configuration period varies depending on the occurrence of SS/PBCH blocks and downlink symbols of the UL-DL time division duplex (TDD) configuration, the duration of the association pattern in number of RACH configuration periods varies. The association period pattern includes one or more association periods, such that it is a periodic pattern between SS/PBCH blocks and PRACH occasions, with duration not exceeding 160 ms.

4.1.2.4.2 Dedicated RACH Configuration The network configures dedicated PRACH preambles through *rach-ConfigDedicated*. Dedicated PRACH preambles are also known as contention free preambles or non-contention preambles. The dedicated RACH configuration is used for reconfiguration with sync procedures such as handover. The network can configure dedicated preambles that are associated with either SS/PBCH blocks or with CSI-RS resources. *rach-ConfigDedicated* includes;

– The PRACH occasions to use for dedicated preambles, which is given by;
  o *rach-ConfigGeneric* for defining the generic PRACH Occasion parameters, such as the time and frequency domain positions of the PRACH occasions.
  o *ssb-perRACH-Occasion* for defining the number of SS/PBCH blocks associated with a PRACH Occasion. Hence it allows the determination of the association period and the association period pattern.
  If the parameters defining the PRACH occasions are absent from the dedicated RACH configuration, the UE uses the random access configuration from rach-ConfigCommon of the first active UL Bandwidth Parts (BWP).
– The dedicated (i.e. contention free) PRACH resources. This is a list of preamble resources associated with either SS/PBCH blocks or CSI-RS resources.
– Parameters to apply for prioritized CFRA procedures.

If the network configures dedicated PRACH resources associated with SS/PBCH blocks, the network can configure up to 64 dedicated PRACH resources. Where the configuration establishes an association between the SS/PBCH block and the preamble index within a PRACH occasion. The network also configures a PRACH mask index, as described in Section 4.1.2.2.3 and Table 4.4, to determine the PRACH occasion to use for the dedicated preamble within a set of PRACH occasions associated with an SS/PBCH block. The association of PRACH occasions with SS/PBCH blocks is performed as described in Section 4.1.2.4.1. When the UE triggers a random access procedure using a dedicated RACH configuration configured with PRACH resources associated with SS/PBCH blocks, the UE identifies the SS/PBCH block index of the candidate beam, and finds the corresponding PRACH occasion and preamble index to use.

If the network configures dedicated PRACH resources associated with CSI-RS resources, the network can configure up to 96 dedicated PRACH resources. Where the configuration establishes an association between the CSI-RS resource and a list of up to 64 PRACH occasions, given by higher layer parameter *ra-OccasionList*, and a preamble index within each PRACH occasion in the list. The indexing of the PRACH occasions is reset to zero at the start of each association period pattern. PRACH occasions are number sequentially;

– First in ascending order of frequency resource index, for the frequency multiplexed PRACH occasions in the same time instance as determined by higher layer parameter *msg1-FDM*.
– Second in ascending order of time resource index within a PRACH slot. Time multiplexed PRACH occasions within a PRACH slot are determined by higher layer parameter *prach-ConfigurationIndex*.
– Third in ascending order of PRACH slot index, where the available PRACH slots are determined by higher layer parameter *prach-ConfigurationIndex*.

Figure 4.17 shows an example of the mapping of a channel-state information reference signals (CSI-RS) resource index (CRI) configured with a list of PRACH occasions (*ra-OccasionList*) to the corresponding PRACH Occasions. When the UE triggers a random access procedure using a dedicated RACH configuration configured with PRACH resources associated with CSI-RS resources, the UE identifies the CRI of the candidate beam, and finds the corresponding PRACH occasions and preamble index to use within the PRACH occasions.

4.1.2.4.3 Beam Failure Recovery RACH Configuration The network configures BFR PRACH preambles through *BeamFailureRecoveryConfig*. These are dedicated PRACH preambles the UE uses to indicate to the gNB a new SS/PBCH block or CSI-RS resource when beam failure is detected at the UE. The network can configure BFR preambles that are associated with SS/PBCH blocks and/or with CSI-RS resources. The network can configure up to 16 candidate beam reference signals (i.e. SS/PBCH blocks and/or with CSI-RS resources) each associated with a PRACH occasions and preamble index within the associated PRACH occasions. The configuration of PRACH resources for BFR associated with SS/PBCH blocks is similar to the configuration of dedicated PRACH resources associated with SS/PBCH blocks described in Section 4.1.2.4.2. The configuration of PRACH resources for BFR associated with CSI-RS resources is similar to the configuration of dedicated PRACH resources associated with CSI-RS resource described in Section 4.1.2.4.2.

BFR is described in Section 4.2.6.

4.1.2.4.4 Power Control Procedure The power control procedure determines the transmission power of the PRACH preamble and for the Msg3 PUSCH transmission. The UE determines a transmission power for the PRACH preamble, in PRACH Occasion i, according to the following equation:

$$P_{\mathrm{PRACH}}(i) = min(P_{\mathrm{CMAX}}(i), P_{\mathrm{PRACH,target}} + PL)$$

where,

- P_{CMAX} is the configured maximum output power of carrier/cell pair used to transmit the PRACH preamble.
- $P_{\mathrm{PRACH, target}}$ is the target preamble receive power at the base station, this is configured by higher layer parameter *preambleReceivedTragetPower*.
- PL is the pathloss (PL) in dB measured based on a downlink reference signal determined as described next.

If the PRACH transmission is not in response to a PDCCH order that triggers a non-CBRA procedure, the PRACH preamble is transmitted in a PRACH Occasion that is associated with an SS/PBCH block index, as described in Section 4.1.2.4.1. That SS/PBCH block index is the downlink reference signal used for pathloss measurement. If the PRACH transmission is in response to a PDCCH order that triggers a non-CBRA procedure, the downlink reference signal for pathloss estimation is the SS/PBCH block index or CRI with which the PDCCH of the PDCCH order is quasi co-location (QCL) Type-D with. The pathloss measurement is given by:

$$PL = referenceSignalPower - \text{higher layer filter RSRP}$$

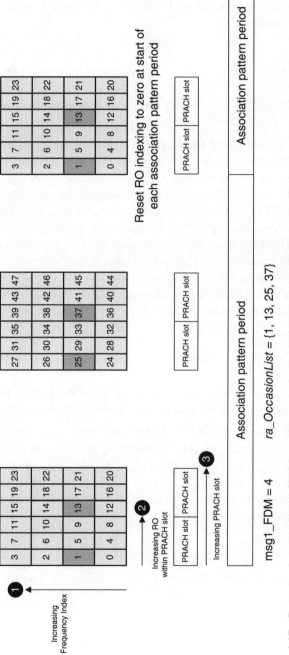

Figure 4.17 Example of association of CSI-RS resource to multiple PRACH occasions within ra_OccasionList.

where:

- In the case where the downlink reference signal is an SS/PBCH block, the *referenceSignalPower* is the average energy-per-resource-element (EPRE) of the Resource Elements (REs) carrying the SSS in dBm, provided by higher layer parameter *ss-PBCH-BlockPower* in the System Information.
- In the case where the downlink reference signal is a CSI-RS, the *referenceSignalPower* is the sum of the higher layer parameter *ss-PBCH-BlockPower* and the higher layer parameter *powerControlOffsetSS*. Where, *powerControlOffsetSS* is the power offset between the CSI-RS RE and SS RE.
- The higher layer filtered RSRP is measured using the corresponding downlink reference signal.

If the UE doesn't receive a RAR for the PRACH transmission within the RAR window as described in Section 4.1.2.2.1, the UE retransmits the PRACH preamble as long as the maximum number of retransmissions is not exceed. When the UE retransmits the PRACH preamble on the same beam as that of the latest PRACH transmission, the UE increments the PREAMBLE_POWER_RAMPING_COUNTER by one, which increments the PRACH transmission power by *powerRampingStep* or *powerRampingStepHighPriority* in dB. Where, *powerRampingStepHighPriority* is the power ramping factor for prioritized Random access procedures. Prioritized random access procedures are triggered by BFR or CFRA handover. For other trigger events of the PRACH procedure, the power ramping factor is given by *powerRampingStep*. When the UE uses a different beam for the retransmission of the PRACH preamble compared to that used for the latest PRACH transmission, the UE doesn't increment PREAMBLE_POWER_RAMPING_COUNTER, i.e. it maintains the same power for the retransmitted preamble.

The transmit power, in dB, of a PUSCH transmission scheduled by the RAR UL grant is given by (Figure 4.18):

Where, P_{CMAX} is the configured maximum output power during the transmission time of the PUSCH scheduled by the UL RAR. The transmit power consists of five components:

- The target PUSCH receive power in dB, which is given by $P_{O\_PUSCH} = P_{O\_NOMINAL\_PUSCH} + P_{O\_UE\_PUSCH}$. For PUSCH scheduled by UL grant, $P_{O\_UE\_PUSCH} = 0$, $P_{O\_NOMINAL\_PUSCH} = P_{O\_PRE} + \Delta_{PREAMBLE\_Msg3}$. Where, P_{O_PRE} is given by higher layer parameter *preambleTragetReceivedPower*, and $\Delta_{PREAMBLE\_Msg3}$ is given by higher layer parameter *msg3_DeltaPreamble*, which represents the offset in dB between the preamble and the PUSCH scheduled by the RAR UL grant.

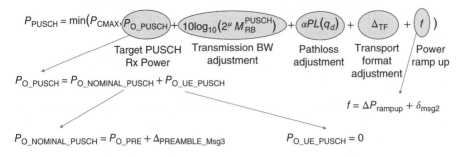

Figure 4.18 Schematics for power control.

- The term, $10\log_{10}(2^\mu M_{RB}^{PUSCH})$, represents the adjustment due to the transmission band-width of the PUSCH in dB. μ is the SCS configuration, and M_{RB}^{PUSCH} is the bandwidtfh of the PUSCH resource allocation in number of RBs.
- The term, $\alpha PL(q_d)$, represents transmit power adjustment due to pathloss in dB. $\alpha \in [0, 1]$ represents the fractional pathloss compensation factor. This is the open-loop power control component, which compensates large scale fading attenuation. When $\alpha = 0$, there is no pathloss compensation. When $\alpha = 1$, there is full pathloss compensation. When α is between 0 and 1, there is partial, or fractional, pathloss compensation. For a PUSCH scheduled by the UL RAR, α is given by the higher layer parameter $msg3\text{-}Alpha$. $PL(q_d)$ is the pathloss calculated using the same RS resource index q_d as the corresponding PRACH transmission.
- The term, Δ_{TF}, represents the adjustment due to the transport format of the PUSCH trans-mission. Δ_{TF} in dB is given by:

$$\Delta_{TF} = \begin{cases} 10\log_{10}\left(\left(2^{1.25\ BPRE}-1\right)\cdot\beta_{offset}^{PUSCH}\right) & \text{if higher layer parameter} \\ & \text{deltaMCS is enabled} \\ 0 & \text{Otherwise} \end{cases}$$

where, BPRE is the bit per RE of the PUSCH transmission, and $\beta_{offset}^{PUSCH} = 1$.
- The term, f represents the closed-loop power control adjustment state in dB. This has two parts; the first part, ΔP_{rampup} in dB, comes from the ramp up of the preamble, from the first PRACH preamble transmission till the last PRACH preamble transmission detected by the base station and corresponding to the transmitted RAR. The second part, δ_{msg2}, is a power adjustment step in dB, in response to the TPC command in the UL grant of the RAR (see Table 4.2). The TPC command is three bits, with eight values corresponding to $\{-6, -4, -2, 0, 2, 4, 6, 8\}$ dB.

4.1.2.5 RACH in Release 16
Release 16 is considering further enhancements for RACH, to reduce latency and overhead. As well as enhancements to better adapt to features being developed in release 16 such as integrated access and backhaul (IAB), and NR in unlicensed bands (NR-U). In this section we briefly discuss some of the RACH enhancements being considered.

4.1.2.5.1 *Two-Step RACH* The four-step RACH procedure described in Section 4.1.2.2.1 requires two round-trip cycles between the UE and the gNB to complete the random access procedure. The two-step RACH procedure combines the messages sent in each direction into a single message as shown in Figure 4.19. In the uplink direction, from the UE to the gNB, MsgA combines the random access preamble (Msg1), and UL scheduling transmis-sion (Msg3) into a single message. Similarly, in the downlink direction, MsgB combines the RAR (Msg2) and the Contention Resolution (Msg4) into a single message. As a result, for two-step RACH there is only one round-trip cycle between the UE and the gNB to complete the random access procedure. This feature is introduced to reduce the latency and control channel signaling overhead. Furthermore, the reduced number of messages benefits NR-U as it reduces the number of LBT attempts.

The channel structure of MsgA and physical layer and higher layer procedure aspects of two-step RACH are currently under consideration for NR release 16.

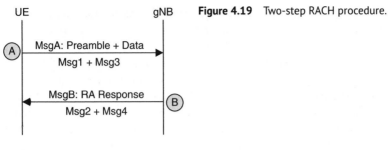

Figure 4.19 Two-step RACH procedure.

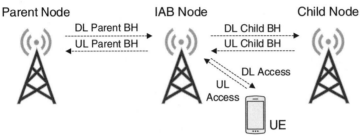

Figure 4.20 Integrated access and backhaul (IAB) backhaul and access links.

4.1.2.5.2 RACH Enhancements for IAB IAB is a feature introduced for NR in release 16 for support of wireless backhaul and relay links sharing the same spectrum as access UEs. The IAB node, can be a parent Node of another child IAB node, and/or it can be a child node of another parent IAB node, in addition to serving its own access UEs as shown in Figure 4.20. The IAB node serves access and backhaul links. In this section, we consider the changes introduced for the RACH design of the backhaul links.

IAB supports the flexibility to configure PRACH occasions with longer periodicities and different offset from the PRACH occasions available for access UEs to orthogonalize access and backhaul links, and time division multiplex backhaul RACH resources across adjacent hops. The IAB backhaul RACH occasions reuses the release 15 RACH configurations (see Figure 3.49 of Section 3.6.4) but adapts these configurations by introducing additional parameters to scale the periodicity and offset when determining the location of the time-domain PRACH occasions. The following parameters are introduced:

– Periodicity scaling factor λ. This scales the PRACH configuration period x as follows

$$x_{IAB} = \lambda x$$

where, $\lambda \in \{1, 2, 4, 8, 16, 32, 64\}$ such as $x_{IAB} \leq 64$ frames.
– Offset for frame containing IAB backhaul ROs Δy. The frame(s) containing IAB backhaul ROs are given by frames whose SFN satisfies:

$$(n_{SFN} \bmod x_{IAB}) = ((y + \Delta y) \bmod x_{IAB})$$

where y is the offset given by the PRACH configuration of Figure 3.49 of Section 3.6.4. $\Delta y \in \{0, 1, \ldots, x_{IAB} - 1\}$
– Offset for the subframe/60 kHz slot, in FR1/FR2 respectively, containing the IAB backhaul ROs Δs. The subframe(s)/slot(s) containing IAB backhaul ROs are given

$(s_n + \Delta s) \bmod L$. Where s_n is the subframe/60 kHz slot number of Figure 3.49 of Section 3.6.4. $\Delta s \in \{0, 1, \ldots, L-1\}$, L is the number of subframes/60 kHz slots in a frame.

An example of the determination of the IAB backhaul ROs for a RACH configuration is shown in Figure 4.21.

As a result of the longer RACH configuration period for the IAB backhaul links (up to 64 frames), the SSB-to-RACH association period and the association pattern period, described in Section 4.1.2.4.1, for IAB backhaul links can be as long as 64 frames.

4.1.2.5.3 RACH Enhancements for NR-U The NR-unlicensed work item in release 16 addresses the support of NR in unlicensed frequency band. As unlicensed spectrum is shared with other radio access technologies there are regulatory requirements for accessing and utilizing this spectrum. The occupied channel bandwidth (OCB), which is defined as the bandwidth containing 99% of the signal power, shall be between 80% and 100% of the declared Nominal Channel Bandwidth [38.889] [51]. The bandwidth of the short sequence PRACH preamble formats of NR release 15 depends on the SCS. The short sequence has a length of 139 subcarriers, with a SCS of 15 kHz, the bandwidth of the PRACH signal is 2.085 MHz. To meet the OCB requirement for 10 MHz Nominal Channel Bandwidth, the PRACH preamble format bandwidth should increase by a factor of four.

3GPP studied different methods to increase the PRACH preamble format bandwidth [38.889]:

– Uniform PRB (Physical Resource Block)-level interlace mapping.
– Non-uniform PRB-level interlace mapping.
– Uniform RE-level interlace mapping.
– Non-interlaced mapping.

In release 16, non-interlaced mapping has been selected to increase the PRACH preamble format bandwidth to meet the OCB requirement. Two flavors are under consideration;

– Using longer Zhadoff-Chu sequence, e.g. a sequence with length 571, and SCS 15 kHz has a preamble format with bandwidth 8.565 MHz. This meets the OCB requirement for a 10 MHz Nominal Channel Bandwidth. Increasing the sequence length further to 1151 leads to a preamble format with bandwidth 17.265 MHz, which meets the OCB requirement for a 20 MHz Nominal Channel Bandwidth.
– Alternatively, the release 15 short sequence of length 139 can be repeated four or eight times to achieve the OCB requirement for a Nominal Channel Bandwidth of 10 or 20 MHz respectively.

Before transmitting on a channel in the unlicensed spectrum, the transmitting device senses the channel using a procedure known as Listen-Before-Talk to confirm that the channel is available. If the channel is available (LBT success), the transmitter proceeds with the transmission, else (LBT failure) that transmission is canceled. The LBT procedure has implications on the random access procedure.

– In case the preamble is not transmitted due to an LBT failure, the PREAMBLE_POWER_RAMPING_COUNTER (Section 4.1.2.4.4) and PREAMBLE_TRANSMISSION_COUNTER (Section 4.1.2.2.1) are not incremented.

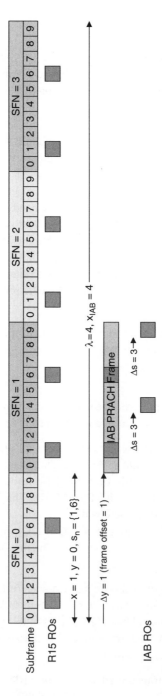

Figure 4.21 Example of IAB backhaul ROs, based on a release 15 PRACH configuration with $x = 1$, $y = 0$, $s_n = \{1,6\}$, with $\lambda = 4$, $\Delta y = 1$, and $\Delta s = 3$.

– As a result of a potential LBT failure for the RAR, the reception of the RAR at the UE could be delayed. Accordingly, the RAR window is increased to minimize the likelihood of a random access procedure failure due to the delay of RAR caused by LBT failure.

Two-step RACH benefits described in Section 4.1.2.5.1, includes reducing the number of channel accesses of the random access procedure from four channel accesses for four-step CBRA to two channel accesses, this benefits NR-U due to the fewer LBT attempts with a reduced number of messages.

4.2 Beam Management (Mihai Enescu, Nokia Bell Labs, Finland, Claes Tidestav, Ericsson, Sweden, Sami Hakola, Juha Karjalainen, Nokia Bell Labs, Finland)

4.2.1 Introduction to Beam Management

As already mentioned, one of the new areas of operation of 5G NR is the deployment in higher frequency bands. As the spectrum utilized by the traditional cellular communications has been filled up, the industry is looking into harvesting the spectrum available at higher frequencies. However, as the carrier frequency is increasing, the propagation gets more challenging as the pathloss between transmitter and receiver is increasing due to the assumption of a fixed antenna size relative to the wavelength. Also, as the carrier frequency increases beyond roughly 10 GHz, diffraction will no longer be a dominant propagation mechanism. Beyond 10 GHz, reflections and scattering will be the most important propagation mechanism for non-line-of-sight propagation links. Furthermore, the penetration loss from propagating into a building tends to increase as the carrier frequency increases. To maintain coverage, this requires a larger number of antennas at both transmission ends. At the transmitter, multiple antennas are needed to steer the transmit power toward the receiver and at the receiver, multiple antennas are needed to enhance the signal coming from a certain direction.

In the multi-antenna solutions traditionally used in cellular systems, each individual antenna element is directly accessible from baseband: e.g. in LTE FD-MIMO with 32 ports, the baseband needs to handle 32 parallel signals. Simply scaling this solution to a larger number of antenna elements, and at the same time to larger carrier bandwidths, is not deemed feasible: the complexity of such an all-baseband architecture would be prohibitively high. An alternative solution would be to implement a purely analog beamforming architecture, where the transmit and receive beams are formed using analog phase shifters. However, such a solution would not be capable of any spatial multiplexing, which is deemed to be central in a 5G system.

For NR, it was deemed important to be able to implement an antenna solution that does not have baseband access to each antenna element, while still maintaining steerability using the entire antenna aperture. We put these facts into perspective in Figure 4.22 where we show a hybrid architecture which is an alternative to the all-baseband and all-RF architectures. In this case the control of MIMO and beamforming is split between baseband and RF. In the hybrid architecture, each RF beam is driven by a transceiver, and multi-stream beam weighting or precoding is applied at baseband to the inputs of the

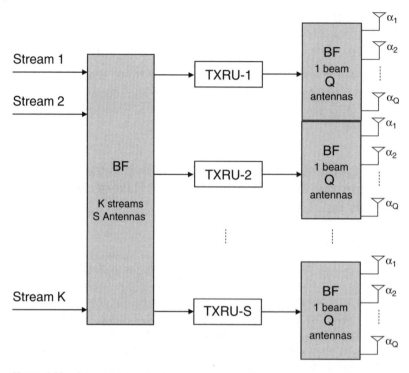

Figure 4.22 One example of hybrid transceiver architecture.

transceivers. Figure 4.22 shows a hybrid architecture for a sub-array/panel configuration, where each RF weight vector is applied to a unique subset of the antenna elements. One advantage of the sub-array configuration is the lack of summation devices behind the antenna elements. A hybrid architecture provides additional flexibility over an all-RF architecture as the baseband transmit portion can be adapted across the signal bandwidth to further optimize performance. Similar tendency in transceiver and antenna architecture can be considered for the UE side as well. Indeed, at the UE, multiple panels may be placed on the sides of the device so that at any point in time, at least one of the panels is able to receive/transmit. It can be argued that the presence of gNB and UE panel antennas is the factor that would require the development of the beam management procedures in NR. What is typically referred as *beam based operation* is related to the utilization of analog beamforming at both the transmitter and receiver: *all* the communication between the gNB and the UE may take place over a spatially filtered link.

We have mentioned above that a larger number of antennas is needed at both the transmitter and receiver. When it comes to the antenna array size, there is however good news: as the carrier frequency increases, the wavelength decreases, and the size of an antenna elements are typically proportional to the wavelength. The result is that as the carrier frequency increases, the number of antennas that can fit into a given fixed area increases significantly. Viewed another way, as the carrier frequency increases the size of an array with a fixed number of antenna elements decreases significantly. Figure 4.23 provides such an intuitive illustration where the same amount of 128 antennas is considered for 3.5 and 28 GHz. The array size shrinks from 343 to 42 mm.

Figure 4.23 128 antenna array size for 3.5 and 28 GHz.

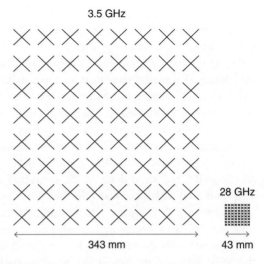

So far, we have mentioned that the beam-based operation is utilized at higher frequency bands, that both the transmitter and receiver are having one or multiple panels, which are able to perform directional transmission. In many cases, both ends of the communication are *only* capable of beam-based operation: omni-directional reception and transmission may not be possible! This opens up quite an amount of challenges which need to be solved and which are described in detail in the next sections. Due to the possibility of the signals to be transmitted from any transmission point (TRP), a set of so-called *quasi co-location (QCL)* rules have been defined to ensure proper reception of such signals. As a result, both the gNB and the UE would benefit from a set of rules facilitating the beam indication. The UE would benefit from knowing the TRP of origin for DL signals and their linkage with previously transmitted DL signals. The indication of such rules is possible based on the Transmission Configuration Indicator (TCI) framework. A similar indication strategy would be needed for the UL transmission, in conjunction with the reception of the DL signals, this being possible through the definitions of a set of spatial relations.

4.2.2 Beam Management Procedures

It is worth starting the discussion in this chapter with a few definitions and clarifications.

Beam management is understood as a set of layer 1 and layer 2 procedures used to acquire and maintain a set of beams at both transmitter and receiver which are used for the transmission of the control and data channels. Beam management consists of a set of procedures which are valid for both the gNB and UE:

- *Beam determination* needs to be performed first by both ends of the transmission. This implies the identification of best beam for transmission and reception of data which is the outcome of the *beam sweeping process*. A key concept of this procedure is *beam correspondence*.
- *Beam measurement* implies obtaining the characteristics (mainly the power) of the received beamformed signals.
- *Beam reporting* leads to the transmission of a beam measurement report to the transmitter.

In the following explanation we will describe in more detail these procedures for both the transmitter and receiver. But first, let us set the stage for the environment and challenges of these beam management procedures.

One of the fundamental elements of beam management is the concept of beam correspondence. This should be seen as a node capability, in other words the gNB or the UE are capable of maintaining beam correspondence if the gNB or the UE are able to transmit with the same beam used for the reception. Beam correspondence at the UE is declared if the UE is able to determine a Tx beam for UL transmission based on the measurement of DL signals utilizing a UE Rx beam. At the same time, beam correspondence at the UE may be declared if the UE is able to determine a Rx beam for DL reception based on the gNB's indication done on the measured beams transmitted by the UE in uplink. The same routine happens at gNB, beam correspondence is declared if the gNB is able to determine a Tx beam for DL transmission based on the measurement of UL signals utilizing a gNB Rx beams and also beam correspondence is declared if the gNB is able to determine a Rx beam for the UL reception based on the measurement at the UE of the gNB's beams. One should not see this possible only in line of sight, as such beam correspondence can also happen based on reflections from scatterers. In Figure 4.24 we depict a transmitter having multiple beams and a UE which also operates with multiple beams. Note that these beams can be used for both transmission and reception. If the UE utilizes the same beam for transmitting and receiving, it is said that it is capable of beam correspondence.

There are a few challenges in the process of maintaining beam pairs. Figure 4.22 shows three different intra-cell downlink beam training use-cases for UE movement. In all cases, it is assumed that UE moves its position linearly with respect to the location of a base station. In the example on the left of Figure 4.25 a linear movement introduces a need for the change of TX beam at base station while using the same RX beam at UE. In the middle example, only RX beam needs to change by using the same TX beam. In the right example, both TX and RX beams need to be changed (Figure 4.25).

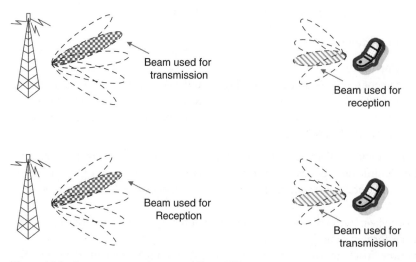

Figure 4.24 Beam correpondence at gNB and UE.

Figure 4.25 Downlink beam training use cases for linear movement of UE (diagonal hash beam = old beam, square dots beam = new beam).

Figure 4.26 Necessity for beam change: movement of the UE (left figure), UE rotation (middle figure), beam blockage (right figure).

Figure 4.26 presents intra-cell beam training use cases for UE rotation and beam blockage. In the left example we have UE rotation which causes the RX beamformer at the UE to change while the TX beam at the gNB is unchanged. In the case of beam blockage, the signal path is blocked by an obstacle leading to a significant drop of signal quality, e.g. tens of decibels in received power, at the receiver. An alternative path, for example due to reflections (blue beams) may become the best link between the BS and UE.

4.2.2.1 Beamwidths

Both the transmitter and receiver are using beams. In the beam management chapter we would typically understand the gNB as the transmitter and the UE as receiver, unless otherwise stated. There is a choice of beamwidths at both transmission ends, however the 5G NR specification does not mention anything about beamwidths, this being an implementation choice for both gNB and UE. In downlink, the main reference signals over which measurements are performed are SSBs and CSI-RS. Both type of reference signals can be transmitted with same beamwidths, but it is typically understood that SSB uses wider beams while CSI-RS uses narrower beams. In the next subsection we are describing procedures used for such beamwidths change. A similar approach is taken for the UE, while the beam width is an implementation specific, the UE may used wider beams for RACH transmission and narrower beams for PUSCH transmission.

4.2.2.2 Beam Determination

We have mentioned so far that both the transmitter and receiver are utilizing beams. There is a set of procedures which are used for beam determination, in downlink, these have been called P1, P2, and P3.

P1 procedure is understood to utilise more of an initial access type of beam determination, where the UE has no prior knowledge of what can be its best serving beam. This procedure supports the selection of the best beam of a TRP and hence of the best TRP as a consequence. For achieving this, each TRP transmits the synchronization signals (SS/PBCH

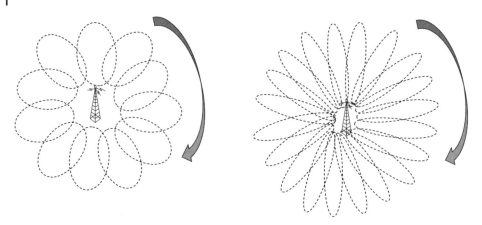

Figure 4.27 Beam sweeping with wider beams (left figure), beam sweeping with narrow beams (right figure).

blocks) in each beam and since the whole angular space needs to be covered, it is expected that such beams are of wider angular range in nature. It is also expected that during the P1 process, the UE also selected its reception beam.

In Figure 4.27 we present two strategies for beam sweeping, with wider and narrower beams. There are a few trade-off points in the beam sweeping strategy. While the wider beams (left figure) would imply that a smaller amount of beams, and hence time, is needed to cover the angular space, these have smaller coverage compared to the case of more narrow beams (right figure). A consequence of utilizing narrow beams is that the system would experience higher overhead. At the end of the P1 procedure it is important that the UE would find one beam that provides good link quality, i.e. with an RSRP over the threshold. Hence, the decision on the number of beams is an important trade-off, more on this procedure being described in Section 4.1.1. While the UE identifies the best DL beams, the same beam may be used by the gNB for the reception of the transmission performed by the UE. Hence in downlink the synchronization signal is transmitted, while the same beam is used for the reception of the random access signals. More on this procedure is described in Section 4.1.2.

P2 procedure is a procedure allowing the so-called beam refinement and hence it is an enabler for gradual adjustment of the gNB Tx beam when the gNB has prior knowledge about a suitable beam. The P2 algorithm may or may not lead to a change in beam width. The P1 beams can be improved in terms of coverage and one needs to perform the transition from wide beams to narrow beams. This is possible by the P2 procedure and it is illustrated in Figure 4.28.

While the P1 procedure can be seen a "global" procedure, where the gNB covers the whole angular space, the P2 procedure may be local and it is based on the identified best wide beam by the UE in the P1 procedure. Indeed, Figure 4.28 shows how the identified wide beam is refined by the gNB, which transmits a set of narrow beams, covering the angular space as the identified beam in the P1. We call this P2 procedure as "local" because the gNB does not necessarily need to cover the whole angular space but rather transmits the beams in the direction where the UE has been identified. Naturally, more

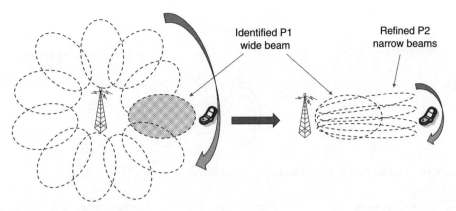

Figure 4.28 Transmission of narrow beams for P2 procedure.

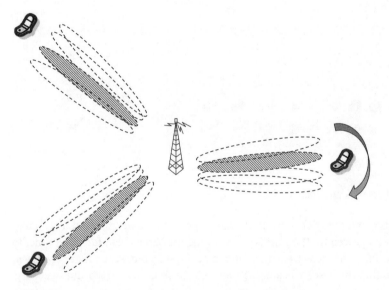

Figure 4.29 Transmission of narrow beams for P2 procedure.

such wide beam directions may be active, depending on the amount of identified UEs, an example being shown in Figure 4.29. The P2 procedure also uses the process of beam sweeping and the same tradeoffs as in the case of P1 procedure do apply in the sense that the time needed for the identification of the best beams is proportional to the number of scanned beams.

The P2 procedure would be based in principle on CSI-RS, being a beam refinement procedure, it is expected that initial beam acquisition has been done based on SS Block identification. Nevertheless, the standard does not mention any specific RS which can be used for this procedure, neither the procedures themselves. On the other hand the CSI-RS can be transmitted periodically, semi-persistent and aperiodically, compared to the SS/PBCH block which is just a periodic signal. Hence the CSI-RS has more configuration flexibility for being configured for P2 procedure.

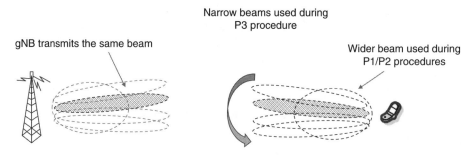

Figure 4.30 Transmission of narrow beams for P3 procedure.

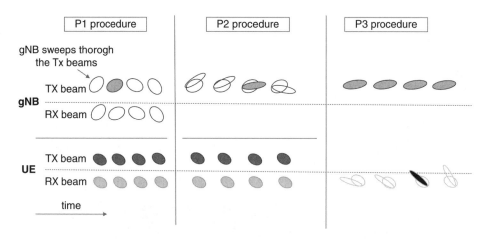

Figure 4.31 Timeline of Px procedures.

P3 procedure is about the UE beams. In P1 and P2 procedures the transmitter is changing the beams. In P3 procedure the gNB is transmitting the same beam for some time, to allow the UE to perform the measurement on the same Tx beam but with different received beams, in order to find the best Rx beam. If P1 and P2 are about the gNB beam sweeping, P3 is about the UE beam sweeping.

Figure 4.30 illustrates the P3 procedure. The gNB would transmit in principle using a narrow beam, the mandatory thing being the fact that same beam is used for consecutive symbols so that the UE may perform a sweeping process with its receive beams. At the end of the P1, P2, and P3 procedures, the gNB has determined the best Tx beam and the UE has determined the best Rx beam. If the gNB has beam correspondence, it has also determined the best Rx beam, and if the UE has beam correspondence, it has identified the best Tx beam (Figure 4.31).

We have so far described the DL based beam determination procedures. When beam correspondence holds, these are sufficient for establishing the best beam pairs between gNB and UE. However, there are situations when such beam correspondence is problematic, and for this case the system needs alternative methods to establish the best beam pairs. These procedures are called U1/U2/U3 and we will describe them in the following.

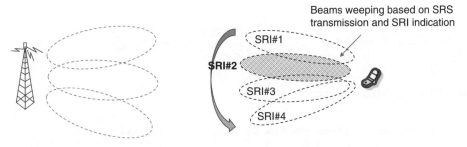

Figure 4.32 UL UE beam identification based on SRS transmission and SRI.

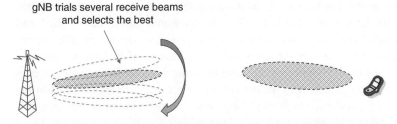

Figure 4.33 UL gNB beam identification based on same beam transmission from UE.

U1 procedure is used to enable gNB measurement of different UE transmitted beams to support the selection of UE Tx beam. Intuitively, this is similar to the P1 procedure, the only difference being that this happens in UL. In Figure 4.32 we describe this procedure where the UE is sweeping through a set of UL beams and the gNB is receiving these. Such UL transmission happens based on the transmission of sounding reference signal (SRS) in UL and the gNB is able to indicate to the UE the best UL beam based on the SRI (Sounding Reference Signal Resource Indicator) which is an SRS index.

U2 procedure is used to enable gNB measurement of different gNB RX beams to possibly change/select the best beam. In this procedure the UE would transmit the same beam for a certain amount of time, allowing for the gNB to perform measurements on different Rx beams and select the best one. In this procedure, depicted in Figure 4.33, the sweeping process takes place at the gNB on its receive beams. From the standards procedure, this procedure is simply based on the configuration of an SRS transmission for several occasions. All the SRS resources in the triggered SRS resource set would be configured with the same spatial relation. The gNB does its own measurements and selection of the best Rx beam at the end of this process, without informing the UE.

U3 procedure is used to enable the gNB measurements on the same gNB Rx beam to change the UE Tx beam in case the UE uses beamforming. As illustrated in Figure 4.34, the gNB uses the same receive beam during several time instances, while the UE can transmit based on multiple beams. The sweeping procedure happens at the UE and as in the case of U1, it is based on SRS transmission and SRI indication of the best transmitted beam. From an specification perspective, the U1 and U3 produces are pretty similar if not the same.

Figure 4.34 UL UE beam selection.

4.2.3 Beam Indication Framework for DL Quasi Co-location and TCI States

4.2.3.1 QCL

Let us start with a bit of fundamental discussion on QCL. In legacy systems, such as early LTE releases prior to Release 11, one of the main assumptions was that the transmitted signals originate from the same TRP. For each reference signal transmitted by the based station, the UE needs to estimate a set of parameters for constructing the channel estimation filter. Such parameters include, e.g. delay and doppler spread of the channel, time and frequency error. Not all the reference signals have the same densities in frequency or periodicities in time to allow standalone channel statistics estimation. In fact, this might not be even efficient in practical implementation where one channel estimation strategy would be to estimate the channel statistics from the reference signal with the highest number of available samples and infer some of those estimated statistics in the derived channel estimation filter coefficients from other reference signals. This holds however only when the signals are originating from the same TRP and have same spatial characteristics. When this is not true, the parameters of the channel estimator need to be obtained for each reference signal. This implies both increased UE complexity but also proper design of that particular reference signal, which leads to increased system overhead.

What is the QCL about? As mentioned above, the channel statistics corresponding to different physically located antenna ports would be completely different and it is beneficial to inform the UE about the *relation* between the different reference signals it receives. These relations are defined using the concept of QCL:

Definition: Two antenna ports are said to be quasi co-located if properties of the channel over which a symbol on one antenna port is conveyed can be inferred from the channel over which a symbol on the other antenna port is conveyed.

Strictly speaking, QCL defines the relation between two reference signals at the UE receiver. In practice, the gNB can only guarantee that the properties of two reference signals are similar if the two reference signals are transmitted from the same TRP.

In LTE, there are two QCL rules: QCL types A and B. In QCL type A the UE may assume that the antenna ports of Common Reference Signal (CRS), CSI-RS, and DM-RS have the same delay spread, Doppler spread, Doppler shift, and average delay. In QCL type B the UE may assume that antenna ports CSI-RS and DM-RS have the same to Doppler shift, Doppler spread, average delay, and delay spread.

NR considers the transmission of any reference signal from any TRP. In some sense this might imply that all the reference signals should be designed so that the UE can estimate the channel with enough accuracy without having to rely on another reference signal. Such a transmission may be thought of as being *self-contained*. Although such an approach is desirable, it would place quite tough requirements on all reference signals.

Instead, NR defines a set of QCL rules that can be signaled to the UE. These QCL rules define what properties are the same between the two reference signals. To reduce the signaling, the individual channel properties are grouped in four groups as follows:

- QCL type A: Doppler shift, Doppler spread, average delay, delay spread
- QCL type B: Doppler shift, Doppler spread
- QCL type C: average delay, Doppler shift
- QCL type D: Spatial Rx parameter

The QCL types A, B and C are applicable for all carrier frequencies, but the QCL type D is applicable only in higher carriers, like FR2 and beyond, where essentially the UE may not be able to perform omni-directional transmission, i.e. the UE would need to form beams.

The QCL types A, B, and C are like the QCL types defined in LTE. For example, if two RSs are QCL type B, the UE may assume that the Doppler shift and the Doppler spread of the two RSs are the same, and the UE may derive the Doppler shift and spread from RS 1 and used that when estimating the channel using RS 2.

Conveying spatial domain QCL properties is a new dimension of NR compared to LTE. As the system needs to support directional transmission, both the gNB and UE are able to create beams, and obviously both ends need the flexibility to change beams at certain points in time to handle, e.g. UE mobility. In some cases, the gNB must inform the UE when it changes its Tx beam, and it may also be necessary to inform the UE about the spatial Rx properties of the signal received using the new Tx beam. During the NR Release 15 design, many parameters have been considered as potential candidates to describe the spatial RX properties: e.g. Angle of Arrival (AoA), Dominant AoA, average AoA, Power Angular Spectrum (PAS) of AoA, average Angle od Departure (AoD), PAS of AoD, transmit/receive channel correlation, transmit/receive beamforming, spatial channel correlation. It was however agreed to refer to the generic term as *spatial Rx* having the understanding that the UE can receive the signals with the same RX beamforming weights. Hence, when we refer to the situation where reference signal A is spatially co-located with reference signal B, it means that the UE may assume that the signals can be received with the same spatial filter.

4.2.3.2 TCI Framework

We have seen so far that the reference signals can be linked to each other with respect to what the UE can assume about their statistics. All these rules need to be placed in a framework and this is the purpose of the Transmission Configuration Indication (TCI) framework which defines pairs of reference signals for QCL indication. The TCI describes which reference signals are used as QCL source, and what QCL properties can be derived from each reference signal. The framework also describes how the TCI states are signaled to the UE.

The following RS can be used a source RS to indicate TX beam for downlink:

- SS/PBCH block: RX beam used to receive certain SS/PBCH block is used as RX beam for the DL transmission
- CSI-RS for beam management
- CSI-RS for CSI acquisition
- CSI-RS for tracking

The following RS can be used a source RS to indicate TX beam for uplink:

- SS/PBCH block: RX beam used to receive certain SS/PBCH block is used as TX beam
- CSI-RS for beam management: RX beam used to receive certain CSI-RS resource is used as TX beam
- CSI-RS for Channel State Information (CSI) acquisition
- SRS: TX beam used to transmit certain SRS resource is used as TX beam

All reference signals except the SS/PBCH block and the periodic CSI-RS requires that a valid TCI state is provided. In cases where QCL Type D is *not* applicable, i.e. in FR1, a TCI state contains only a single reference signal, and that reference signal provides the large-scale channel properties corresponding to QCL Type A, Type B, or Type C. For cases when QCL Type D *is* applicable, i.e. in FR2, the TCI state contains two reference signals, where one of the reference signals provides the large-scale channel properties corresponding to QCL Type A, B, or C, and the second reference signal provides the large-scale channel properties corresponding to QCL Type D.

The TCI framework assists the reception of CSI-RS for CSI acquisition, CSI-RS for beam management, DM-RS for PDCCH demodulation and DM-RS for PDSCH demodulation. For each such *target* reference signal, only certain reference signals are allowed in the TCI state conveying the QCL information.

The reference signals allowed in the TCI state are provided in the tables below for the different types of target reference signals. For each target reference signal, each row in the tables contains two reference signals, where only the first reference signal is present in the TCI state for operation in FR1, whereas both reference signals are present for operation in FR2.

4.2.3.2.1 *TCI States: QCL Associations*

For a periodic tracking reference signal (TRS) there are two possible configurations TCI configurations, described in Table 4.5.

Let's discuss in detail the above two configurations. For configuration 1, the QCL Type C properties, i.e. average delay and Doppler shift, can be inferred from an SS/PBCH Block for the reception of the periodic TRS. In case QCL Type D is applicable, RS 2 is also configured in the TCI state, and the UE can utilize the same spatial Rx filter as used for the reception of the SS/PBCH block. The case in configuration 2 is more interesting: for operation in FR1 the UE infers the average delay and the Doppler shift of the periodic TRS from an SS/PBCH Block. For operation in FR2, the UE gets the average delay and Doppler shift from the SS/PBCH Block and the spatial RX properties from a CSI-RS for beam management. This means that the UE may adjust its Rx beam based on measurements on a CSI-RS

Table 4.5 TCI states for periodic TRS.

TCI state configuration	DL RS 1	qcl-Type1	DL RS 2 (if configured)	qcl-Type2 (if configured)
1	SS/PBCH Block	QCL-TypeC	SS/PBCH Block	QCL-TypeD
2	SS/PBCH Block	QCL-TypeC	CSI-RS for beam management	QCL-TypeD

Table 4.6 TCI states for aperiodic TRS.

TCI state configuration	DL RS 1	qcl-Type1	DL RS 2 (if configured)	qcl-Type2 (if configured)
1	Periodic TRS	QCL-TypeA	Periodic TRS	QCL-TypeD

Table 4.7 TCI states for CSI-RS for CSI acquisition.

TCI state configuration	DL RS 1	qcl-Type1	DL RS 2 (if configured)	qcl-Type2 (if configured)
1	TRS	QCL-TypeA	SS/PBCH Block	QCL-TypeD
2	TRS	QCL-TypeA	TRS	QCL-TypeD
3	TRS	QCL-TypeA	CSI-RS for beam management	QCL-TypeD
4	TRS	QCL-TypeB		

Table 4.8 TCI states for CSI-RS for beam management.

TCI state configuration	DL RS 1	qcl-Type1	DL RS 2 (if configured)	qcl-Type2 (if configured)
1	TRS	QCL-TypeA	TRS	QCL-TypeD
2	TRS	QCL-TypeA	CSI-RS for beam management	QCL-TypeD
3	SS/PBCH Block	QCL-TypeC	SS/PBCH Block	QCL-TypeD

for beam management, which may be transmitted with a narrower beam width than the SS/PBCH block.

For the aperiodic TRS, Table 4.6, the only valid TCI state contains one or two periodic TRSs. This means that the aperiodic TRS cannot exist in the system without a periodic TRS. While this is the Release 15 operation, it is also likely that in the future, aperiodic TRS could be transmitted without relying on the periodic TRS, from the QCL perspective this being a straight forward change.

For a CSI-RS used for CSI acquisition the possible configurations TCI configurations described in Table 4.7:

For a CSI-RS used for beam management the possible configurations TCI configurations described in Table 4.8:

For a DM-RS for PDCCH/PDSCH demodulation the possible configurations TCI configurations described in Table 4.9:

4.2.3.2.2 TCI Signaling Mechanisms The TCI states provide the UE with the QCL information necessary to receive the various reference signals. Unlike LTE, the QCL information contained in the TCI states are indicated to the UE via explicit signaling, thus, the UE must be informed about the QCL relation between any pair of reference signals.

Table 4.9 TCI states for DM-RS for PDCCH/PDSCH DM-RS.

TCI state configuration	DL RS 1	qcl-Type1	DL RS 2 (if configured)	qcl-Type2 (if configured)
1	TRS	QCL-TypeA	TRS	QCL-TypeD
2	TRS	QCL-TypeA	CSI-RS for beam management	QCL-TypeD
3	CSI-RS for CSI	QCL-TypeA	CSI-RS for CSI	QCL-TypeD

The signaling mechanism is different for different target reference signals:

- For CSI-RS, the TCI state is provided via RRC for periodic CSI-RS, via MAC CE for semi-persistent CSI-RS, and via a combination of RRC, MAC CE, and DCI for aperiodic CSI-RS. This is true for all types of CSI-RS: CSI-RS for CSI acquisition, CSI-RS for beam management and TRS.
- For PDCCH DM-RS, the TCI state is provided via a combination of RRC and MAC CE
- For PDSCH DM-RS, the TCI state is provided via a combination of RRC, MAC CE, and DCI.

It is obvious that in a beam-based system we need the flexibility of fast beam switching, such configuration being possible only by DCI. The DCI needs to select from a list of candidates, while the size of the DCI is proportional with the number of candidates, however, if the amount of candidates is large, then the DCI bits needs for this purpose become prohibitive. This is why, if there is a desire to configure a large amount of reference signals linkages, these cannot be really handled by the DCI as such but an intermediate selection step is needed. In other words, the DCI cannot have an efficient size if reading directly from a large RRC configured list, but it needs an intermediate down-selection step in the form of a MAC CE. This type of hierarchy is needed in order to allow a larger amount of RRC configured pairs of reference signals.

All the TCI states are defined in the PDSCH configuration, and any subsequent use of the TCI states refers to this pool. One example of TCI states configuration, and its use as QCL sources for some reference signals are depicted in Figure 4.35.

The most straightforward way to configure a TCI state as a source for a reference signal is to use RRC signaling. This method can be used for periodic CSI-RS, and also for PDCCH.

A semi-persistent CSI-RS is activated using MAC CE, as described in Section 4.33. In the activation command, the network also provides the UE with a TCI state that the UE would use as QCL source for the reception of that CSI-RS.

As described in Section 4.33, an aperiodic CSI-RS is scheduled by the CSI request field in the triggering DCI. The CSI request field points to a report configuration, which contains a CSI-RS resource set, and a list of TCI states, one TCI state per CSI-RS resource in the CSI-RS resource set. Different CSI requests points to different report configurations, which provides quite flexible means to provide a CSI-RS with a TCI state. One example of this indication is shown in Figure 4.36.

Note that the description above is somewhat simplified: the actual configuration contains more levels. Also, the list of report configurations can be larger than what can be accessed

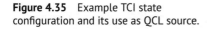
Figure 4.35 Example TCI state configuration and its use as QCL source.

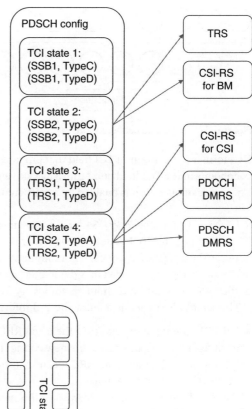

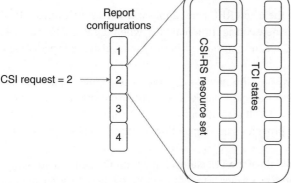

Figure 4.36 The TCI states for aperiodic CSI-RS is signaled via the CSI request field in DCI. The CSI request points to a report configuration, which contains the CSI-RS resources and the corresponding TCI states.

using the up to six bits in the CSI request field, in which case MAC CE is used to down select among the configured report configurations. More details can be found in Section 4.33.

For PDCCH DM-RS, a list of TCI states is configured in the CORESET. If the list has only a single element, the UE directly uses that TCI state as the source for its PDCCH DM-RS reception. If the list has more than one element, MAC CE is used to activate one of the TCI state, which the UE uses to receive the PDCCH DM-RS.

There are two ways to inform the UE about the TCI state to use for PDSCH. In one mode of operation, the TCI state of the PDSCH DM-RS can be tied to the PDCCH DM-RS TCI state. This is the only possible operation when the PDSCH is scheduled using DCI format 1_0.

Alternatively, it is possible to convey the TCI state of the PDSCH DM.RS explicitly in DCI in case there is a need to have different TCI states for PDCCH and PDSCH reception.

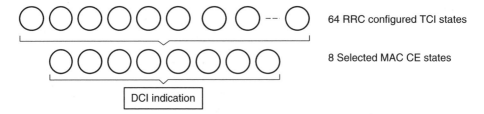

Figure 4.37 TCI states configuration.

DCI format 1_1 contains a TCI field that provides the UE with the TCI state to use, and the codepoints in the TCI field point into a list activated by MAC CE, as depicted in Figure 4.37. This mode of operation is enabled by a configuration option in the scheduling PDCCH.

4.2.3.2.3 TCI Timeline Rules So far, we have mentioned the signaling framework and the content of possible TCI states. No signaling is instantaneous: there will always be a gap between the time the network sends a command and the time the UE applies it. A set of rules and assumption s are needed in order to put together a proper timeline for the applicability of the QCL information without an ambiguity for the gNB or the UE.

The set of rules depends on how the TCI information is signaled:

- For TCI state signaling via RRC, the usual RRC rules applies: when a synchronized RRC reconfiguration is performed, the new configuration applies when the PRACH preamble is transmitted, and when an RRC reconfiguration without synchronization is performed, the new configuration applies ....
- For TCI state signaling via MAC CE, the new configuration is applied 3 ms after the HARQ ACK of the corresponding transport block is received. Note that this activation time if much shorter than, e.g. the SCell activation time. The 3 ms activation delay facilitates UE processing, and it also provides some protection in case the HARQ ACK is not safely received at the gNB.
- The timeline of the DCI-based update of the TCI state update is more complicated. This is mostly caused by the fact that there was a desire to reduce the activation delay as much as possible. Still, to apply a new TCI state signaled in DCI, the UE would need time to adjust its Rx beam, which has led to the introduction of a minimum permissible scheduling offset for PDSCH and aperiodic CSI-RS. Furthermore, the rules are different for PDSCH, CSI-RS for CSI acquisition and CSI-RS for beam management.

In Figure 4.38 we present a diagram of several PDSCH timeline configurations depending on the network and UE parameterization. At the core of these options we have the presence of the TCI signaling field in the DCI and the scheduling offset which can be larger or smaller than the scheduling threshold reported by the UE.

- The *presence or not of the TCI in the DCI* relates to the fact that there is a need for a set of default assumptions when the TCI is not included in DCI, this being a light operation mode when beam indication is not needed. For this to work, the reference signals/beam associations are hardcoded.
- The *scheduling threshold* accounts for the time needed for the UE to decode the DCI and potentially perform a beam change which could be requested by the network in DCI.

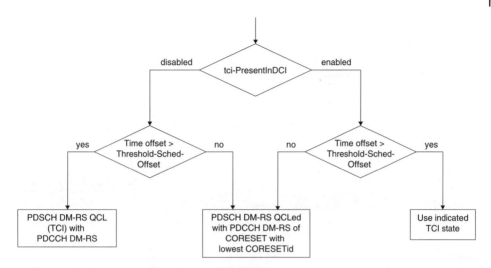

Figure 4.38 TCI timeline configurations for PDSCH.

Various UEs may have different such scheduling thresholds as some devices can change the Rx beams faster than others. If there is no enough time for such Rx beam change, a more conservative TCI operation needs to be assumed, like no Rx beam changes to be performed by the UE and a same Rx beam utilization for the PDSCH reception as it was used for the PDCCH reception. If there is enough time and the actual data scheduling comes after the necessary time for changing the Rx beam, then the UE would utilize the beams indicated by the TCI state in the DCI. This is the most versatile operation mode, allowing for dynamic beam indication.

A similar framework controls the TCI state handling for aperiodic CSI-RS, Figure 4.39. There is a scheduling threshold that accounts for the time it takes for the UE to decode the DCI and potentially change its Rx beam. The NW takes this scheduling threshold into account when it configures the CSI-RS resource set: each CSI-RS resource set contains a triggering offset that determines in which slot the CSI-RS resources are transmitted. If the triggering offset is larger than the scheduling threshold, the UE will apply the TCI state associated with the CSI-RS resource. If the triggering offset for CSI-RS for CSI acquisition is smaller than the scheduling threshold, the UE will apply a default QCL assumption. For CSI-RS for beam management, there is no such default QCL assumption specified.

4.2.4 Beam Indication Framework for UL Transmission

Downlink signals are transmitted under the QCL framework, or when not available, based on a set of default assumptions with respect to the QCL properties, especially QCL type D in FR2. A similar operation mode is needed in UL where from the spatial properties perspective it is important for the gNB to be aware of what UL beams the UE has used for the transmission of UL channels. In a directional transmission as it is the case in FR2, the gNB needs to be have the knowledge of the direction in which it "needs to listen." While the DL Tx beam indication is done based on the TCI framework, the UE TX beam indication for

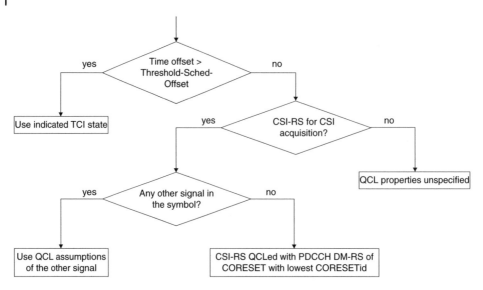

Figure 4.39 TCI timeline configurations for CSI-RS.

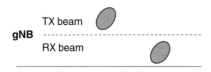

Figure 4.40 UL Beam indication example.

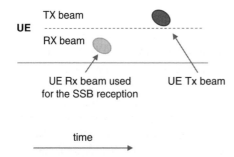

uplink transmissions is performed through the *spatial relations* that provide a reference RS based on which the UE determines the TX beam for the target UL signal to be transmitted. As discussed in Section 4.2.3.2., the following reference signals can be used as a reference RS to indicate TX beam for uplink:

- *SS/PBCH block*: UE RX beam used to receive certain SS/PBCH block is used as UE TX beam (Figure 4.40).
- *CSI-RS*: UE RX beam used to receive certain CSI-RS resource is used as UE TX beam.
- *SRS*: UE TX beam used to transmit certain SRS resource is used as UE TX beam.

Spatial relations are individually configured for the different types of uplink signals. For the PUCCH, a spatial relation is provided per PUCCH resource. The UE can be configured

Figure 4.41 Different Beams configured for a PUCCH resource, a single beam is active at a time.

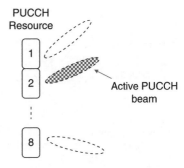

with up to eight spatial relations per each PUCCH resource, however, only one can be active at a time, Figure 4.41. The fact that only one is active at a time is because the UE is envisioned so far as being able to have a single active UL transmission at a time;if in the future multiple UL transmissions may be able in the same time, then more active spatial relations could be enabled. From a signaling perspective, the configuration of the eight spatial relations is provided in RRC level and dynamic activation of the spatial relation info in MAC CE thus supporting dynamic TX beam switch for the PUCCH. The fact that the amount of (pre-)configured spatial relations is high, means in fact that the UE would have an increased amount of choices for performing UL transmission. We can see here the system flexibility where the number of beams is thought to be high, or low, depending on the deployment and on the architecture of the gNB in terms of amount of beams which can be constructed. A high number of beams in DL means that the UE can be scheduled in a dynamic way by switching between beams, another possibility being that the high amount of beams arises from the fact that the gNB may operate with narrow beams.

4.2.4.1 SRS Configurations

The SRS resource can be configured to be transmitted periodically, in semi-persistent manner or triggered to be transmitted in aperiodic manner. Different approaches for providing spatial relation, depicted also in Figure 4.42, are defined according to time-domain behavior of the SRS resource:

- Periodic SRS resource can have spatial relation defined by DL RS which is periodic or semi-persistent and another periodic SRS.
- Semi-persistent sounding reference signals (SP-SRS) resource can have spatial relation defined by DL RS which is periodic or semi-persistent, periodic or another semi-persistent SRS.
- Aperiodic SRS resource can have spatial relation provided by periodic, semi-persistent or aperiodic DL RS or SRS.

The possible linkages between periodic, aperiodic, and semi-persistent SRS and the periodic, aperiodic, and semi-persistent downlink RS are natural. For example, a periodic SRS can be linked to a periodic DL RS or a semi-persistent DL RS (such an RS being similar to the periodic one except that it can be enabled/disabled) but it cannot be linked to an aperiodic DL RS. Indeed, since the nature of the aperiodic DL RS is that it is a one time transmission, this cannot control the spatial RS for a periodic SRS.

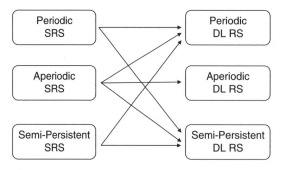

Figure 4.42 Possible spatial relations configurations of periodic, aperiodic and semi-persistent SRS to periodic, aperiodic and semi-persistent downlink RS.

4.2.4.2 Signaling Options for SRS Used for UL Beam Management

For a periodic SRS resource only an RRC level configuration of the spatial relation is supported which means that no dynamic beam switching is possible. For the periodic SRS resource, the reference RS can be SS/PBCH block, periodic, or semi-persistent CSI-RS or another periodic SRS.

Regarding SP-SRS, gNB configures by RRC signaling a reference RS for the SRS resource, which can be updated using a MAC CE activation command. The SP-SRS activation command for an SRS resource set overrides RRC configured or earlier activated spatial relations by providing a list of reference RSs, one per each resource within the set. Thus, a dynamic beam switching is supported for SP-SRS. The reference RS for the SP-SRS can be SS/PBCH block, periodic, or semi-persistent CSI-RS, periodic or semi-persistent SRS.

For an aperiodic SRS resource, a spatial relation can be provided via higher layer configuration only. The reference RS can be SS/PBCH block, periodic CSI-RS, semi-persistent CSI-RS, aperiodic CSI-RS, periodic SRS, semi-persistent SRS, or aperiodic SRS. Similar to periodic SRS, no dynamic beam switch is supported for the aperiodic SRS resource.

Beam-switching mechanism and applicable reference RS for SRS resource of different time-domain behavior is summarized in Table 4.10.

4.2.4.3 Beam Reporting from a UE with Multiple Panels

A UE can be equipped with one or multiple panels, these panels being placed on different sides of the device in order to guarantee the signal reception in any device position. The

Table 4.10 Beam switching mechanism and applicable reference RS for SRS resource of different time domain.

Time-domain behavior	Beam switch mechanism	Reference RS
Periodic SRS	RRC (semi-static)	SS/PBCH block, periodic CSI-RS, semi-persistent CSI-RS, periodic SRS
Semi-persistent SRS	MAC CE (dynamic)	SS/PBCH block, periodic CSI-RS, semi-persistent CSI-RS, periodic SRS, semi-persistent SRS
Aperiodic SRS	RRC (semi-static)	SS/PBCH block, periodic CSI-RS, semi-persistent CSI-RS, aperiodic CSI-RS, periodic SRS, semi-persistent SRS, aperiodic SRS

purpose of the panel specific UL beam selection is that the gNB can indicate the beam(s) of certain panel(s) to be used for the uplink transmission. This is intended for primary operation mode in FR2 where spatial source RSs for uplink signals are DL RSs. In other words, downlink RSs are used to determine UL TX beam utilizing TX/RX beam correspondence at the UE. That is to avoid SRS based beam search, selection and refinement schemes that are resource hungry and thus introduce high system overhead.

To facilitate panel specific UL beam selection using DL RSs as spatial sources for UL TX beams, UE panel specific beam reporting would be beneficial where UE would insert an identifier (that reflects panel visible to gNB) to each reported L1-RSRP measurement. In another alternative, the gNB could request the beam measurement result specifically from a certain UE panel by providing the identifier for which the measurements are to be provided. To enable that, a measurement configuration and reporting specifically for panel aware uplink TX beam selection separate from the downlink beam selection is seen as beneficial.

One further related aspect which was not discussed thoroughly in Release15 is the UE's potentially needed transmission power backoff as a function of spatial direction. Certain DL beams could have a human body as a blocker between the TX and corresponding (UE) RX beam. Depending on the propagation conditions, these beams could still have good observed L1-RSRP (e.g. even when assuming body loss of 3 dB) resulting in low pathloss (PL) estimate. Due to UE requiring additional MPR (maximum power reduction), P-MPR, to meet the emission related requirements (e.g. electromagnetic energy absorption requirements), the actual achievable PL for a given UL resource allocation will be reduced compared to the level that could be estimated based on the L1-RSRP. Thus, for the accurate DL and UL beam selection based on L1-RSRP beam measurements, the gNB should get separate information which DL RSs are feasible to determine good downlink "beam pair links," i.e. gNB TX beam and UE RX beam, and which DL RSs are feasible to determine good uplink "beam pair links," i.e. UE TX beam and gNB RX beam. Also, for that purpose (in addition to panel aware measurement) it would be beneficial if the UE would be able to provide separately measurement results on DL RSs:

1) for DL beam selection.
2) for UL beam and panel selection where UE would determine reported DL RS indices (together with measurement result and an [panel] identifier) that are good to determine UL TX beam where determination could be based e.g. on power headroom calculated per DL RS or some similar metric that considers achievable EIRP per UL TX beam.

4.2.5 Reporting of L1-RSRP

Prior to the introduction of NR Rel-15 specification, joint TX beam selection and CSI reporting procedure has been discussed and specified in the framework of LTE-A Rel-13 TS 36.213 [9]. Figure 4.43 shows an example of Class B CSI reporting in LTE-A Rel-13. A network configures multiple beamformed non-zero-power channel-state information reference signals (NZP-CSI-RS) resources in spatial domain, i.e. azimuth and elevation, where up to eight NZP-CSI-RS antenna ports per resource supported, UE to perform CSI measurements, i.e. L1-RSRP, Precoder Matrix Indicator (PMI), Channel Quality Indicator

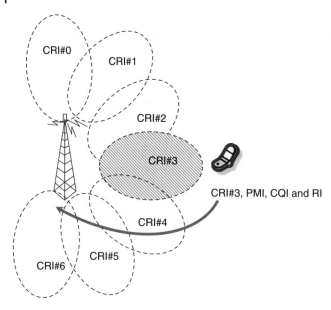

Figure 4.43 An example of Class B CSI reporting in LTE-A Rel-13.

(CQI), and RI, as well as the CSI reporting including CSI resource selection, i.e. DL TX beam. An interested reader is advised to see further details on CSI acquisition, PMI, RI and CQI in Section 4.3. Based on the multiple configured beamformed NZP-CSI-RS resources, the UE selects a single NZP CRI, i.e. DL TX beam in terms of measured L1-RSRP, as well as computes multiple CSI hypothesis per resource related to reported PMI, RI and CQI values. From the perspective of beam management in NR, the most important reporting parameters are L1-RSRP and CRI describing the received downlink TX beam power associated with a specific resource index, respectively. In addition to NZP-CSI-RS resources, NR Rel-15 also supports the use of SSB resources for beam management.

As discussed in Section 4.3, NR Rel-15 enables UE to be configured with $N \geq 1$ CSI Report Settings and $M \geq 1$ Resource Setting TS 38.214 [44]. The CSI Report Setting defines the content of the CSI report, the CSI hypothesis which shall be assumed as well as the time-domain behavior of the CSI report. The resource setting defines CSI measurement resources, e.g. resource type, resource allocation, time nature of resource, for a specific BWP.

To enable DL CSI measurements for beam management, NR Rel-15 supports two different CSI resources settings, i.e. NZP-CSI-RS and SSB TS 38.214 [44]. Referring to discussion in CSI-RS Section 3.7, the time–domain behavior of the CSI-RS resources for beam management can be configured by higher layer signaling to be aperiodic, periodic, or semi-persistent. The supported NZP-CSI-RS RE-pattern and antenna port configurations for beam management are discussed in CSI-RS Section 3.7.

Rel-15 provide supports for both UE non-group and group-based beam reporting schemes TS 38.214 [44]. Figure 4.44 shows an example of non-beam group based reporting with NZP-CSI-RS resources. Network can configure up to four CRIs with L1-RSRP values to be reported. Here, for simplicity, only two CRIs are shown when network configures beam reporting to be non-group based reporting, UE is not assumed to receive simultaneously

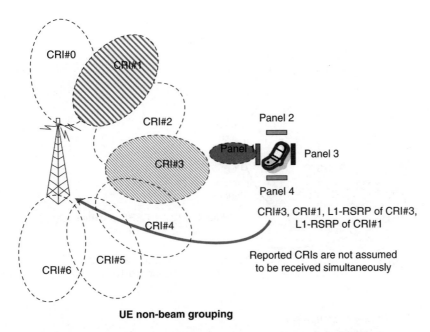

CRI#0

CRI#1

CRI#2

Panel 2

Panel 1

Panel 3

CRI#3

Panel 4

CRI#3, CRI#1, L1-RSRP of CRI#3,
L1-RSRP of CRI#1

CRI#4

Reported CRIs are not assumed
to be received simultaneously

CRI#5

CRI#6

UE non-beam grouping

Figure 4.44   An example of non-group beam reporting schemes in NR Rel-15.

reported CRIs associated with L1-RSRP values. When TX beam switch occurs among reported CRIs, some extra time need be reserved enable UE change it RX beam and/or antenna panel accordingly. As a result of this, the scheduling flexibility of the network is limited and the beam group-based reporting can be used.

Figure 4.45 shows an example of beam group based reporting with NZP-CSI-RS resources with a single UE beam group. When only single beam group is configured, UE is assumed to receive simultaneously up four CRIs. Here, for simplicity, only two CRIs are shown. Since multiple CRIs can be received with a single UE beam group, no extra time needs to be reserved when TX beam change occurs within the beam group. As a result of this, the scheduling restrictions of a network can be reduced.

To enhance further the spatial multiplexing capability of simultaneous downlink multi-beam transmission, network can configure two separate UE beam reporting groups with one CRI in each group, as shown in Figure 4.46. Here, it is assumed that UE has a capability to use at least two active antenna panels for the simultaneous reception. As result of this, two different downlink TX beams associated with CRIs can be received simultaneously. It is worth noting that Rel-15 specification is UE antenna panel and beam agnostic as well as the number of active UE antenna panels is up to UE reception capability.

Regarding to CSI report setting for beam management, both non- and differential based reporting for both non-group and group-based beam reporting are supported in NR Rel-15 TS 38.214 [44]. It is worth noting that a reporting format is reused among non- and group-based schemes. Differential reporting aims at reducing CSI reporting overhead by utilizing differential encoding. When the number of reported CRIs is larger than one, differential reporting is used. Figure 4.47 shows an example of non- and different reporting schemes. As can be seen, seven bit-length field is reserved to indicate quantized

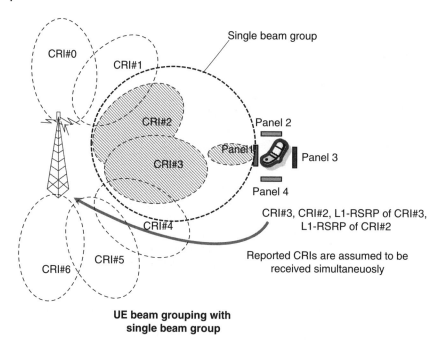

**UE beam grouping with
single beam group**

Figure 4.45 An example of group-based beam reporting scheme with single beam group in NR Rel-15.

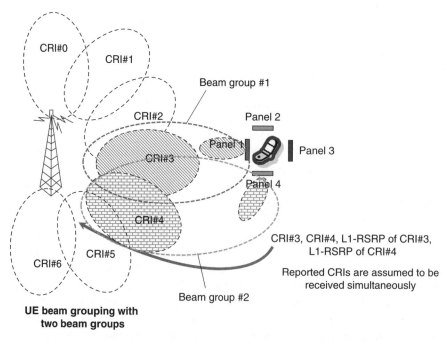

Figure 4.46 An example of group-based beam reporting scheme with two beam groups in NR Rel-15.

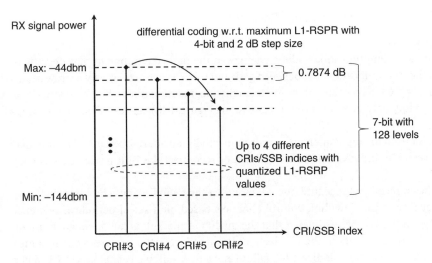

Figure 4.47 An example of non- and differential reporting in Rel-15.

Table 4.11 Summary for different combinations CSI reporting activation with different CSI-RS configurations in Rel-15 [TS 38.214].

CSI-RS configuration	Periodic CSI reporting	Semi-persistent CSI reporting	Aperiodic CSI reporting
Periodic CSI-RS	No dynamic activation	For PUCCH based reporting, MAC CE activates. For PUSCH based reporting, DCI activates.	DCI can activate reporting. Additionally, MAC CE can also active it.
Semi-Persistent CSI-RS	Not supported	For PUCCH based reporting, MAC CE activates. For PUSCH based reporting, DCI activates.	DCI can activate reporting. Additionally, MAC CE can also active it.
Aperiodic CSI-RS	Not supported	Not supported	DCI can activate reporting. Additionally, MAC CE can also active it.

measured L1-RSRP between the largest and smallest L1-RSRP value (-140 to -44 dBm). Additionally, four bit-length fields are reserved to indicate differentially coded L1-RSRP value with respect to the maximum value with two-bit step-size. Furthermore, network can configure up to four CRIs or secondary synchronization/physical broadcast channel block resource indicators (SSBRIs) associated with L1-RSRP values to be reported.

In Rel-15, the network can configure CSI reporting for beam reporting with different types of time periodicities, i.e. periodic/semi-persistent/aperiodic in conjunction with different UL channels, i.e. PUSCH/PUCCH. Table 4.11 provides a summary for different combinations of CSI reporting activation with different CSI-RS configurations. As shown, CSI-RS measurement resources can be periodic/semi-persistent/aperiodic. The payload size of the beam report can vary according to the number of reported and measured CSI resources as well as whether differential or non-differential coding is used.

4.2.6   Beam Failure Detection and Recovery

4.2.6.1   Overview

In beam-based communication sudden changes in the radio environment may degrade or interrupt the communication link between gNB and UE. The use of relatively narrow beams at UE and/or gNB makes the link prone to quality degradation due to blockage where e.g. an obstacle suddenly interrupts the transmission (Figure 4.48), as described in a previous section.

To recover from the rapid interruptions of connectivity, an alternative candidate link may exist between UE and gNB and to re-establish or reconnect the BFR procedure has been specified.

In the beam recovery procedure, the UE monitors the radio link by estimating the hypothetical quality of the downlink control channels based on a set of periodical reference signals. When the UE has estimated that the quality of the link is not adequate to maintain reliable communication, the UE declares beam failure. After declaring beam failure, the UE initiates recovery to indicate the failure and a new suitable beam to the gNB. When the UE-gNB link fails, it is understood that both the DL and the UL transmission are not possible on that particular link. Hence, one may wonder how the recovery indication may be performed if such links are not available anymore? This will be explained in the following sections.

BFR is a combination of L1 and L2 procedures. L2 part of the beam failure detection (BFD) and recovery procedures are covered in MAC specification 38.321 [43] and L1 procedures in 3GPP 38.213 [9] Physical layer Procedures specification. The BFR procedure in 38.213 is referred as *Link Recovery Procedures*.

On a high level, the BFR procedure consists of the following steps, Figure 4.49:

- BFD
- New candidate beam identification
- BFR request
- Recovery response

BFD and Radio Link Monitoring (RLM) have some similarities but are separate UE procedures. This is further discussed in Section 4.4.3.

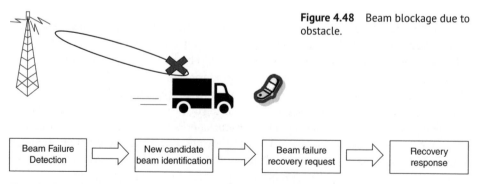

Figure 4.48   Beam blockage due to obstacle.

Figure 4.49   Beam failure recovery procedure steps.

4.2.6.2 Beam Failure Detection

BFD and declaration is a combined L1/L2 procedure where L1 provides the MAC layer indications of beam failure instances (BFIs). The MAC layer counts the indications and declares failure when configured maximum number of BFI indications has been reached.

4.2.6.2.1 *Configuration of BFD Reference Signals*

BFD is performed by evaluating the quality of the downlink reference signals, referred as beam failure detection reference signals (BFD-RS). BFD-RS are configured by the network using either explicit or implicit configuration, similar to RLM. In 38.213 [9] the BFD-RS are referred as set $q0$.

In *explicit configuration*, the network indicates UE the CSI-RS resource indices for BFD using RRC signaling. The explicit configuration is similar to Radio Link Monitoring Reference Signal (RLM-RS) configuration and it is possible to configure both BFD-RS and RLM-RS jointly, using explicit configuration (in the same message) to reduce the configuration overhead. It should be noted that for BFD, specific restrictions on the failure detection signals apply: for BFD only periodical CSI-RSs that are quasi co-located with the PDCCH DM-RS can be used.

In *implicit configuration*, the UE is not provided by the network a list of failure detection resources for BFD. The RS for BFD are determined based on the activated TCI states for PDCCH reception. When a specific TCI state for PDCCH is active, the UE determines the RS indicated by the TCI state to be used as BFD-RS i.e. the RS index is included in the set q0. If a TCI state includes two RSs, the UE selects the RS which is configured with QCL-typeD if the QCL type is configured. As QCL-typeD indicates UE the spatial RX assumption (i.e. UE is provided with information how to set its RX beam) it is configured when UE is assumed to use TX/RX beam forming. The BFD-RS is updated by UE based on the PDCCH beam indication by the network, i.e. when the network activates a new TCI state for PDCCH reception, the UE updates the set of RS used for BFD.

Regardless of the configuration method used, only periodical CSI-RS that are quasi co-located with the PDCCH DM-RS can be used for BFD.

4.2.6.2.2 *Beam Failure Instance Indication*

For BFD, the quality of each BFD-RS in set q0 is individually compared against the threshold Q_{out_LR}, which maps to 10% BLER of a hypothetical PDCCH. Similarly, as discussed in Section 4.4.1.1 for RLF, the evaluation of link condition is based on the estimated quality of a hypothetical downlink control channel based on the configured BFD-RS. The quality threshold Q_{out_LR} corresponds to the out-of-sync (OOS) threshold, Q_{out}, used for RLM which e.g. 10% BLER.

A BFI indication is provided to MAC when all the quality of all configured BFD reference signals are below the configured threshold Q_{out_LR}. If the quality of least one reference signal is above the configured threshold Q_{out_LR}, no failure instance indication is provided to MAC. The conditions are depicted in Figure 4.50

In non-discontinuous reception (DRX) mode, BFI indication interval is derived from the configured periodicity of the reference signals used for BFD. The UE determines the indication interval to be maximum between the shortest periodicity of configured RS(s) in the set q0 and 2 ms. Thus, the BFI indication interval has a lower bound of 2 ms. In DRX mode, the indication interval is lower bounded by the DRX cycle length.

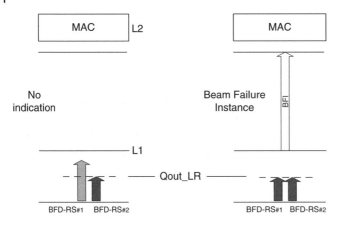

Figure 4.50 Beam failure instance indication.

4.2.6.2.3 ***Declaring Beam Failure*** Beam failure is declared by MAC layer based on the BFIs described in the previous section. The MAC layer implements a counter, a BFI counter, to count the failure indications. To prevent the counting of BFI indications from too long period and due to the lack of in-sync indications, as in RLM, the operation of the BFI counter is supervised by a BFD timer. Each time the MAC layer receives a BFI indication the BFD timer is restarted, and a BFI counter is increased by 1. When the timer expires, the BFI counter is reset to zero (0).

The BFD can be managed by configuring different timer values. The values are multiples of BFI indication intervals. As an example, configuring a timer value of one indication interval means that UE counts only consecutive BFI instances for failure detection. Setting the timer value to be multiple indication intervals allows some hysteresis in the BFD procedure, i.e. the BFI counter is not reset in case the BFI indications are not consecutive. This means that beam failure will be declared earlier.

When the BFI counter reaches the network configured maximum value a beam failure is declared. An example of BFD procedure is illustrated in Figure 4.51.

4.2.6.3 New Candidate Beam Selection

Once the UE has declared beam failure, it initiates the recovery procedure to indicate the gNB the failure and a new suitable beam for recovering the failed link. BFR procedure reuses the random-access procedure and both contention-based (CBRA) and contention free (CFRA) mechanisms are supported in NR.

For CFRA BFR, the gNB provides the UE with a list of candidate beams for new beam identification. Each candidate beam (a downlink reference signal) is associated with dedicated CFRA preamble and by transmitting the selected dedicated preamble, the UE indicates to the gNB both that beam failure has been declared and a new candidate beam for recovery As the preamble is dedicated, it also identifies the UE to gNB. The candidate beams which can be used for CFRA BFR can be SSB or CSI-RS signals or both.

Although the BFD metric is -based on the perceived reception quality of a hypothetical PDCCH, i.e. the evaluation takes into account possible interference component, the new candidate beam is selected based only on received signal strength. The candidate beam is

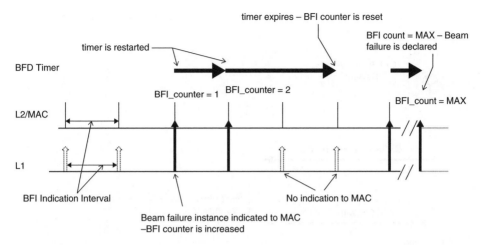

Figure 4.51 Illustration of beam failure detection procedure.

selected by MAC layer based on L1-RSRP measurements provided by the physical layer. If the UE has been configured with candidate beam list (the set of *q1*), it first checks if the L1-RSRP of any of the CFRA candidate beams is above the network configured selection threshold (*rsrp-ThresholdSSB*). If multiple CFRA candidates are above the threshold, the UE selects one of the candidates. If no CFRA candidates are above the threshold, the CBRA based recovery is used. Also, if UE is not configured with CFRA based recovery the CBRA BFR is used by default.

In CBRA recovery the UE performs CBRA where it indicates to the gNB an SSB by transmitting a corresponding preamble. The CBRA recovery is a normal RACH procedure where the SSB is selected based on L1-RSRP measurements. The UE selects one SSB with L1-RSRP above RSRP threshold (for CBRA) and if no SSBs has RSRP above the threshold, the UE can select any SSB as new candidate, as long as the cell is suitable. In MAC layer the selection procedure prioritizes the candidate beams that can be indicated using dedicated preambles.

In addition, the use of CFRA preambles for candidate indication, can be configured to be supervised by a *beamFailureRecoveryTimer*. When configured, and when the timer is running, the UE may use CFRA signaling for candidate indication. When the timer has expired, the UE cannot not use CFRA candidates for BFR. Regardless whether the timer is running or has expired, the UE has always the option to use CBRA BFR if the quality of the CFRA candidates are not above the quality threshold.

4.2.6.4 Recovery Request and Response

4.2.6.4.1 CFRA BFR Although the beam failure recovery request (BFRQ) is modeled as part of the RACH procedure, there is one notable difference between CFRA for BFR and normal CF random access regarding the monitoring of gNB response. In normal RACH procedure the UE monitors the RAR which is scrambled with RA-RNTI. However, in CFRA BFR, the gNB response is scrambled with C-RNTI. The CFRA BFR procedure is depicted in Figure 4.52.

For receiving network response on CFRA BFR request, a dedicated CORESET is configured by the network. This CORESET, CORESET-BFR, is monitored according to the search

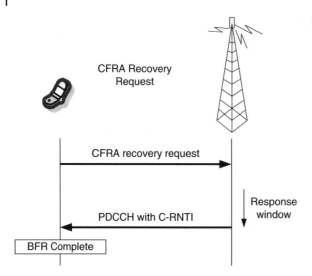

Figure 4.52 CFRA BFR procedure.

space configuration for BFR. CORESET-BFR is only monitored when BFR is performed using dedicated signals. When the UE has selected and transmitted CFRA preamble, the UE monitors the gNB response during a monitoring window.

When the UE has successfully received DCI scrambled with its C-RNTI it considers the BFR to be successful. After receiving the network response successfully, it continues to monitor PDCCH and PDSCH using the QCL assumption of indicated candidate beam, until a new TCI State for PDCCH (a new beam for PDCCH) is indicated (activated/reconfigured). Also, for the transmission of the uplink control channel the same spatial filter as was used for indicating the new candidate beam is used by UE until reconfigured by gNB.

4.2.6.4.2 CBRA BFR When BFR is performed using CBRA, no specific response is provided by the network and the recovery procedure is carried out as normal contention-based RACH procedure. The CBRA BFR procedure is depicted in Figure 4.53. As opposed to CFRA BFR where preamble explicitly indicates the recovery and new candidate beam, in CBRA BFR no specific indication is provided by UE to gNB during the recovery procedure. CBRA recovery complements CFRA based recovery and works as fall-back mechanism if CFRA candidates are not suitable, i.e. cannot be selected. Also, when CFRA recovery is not configured the CBRA is used by default.

If UE does not receive network response for the transmitted preamble (determined by the response monitoring window), for the retransmission of recovery request it reruns the CFRA/CBRA selection procedure.

4.2.6.5 Completion of BFR Procedure
From UE perspective, the BFR is completed successfully when UE receives gNB response in CFRA BFR or UE successfully completes CBRA procedure.

When UE has made configured maximum number of recovery attempts (either CFRA or CBRA) it determines that BFR is unsuccessful. Failure in RACH is indicated to RRC and the indication is modeled as RACH failure (random access problem) from MAC to

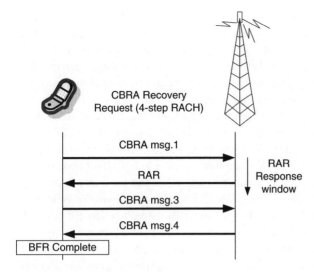

Figure 4.53 CBRA BFR procedure.

RRC. The unsuccessful BFR leads to declaration of RLF and subsequent RRC level recovery procedure. When UE successfully completes the BFR (CFRA or CBRA), no indication is provided to RRC.

4.3 CSI Framework (Sebastian Faxér, Ericsson, Sweden)

Multi-antenna techniques are an integral part of the NR system design, and the gNB is typically deployed with a large antenna array, which enables it to apply both beamforming for improving the received signal strength as well as spatial multiplexing for increasing the rank and achievable data rate of the transmission. However, in order to utilize the increased spatial degrees of freedom offered by the arrays, the gNB needs CSI for the UEs which it intends to serve. Generally speaking, the CSI is needed for two things, how to determine the precoding of the ports of the antenna array and how to set the link adaptation, i.e. selecting a proper Modulation and Coding Scheme (MCS) of the PDSCH transmission. Different types of precoding strategies require different types of CSI. For instance, the requirement of CSI accuracy for MU-MIMO transmission is much larger than for SU-MIMO transmission as the inter-user interference has to be suppressed at the transmitter side for the former case. For different antenna array sizes, the preferred CSI acquisition mechanism may also differ. For instance, for a small number of transmit antennas it may be preferred to sound each antenna with a separate, non-precoded, CSI-RS which are measured upon by all UEs in the cell where the UEs perform a full channel measurement and feeds back a CSI report corresponding to the entire antennas array, while for larger antenna sizes that approach would result in too large a reference signal overhead and instead UE-specifically beamformed CSI-RS operation is required where the reference signals are precoded using some prior or rough channel knowledge. There is also the possibility to apply coordinated multi-point transmission (CoMP, under development in Release 16) where neighboring TRPs coordinate either their scheduling, precoding, and/or link adaptation in order to improve performance. This type of operation requires yet another type of CSI measurement and report. In fact, as NR features only a single transmission scheme based on UE-transparently precoded DMRS, the actual transmission strategy in terms of utilized antenna array, deployment and precoding can be extremely flexible.

To enable the plurality of transmission and precoding strategies that are possible to implement with the standard, NR features a flexible and modular CSI framework. One of the key principles behind the design of the NR CSI framework was to decouple the CSI reports from the actual applied PDSCH transmission and treat them as independent entities. This enables the gNB to employ different CSI acquisition strategies and for instance combine different kinds of CSI reports or evaluate multiple CSI transmission hypotheses without being tied to a certain transmission mode. That is, the gNB is free to choose if it should apply the CSI recommendation by the UE or not. It also results in a forward-compatible design where the existing reporting structures can enable new deployments and use cases. Another motivation for the modular structure of the CSI framework is to enable different types of CSI content to be derived from the same measurement resources and to enable resources to be shared, or pooled, meaning that a fixed set of resources can be dynamically allocated to different users at different times, in an efficient manner.

A single framework is used for both CSI reporting and for beam measurement and reporting procedures for beam management (as discussed in Section 4.2), and in fact, a beam reporting can be seen as just a special kind of CSI report. In addition, the CSI framework is tightly integrated with carrier aggregation (CA) operation, supporting measurements and reporting both for the same and across carriers.

In the following sections, the constituent part of the NR CSI framework will be elaborated.

4.3.1 Reporting and Resource Settings

On the top level of the CSI framework (Figure 4.54), the UE can be configured with $N \geq 1$ CSI Report Settings (*CSI-ReportConfig*) and $M \geq 1$ Resource Setting (*CSI-ResourceConfig*). The CSI Report Setting defines to content of the CSI report, such as what CSI parameters to fed back, their frequency-granularity, the CSI hypothesis which shall be assumed as well as the time-domain behavior of the CSI report. A CSI Report Setting is also associated with a certain carrier and thus multiple CSI Report Settings needs to be configured to enable multi-carrier CSI reporting for CA.

The Resource Setting on the other hand can be seen as a container to logically group the resources for channel and/or interference measurement that is intended to be used for a certain purpose. The Resource Setting is associated with measurement resources for a certain BWP. Two types of Resource Settings can be configured, either based on NZP CSI-RS resources or on CSI-IM resources. As described in Section 3.7, the NZP CSI-RS is a reference signal which enables the UE to estimate the coherent channel of the antenna port over which the CSI-RS is conveyed, and the configuration of CSI-RS resource indicates both which REs the RS is transmitted on as well as the RS sequence used. The CSI-IM resource on the other hand defines only a number of REs where the UE is supposed to perform a

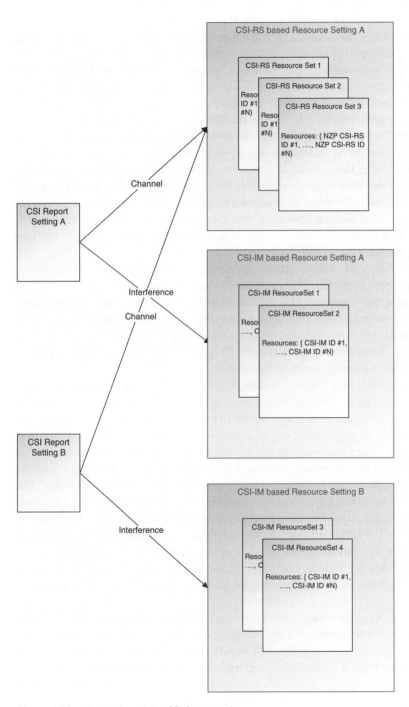

Figure 4.54 Illustration of the CSI framework.

non-coherent measurement of the received interference power. It is also possible to include SSB Resources in a Resource Setting to enable beam reporting.

A Resource Setting is also associated with a time-domain behavior, which can be either periodic, semi-persistent or aperiodic, and all CSI-RS/CSI-IM resource with a Resource Setting have the same time-domain behavior. Periodic CSI-RS/CSI-IM resource are assumed to be always active and present once they have been configured to the UE via dedicated RRC signaling, and a periodic CSI-RS/CSI-IM resource is therefore associated with a periodicity and slot offset, indicating the time-domain position. A semi-persistent (SP) CSI-RS/CSI-IM resource is also configured with a periodicity and slot offset similar to the periodic resource, but is assumed to be inactivated when first configured by RRC. To activate the CSI-RS/CSI-IM resource, an activation command needs to be transmitted using a MAC CE message, and upon receiving such an activation command, the UE assumes that the SP CSI-RS/CSI-IM resource is active and present until a MAC CE deactivation command is received. The detailed message structure is further explained in Section 2.4.5. An aperiodic CSI-RS/CSI-IM resource is only present when explicitly triggered by DCI and constitutes a one-shot measurement.

Typically, periodic and semi-persistent resources are used when the resources are intended to be shared by multiple UEs in the cell, i.e. to encompass cell-specific measurements. Or, to give the opportunity for the UE to perform filtering of the measurements across multiple time-instances to improve the measurement accuracy. Aperiodic resources, on the other hand, are primarily used when each resource is UE-specific, such as when applying UE-specific beamforming on the CSI-RS, conducting aperiodic beam sweeping, or to evaluate a short-term transmission hypothesis.

The CSI-RS/CSI-IM resources are not directly defined in the Resource Setting but are first logically grouped into $S \geq 1$ CSI-RS Resource Sets (or $S \geq 1$ CSI-IM Resource Sets). In fact, all available CSI-RS/CSI-IM resources are first defined and configured as lists on a top-level in *CSI-MeasConfig*, which can be seen as a resource pool. The CSI-RS/CSI-IM resource indices are then configured in the definition of the Resource Sets, which implies that a CSI-RS/CSI-IM resource may be present in multiple Resource Sets, which enables for instance efficient sharing of resources from the resource pool between different UEs. Only a single Resource Set within a Resource Setting is used for measurement for a certain CSI report, and in fact, periodic/semi-persistent Resource Settings can only contain a single Resource Sets (i.e. $S = 1$). In that case, the Resource Setting structure serves no functional purpose, however for aperiodic Resource Settings, multiple (up to $S = 16$) Resource Sets can be defined and which Resource Set is to be used for a certain aperiodic measurement and report is indicated in the aperiodic trigger state, which will be further described in Section 4.3.3.

The Resource Settings thus described the measurement resources while the CSI Report Settings describe the CSI calculations and reporting format which are to be applied on the measurements. To accomplish this each CSI Report Setting refers, or links to, one or more Resource Setting(s) for either channel or interference measurement. This can be done in a number of ways depending on which mode of CSI is configured and a CSI Report Setting can be linked with either one, two or three Resource Settings.

- For beam management operations using P2 or P3 beam sweeping, only channel measurements are needed and thus the corresponding CSI Report Setting is linked with a single NZP CSI-RS based Resource Setting for channel measurement.
- For other types of, regular, CSI reports, both channel and interference measurement are required, and the corresponding CSI Report Setting is linked with one NZP CSI-RS based Resource Setting for channel measurement and one CSI-IM based Resource Setting for interference measurement.
- A new functionality in NR, compared to LTE, is the support of NZP CSI-RS based interference measurement (as will be explained further in Section 4.3.2). In this case, the corresponding CSI Report Setting is additionally linked with one NZP CSI-RS based Resource Setting for interference measurement (in addition to the two linked Resource Settings for the regular CSI report), meaning that the UE measures interference on both CSI-IM resource (typically inter-cell interference) and NZP CSI-RS resources (typically intra-cell interference). It is also possible to only configure a single NZP CSI-RS based Resource Setting for interference measurement purpose, whereby the UE needs to estimate inter-cell interference from the residuals of the NZP CSI-RS.

When only CSI-IM based interference measurement is utilized, there is a one-to-one mapping between the NZP CSI-RS resources in the Resource Set for channel measurement and the CSI-IM resources in the Resource Set for interference measurement that is used for a certain CSI report. Each pair of (NZP CSI-RS resource, CSI-IM resource) thus constitute a channel/interference hypothesis. Typically, though, when multiple such resources are configured in the Resource Set, the gNB is only interested in probing different channel hypotheses whereby the multiple CSI-IM resource in the Resource Set are duplicates of the same resource.

For NZP CSI-RS based interference measurement, it is only possible to configure a single NZP CSI-RS resource in the Resource Set for channel measurement and a single CSI-IM, this implies that the UE cannot perform any channel/interference hypothesis selection. However, multiple NZP CSI-RS resources for interference measurement can be configured, where each resource typically corresponds to the precoded channel of a hypothetical co-scheduled UE. In this case, the multiple configured resources shall all be accounted for as part of a single channel/interference hypothesis, i.e. there is a one-to-many mapping between channel and interference measurement resources.

The CSI Report Settings are also configured with a time-domain behavior, which can be either periodic, semi-persistent, or aperiodic, and defines on which time occasions the associated CSI report is to be transmitted.

Periodic CSI reports can only be transmitted on PUCCH, as resources for PUSCH needs to be dynamically indicated whereas the PUCCH resource to be used to convey the periodic CSI report can be preconfigured. As PUCCH resources are defined for each UL BWP, which can be dynamically switched, each periodic CSI Report Setting is configured with an associated PUCCH resource for each UL BWP candidate. Typically, periodic CSI reports are configured with a relatively longer reporting periodicity and is only intended to give a coarse, wideband, CSI to be used for initial packet scheduling or PDCCH link adaptation.

Once the UE has more data to be transmitted, finer granular CSI can be requested using semi-persistent or aperiodic reporting. Similar to periodic CSI-RS/CSI-IM resources, a periodic CSI report is activated once the UE receives the corresponding RRC configuration.

Semi-persistent CSI reports can be transmitted either on PUCCH or PUSCH. PUCCH-based semi-persistent channel state information (SP-CSI) reporting, is similar to periodic CSI reporting except that the SP-CSI reports needs to be activated (and deactivated) via a MAC CE command. PUSCH-based SP-CSI reporting on the other hand is activated by DCI, implying that the resource allocation can be changed dynamically which enables the gNB to perform more adaptive link adaptation of the CSI report or adapt to varying CSI payload. This is not possible to the same extent for PUCCH-based CSI reporting since the PUCCH resource is pre-configured.

Aperiodic CSI reports can only be carried on the PUSCH and constitutes a one-shot report which is typically triggered whenever the gNB requires up to date fine granular CSI. PUSCH has much larger payload capacity compared to PUCCH and can therefore carry the more heavy-weight subband CSI and even multiple CSI report in case CA operation is used (the longer PUCCH Formats 3&4 also has capacity for subband CSI reporting, but to a much lower extent than PUSCH). In typical operations, aperiodic CSI reports are triggered by an UL DCI which also schedules UL-SCH data transmission whereby the CSI reports are multiplexed with the data on the same PUSCH transmission, but it is also possible to transmit only CSI without data (by setting the UL-SCH flag in DCI Format 0_1 to zero).

The time-domain behavior of the CSI Report Setting must have the same or more dynamic time-domain behavior than the associated Resource Settings. That is, periodic CSI reporting can only be done based on periodic CSI-RS/CSI-IM resources, whereas semi-persistent reporting can be done on either periodic or semi-persistent resources and aperiodic reporting can be done on resources with any time-domain behavior. This makes sense, since otherwise the report may need to be done on resources which have not yet been activated or triggered, which would result in an error case. To reduce the number of possible configuration combinations, there is a restriction that all associated Resource Settings for a CSI Report Setting needs to have same time-domain behavior.

A summary of possible combinations of time-domain behaviors of CSI Report Setting and Resource Setting and the associated triggering mechanisms are presented in Table 4.12.

Table 4.12 Summary of possible combinations of time-domain behaviors of CSI report setting and resource setting and the associated triggering mechanisms.

		Periodic CSI reporting	Semi-persistent CSI reporting	Aperiodic CSI reporting
Time-domain behavior of Resource Setting	Periodic CSI-RS	No dynamic triggering/activation	PUCCH-based: MAC CE PUSCH-based: DCI	DCI
	Semi-persistent CSI-RS	Not supported	PUCCH-based: MAC CE PUSCH-based: DCI	DCI
	Aperiodic CSI-RS	Not supported	Not supported	DCI

4.3.2 Reporting Configurations and CSI Reporting Content

In the CSI Report configuration, the content of the CSI report is defined. In general, there are two types of approaches that can be used when feeding back CSI, *explicit* or *implicit* CSI feedback.

In an explicit feedback methodology, the UE strives for feeding back a representation of a channel state as observed by the UE on a number of antenna ports, without taking into account how the reported CSI is processed by the eNB when transmitting data to the UE. Similarly, eNB also does not know how such a hypothetical transmission is processed by the UE on the receiver side. The channel part of the feedback report could consist of quantized coefficients of the $N_R \times N_T$ channel matrix $\boldsymbol{H}$, the TX-side correlation matrix $\boldsymbol{H}^H \boldsymbol{H}$ or its eigenvectors, or a quantity derived thereof (such as a measured RSRP).

In an implicit feedback methodology, on the other hand, a UE feeds back a desired transmission hypothesis as well as consequences of that transmission hypothesis. This is the approach used for both NR and LTE CSI reporting, where the transmission hypothesis is indicated as a desired precoder matrix $\boldsymbol{W}$ selected from a set of candidate precoder matrices (a precoder codebook) which is applied to the measured CSI-RS ports to form the hypothetical transmission. In conjunction with this, the UE typically reports an appropriate MCS for the transmission hypothesis, which thus takes receiver processing into account.

4.3.2.1 The Different CSI Parameters

NR defines a number of different CSI parameters that make up the CSI reporting quantity and where a certain number of limited combinations of CSI parameters can be configured to be reported, where each combination is associated with a certain transmission hypothesis. In the following subsections we will stay in the logic of implicit feedback described above, where (Figure 4.55), CSI-RS resources are transmitted by the gNB to the UE and CSI components (CRI, PMI, RI, LI, CQI) are fed back to the gNB.

4.3.2.2 CSI-RS Resource Indicator (CRI)

First of all, a certain CSI is only calculated conditioned on a single NZP CSI-RS resource for channel measurement (which may be coupled with a CSI-IM resource according to one-to-one mapping or a multiple NZP CSI-RS resources for interference measurement). So, if the number of NZP CSI-RS resources in the Resource Set for channel measurement is larger than one, the UE first performs a down selection and indicates a preferred CSI-RS resource using a *CSI-RS Resource Indicator (CRI)*. Typically, the multiple CSI-RS resources in the Resource Set are precoded differently and correspond to different beam directions. The CRI can thus be seen as a preferred beam indicator. In Figure 4.2, the UE is configured

Figure 4.55 Illustration of the CSI feedback loop.

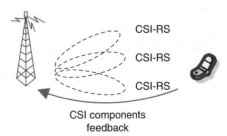

CSI-RS

CSI-RS

CSI-RS

CSI components
feedback

Table 4.13 Limitation on number of CSI-RS ports per CSI-RS resource.

Number of CSI-RS resources K_s in a resource set	Maximum number of ports per CSI-RS resource
1	32
2	16
3-8	8

with three CSI-RS resources on three beams, while the second beam is identified as the best one. The UE can either determine the best CSI-RS resource (i.e. best beam) by calculating a complete CSI report for each candidate CSI-RS resource, however this may be costly in terms of processing and the UE may instead use a simplified approach such as determining the CRI based on received signal strength of the CSI-RS resources.

For beam reporting (i.e. L1-RSRP), the number of CSI-RS resources in the Resource Set can be $K_s \leq 16$ and each CSI-RS resource can in that case contain at most two antenna ports. For other types of CSI, which is more complex to determine, between $1 \leq K_s \leq 8$ CSI-RS resource can be configured in a Resource Set and the limitation on how many ports each resource can contain is according to Table 4.13. The CRI is thus defined as the local index of the preferred CSI-RS resource in the Resource Set and can be conveyed with $\lceil \log_2 K_s \rceil$ bits. If only a single CSI-RS resource is configured in the Resource Set, CRI reporting is omitted.

4.3.2.3 SSB Resource Indicator

For L1-RSRP reporting, it is also possible to configure the measurement resources as an SS/PBCH Block. In that case, similarly to as for NZP CSI-RS, the UE can be configured to measure on an SSB Resource Set and feed back a preferred SSB Resource using a SSBRI.

4.3.2.4 Precoder Matrix Indicator (PMI) and Rank Indicator (RI)

The NR CSI feedback is designed around the UE indicating a desired precoder matrix $W \in \mathbb{C}^{N_P \times \upsilon}$, from a precoder codebook, where N_P is the number of CSI-RS ports and υ is the rank of the precoder matrix. The codebooks (NR supports a number of different precoder codebooks) are defined separately for each transmission rank υ and the desired precoder matrix is jointly indicated with the Rank Indicator (RI) and the PMI, where the latter indicates the precoder within the rank-specific codebook.

The selection of an appropriate transmission rank by the UE depends on both the rank of the propagation channel itself, as well as on the Signal-to-Noise and Interference Ratio (SINR) level experienced by the UE. The rank of the propagation channel depends on the number of multi-path components and their angular distribution. A pure line-of-sight channel consisting of a single strong tap is typically of rank-2, due to the two orthogonal polarizations of the radio channel, whereas a channel consisting of multiple taps can be up to rank $\min(N_P, N_{RX})$. However, even if the channel itself is full rank, it may not be optimal for the UE to select a high-rank transmission with multiple layers since the SINR per layer is reduced due to transmission power sharing between layers and inter-layer interference. Typically, the SINR per layer needs to be sufficiently large, i.e. in the logarithmic region of the Shannon capacity ($\log_2(1 + \text{SINR}) \approx \log_2(\text{SINR})$) rather than in the linear region ($\log_2(1 + \text{SINR}) \approx \text{SINR}$) for increasing the transmission rank to be beneficial.

The number of candidate precoders in the precoder codebook scales with the number of antenna ports, since the beam pattern becomes more narrow as the number of antennas increases and thus more beams are needed to cover the entire intended angular serving area. This means that for a small number of antenna ports, the precoders can be explicitly listed in a table and the PMI can simply be an index pointing into this table. This is the case for a 2 Tx codebook, which only consists of four candidate precoders for rank-1 and two candidate precoders for rank-2. However, for a larger number of antenna ports, listing all precoders explicitly becomes infeasible and instead the precoder codebooks are expressed using formulas that depend on a number on indices. In order to conserve the feedback overhead, the reporting granularity in frequency for these indices can also vary. For instance, on a high-level,[1] the precoder matrices $W \in \mathbb{C}^{N_P \times v}$ in the NR codebooks can be described as being decomposed into two matrix factors which are each matrix factor is indicated using separate indices i_1 and i_2 respectively, such that $W(i_1, i_2) = W_1(i_1)W_2(i_2)$, where $W_1 \in \mathbb{C}^{N_P \times K}$ is precoder matrix targeting the long-term channel which is selected on a wideband basis using the index i_1 while $W_2 \in \mathbb{C}^{K \times v}$ targets the short-term/frequency-selective properties of the channel and can be reported on a per-subband basis using the index i_2. Here, K represents the size of the "conversion dimension" and defines how the precoder space is divided between the wideband and per-subband parts. The motivation for such a design is to have K sufficiently small so that lions share of the precoder space can be selected by the wideband precoder part, which only needs to be reported once, or at least with far lower periodicity than the subband part, so that the per-subband overhead is kept small.

To make it more complicated, the wideband and subband precoder indices i_1 and i_2 are not in fact integer indices themselves, they are in turn comprised of sub-indices, (i.e. they can be expressed as vectors $i_1 = [i_{1,1}, i_{1,2}, \ldots], i_2 = [i_{2,1}, i_{2,2}, \ldots]$). The set of these sub-indices is what constitutes the reported PMI and each constituent sub-index is mapped to a number of bits and reported in uplink control information (UCI) as part of the CSI report.

The actual structure of the precoder codebooks in more detail are given in Section 4.3.7.

4.3.2.5 Channel Quality Indicator (CQI)

The CQI indicates a desired MCS, given a hypothetical PDSCH transmission where the reported precoder matrix (as indicated by the PMI and RI) is applied to the CSI-RS to form a set of hypothetical PDSCH layers. This informs the gNB that if it applies the reported precoder matrix directly for PDSCH precoding, it can set the MCS according to the reported CQI. Although, if the gNB decides not to schedule the exact transmission hypothesis recommended by the UE, by for instance performing MU-MIMO transmission or adapting the precoder or rank, the recommended CQI cannot be directly applied. It is therefore typical that the CQI is mapped to a SINR value, which can then be used as input for link adaptation.

When determining the CQI, the UE makes an assumption of a hypothetical PDSCH transmission scheduled with a single transport block and occupying the so-called *CSI reference resource*. The CSI reference resource defines a set of assumption the UE should make on the hypothetical PDSCH transmission, such as the bandwidth, reference signal overhead and

---

1 All precoders cannot be decomposed in this fashion, but it is useful to illustrate the principle.

what precoding is used. The CSI reference resource is also associated with a time instance for which the CSI should correspond. Essentially, the CSI reference resource points out a downlink slot as the timing reference, implying that no later channel/interference measurements than that slot shall be used as input for the CSI report.

Based on this hypothetical scheduling, the UE selects the highest CQI index (each CQI index corresponding to a modulation order and code rate) which would result in the hypothetical scheduling being received with a transport block error probability of less than target BLER. In typical operation, the target BLER is equal to 10%. However, to support Ultra Reliable low latency Communications (URLLC) with very high reliability requirements, the UE can be configured to report CQI using a special CQI table targeting low code rate and in that case the target BLER is equal to 10^{-5}. The CQI tables are intended to be a subset of the corresponding MCS tables, four-bit CQI tables are used whereas the MCS tables are five-bits, implying that the CQI table is constructed by roughly taking every other entry from the MCS table. There thus exists a corresponding CQI table for each MCS table, i.e. there are three CQI tables in Rel-15: one for max 64QAM modulation, one for max 256QAM and the aforementioned one for URLLC. The 64QAM CQI table is given in Table 4.14.

For the corresponding PDSCH scheduling, only a single wideband MCS value can be indicated that applies to the entire scheduled transport block. However, if the gNB utilizes frequency-selective scheduling, it can be of interest to the gNB to know which subbands experience good conditions (either by having a frequency-selective fading peak or experiencing a low interference level). This can be achieved by reporting subband CQI. If subband CQI is configured, the UE in addition to a wideband CQI, reports a differential subband CQI

Table 4.14 The four-bit 64QAM CQI table.

CQI index	Modulation	Code rate x 1024	Efficiency
0		out of range	
1	QPSK	78	0.1523
2	QPSK	120	0.2344
3	QPSK	193	0.3770
4	QPSK	308	0.6016
5	QPSK	449	0.8770
6	QPSK	602	1.1758
7	16QAM	378	1.4766
8	16QAM	490	1.9141
9	16QAM	616	2.4063
10	64QAM	466	2.7305
11	64QAM	567	3.3223
12	64QAM	666	3.9023
13	64QAM	772	4.5234
14	64QAM	873	5.1152
15	64QAM	948	5.5547

Table 4.15 Definition of subband differential CQI.

Sub-band differential CQI value	Offset level
0	0
1	1
2	≥ 2
3	≤ -1

value for each configured subband. A 2-bit differential subband CQI value reported for each subband, according to:

- Sub-band Offset level (s) = sub-band CQI index (s) − wideband CQI index.

The mapping from the two-bit sub-band differential CQI values to the offset level is shown in Table 4.15.

4.3.2.6 Layer Indicator (LI)

The Layer Indicator (LI) is intended to be used in FR2 operation when phase tracking reference signal (PT-RS) is transmitted in conjunction with the PDSCH in order to assist the UE in tracking and compensating for the phase noise, which is much more prevalent for higher carrier frequencies. To optimize the phase noise tracking, the PT-RS should ideally be transmitted with the same precoding as the strongest layer of the PDSCH, in order to maximize its SINR. However, in the PMI report, the preferred precoder matrix from the precoder codebook is indicated and there is no information about which column of the precoder matrix (which defines the precoding of a layer) is received with the highest SINR. The CQI cannot be used to infer this information either, since one codeword is used for up to rank-4, and thus one CQI value is reported corresponding to the up to four layers jointly.

Therefore, the LI was introduced to enable the UE to indicate which columns of the precoder indicated by the reported PMI corresponds to the strongest layer. Based on this information, the gNB can permutate the columns of the reported precoder before applying the precoder for PDSCH transmission to the UE (remember, the gNB does not need to follow the precoding recommendation and the actual precoding of PDSCH is transparent to the UE), so that indicated precoder column is mapped to the first DM-RS port where also the PT-RS is transmitted.

4.3.2.7 Layer-1 Reference Signal Received Power (L1-RSRP)

The L1-RSRP measurement quantity is used for beam reporting, where the linear average of the received power on the REs occupied by the reference signal (a CSI-RS or an SSB) is reported. L1-RSRP details are presented in Section 4.2.6.

4.3.2.8 Reporting Quantities

A limited number of different combinations of the described CSI parameters can be configured as the content of the CSI report by setting the parameter *reportQuantity*. Each possible value of report quantity is also associated with a certain CSI hypothesis assumption.

4.3.2.8.1 "none" In some cases, the gNB may only want to aperiodically trigger an aperiodic CSI-RS transmission to allow the UE to perform a measurement and adjust some of its reception parameters, without transmitting an associated CSI report. Specifically, there are two use cases:

1. Aperiodic transmission of the Tracking Reference Signal (TRS, or CSI-RS for tracking), which enables the UE to perform time/frequency synchronization refinement
2. Aperiodic Rx beam sweeping, a so called P3 beam sweep, which enables the UE to adjust its Rx spatial received filter (see Section 4.2.2 for beam management procedures).

Neither of these operations require CSI feedback to be transmitted back to the gNB, but since the aperiodic CSI-RS triggering is performed with the same framework as an aperiodic CSI request, with a CSI Resource Setting always being associated with a CSI Report Setting, the decision was made to introduce and "empty" CSI Report by setting *reportQuantity* to "none" rather than introducing a separate mechanism to trigger an aperiodic CSI-RS which is not associated with any CSI report.

4.3.2.8.2 "cri-RI-PMI-CQI" This value of report quantity corresponds to the conventional form of PMI/RI/CQI reporting familiar from LTE and is by far the most common CSI report configuration. With this report setting, the UE indicates a preferred precoder matrix from the codebook via first selecting the rank with the RI and then indicating the precoder matrix with the PMI. Conditioned on the reported precoder matrix, the UE feeds back a corresponding CQI. If the UE is configured with multiple CSI-RS resources for channel measurement, it first selects the preferred CSI-RS resource with the CRI. In many cases however, the UE is only configured with a single CSI-RS resource whereby the CRI reporting is omitted.

4.3.2.8.3 "cri-RI-LI-PMI-CQI" This reporting quantity is an extension of the 'cri-RI-PMI-CQI', where additionally the LI is reported. As mentioned, such a configuration is appropriate in FR2 operation where PT-RS is used and where the reported LI can assist the gNB in determining the optimal PT-RS port mapping.

4.3.2.8.4 "cri-RI-i1" The 'cri-RI-i1' reporting mode is intended to be used with hybrid beamformed/non-precoded CSI acquisition operation, Figure 4.56. That is, the gNB would primarily rely on UE-specifically beamformed CSI-RS, with a few ports per CSI-RS resource, for the CSI acquisition. But in order to know how to beamform the CSI-RS for each UE, the gNB would intermittently transmit a non-precoded CSI-RS resource

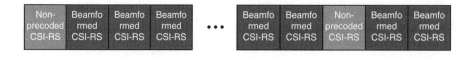

Time [ms]

Figure 4.56 Hybrid non-precoded/beamformed CSI-RS feedback.

with many, perhaps 32, ports. Based on this, the UEs could report a PMI corresponding to the full 32 ports, which would indicate a preferred beam direction. The intention is then that the beamforming of the CSI-RS can be based on the wideband/long-term part of the reported precoder, i.e. the W_1 matrix as indicated by the index i_1. That is, the reported W_1 matrix is used to virtualize the UE-specifically beamformed CSI-RS ports. Then, the short-term/subband CSI (corresponding to the W_2 part) is determined from the beamformed CSI-RS. If such an operation is desired, it is not necessary for the UE to report the W_2 part of the precoder in the non-precoded CSI report, as only the W_1 part is needed. Hence, the i_2 index may can be omitted to conserve reporting overhead. However, the overhead saving with this reporting mode is rather slim and the reporting mode is associated with an optional UE capability, which makes the practical use of this *reportingQuantity* limited.

4.3.2.8.5 *"cri-RI-i1-CQI"* In this reporting mode, the UE reports a W_1 matrix and a CQI assuming semi-open loop CSI hypothesis. That is, the UE assumes that the gNB randomly cycles through the possible W_2 matrices in order to achieve transmit diversity. Such a precoding strategy can be useful if the short-term properties of the channel changes so rapidly so that the W_2 report would be outdated at the time the gNB would apply the precoding. This may happen in case of high Doppler such as when the UE is moving at a high speed. However, in many cases the long-term properties of the channel (as captured by the W_1 matrix) would still be valid even though the short-term properties vary too quickly. Hence, similarly to the "hybrid CSI" reporting mode, the UE only reports the i_1 part of the PMI, indicating a desired W_1 matrix. However, a CQI is still needed in this case in order to properly set the link adaptation, and to generate the CQI, the UE needs an assumption of what precoding should be assumed even for the W_2 part. Since any transmit diversity schemes in NR is transparent to the UE however, the UE does not know exactly what precoding the gNB will apply. Therefore, it is assumed for CQI calculation purpose that a random W_2 matrix is selected for each precoding resource block group (PRG), to mimic that the gNB would apply per-PRG precoder cycling. Here, the assumed PRG size for CQI calculation purpose is explicitly configured in the CSI Report Setting and may not be the same as the actual PRG size used for PDSCH DMRS (in order to decouple the CSI feedback from data transmission and as the actual PRG size dynamically vary with DCI indication). It is also possible for the gNB to indicate which subset of the possible W_2 precoders the UE should randomly pick from via indication of a codebook subset restriction of the i_2 index.

4.3.2.8.6 *"cri-RI-CQI"* This reporting mode is known as "non-PMI feedback" and thus operates without the use of a PMI codebook. It is intended to be used with reciprocity-based operation where the gNB can derive the desired PDSCH precoding based on SRS measurements in the uplink. However, the interference conditions experience by the UE cannot be known to the gNB based on UL channel measurements, only the channel part of the CSI, and hence, the gNB needs to know a proper rank selection and CQI. Therefore, in this reporting mode, it is assumed that the gNB precodes the CSI-RS in a similar fashion as PDSCH. The UE then only needs to select a desired RI and determined a corresponding CQI. In other words, the UE simply measures the corresponding beamformed CSI-RS antenna ports and assumes that an identity matrix is used as the precoding matrix over

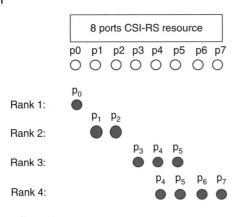

Figure 4.57 Example of port indication for non-PMI feedback.

the corresponding CSI-RS ports, as the actual precoding has already been applied at the transmit side. Which CSI-RS port shall be used for a certain rank hypothesis is indicated in the CSI report setting via the *non-PMI-PortIndication* field. This serves two purposes. Firstly, it assures that both rank-nested (e.g. where layer v corresponds to CSI-RS port v irrespective of rank hypothesis) and more generally non rank-nested (e.g. where layer 0 for rank-1 is transmitted on port 0, layers 0-1 for rank-2 is transmitted on ports 1–2, etc) precoding can be applied. An example port indication is given Figure 4.57.

Typically, such as when SVD-based precoding is used, the rank-nested property holds. However, for more advanced precoding schemes, such as SLNR or MMSE precoding, that is not necessarily the case. Secondly, the port indication allows a single CSI-RS resource to be shared among multiple UEs so that a single UE only uses a subset of the CSI-RS ports within the resource.

4.3.2.8.7 "cri-RSRP" and "ssb-Index-RSRP"
These two types of reporting are used for beam reporting. As their names imply, the UE is required to compute RSRP measured on CSI-RS or SS Blocks. In principle, the UE may be configured with up to 64 CSI-RS and/or SSB resources for which RSRP needs to be computed/updated. However, in order to reduce complexity, such RSRP update is dependent also if the CSI-RS and/or SSBs may be grouped. The grouping means that CSI-RS or SSBs received with the same spatial filter by the UE (hence the same beam) may be considered as part of the same group. The use of grouping is controlled by the gNB and hence if grouping is enabled, the gNB would indicate this to the UE by configuring the higher layer parameter *groupBasedBeamReporting* set to "enabled." If such configuration happens, the UE is not required to update measurements for more than 64 CSI-RS and/or SSB resources, and the UE shall report in a single reporting instance two different CRI or SSBRI for each report setting, where CSI-RS and/or SSB resources can be received simultaneously by the UE either with a single spatial domain receive filter, or with multiple simultaneous spatial domain receive filters. If on the other hand the higher layer parameter *groupBasedBeamReporting* set to "disabled," in this case the UE is not required to update measurements for more than 64 CSI-RS and/or SSB resources, and the UE shall report in a single report *nrofReportedRS* (higher layer configured) different CRI or SSBRI for each report setting.

Table 4.16 Nominal subband sizes.

Bandwidth part (PRBs)	Subband size (PRBs)
<24	N/A
24–72	4, 8
73–144	8, 16
145–275	16, 32

4.3.2.9 Frequency-Granularity

The CSI Report Setting also defines which part of the bandwidth the CSI should correspond to, and in addition, what granularity in frequency the CSI should have. To accomplish this, the bandwidth of a BWP is divided into a number of subbands according to Table 4.16: Nominal subband sizes. That is, one out of two possible subband sizes can be configured and the candidate subband sizes depends on the bandwidth of the BWP. Since subband CSI feedback overhead scales linearly with the number of subbands, the subband sizes were chosen so the largest number of subbands did not exceed 19. In addition, it is a desirable property that the subband sizes are both an integer multiple of the PRG size as well as an integer multiple of the Resource Block Group (RBG) size to avoid misalignment, since a UE can be configured to report a PMI and a CQI for each subband. If the subband size was not an integer multiple of the PRG size (which is either two or four PRBs), the subband boundary could occur in the middle of a PRG and hence the subband PMI cannot be applied directly to the precode the PDSCH DMRS and one of the subband PMIs overlapping the PRG at the boundary would have to be selected, which may not be the optimal selection. Similarly, if the subband boundary and RBG boundary is not aligned, there could be complications when determining the MCS based on the subband CQI(s) if frequency-selective scheduling is used. Another requirement to avoid misalignment is that that the PRG, RBG, and subband grids start from the same reference point. That is, they are all aligned to the starting PRB of the carrier rather than the starting PRB of the BWP. This implies that there may potentially be edge subbands (where the actual subband size is smaller than the nominal subband size) at both the start and the end of the BWP.

Based on this division of the BWP into subbands, the CSI reporting band for the CSI report is defined as an arbitrary subset of subbands of the BWP, which is indicated as a bitmap where each bit corresponds to one subband, Figure 4.58. The UE should only take the subbands in the CSI reporting band into account when determining the CSI. This accomplishes so called *partial band* CSI reporting. For instance, different services may be multiplexed in different parts of the band, and hence, different partial bands may experience different interference conditions or have different requirements of the CSI content. It is therefore useful for the gNB to be able to control exactly which part of the bandwidth the UE shall measure and report CSI for. This is in contrast with LTE, where the CSI report always corresponds to the entire carrier bandwidth.

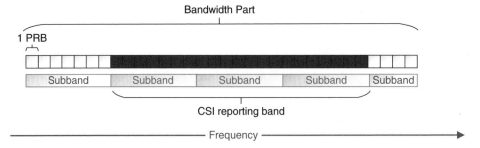

Figure 4.58 Illustration of CSI reporting band as subset of subbands within the BWP.

The CSI Report Setting also defines the respective frequency-granularity of the PMI and CQI, which can be either wideband or subband. For wideband PMI/CQI, a single PMI/CQI corresponding the entire CSI reporting band is reported whereas for subband PMI/CQI, a separate PMI/CQI is reported for each constituent subband in the CSI reporting band. As described in the previous sections, the subband CQI is differentially encoded against a wideband CQI and if subband CQI is configured the wideband reference CQI per codeword is reported in addition. Similarly, for subband PMI, only part of the PMI (the W_2 matrix corresponding to the i_2 index) is reported per subband in addition to a single wideband PMI (the i_1 index).

As a general classification, NR categorizes a CSI Report Setting into wideband and subband frequency-granularity as follows: wideband PMI/CQI reporting, beam reporting, hybrid CSI report, semi-open loop reporting and non-PMI feedback (with wideband CQI) is classified as wideband frequency-granularity CSI, whereas the other configurations of a CSI Report Setting is classified as having a subband frequency-granularity. Only CSI Report Settings with wideband frequency-granularity is allowed to be periodically reported on short PUCCH.

Which of subband or wideband CSI is used is a trade-off between CSI accuracy and UCI overhead. Depending on what UL coverage the UE has, different amount of bits can reliably be fed back. Thus, a UE with good UL coverage could be configured with subband PMI/CQI reporting whereas UEs with poor UL coverage would typically be configured with wideband PMI/CQI.

4.3.2.10 Measurement Restriction of Channel and Interference

In the CSI Report Setting, it is also possible to configure measurement restriction in the time-domain for channel and interference resources respectively. If measurement restriction is configured, the UE is only allowed to use the latest occurrence of the CSI-RS/IM for channel/interference measurement into account when deriving the CSI. That is, the UE is restricted from temporally averaging the measurement of the resources. Otherwise, the UE would typically perform such averaging in order to improve the channel/interference estimation performance. However, if the gNB intends to apply for instance different beamforming weights when precoding the CSI-RS in different time occasions, the averaging would ruin the CSI-RS measurement, and thus the gNB needs a mechanism to turn off the averaging at the UE. Measurement restriction thus enables the gNB to re-use a periodic or semi-persistent CSI-RS/IM resource in time between different

UEs so that the same resource can be shared. In some sense, this is an alternative to using aperiodic CSI-RS/IM.

4.3.2.11 Codebook Configuration

If PMI reporting is applied, which PMI codebook is used among with the parameters for the codebook is configured in the CSI Report Setting. Four codebooks are defined for NR, namely:

- Type I Single-Panel codebook
- Type I Multi-Panel codebook
- Type II codebook
- Type II Port Selection codebook

The Type I codebooks targets SU-MIMO operation and offers a "regular" spatial resolution with a relatively low overhead, whereas the Type II codebooks targets MU-MIMO operation and gives a much finer spatial resolution which is required for intra-cell interference suppression. This comes at the cost of substantially larger overhead. The Type I codebooks can be further configured in two CodebookModes, which offer slightly different performance and overhead. The Type II codebooks, on the other hand, can be more precisely tuned and offers several knobs to adjust the spatial resolution and overhead. The codebooks are more thoroughly explained in Section 4.3.7.

A CSI Report Setting can also be configured with so called codebook subset restriction as well as rank restriction. The codebook subset restriction indicates to a UE that some precoders in the codebook are not allowed to be selected for the PMI report. That is, some precoders are "forbidden." Configuring codebook subset restriction can be useful to control inter-cell interference. For instance, the gNB may know that some precoders, which correspond to certain transmission directions, cause a large amount of interference to neighboring cells. By restricting the UE from selecting those precoders, the UE will be forced to select different precoders which have better inter-cell interference properties.

4.3.2.12 NZP CSI-RS Based Interference Measurement

A new feature of NR is the support of NZP CSI-RS based interference measurements. The intention with this new interference measurement mode is to provide CQI taking the intra-cell MU-MIMO inter-user interference into account. In typical operation, the gNB would determine how many UEs N to co-schedule and the intended precoders $\{W_n\}_{n=0}^{N-1}$ for each UE n by either UL sounding using reciprocity or by requesting a first set of CSI reports from the UEs (which then indicates the desired PMIs). Conditioned on this scheduling hypothesis, a second set of UE-specifically beamformed CSI-RS are then transmitted, where the CSI-RS for a UE is precoded with the intended precoder W_n and each UE is allocated a separate CSI-RS resource (since the CSI-RS ports are already precoded with the intended PDSCH precoder, NZP CSI-RS based interference measurement is typically used in conjunction with non-PMI feedback).

For a certain UE, its allocated CSI-RS resource is included in the CSI-RS resource set for channel measurement whereas the CSI-RS resources for the other, co-scheduled, UEs are included in the CSI-RS Resource Set for interference measurement. The UE then assumes that each CSI-RS port of the resources for interference measurement corresponds to an interference layer which shall be taken into account when calculating the CQI. Typically, the UE is also configured with a CSI-IM resource to capture the inter-cell interference. While it is not explicitly defined in the specification how the UE shall combine the interference estimates from the multiple resources, to simply accumulate the interference estimate is a likely implementation.

Since different CSI-RS resources may have different PDSCH EPRE to NZP CSI-RS EPRE ratio, this needs to be taken into account when calculating the accumulated interference. That is, if $\boldsymbol{H}_i$, $i = 1, \ldots, N-1$ are estimated channels for the NZP CSI-RS resources for interference measurement, $P_c^{(i)}$ is the ratio of PDSCH EPRE to NZP CSI-RS EPRE for resource i and $\boldsymbol{R}_{\mathrm{CSI-IM}} = \frac{1}{K} \sum_{k=0}^{K-1} \boldsymbol{y}_k \boldsymbol{y}_k^H$ is the interference covariance estimate from the CSI-IM (where $\boldsymbol{y}_k$ is the received signal on CSI-IM RE k and K is the number of REs in CSI-IM or averaging region), the total interference covariance matrix used for CQI calculation could for instance be calculated as

$$R_{\mathrm{TOT}} = \boldsymbol{R}_{\mathrm{CSI-IM}} + \sum_{i=1}^{N-1} P_c^{(i)} \boldsymbol{H}_i \boldsymbol{H}_i^H$$

4.3.3 Triggering/Activation of CSI Reports and CSI-RS

While periodic CSI-RS/IM resource and CSI reports are always assumed to be present and active once configured by RRC, aperiodic and semi-persistent CSI-RS/IM resources and CSI reports needs to be explicitly triggered or activated.

4.3.3.1 Aperiodic CSI-RS/IM and CSI Reporting

For both aperiodic CSI-RS/IM resources and aperiodic CSI reports, the triggering is done jointly by transmitting a DCI Format 0-1. This is the DCI format which schedules PUSCH transmission where the aperiodic CSI report is to be carried. The DCI Format 0_1 contains a *CSI request* field which can be configured to be between 0 and 6 bits wide. That is, it can contain at most $S_c = 2^6 = 64$ codepoints. If this field is set to all zeros, no CSI is requested, and the DCI only schedules a regular PUSCH transmission containing UL data. A non-zero codepoint on the other hand points to a so-called *aperiodic trigger state* configured by RRC. An aperiodic trigger state is defined as a list of up to 16 aperiodic CSI Report Settings, identified by a CSI Report Setting ID, (but typically, a much lower number of report settings is used) for which the UE simultaneously should calculate CSI for and include in the scheduled PUSCH transmission. If a CSI Report Setting is linked with periodic/semi-persistent Resource Setting(s), no further information is needed since there is only one Resource Set included in the Resource Setting for channel/interference measurement. However, if the CSI Report Setting is linked with aperiodic Resource Setting (which can comprise multiple Resource Sets), which CSI-RS/IM Resource set should be used for measurement must be indicated. Hence, this allows the gNB, for a given CSI Report Setting, to dynamically switch which CSI-RS/IM resource shall be used for measurement each time the aperiodic report is triggered, by indicating different aperiodic trigger states. This means that the aperiodic NZP CSI-RS Resource Set for channel measurement, the aperiodic CSI-IM Resource Set for interference measurement (if used) and the aperiodic NZP CSI-RS Resource Set for interference measurement (if used) to use for a given CSI Report Setting is also included in the

DCI codepoint

Aperiodic trigger state

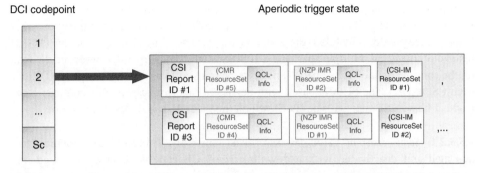

Figure 4.59 Illustration of aperiodic trigger states and mapping from a DCI codepoint.

aperiodic trigger state definition. For aperiodic NZP CSI-RS, the QCL source to use (i.e. the TCI state) is also configured in the aperiodic trigger state, which enables the gNB to dynamically switch Rx beam assumptions for the CSI-RS. The aperiodic trigger state definition is illustrated in Figure 4.59.

It is possible to configure up to 128 aperiodic trigger states via RRC. However, the number of codepoints of the CSI request bitfield in DCI only ranges between 0–64. Therefore, it is possible that more trigger states are configured in RRC than can be indicated with the DCI field. If that is the case, i.e. M aperiodic trigger states are configured in RRC but the CSI request bitfield (with bitwidth $N_{TS} = 0, \ldots, 6$) only contains $S_c = 2^{N_{Ts}} - 1 < M$ non-zero codepoints, an intermediary sub-selection, or mapping, between the S_c codepoints and the M RRC configured trigger states needs to be performed. This sub-selection is performed by transmitting a MAC CE sub-selection command, as is explained in more detail in Section 2.4.5.

The aperiodic CSI-RS/IM is essentially a one-shot measurement which is only present for a single time instance and is only used to determined CSI for a single aperiodic report. The position, in time, of the aperiodic CSI-RS/IM is defined as a slot offset relative to the slot where the DCI which trigger it was received. The slot offset is defined on a CSI-RS resource set level. For aperiodic CSI-IM, there is no explicit slot offset defined but rather it is assumed that the CSI-IM and CSI-RS is present in the same slot to enable efficient CSI processing at the UE.

4.3.3.2 Semi-Persistent CSI-RS/IM and CSI Reporting

While aperiodic CSI-RS/IM and aperiodic CSI reporting is jointly triggered by the CSI request field in DCI, semi-persistent CSI-RS/IM resources and semi-persistent CSI reports are independently activated. This is because semi-persistent CSI-RS/IM resources does not necessarily need to be used together with semi-persistent CSI reporting (but can be used with aperiodic CSI reports), and vice versa, semi-persistent CSI reports can utilize periodic CSI-RS/IM resources. It is also possible that the same semi-persistent CSI-RS or CSI-IM resource is shared between multiple CSI Report Settings.

Semi-persistent CSI reporting can either be PUCCH-based or PUSCH-based, where each option is targeting different use cases. PUCCH-based semi-persistent reporting is more akin to periodic CSI reporting but with the additional functionality to turn the reporting ON or OFF via L2 signaling and is intended for more lightweight CSI. PUSCH-based semi-persistent reporting on the other hand can be seen as a multi-shot aperiodic CSI report. For instance, if the gNB knows that the UE has a lot of data in its DL buffer and knows that it will be frequently scheduled, the gNB can trigger a semi-persistent report on PUSCH instead of triggering multiple aperiodic CSI reports on PUSCH in order to reduce the load on the control channel (only one DCI needs to be transmitted instead of a separate DCI for each CSI report). The PUSCH-based semi-persistent reporting can then encompass more payload-heavy CSI, such as Type II CSI, which the PUCCH-based reporting cannot.

4.3.3.2.1 *Semi-Persistent CSI-RS/IM* Semi-persistent CSI-RS and CSI-IM resources are activated and deactivated with a downlink MAC CE command. Once activated, the resources are assumed to be present and available for measurement until a deactivation command is received. This activation/deactivation is synchronous and is assumed to take effect 3 ms after the slot where the HARQ-ACK acknowledging the PDSCH which carried the MAC CE was transmitted by the UE. Both semi-persistent CSI-RS and CSI-IM are activated/deactivated with the same MAC CE message and the activation/deactivation is performed on a resource set level. This is explained in more detail in Section 2.4.5.

4.3.3.2.2 *Semi-Persistent CSI Reporting on PUCCH* PUCCH-based semi-persistent reporting is activated/deactivated using a similar mechanism as semi-persistent CSI-RS and CSI-IM resources. That is, activation/deactivation is performed by transmitting a MAC CE message.

4.3.3.2.3 *Semi-Persistent CSI Reporting on PUSCH* PUSCH-based semi-persistent CSI reporting is activated by transmitting a DCI Format 0_1. To differentiate the triggering of an aperiodic CSI report with the activation of a semi-persistent CSI report, different RNTIs is used to scramble the CRC of the DCI. For aperiodic CSI reporting, the regular C-RNTI is used while for semi-persistent reporting, a separately configured SP-CSI RNTI is used. When the UE receives a DCI Format 0_1 CRC-scrambled with SP-CSI RNTI, it can thus interpret the CSI request field differently. For this case, a separate list *of semi-persistent trigger states* is defined in RRC, however their definition is much simpler than for their aperiodic counterparts: a semi-persistent trigger state only contains a mapping to a single (semi-persistent) CSI Report Setting. Due to this, there is no MAC CE sub-selection of trigger states required and a codepoint of the CSI request field essentially maps directly to a semi-persistent CSI Report Setting configured in RRC. A DCI CRC-scrambled with SP-CSI RNTI is only used for SP-CSI activation/deactivation of a single semi-persistent CSI Report Setting, i.e. UL data is not multiplexed on the semi-persistent PUSCH (semi-persistent scheduling of UL data is of course possible, but that is a separate feature).

To determine if the DCI contains an activation command or a deactivation command, the UE performs *validation* by verifying that a combination of DCI fields (which are not required for indicating scheduling information for the CSI report) are set according to pre-defined values. To validate a deactivation command, the bitfields must be set according to Table 4.17 while for validating an activation command, the bitfields must be set according to Tables 4.17 and 4.18.

Table 4.17 Bitfields used for validation of SP-CSI deactivation command.

	DCI format 0_1
HARQ process number	set to all '0's
Redundancy version	set to '00'

Table 4.18 Bitfields used for validation of SP-CSI activation command.

	DCI format 0_1
HARQ process number	set to all '0's
Modulation and coding scheme	set to all '1's
Resource block assignment	If higher layer configures RA type 0 only, set to all '0's;
	If higher layer configures RA type 1 only, set to all '1's;
	If higher layer configures dynamic switch between RA type 0 and 1, then if MSB is'0', set to all '0's; else, set to all '1's
Redundancy version	set to '00'

4.3.4 UCI Encoding

The reported parameters of the CSI report(s) are encoded in UCI and mapped to PUSCH or PUCCH and the encoding format used is different depending both on the physical channel used and the frequency-granularity of the CSI report(s). The reason for the different encoding schemes is that the payload size of the CSI generally varies with UE's selection of CRI and RI. That is, the codebook size for PMI reporting is different for different ranks, especially for Type II CSI reporting and subband PMI reporting in general, where it can vary drastically. Similarly, as one codeword is used up to rank-4 and two codewords are used for higher ranks, the number of CQI parameters (which is given per codeword) included in the CSI report will vary depending on the selection of rank.

For PUCCH-based CSI reporting with wideband frequency-granularity, the variation of PMI/CQI payload depending on the selected rank is not too large and therefore a single packet encoding of all CSI parameters in UCI is used. Since the gNB needs to know the payload size of the UCI in order to try to decode the transmission, the UCI is padded with a number of dummy bits corresponding to the difference between the maximum UCI payload size (i.e. corresponding to the RI which results in the largest PMI/CQI overhead) and the actual payload size of the CSI report. This ensures that the payload size is fixed irrespective of UE's RI selection. If this measure was not taken, the gNB would have to blindly detect the UCI payload size and try to decode for all possible UCI payload sizes, which is not feasible.

However, for PUCCH-based CSI with subband frequency–granularity as well as PUSCH-based CSI reporting, always padding the CSI report to the worst-case UCI payload size would result in too large overhead. For these cases, the CSI content is instead divided into two CSI Parts, CSI Part1 and CSI Part 2, where CSI Part 1 has a fixed payload size (and can be decoded by the gNB without prior information) but where CSI Part 2 has a variable payload size. The information about the payload size of CSI Part 2 can be derived from the CSI parameters in CSI Part 1. That is, the gNB first decodes CSI Part 1 to obtain a subset of the CSI parameters, based on these CSI parameters, the payload size of CSI Part 2 can be inferred, and CSI Part 2 can be subsequently decoded to obtain the remainder of the CSI parameters.

For PUCCH-based subband CSI reports and PUSCH-based reports with Type I CSI feedback, the CSI Part 1 contains RI (if reported), CRI (if reported) and CQI for the first codeword while CSI Part 2 contains PMI the CQI for the second codeword when RI > 4.

For Type II CSI feedback on PUSCH, CSI Part 2 in addition contains an indication of the number of "non-zero wideband amplitude coefficients" per layer. The wideband amplitude coefficient is part of the Type II codebook and depending on if a coefficient is zero or not, the PMI payload size will vary, which is why an indication of the number of non-zero such coefficients needs to be included in CSI Part 1.

For beam reporting of L1-RSRP, a single part encoding is also used since the payload size cannot vary.

4.3.4.1 Collision Rules and Priority Order

It can happen that two or more CSI report transmissions "collide," in the sense that they are scheduled to be transmitted simultaneously (for instance a periodic and an aperiodic). It may also occur that a number of CSI reports scheduled to be transmitted simultaneously result in too large a payload size so cannot fit in the UCI container (for instance due to that HARQ-ACK and/or SR additionally needs to be multiplexed). For these situations, some CSI reports may have to be dropped or omitted. To know which CSI reports to prioritize in this case, a number of prioritization rules are defined.

CSI reports are first prioritized according to their *time-domain behavior* and *physical channel*, where more dynamic reports are given precedence over less dynamic reports and PUSCH has precedence over PUCCH. That is, an aperiodic report has priority over a semi-persistent report on PUSCH, which in turn has priority over a semi-persistent report on PUCCH, which has priority over a periodic CSI report. This means that if an aperiodic report is scheduled at the same time where a periodic report is to be transmitted, the periodic report is dropped and not reported.

If multiple CSI reports with the same *time-domain behavior* and *physical channel* collide, the reports are further prioritized depending on CSI content, where beam reports (i.e. L1-RSRP reporting) has priority over regular CSI reports. The motivation is that the CSI report is typically conditioned on a serving beam, so if the beam is not correct the CSI report is useless anyway.

If there is still need for differentiation, the CSI reports are further prioritized based on for which serving cell the CSI corresponds (in case of CA operation). That is, CSI corresponding to the PCell has priority over CSI corresponding to Scells. Finally, in order to avoid any ambiguities in which CSI report is to be transmitted, the CSI reports are prioritized based on the *reportConfigID*.

The above priority rules are applied so that only a single CSI report is transmitted in case of CSI collision, with the exception of if multiple PUCCH-based CSI reports collide. In this case, it is possible to configure the UE with a larger "multi-CSI" PUCCH resource, where several CSI reports can be multiplexed in case of collision. In this case, as many CSI reports is possible without exceeding a maximum UCI code rate is transmitted in the "multi-CSI" PUCCH resource.

4.3.4.2 Partial CSI Omission for PUSCH-Based CSI

For PUSCH-based CSI reporting and Type II CSI reporting in particular, the CSI payload size can vary quite dramatically depending on the RI selection. For instance, for Type II reporting, the PMI payload for $RI = 2$ is almost double to that of $RI = 1$. Since the RI selection is not known to the gNB prior to scheduling an aperiodic CSI report on the PUSCH, the gNB has to allocate a PUSCH resources (i.e. in frequency and time domain) by using a best guess of the RI selection the UE will make, perhaps by looking at historic RI reports. Thus, it can happen that the gNB has allocated PUSCH resources with the assumption that the UE will report $RI = 1$, but the UE actually reports $RI = 2$. IN that case, it may be so that the CSI payload will not fit in the PUSCH container, i.e. the code rate will be too large or even the un-coded systematic bits will not fit. Instead of dropping the entire CSI report in this case, which would be quite wasteful, NR introduces a schemes of partial CSI omission, where a portion of the CSI (which can provide some utility to the gNB and at least give information about the RI selection so that the gNB can allocate a proper PUSCH resource for the next aperiodic CSI request) can still be reported.

This is accomplished by ordering the CSI content in CSI Part 2 in a particular fashion. If multiple CSI reports are transmitted in the PUSCH, the wideband CSI components (i.e. the wideband PMI and CQI) for all the reports are mapped to the most significant bits of the UCI. Then, the subband CSI for each report are mapped according to the previously described priority rules, where the subband CSI for even numbered subbands are mapped first, followed by subband CSI for the odd numbered subbands, as is illustrated in Figure 4.60: Example of partial CSI Part 2 omission.

If the resulting code rate of the UCI is above a threshold, a portion of the least significant UCI bits are omitted, until the code rate falls below the threshold. This means that subband CSI for odd numbered subbands for a report are omitted first. The motivation is that the gNB in this case would have subband PMI and CQI for ever other subband in the frequency domain and can therefore interpolate the PMI/CQI between two reported subbands to try to estimate the missing PMI/CQI values for the subband in the middle. While this will not result in perfect reconstruction, it is better than omitting CSI n entire chunk of consecutive subbands.

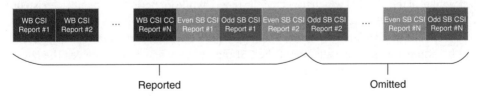

Reported   Omitted

Figure 4.60 Example of partial CSI Part 2 omission.

4.3.5 CSI Processing Criteria

In LTE, the concept of a CSI process was introduced in Rel-11 for the purpose of CoMP, i.e. feedback of several CSI corresponding to multiple TRPs. Each CSI process was associated with a specific kind of reporting configuration (i.e. CSI content and measurement resource) and the UE is assumed to always be able to provide CSI for all its supported CSI processes on a carrier. Thus, for LTE, the CSI computation capability reported by the UE is the support of a number of reporting configurations. However, such a tight coupling between CSI capability and configured reporting configurations was not suitable for the more flexible NR CSI framework. Instead, the NR CSI computation capability separates the number of supported *configured* CSI Report Settings and the number of supported *simultaneous* CSI calculations. That is, the concept of CSI process is generalized in NR with the introduction of the Channel State Information processing Unit (CPU), where the number of CPUs is equal to the number of simultaneous CSI calculations supported by the UE. The CPU can be seen as a generic CSI calculation engine which can process any kind of CSI report. The CPUs are a pool of computational resources. For instance, the UE can indicate support for four configured CSI Report Settings but only support a single simultaneous CSI calculation (i.e. supporting a single CPU). This means that the gNB can trigger any of the four different CSI reports, but has to multiplex the different CSI reports calculations in time. The different configured CSI report settings may for instance correspond to different codebook configurations (i.e. Type I and Type II codebooks), different types of beam reports (e.g. P2 and P3), different CSI hypotheses used in CoMP operation or CSI reports corresponding to different carriers.

The framework works as follows. When calculation of a CSI report is about to proceed, i.e. either when the UE gets triggered with an aperiodic CSI report or when the computation starts for a periodic or semi-persistent CSI report, the CSI report is allocated to one or multiple available CPU(s). If there are not enough CPUs available due to the fact that the UE is already processing other CSI reports, the CSI reporting to be allocated does not have to be calculated by the UE and it can instead report stale CSI, such as a previously calculated CSI report stored in memory or simply padding the CSI report with dummy bits. The CSI report is not dropped in this case, but some content is always transmitted in order to not change the rate matching procedures for the PUSCH or PUCCH transmission, which could be error prone. In practice, the gNB should strive for only triggering/configuring as many CSI reports as the UE is capable of handling so that stale CSI does not need to be reported by the UE.

Each CSI report that is committed for calculation by the UE thus occupies a number O_{CPU} CPUs from a starting allocation time until the last symbol of the physical channel (i.e. PUCCH or PUSCH) carrying the CSI report has finished transmitting, whereby the $O_{CPU}^{(n)}$ CPUs are released. For aperiodic CSI report, the starting allocation time of the CPU(s) is the last symbol of the PDCCH which triggered the report, while for periodic and semi-persistent CSI reports, the CPUs are allocated from the time of the occurrence of the latest CSI-RS/IM resource used to calculate the report. That is, for periodic/semi-persistent reports, the UE can be assumed to start calculation of the CSI report as soon as it has received the latest occurrence of the measurement resource.

The number of CPUs O_{CPU} occupied by a certain CSI report depends on the content of the report. For non-beam related CSI reports (i.e. when the reportQuanity is not equal to "cri-RSRP," "ssb-Index-RSRP," or "none"), the CSI report occupies as many CPUS as the number of CSI-RS resources in the CSI-RS resource set for channel measurement. This is because a UE may, in the worst case, need to calculate a complete CSI report for each CSI-RS resource in parallel in order to determine which CSI-RS resource is optimal and shall be selected with the CRI (of course, a UE implementation may use simpler approaches to determine the CRI, such as comparing the signal strength of the resources). For beam-related reports, on the other hand, the required computations are not as complex and only a single CPU ($O_{CPU} = 1$) is occupied, even if multiple CSI-RS resources are included in the CSI-RS resource set for channel measurement. The gNB also has the possibility to trigger an aperiodic TRS using the triggering mechanisms of the CSI framework, however this does not occupy any CPUs and it is instead assumed that the UE has dedicated resources for TRS processing.

If multiple CSI reports are about to be allocated to CPUs on a given OFDM symbol, they are ordered according to the set of priority rules described in Section 4.3.4. That is, if N CSI reports start occupying their respective CPUs on the same OFDM symbol on which $N_{CPU} - L$ CPUs are unoccupied, where each CSI report $n = 0, \ldots, N - 1$ corresponds to $O_{CPU}^{(n)}$, the UE is not required to update the $N - M$ requested CSI reports with lowest priority where $0 \leq M \leq N$ is the largest value such that $\sum_{n=0}^{M-1} O_{CPU}^{(n)} \leq N_{CPU} - L$ holds.

The concept of CPU occupation is illustrated in Figure 4.61: Illustration of CPU occupation, where CPU#1 gets allocated by the P CSI report in slot 0, which is the slot of the latest NZP CSI-RS occurrence (no later than the CSI reference resource) used by the P CSI report. While the P CSI reports is calculated, the UE gets triggered with two consecutive aperiodic CSI reports, which is allocated to CPU #2. After both CPUs are released, the UE gets triggered with two simultaneous aperiodic CSI reports, which respectively occupies CPU#1 and CPU#2. Before these CSI reports have finished calculating, the UE gets triggered with another aperiodic CSI report. However, since there are no CPUs available, that CSI report is not computed by the UE and instead stale or dummy CSI is reported.

4.3.6 CSI Timeline Requirement

When aperiodic CSI reports are triggered by PDCCH, not only does the UE need to have available computational resources to calculate the report, as was described in the previous section, it also needs enough time to perform the computation. In LTE, scheduling of PUSCH, where aperiodic CSI is carried, used a fixed scheduling offset of four subframes, corresponding to 4 ms and it was assumed that any triggered aperiodic CSI report could be calculated during this time period and no additional timing requirement for CSI was needed. However, since NR features both a more diverse set of CSI content with different computational complexities, as well as flexible scheduling offset of the uplink transmissions to carry the CSI report, a separate CSI timeline requirement is needed to ensure that the PUSCH carrying the CSI is not scheduled too aggressively.

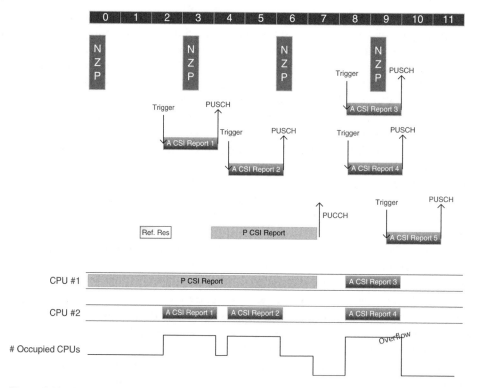

Figure 4.61 Illustration of CPU occupation.

The CSI reports (i.e. each CSI report Setting) is classified into three Latency Classes., each with different timing requirements. *Low Latency CSI* is classified as CSI which fulfills the following criteria:

- Wideband frequency-granularity (as defined in Section 4.3.2)
- A single CSI-RS resource (i.e. no CRI reporting) with at most four CSI-RS ports
- PMI reporting with Type I SinglePanel codebook or non-PMI reporting

The remaining types of CSI content, excluding beam reporting, is classified as *High Latency CSI* while beam reporting is defined as an own CSI Latency Class.

Two timing requirements for aperiodic CSI reporting is defined in NR. The first requirement is defined as the minimum number of OFDM symbols Z between the last symbol of the PDCCH triggering the aperiodic CSI report and the first symbol of the PUSCH which carries the CSI report. During this time, the UE needs to be able to decode the PDCCH, perform possible CSI-RS/IM measurements (if it does not already have an up-to-date previous channel/interference measurement stored in its memory), perform possible channel estimation, calculate the CSI report, and perform UCI multiplexing with UL-SCH. However, if aperiodic CSI-RS/IM is used with the report, this first requirement alone does not guarantee that that the UE has sufficient time to compute the CSI, since the aperiodic CSI-RS could potentially be triggered close to the PUSCH transmission. Therefore, the second requirement is defined as the minimum number of OFDM symbols Z' between the last symbol of

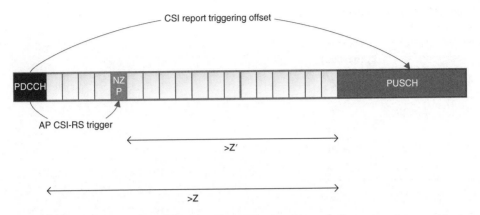

Figure 4.62 Illustration of the CSI timeline requirement (Z,Z') symbols.

Table 4.19 Timing requirement for low latency CSI, high latency CSI and beam reports.

μ	Low latency CSI (symbols)		High latency CSI (symbols)		Beam reporting (symbols)	
	Z_1	Z'_1	Z_2	Z'_2	Z_3	Z'_3
0	22	16	40	37	22	X_1
1	33	30	72	69	33	X_2
2	44	42	141	140	$\min(44, X_3 + KB_1)$	X_3
3	97	85	152	140	$\min(97, X_4 + KB_2)$	X_4

the aperiodic CSI-RS/IM used to calculate the report and the first symbol of the PUSCH which carries the CSI report. This is illustrated in Figure 4.62.

The numerical values for timeline requirements (Z,Z') for the three CSI Latency classes is given in Table 4.19: Timing requirement for Low Latency CSI, High Latency CSI and Beam reports, for each subcarrier configuration μ (where μ is counted in the smallest SCS if different numerologies are sued for PDCCH, PUSCH and/or CSI-RS). In practice, the only difference between the Z and Z' timing requirements is that the Z requirement should additionally encompass DCI decoding time, which is why the Z is typically a few symbols larger than the corresponding Z' value. For Low/High Latency CSI Classes, the timeline requirement is fixed in specification while it for beam reporting depends on the UE's reported capability on beam report timing (X_μ) and beam switch timing (KB_i). If the CSI report is multiplexed with UL-SCH, some additional OFDM symbols are added to the requirement to account for UCI and UL-SCH multiplexing time. If more than one CSI Report Setting is aperiodically triggered by a PDCCH, the largest (Z,Z') values are used for all the CSI reports. That is, if one High Latency and one Low Latency CSI report is triggered, the High Latency CSI timing requirement is used.

If the Z-criterion (or Z'-criterion) is not fulfilled and the gNB triggers the PUSCH too close to the PDCCH (or the aperiodic CSI-RS/IM), the UE can simply ignore the scheduling DCI

Table 4.20 Ultra-low latency CSI timing requirement.

	Z_1 (symbols)	
μ	Z_1	Z'_1
0	10	8
1	13	11
2	25	21
3	43	36

if the UE is not also scheduled with UL-SCH or HARQ-ACK and not transmit anything. If UL-SCH or HARQ-ACK needs to be multiplexed on the PUSCH however, the UE still transmits the PUSCH but pads the CSI report with dummy bits or transmits a stale CSI report.

Additionally, there is a special requirement for ultra-low latency CSI with a very short timeline. This timing requirement can only be applied if a *single* Low Latency CSI report is triggered, without multiplexing with either UL-SCH or HARQ-ACK and when the UE has all of its CPUs unoccupied. The UE can then allocate all of its computational resources to compute this CSI in a very short time. Hence, all of the CPUs become occupied for this duration and the UE cannot calculate any other CSI. The ultra-low latency timing requirement is given in Table 4.20.

4.3.7 Codebook-Based Feedback

The initial studies of NR have considered multiple forms of CSI acquisition schemes which were based on implicit feedback, where the parameters indicating the CSI are based on a set of hypotheses associated with one particular UE, explicit feedback using both quantized and unquantized CSI feedback and reciprocity-based feedback. The implicit and reciprocity-based schemes are also used in LTE, where implicit CSI is computed at the UE based on DL channel sounding where the UE is reporting CQI, PMI, RI, while for reciprocity based CSI is obtained at the UE based on UL channel sounding. There are many trade-offs between these main CSI acquisition schemes such as UE computation complexity, feedback overhead, resolution of the computed CSI, etc. The first step of NR design contains implicit and reciprocity based CSI computation while explicit CSI computation remains something to be studied in the future.

In this section, the precoder codebooks supported by Release 15 NR are described in more detail.

Codebook-based precoding can be seen as a type of vector quantization of the channel experienced by the UE, where the precoder codebook, which is a set of precoder matrices, is designed taking into account typical cellular propagation channels and antenna deployments.

Four codebooks are defined for Release 15 NR:

- Type I Single-Panel codebook
- Type I Multi-Panel codebook
- Type II codebook
- Type II Port Selection codebook

And in addition, an enhance Type II codebook is introduce in NR Release 16. All codebook types (expect the port selection codebook) are designed based on 1D/2D-DFT vectors and hence implicitly assumes that a uniform linear or planar array is employed at the gNB. Since a wide variety of 2D antenna array dimensions can be used, the codebooks are configurable and scalable. That is, the antenna port layout (of a panel) in vertical and horizontal dimensions (N_1 and N_2 respectively) is explicitly configured as part of the codebook configuration. For the multi-panel codebook, the number of panels N_g is also configured. It is also implicitly assumed that dual-polarized antenna arrays are used, implying that the total number of CSI-RS antenna ports that is used by the codebook is $P = 2N_g N_1 N_2$, where $N_g = 1$ for the single panel and Type II codebooks. Up to 32 antenna ports in supported for the NR codebooks and the supported antenna port layouts (for the single-panel and Type II codebooks) are illustrated in Figure 4.63.

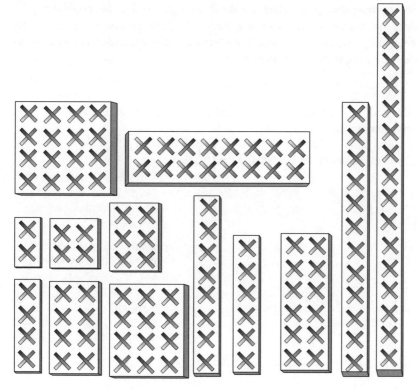

Figure 4.63 Supported 1D and 2D antenna port layouts for the single panel codebooks.

The Type I codebooks targets SU-MIMO operation and offers a "regular" spatial resolution with a relatively low overhead, whereas the Type II codebooks targets MU-MIMO operation and gives a much finer spatial resolution which is required for intra-cell interference suppression. What is meant by spatial resolution will be further elaborated in the next section.

4.3.7.1 Motivation for the Use of DFT Codebooks

The use of codebooks based on DFT vectors arise from the fact that cellular propagation channels are often spatially correlated and that uniform linear/planar arrays are typically employed at the gNB. For instance, consider a propagation channel with only one propagation path such that the channel at a point r_0 in space and time t is $h_0(t)$, i.e. $h(r_0, t) = h_0(t)$. Assume further that the distance from the transmitter or closest scatterer to r_0 is sufficiently large, so that the propagation in the immediate area around r_0 may be described as a plane wave. Then, the channel can be expressed as

$$h(r, t) = h_0(t_0) \cdot e^{j(k \cdot (r - r_0) - \omega t)}$$

in the vicinity around r_0, where $h_0(t_0)$ is the channel at time zero, $k = \frac{2\pi}{\lambda} \begin{pmatrix} \sin\theta\cos\phi \\ \sin\theta\sin\phi \\ \cos\theta \end{pmatrix}$ is the

wave vector of the propagation path (where λ is the wavelength, θ the zenith angle and ϕ the azimuth angle of the propagation path) and ω is its angular frequency. As we are interested only in the spatial properties, we can set $t = 0$. We further look at the channel along a single direction, specified by the vector $r = r_0 + r\hat{r}$, so that

$$h(r) = h_0(t_0) \cdot e^{jk \cdot (r - r_0)} = h_0(t_0) \cdot e^{j \cdot r \cdot (k \cdot \hat{r})} = h_0(t_0) \cdot e^{j \cdot r \cdot \frac{2\pi}{\lambda}(\hat{k} \cdot \hat{r})}.$$

The wave field and the position vector are illustrated in Figure 4.64: Illustration of position vector in plane wave field.

For simplicity of explanation, let $\theta = 90°$ and $\hat{r} = \begin{pmatrix} 1 \\ 0 \\ 0 \end{pmatrix}$, then

$$h(r) = h_0(t_0) \cdot e^{j \cdot r \cdot \frac{2\pi}{\lambda} \cdot \cos\phi}.$$

Figure 4.64 Illustration of position vector in plane wave field.

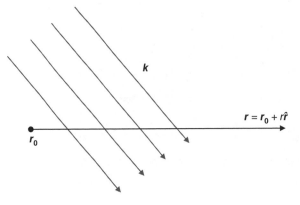

Let us further normalize the position with the wavelength, for convenience, so that $r_\lambda = \frac{r}{\lambda}$. If we take the Fourier transform of the channel, with respect to the position r_λ we get the expression

$$H(\Phi) = \mathcal{F}\{h(r_\lambda)\} = h_0(t_0) \cdot \delta(\Phi - \cos\phi),$$

where Φ is the Fourier transform variable of r_λ and $\delta(.)$ denotes the Dirac distribution. Thus, in the transform domain, Φ relates to the cosine of the propagation path angle. As seen, the channel is infinitely sparse in the Fourier transform domain.

Let's further assume that we sample the positions using uniform sampling with distance $d_\lambda \lambda$ between each point. The channels for each sample point $n = 0, 1, \ldots$ is then expressed as

$$h_n = h(r_\lambda = n \cdot d_\lambda) = h_0(t_0) \cdot e^{j \cdot n \cdot d_\lambda \cdot 2\pi \cdot \cos\phi},$$

That is, the channel as a function of the spatial sampling index n is a DFT vector! Thus, the optimal precoder for such a channel is also a DFT vector with a steering angle at the conjugate direction.

In this analysis, setting the sampling positions in space in such a way of course corresponds to an infinitely large uniform linear array (ULA), using an element separation of d_λ wavelengths. By expressing it as a sampling problem, though, one may trivially understand that the *angular spectrum* of the channel is attained by applying the discrete-time Fourier transform on h_n, which may be calculated using the Poisson summation formula as

$$H_{DT}(\Phi_{DT}) = \mathrm{DTFT}(h_n) = \frac{1}{d_\lambda} \sum_k H\left(\Phi_{DT} - \frac{k}{d_\lambda}\right) = \frac{h_0(t_0)}{d_\lambda} \sum_k \delta\left(\Phi_{DT} - \frac{k}{d_\lambda} - \cos\phi\right).$$

Since the quantity of Φ_{DT} is in cosine, which takes values between $[-1, 1]$, this is also the relevant interval at which to analyze the angular spectrum. Due to the space-sampling, the space-continuous Fourier transform of the channel will be periodically copied, as is illustrated in Figure 4.65: Illustration of time-discrete Fourier transform of space-sampled channel. To avoid spectral aliasing, that is, to avoid introducing *grating lobes*, all of the copies in the sum (for $k \neq 0$) must lie outside the interval $[-1, 1]$. The worst case happens when $\cos\phi = 1$ (or -1), then $\frac{1}{d_\lambda} \geq (1 - (-1)) = 2$ to avoid aliasing. That is, $d_\lambda \leq \frac{1}{2}$ must hold, i.e. the array must be sampled at least with half-wavelength distance in order to avoid grating lobes.

Finally, we assume that the sampled points in space are not infinite, but is limited to N, corresponding to a ULA of N antennas. This can familiarly be expressed as a multiplication

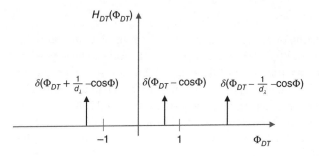

Figure 4.65 Illustration of time-discrete Fourier transform of space-sampled channel.

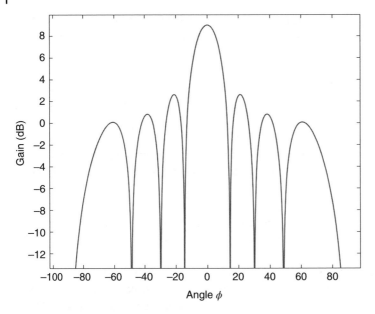

Figure 4.66 Illustration of beam pattern.

of the infinite-space signal h_n with a rectangular window function

$$w_n = \begin{cases} 1, & n < N \\ 0, & n \geq N \end{cases}$$

so that $h_{n,\,ULA} = h_n \cdot w_n$. The DTFT of $h_{n,\,ULA}$ is thus the convolution of $H_{DT}(\Phi_{DT})$ and DTFT of w_n, which is the Dirichlet kernel $D(\Phi_{DT})$, resulting in

$$H_{ULA,DT}(\Phi_{DT}) = \text{DTFT}(h_n) * D(\Phi_{DT}) = \frac{h_0(t_0)}{d_\lambda} \sum_k D\left(\Phi_{DT} - \frac{k}{d_\lambda} - \cos\phi\right).$$

Plotting the Dirichlet kernel as a function of the angle, as Figure 4.66: Illustration of beam pattern, in will result in the well-known "beam" pattern.

A general propagation channel of course consists of multiple propagation paths, i.e. can be described as

$$H(f) = \sum_{i=1}^{M} c_i \boldsymbol{a}^T(\theta_i) e^{-j2\pi f \tau_i}$$

In this case, a precoder consisting of a single DFT beam will not be able to capture the full channel energy, and instead, multiple DFT beam components needs to be applied, so as to illuminate the multiple propagation paths of the channel. Luckily, it is often enough to apply relatively few DFT beam components to capture most of the channel energy. This is illustrated in Figure 4.67: Fraction of channel energy captured in the N strongest DFT beams, where a C.D.F. of the fraction of the channel energy captured in the $N_{beams} = \{1, 2, 3, 4, 5, 6, 7, 8\}$ strongest components of a DFT basis is shown. For a 1x16 antenna array using the 3GPP 3D UMi channel model. Since an ULA with 16 antennas is used, the DFT basis

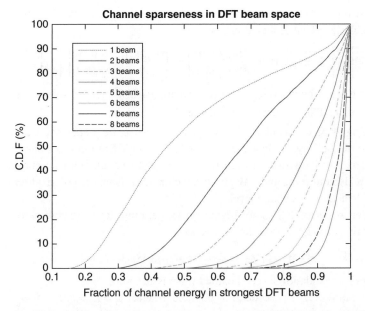

Channel sparseness in DFT beam space

Legend:
- 1 beam
- 2 beams
- 3 beams
- 4 beams
- 5 beams
- 6 beams
- 7 beams
- 8 beams

(Y-axis: C.D.F (%); X-axis: Fraction of channel energy in strongest DFT beams)

Figure 4.67 Fraction of channel energy captured in the N strongest DFT beams.

comprises 16 beams. Each point in the C.D.F corresponds to a user position. An immediate observation is that for the higher percentiles, almost all of the channel energy is captured in the strongest DFT beam. This of course corresponds to the (outdoor) LOS users with a specular LOS path. However, for the median user, only ~45% of the channel energy is captured in the strongest beam, while for the 5th percentile user (with large angular spread), it's only around 20%. If a couple of more beams are included in the basis, the fraction of energy captured increases significantly and with three beams, 80% of channel energy is captured for the median user while 55% is captured for the 5th percentile user.

4.3.7.2 DL Type I Codebook

4.3.7.2.1 Single-Panel Codebook The NR Type I Singe-panel codebook uses a similar structure as the LTE codebooks for > = 4 antenna ports. That is, it is a constant modulus DFT codebook tailored for a dual-polarized 2D Uniform Planar Array (UPA). Intuitively, the construction of the codebook can be explained in the following manner.

A DFT precoder where the precoder vector used to precode a single-layer transmission using a single-polarized ULA with N antennas can be defined as

$$\boldsymbol{w}_{1D}(k) = \frac{1}{\sqrt{N}} \begin{bmatrix} e^{j2\pi \cdot 0 \cdot \frac{k}{QN}} \\ e^{j2\pi \cdot 1 \cdot \frac{k}{QN}} \\ \vdots \\ e^{j2\pi \cdot (N-1) \cdot \frac{k}{QN}} \end{bmatrix},$$

where $k = 0, 1, \ldots QN - 1$ is the precoder index and Q is an integer oversampling factor. A corresponding precoder vector for a two-dimensional UPA can be created by taking the

Kronecker product of two precoder vectors as $w_{2D}(k, l) = w_{1D}(k) \otimes w_{1D}(l)$. Extending the precoder for the dual-polarized UPA is then done as

$$w_{2D,DP}(k, l, \phi) = \begin{bmatrix} 1 \\ e^{j\phi} \end{bmatrix} \otimes w_{2D}(k, l) = \begin{bmatrix} w_{2D}(k, l) \\ e^{j\phi} w_{2D}(k, l) \end{bmatrix} = \begin{bmatrix} w_{2D}(k, l) & 0 \\ 0 & w_{2D}(k, l) \end{bmatrix} \begin{bmatrix} 1 \\ e^{j\phi} \end{bmatrix},$$

where $e^{j\phi}$ is a co-phasing factor between the two orthogonal polarizations, that for rank-1 precoding is selected from a QPSK alphabet $\phi \in \left\{ 0, \frac{\pi}{2}, \pi, \frac{3\pi}{2} \right\}$.

The optimal co-phasing $e^{j\phi}$ between polarizations typically vary over frequency while the optimal beam direction $w_{2D}(k, l)$ typically is the same over the whole CSI reporting band. Thus, the precoder matrix is split up into a W_1 matrix factor comprising the beam direction which is selected on a wideband level, and a W_2 matrix factor comprising the polarization co-phasing, which is selected on a subband level.

To create the precoder matrices $W_{2D, DP}$ for higher ranks, i.e. multi-layer transmission, several DFT precoders $w_{2D,DP}(k, l, \phi)$ are concatenated as

$$W_{2D,DP} = \begin{bmatrix} w_{2D,DP}(k_1, l_1, \phi_1) & w_{2D,DP}(k_2, l_2, \phi_2) & \cdots & w_{2D,DP}(k_R, l_R, \phi_R) \end{bmatrix},$$

where R is the number of transmission layers, i.e. the transmission rank.

As an example, for the rank-2 DFT precoder, it is possible to select $k_1 = k_2 = k$, $l_1 = l_2 = l$ and $\phi_2 = \phi_1 + \pi$ meaning that

$$W_{2D,DP} = \begin{bmatrix} w_{2D,DP}(k, l, \phi_1) & w_{2D,DP}(k, l, \phi_1 + \pi) \end{bmatrix}$$

$$= \begin{bmatrix} w_{2D}(k, l) & 0 \\ 0 & w_{2D}(k, l) \end{bmatrix} \begin{bmatrix} 1 & 1 \\ e^{j\phi_1} & -e^{j\phi_1} \end{bmatrix} = W_1 W_2.$$

Depending on the codebook configuration, the codebook can also contain a subband beam selection component, where four adjacent DFT beams, e.g. $[b_1 \ b_2 \ b_3 \ b_4] = [w_{2D}(k_0, l_0) \ w_{2D}(k_0 + 1, l_0) \ w_{2D}(k_0 + 2, l_0) \ w_{2D}(k_0 + 3, l_0)]$ is selected in the W_1 matrix and a beam selection component is added to the W_2 matrix, as is illustrated in Figure 4.68.

Thus, the Type I codebook always selects a single DFT beam for the precoding of each layer, as is illustrated in Figure 4.69 below. This corresponds to directing the transmission

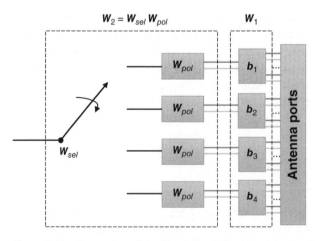

Figure 4.68 Illustration of the Type I Single-Panel codebook with subband beam selection.

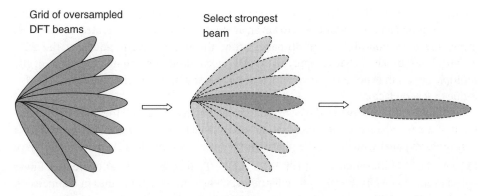

Grid of oversampled DFT beams

Select strongest beam

Figure 4.69   Illustration of the Type I codebook.

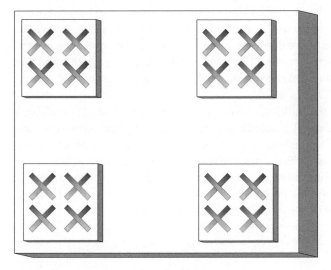

Figure 4.70   Illustration of multi-panel port layout.

toward the strongest propagation path of the UE's channel, and is typically enough to get good SU-MIMO performance, even though for many users the strongest propagation path may only capture a relatively low fraction of the total channel energy.

4.3.7.2.2 Multi-Panel Codebook   When designing the NR CSI framework, consideration was made to also consider the case where the gNB is equipped with multiple, possibly uncalibrated, antenna panels. The motivation for using multiple panels instead of a fitting all the antenna elements into a single calibrated panel is to decrease the implementation complexity. An example of a multi-panel array is given in Figure 4.70.

The Type I Single-panel codebook, as described in the previous section, contains DFT precoder vectors with linearly increasing phases over the antenna ports in each spatial dimension. As we saw in the previous section, such a codebook design implicitly assumes an antenna setup of phase-calibrated and equally spaced antenna ports in each dimension. In that case, the precoders of the codebook perfectly matches the array response assuming a pure line-of-sight channel and gives a good representation of the dominant channel path for other propagation conditions. However, in the case of an uncalibrated multi-panel array, and/or, a non-uniform multi-panel array, the implicit assumptions of the DFT codebook are broken. That is, applying a DFT precoder across antenna elements of the multiple panels may not result in an efficient representation of the channel response. This is due to several factors.

Firstly, the spacing between the last antenna element of a panel and the first antenna element of the next panel is different from the antenna element spacing within a panel for a non-uniform panel array. Thus, the phase shift between said antenna elements would have to be $e^{\frac{j2\pi k(1+\Delta_{panel})}{N}}$ rather than $e^{\frac{j2\pi k}{N}}$ (as it is for the DFT precoder) in order to create a linear phase front, where k is the DFT precoder index, N the number of antennas in a dimension and Δ_{panel} the additional distance between panels compared to the distance between panels in a uniform multi-panel array. This phase difference could of course be compensated directly for in the codebook if the panel distance was, however, the distance between panels is up to implementation and should not be specified by 3GPP. Secondly, there may exist an additional phase offset between antenna panels due to, for instance, different LO phase state or frequency offset. In the worst case, the phase offset may be completely random and thus uniformly distributed in $[0, 2\pi]$. Thus, a special codebook design for this type of antenna setup is needed.

The codebook design for the multi-panel codebook essentially builds on applying a precoder $\boldsymbol{w}_{SP}$ from the single-panel codebook on each constituent panel, where the same single-panel precoder $\boldsymbol{w}_{SP}$ is applied on each antenna panel of the multi-panel antenna array. Additionally, to compensate for the possible calibration error between panels as well as the phase offset due to the non-uniform panel placement, a per-panel cophasing factor φ_n is applied to each panel n. This implies that the precoding vector for a layer can be described as for a multi-panel codebook consisting of $N_g = 4$ panels:

$$\boldsymbol{w}_{MP} = \begin{bmatrix} \boldsymbol{w}_{SP} \\ \varphi_1 \boldsymbol{w}_{SP} \\ \varphi_2 \boldsymbol{w}_{SP} \\ \varphi_3 \boldsymbol{w}_{SP} \end{bmatrix}$$

4.3.7.3 DL Type II Codebook

While feeding back spatial information about only the strongest propagation path of the channel, as for the Type I codebooks, is sufficient for SU-MIMO operation, it is not good enough for MU-MIMO. The reason is that MU-MIMO precoding needs to know about the full channel space in the spatial domain in order to suppress interference toward co-scheduled UEs. Thus, information about more propagation paths than the strongest one is needed, i.e. more than one DFT beam needs to be indicated. However, when the precoder vector for a layer is constructed by multiple DFT beams, the gNB needs to know *how* to linearly combine the multiple beams in order to for the signal to add up coherently at the UE and improve the SINR. This implies that linear combining coefficients need to be provided in addition to simply indicating the preferred beam directions. Due to the frequency-selective nature of the propagation channel, at least the phase of the linear combining coefficients can vary quite rapidly across the subband, which means that associated overhead for feeding back these coefficients can be quite large. Essentially, the Type II codebook can be seen as a form of transform compression, where the principal eigenvectors of the UEs channel are transformed into beam domain using a spatial basis, the $2L$ dominant (dual-polarized) basis vectors are selected, and linear combining coefficients for these basis vectors are calculated and reported, where $L = 2, 3, 4$ is the number of spatial beams included in the precoder.

The NR Type II Codebook design consists of two components, spatial basis selection and basis linear combination. The spatial basis is constructed from columns of a dual-polarized 2D-DFT matrix (again, assuming UPA structure of antenna ports) in order to correspond to different beam 2D directions. The precoder vector for a layer is then formed by linearly combining the basis vectors, i.e. weighting them together using different amplitude and phase weights. The precoding vectors use the same dual-stage $W = W_1 W_2$ structure as the Type I codebooks, where W_1 is selected wideband while W_2 is selected per subband. The basis/beam selection is performed in W_1 while selection of beam phase weights is done frequency-selectively in W_2. Wideband beam amplitude weights are also included in W_1 and in addition, differential subband amplitude weights can be included in W_2.

In order to achieve optimal quantization performance of the coefficients it is important that an orthogonal DFT basis is used. However, this basis can be rotated, so as to better align the orthogonal basis vectors to the propagation directions of the channel.

To express the basis selection in W_1, we can first define a dual-polarized rotated 2D-DFT beam space transformation matrix $\widetilde{B}_{N_V, N_H}(q_V, q_H)$ as

$$
\begin{aligned}
\widetilde{B}_{N_V, N_H}(q_V, q_H) &= \begin{bmatrix} B_{N_V, N_H}(q_V, q_H) & 0 \\ 0 & B_{N_V, N_H}(q_V, q_H) \end{bmatrix} \\
&= \begin{bmatrix} (R_{N_H}(q_H) D_{N_H}) \otimes (R_{N_V}(q_V) D_{NV}) & 0 \\ 0 & (R_{N_H}(q_H) D_{N_H}) \otimes (R_{N_V}(q_V) D_{NV}) \end{bmatrix} \\
&= \begin{bmatrix} b_0 & b_1 & \cdots & b_{N_V N_H - 1} & 0 & 0 & \cdots & 0 \\ 0 & 0 & \cdots & 0 & b_0 & b_1 & \cdots & b_{N_V N_{H-1}} \end{bmatrix},
\end{aligned}
$$

where D_N is a size $N \times N$ DFT matrix, i.e. the elements of D_N are defined as $[D_N]_{m,n} = \frac{1}{\sqrt{N}} e^{\frac{j 2\pi mn}{N}}$. The orthogonal 2D beams may thus be indexed by the orthogonal beam indices $0 \leq n_1 < N_1 - 1, 0 \leq n_2 < N_2 - 1$. Further, $R_N(q) = \text{diag}\left(\begin{bmatrix} e^{j2\pi \cdot 0 \cdot \frac{q}{N}} & e^{j2\pi \cdot 1 \cdot \frac{q}{N}} & \cdots & e^{j2\pi \cdot (N-1) \cdot \frac{q}{N}} \end{bmatrix} \right)$ is a size $N \times N$ rotation matrix, defined for $q = 0, 1, \ldots, O - 1$, where O is an oversampling factor of the DFT basis. Multiplying D_N with $R_N(q)$ from the left creates a rotated DFT matrix with entries $[R_N(q) D_N]_{m,n} = \frac{1}{\sqrt{N}} e^{\frac{j 2\pi m(n+q)}{N}}$. Rotating the beam space basis has an effect similarly to oversampling a codebook, for example, if the channel is a pure LOS channel and the angle of the LOS ray if perfectly aligned with a constituent beam in the beam space, the channel matrix can be described by only one beam coefficient. However, if the angle of the LOS ray lies in between two beams in the beam space, two beam coefficients are required to express the channel, doubling the amount of overhead needed.

By utilizing this structure for the DFT basis selection, the basis vectors are always orthogonal, as is illustrated in Figure 4.71: An example of rotated orthogonal beams expressed as oversampled DFT beams.

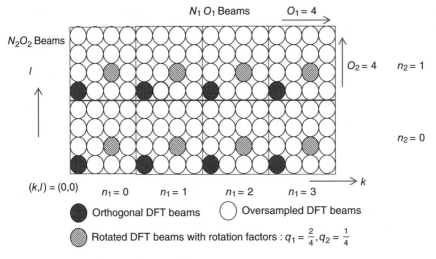

Figure 4.71 An example of rotated orthogonal beams expressed as oversampled DFT beams.

A selected beam matrix consists of columns from $\widetilde{\boldsymbol{B}}_{N_V,N_H}(q_V,q_H)$, where L beams are selected, as

$$B_{I_S} = \begin{bmatrix} \boldsymbol{b}_{I_S(0)} & \boldsymbol{b}_{I_S(1)} & \cdots & \boldsymbol{b}_{I_S(L-1)} & 0 & 0 & \cdots & 0 \\ 0 & 0 & \cdots & 0 & \boldsymbol{b}_{I_S(0)} & \boldsymbol{b}_{I_S(1)} & \cdots & \boldsymbol{b}_{I_S(L-1)} \end{bmatrix},$$

where I_s denotes the selected beam indices

$$I_S = [i_0, i_1, \ldots, i_{L-1}] = [N_2 n_1^{(0)} + n_2^{(0)}, N_2 n_1^{(1)} + n_2^{(1)}, \ldots, N_2 n_1^{(L-1)} + n_2^{(L-1)}].$$

Thus, feedback of basis selection comprises signaling the following quantities:

- Beam space rotation indices q_1, q_2
- Selection of L beams: $n_1^{(0)}, n_2^{(0)}, n_1^{(1)}, n_2^{(1)}, \ldots n_1^{(L-1)}, n_2^{(L-1)}$

The selected beams in the basis are linearly combined with phase and amplitude weighing to form the resulting precoder. The total precoder for rank 2 (as an example) can then be expressed as

$$W^{(2)} = B_{I_S} \widetilde{W} = \begin{bmatrix} \widetilde{w}_{0,0} & \widetilde{w}_{0,1} \\ \widetilde{w}_{1,0} & \widetilde{w}_{1,1} \end{bmatrix}$$

where

$$\widetilde{w}_{r,l} = \sum_{i=0}^{L-1} \boldsymbol{b}_{I_S(i)} \cdot p_{r,l,i}^{(WB)} \cdot p_{r,l,i}^{(SB)} \cdot c_{r,l,i}, r = 0, 1$$

is the precoder weights for layer l for the antennas on polarization r. The phase combining coefficients $c_{r,l,i}$ can take values from an 2^N-PSK constellation, where $N = 2, 3$ is configurable as part of the codebook configuration

$$c_{r,l,i} \in \left\{ 1, \exp\left(\frac{j2\pi}{2^N}\right), \ldots, \exp\left(\frac{j2\pi \cdot (2^N - 1)}{2^N}\right), \ldots \right\}, \forall r, l, i$$

but where the strongest coefficient for each layer (indexed by (r_l^*, i_l^*)) is not reported and instead assumed to be set to $c_{r_0^*,0,i_0^*} = c_{r_1^*,1,i_1^*} = 1$ since only the relative phases between the beams for each layer matters. The index of the strongest coefficient for each layer however needs to be reported so that the gNB knows which coefficient is set to one.

The amplitude coefficients are factorized into a wideband part $p_{r,l,i}^{(WB)}$ and a subband part $p_{r,l,i}^{(SB)}$ in order to reduce overhead. The WB amplitude coefficients $p_{r,l,i}^{(WB)}$ are linearly quantized with 3 bits in the dB domain but where a zero state is included, so that $(p_{r,l,i}^{(WB)})^2 \in \{0\,\text{dB}, -3\,\text{dB}, -6\,\text{dB}, -9\,\text{dB}, -12\,\text{dB}, -15\,\text{dB}, -18\,\text{dB}, 0\,(\text{lin})\}$.

Including a zero state is useful to "turn off" weaker beams. That is, if a UE is configured with L = 4 beams but only finds that L = 3 beams needs to be included in the precoder for optimal performance, the weakest beam can be turned off and the phase and subband amplitude coefficients for that beam does not need to be reported, which reduces the PMI overhead. This results in that the PMI payload can vary dynamically, even for the same selected RI. As mentioned in Section 4.3.4, this requires the number of non-zero wideband amplitude coefficients to be indicated in CSI Part 1 in order for the CSI Part 2 payload to be known.

The subband amplitude reporting can be turned ON or OFF as part of the codebook configuration, if subband amplitude reporting is turned off, only the WB amplitude is reported and the subband amplitude coefficients are fixed to $p_{r,l,i}^{(SB)} = 1$. If subband amplitude reporting is turn on, $p_{r,l,i}^{(SB)} \in \{1, \sqrt{0.5}\}$, i.e. one-bit differential subband amplitude quantization is used. To reduce the subband overhead when subband amplitude reporting is turned ON, an unequal allocation of quantization bits for the coefficients among the stronger and weaker beams is applied, so that the weaker beams of a layer are quantized with more coarse resolution.

The principle of the Type II codebook is illustrated in Figure 4.72. That is, on a high-level it can be described as the EU performs beams election of the L strongest beams from the DFT basis, scales the power of each beam to correspond to the power distribution of the

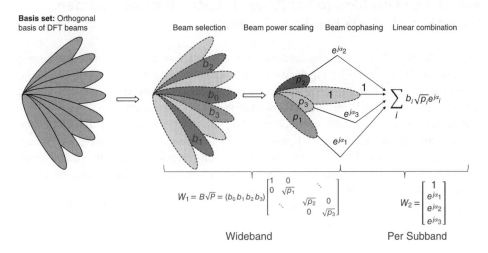

Figure 4.72 Illustration of the Type II codebook with wideband amplitude.

Table 4.21 Example overhead calculation for Type II CSI feedback, assuming 13 subbands, 32 CSI-RS ports, and 8-PSK co-phasing.

W1 overhead

L	Beam selection: $\left\lceil \log_2\left(\dfrac{N_1 N_2}{L}\right)\right\rceil$ bits	Rotation: $\lceil\log_2(O_1 O_2)\rceil$ bits	Strongest beam indication $\lceil\log_2(2L)\rceil$ bits per layer	WB Amplitude: $3\cdot(2L-1)$ bits per layer	Total W1 payload Rank 1	Total W1 payload Rank 2
2	7	4	2	9	22	33
3	10	4	3	15	32	50
4	11	4	3	21	39	63

W2 overhead

L	SB Amplitude (if applied): $(2L-1)$ bits per layer	SB Phase: $3\cdot(2L-1)$ bits per layer	Total W2 payload Rank 1	Total W2 payload Rank 2	Total Rank 1 payload for 10 SB	Total Rank 2 payload for 10 SB
2	(3)	9	9 (12)	18 (24)	112 (142)	213 (273)
3	(5)	15	15 (20)	30 (40)	182 (232)	350 (450)
4	(7)	21	21 (28)	42 (56)	249 (319)	483 (623)

propagation paths of the channel and finally co-phasing the beams so that the signal adds up constructively at the UE.

The resulting overhead for the Type II codebook depends on both the codebook configuration (number of beams L, Q-PSK or 8-PSK co-phasing and SB amplitude ON or OFF), number of antenna ports and port layout as well as the number of subbands in the CSI reporting band. An example of overhead calculation is given in Table 4.21.

4.4 Radio Link Monitoring (Claes Tidestav, Ericsson, Sweden, Dawid Koziol, Nokia Bell Labs, Poland)

In RRC_CONNECTED state, the network maintains the connection between the serving cell and the UE, e.g. by collecting measurements, and providing the UE with updated configuration when needed. Ensuring that the radio link quality between the network and the UE is, under normal circumstances, the responsibility of the network.

However, in rare cases, the network may lose the ability to maintain the connection between the serving cell and the UE, e.g. due to bad coverage or a failed handover. Under such circumstances, the UE may be trapped in a non-reachable state, since the network cannot contact the UE to restore the communication link. In these cases, it may be advantageous for the UE to reset the RRC connection and restart the communication.

The standardized procedure for this type of connection reset is called *radio link failure* (RLF). RLF is a procedure where the UE autonomously determines that the radio link is lost. The UE does this by checking whether certain conditions are fulfilled, in which case it declares RLF and performs the associated actions.

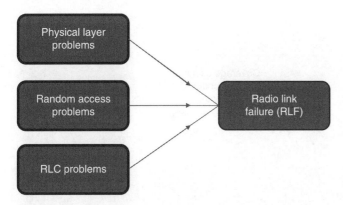

Figure 4.73 Reasons for radio link failure. The UE may declare RLF when it detects physical layer problems, random access problems or RLC problems.

There are three types of problems, Figure 4.73, that may cause the UE to declare RLF:

Physical layer problem: The UE monitors the physical layer quality of the connection with the SpCell. This is performed using measurements on a DL reference signal, a procedure called *radio link monitoring* (RLM). RLM and the associated RLF conditions are described in Section 4.4.1.1

Random access problem: As described in Section 4.1.2.4, during the random access procedure, the UE repeatedly transmits PRACH preambles until it detects a DL response. If the UE has performed the configured maximum number of PRACH transmissions without receiving any response, it may under some circumstances declare RLF. Random access problems are described in Section 4.4.1.2.

Radio Link Control (RLC) problem: In acknowledged mode, the RLC protocol relies on retransmissions to achieve error-free communication. In case the maximum number of retransmissions of a certain RLC PDU was reached without succeeding to deliver it to the receiving end, the UE declares RLF. RLC failure is briefly described in Section 4.4.1.3.

When any of these three situations occur, the UE may declare RLF. The actions after RLF depend on whether the RLF occurred in the Master Cell Group (MCG) or the Secondary Cell Group (SCG) and on the configuration details of the connection.

As will be explained in this chapter, RLF results in loss of connectivity and a long service interruption, as well as a significant signaling overhead on the Uu and Radio Access Network (RAN)-internal interfaces. Therefore, RLF should only be declared when there is very little chance that the radio link can be restored through other means. To avoid premature RLF declaration, the procedure is governed by a set of parameters that give the network ample means to control under what circumstances the UE declares RLF.

4.4.1 Causes of Radio Link Failure

4.4.1.1 Physical Layer Problem

A physical layer problem occurs when the quality of the serving link is below an acceptable level. In NR, the network may rely on beamforming to communicate with the UE.

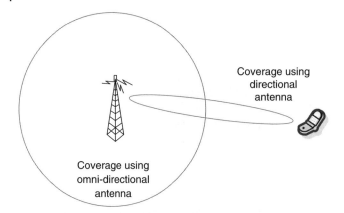

Figure 4.74 When the network can reach the UE using UE-specific beamforming, the UE should not declare RLF. A quality estimate based on the omni-directional transmission may be too pessimistic.

This needs to be considered also in the RLM procedure: as long as the quality of the beam-formed signals are adequate, the UE shall not declare RLF. In principle, the UE shall not declare RLF as long as the quality of any beam the network may choose to reach the UE is operational. Such a situation is illustrated in Figure 4.74.

As the beamforming is UE-specifically configured, the UE does not have a priori information on which signal to monitor: there is a need to configure the UE with additional details on how to perform the monitoring and NR includes two methods to provide the UE with those details.

In both methods, the UE is provided with the identity of one or more reference signals that the UE will use to estimate the quality of the radio link. Each of these radio link monitoring reference signals (RLM RSs) can be either an SS/PBCH block or a periodic single-port CSI-RS resource. Preferably, the provision of adequate reference signals is performed in the first part of the configuration of the RRC connection, so that the UE can start performing RLM already early during the connection.

The fact that the network will rely on UE-specific beamforming implies that different beams may be used to provide coverage in different parts of the cell. To avoid that UE movement causes RLF, the UE would have to monitor different reference signals as it moves across the coverage area of different beams. One option is that the network updates the RLM RS as the UE moves: clearly, the network would anyway have to track the UE to maintain reachability, using the mechanisms described in Section 4.2. Based on the information, acquired from, e.g. measurement reports from the UE, the network could determine that the RLM RS needs to be updated, since the quality measured from the current RLM RS would otherwise lead to RLF. However, such mechanism would require RRC signaling to handle UE-mobility within the cell. Due to the overhead and latency associated with the RRC signaling, this is undesirable.

To alleviate the overhead associated with RRC signaling, the UE can be explicitly configured with a set of RLM RSs. The network would thus configure the UE with not only an RLM RS corresponding to the currently used beam, but also with the RLM RSs transmitted in beams that are close to the current beam. This situation is illustrated in Figure 4.75. Only

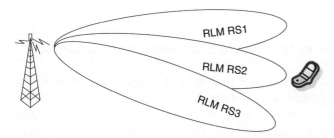

Figure 4.75 UE can be configured with multiple RLM RSs. Only if the quality estimated from all RLM RSs indicate bad quality, the UE would consider there is a radio link problem.

Table 4.22 The maximum number of RLM RSs that can be configured for different maximum number of SS/PBCH blocks.

Maximum number of SS/ PBCH blocks in half-frame	Maximum number of RLM RSs
4	2
8	4
64	8

The maximum number of SS/PBCH blocks is determined per frequency band, as described in Section 4.1.

if the quality estimated from all the RLM RSs indicate bad quality, the UE would consider there is a physical layer problem.

The maximum number of RLM RSs the UE can be configured with depends on the maximum number of SS/PBCH blocks in a half-frame that can be deployed in the cell. This is summarized in Table 4.22.

If SS/PBCH blocks are used as RLM RS, it would be advantageous to configure the UE with all the SS/PBCH blocks in the cell as RLM RSs. It is foreseen that in some cases, a large number of SS/PBCH blocks will be used to cover the cell. For example, at mmW frequencies, the antenna arrays may be able to provide very large array gains, extending the coverage. In this case, to avoid that the SS/PBCH blocks limit the coverage, the SS/PBCH blocks may have to be transmitted over many beams. With many SS/PBCH beams, the maximum number of RLM RSs, as given in Table 4.22, may not be enough to prevent the need of RRC reconfigurations to update the set of RLM RSs due to UE mobility within the cell.

To avoid the need for frequent RRC configurations in deployments where the number of RLM RSs is too low to cover the cell, an alternative method was introduced to provide the UE with the RLM RSs. This *implicit* method to provide the UE with the configuration details for the RLM RS uses the fact that the UE must be provided with a QCL source for the PDCCH reception, as described in Section 4.2.3. The network provides the UE with this QCL source via the *TCI states*, using a combination of RRC and MAC CE signaling. To be more precise, the TCI state is associated with the CORESET, which contains some of the parameters of the PDCCH candidates. As at most three CORESETs can be defined, the same limitation applies also to the number of RLM RSs, when they are implicitly configured.

As described in Section 4.2.3.2, each TCI state may contain one or two RSs. If the TCI state contains only one RS, the UE uses that RS for RLM. However, if QCL Type D is applicable, the TCI state contains two RSs, and in this case the UE should use the RS corresponding to QCL Type D for RLM.

Based on the configured PDCCH candidates, the UE can then determine 1, 2, or 3 RLM RSs, depending on how many CORESETs are configured. For the case when the number of RLM RS is restricted to two and more than two CORESETs are configured, UE will select two of the RLM RS based on pre-determined rules. As previously noted, since the TCI states can be updated using MAC CE, RRC signaling is not required to update the RLM RS as the UE moves in the cell. The signaling overhead associated with the update of RLM RSs during intra-cell mobility is thus manageable.

The quality estimation of the individual RLM RSs is performed in a similar way as in LTE. The UE compares the quality of each RLM RS to an internal threshold Q_{out}, and if the quality of *all* RLM RSs is worse than Q_{out}, the physical layer in the UE indicates *out-of-sync* to higher layers. This indication occurs inside the UE: the separation between the physical layer and higher layers is made to facilitate flexible quality estimation in the UE and at the same time allow adequate network control of the properties of the RLM procedure. The UE also compares the quality of each RLM RS to another internal threshold Q_{in}, and if any of the RLM RSs is better than Q_{in}, the UE indicates *in-sync* to higher layers. It may happen that neither the in-sync or the OOS conditions are fulfilled, in which case the UE indicates nothing to higher layers.

As noted above, the threshold used to define the in-sync and the OOS conditions are UE internal. One option would have been to define the thresholds Q_{in} and Q_{out} as two specific SINR thresholds. However, such a definition was not deemed suitable, since the ability of the network to reach the UE is determined by the performance of the PDCCH and PDSCH, and the mapping between the SINR and the error rates of the DL channels depends on what receiver the UE implemented.

To allow different (improved) UE receiver implementations to be accounted for, the thresholds Q_{out} and Q_{in} are determined in a receiver agnostic manner, using the concept of a *hypothetical PDCCH*:

- The UE would consider the quality worse than Q_{out} if the BLER of a hypothetical PDCCH would be larger than, e.g. 10%.
- The UE would consider the quality better than Q_{in} if the BLER of a hypothetical PDCCH would be smaller than, e.g. 1%.

The network can configure the values of the BLER thresholds, where the default thresholds 10% and 1% are the same as in LTE. To ensure consistent behavior between different UEs, the parameters of the hypothetical PDCCHs are defined in [40]. Also, to ensure sufficient hysteresis between in-sync and OOS, the hypothetical PDCCH configuration used for the definition of Q_{out} is different from the hypothetical PDCCH used for the definition of Q_{in}.

The setup is illustrated in Figure 4.76.

By relying on an indirect definition, any improvement in the UE's ability to decode PDCCH would directly lead to fewer OOS indications, reducing the risk for RLFs.

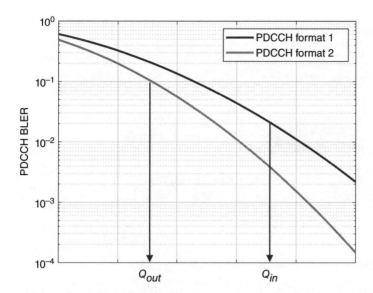

Figure 4.76 The Q_{out} and Q_{in} thresholds are defined indirectly by relying on hypothetical PDCCHs.

In non-DRX mode, the physical layer in the UE evaluates the radio link quality once every T_{ind} ms, where

$$T_{ind} = \max(\min_i T_i, 10)$$

Here, T_i is the period of RLM RS i: when the period of the RLM RSs are longer, the UE evaluates the radio more seldomly. In DRX, the UE tries to save power, and would like to turn off as much as possible of its hardware for as long as possible. Turning on hardware just to evaluate the radio link quality is thus undesirable. To enable significant UE power saving, the indication interval T_{ind} in DRX is therefore increased to

$$T_{ind} = \max(\min_i T_i, T_{DRX})$$

where T_{DRX} is the DRX period.

The physical layer in the UE provides the in-sync and OOS to higher layers. Based on these indications, higher layers determine if and when the UE declares RLF. The procedure is controlled by a set of RLF counters and timers, configured by the network:

- N310 – Specifies the number of consecutive OOS indications after which the T310 timer is started. It can take values from 1 up to 20.
- T310 – This is the timer controlling when RLF is declared. It can take values between 0 and 6 s with intermediate values of tens or hundreds of milliseconds
- N311 – Specifies the number of consecutive in-sync indications for the SpCell received from lower layers, which, if received in time, will prevent the UE from declaring RLF. It can take values between 1 and 10.
- T311 – This is the timer that specifies the maximum time for which the UE may try to find a suitable cell where it would attempt to re-establish its RRC connection after RLF has been declared. It can take values between 1 and 30 seconds.

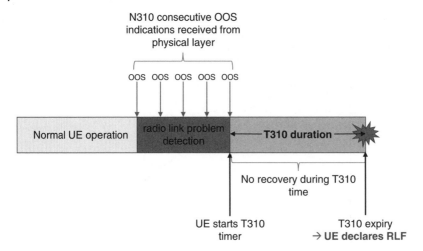

Figure 4.77 RLF declaration based on detection of physical layer problems.

Parameters T310, N310, and N311 are used to control the RLF declaration procedure while T311 is related to RRC connection re-establishment procedure, which is triggered after RLF is declared and is explained in detail in Section 4.4.2.1.

The procedure presented in Figure 4.77 has the following steps:

- Physical layer problems are detected based on RLM RS monitoring and OOS indications are sent from the physical layer to RRC layer in the UE
- In case N310 consecutive (i.e. uninterrupted by any in-sync [IS] indication) OOS indications are received from the physical layer, the UE starts the T310 timer.
- The UE keeps on monitoring RLM RS(s) for the duration of T310. At this stage, in order to prevent RLF declaration, UE would have to receive N311 consecutive (i.e. uninterrupted by any OOS indication) in-sync indications from the physical layer.
- If less than N311 consecutive in-sync indications are received when the T310 timer is running, the UE declares RLF.

As can be seen, the procedure is designed in such a way that the T310 timer responsible for triggering the RLF procedure is started only in case the UE experiences bad radio conditions for a specific time configured by the network. If any in-sync indication is received between OOS indication, which translates into UE having periods of time where it experiences good radio conditions, T310 will not be started. On the other hand, once the timer T310 has been started, it will normally not be stopped unless the UE receives a certain number of consecutive in-sync indications meaning that the radio conditions improved rather permanently and do not fluctuate between the conditions where in-sync and OOS indications are generated by the UE inter-changeably. Of course, the standard gives a network operator the flexibility to configure even the extreme values for N310 or N311 where the T310 would be started or stopped after receiving only a single OOS or in-sync indication. Examples of situations where RLF is not declared by the UE, because the RLF triggering criteria are not met, are presented in Figures 4.78 and 4.79.

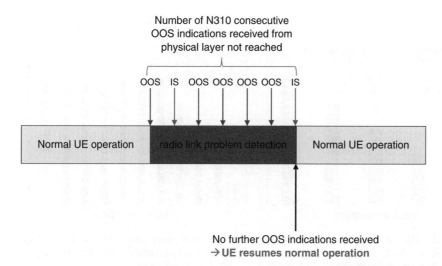

Figure 4.78 Example of a situation where RLF is not declared due to the number of consecutive out-of-sync indications being less than N310.

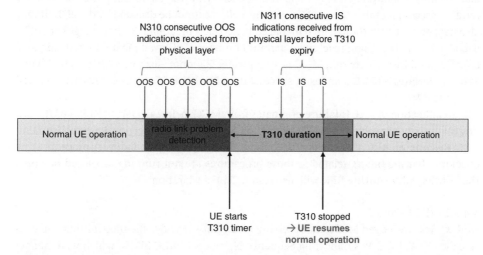

Figure 4.79 Example of a situation where RLF is not declared due to that N311 consecutive in-sync indications being received prior to T310 timer expiry.

4.4.1.2 Random Access Failure

Another issue which may lead a UE to declare RLF is unsuccessful random access procedure (see Section 4.6) performed by the UE in RRC Connected state. As explained in Section 4.6, to control an interference during Random Access procedure, UE does not directly transmit the random access preamble with the maximum possible power but starts with lower power and ramps it up during subsequent preamble transmissions. This way, the following transmissions have a higher chance of being successfully decoded by the gNB, which replies with a RAR message to the UE. However, in case UE finds itself in

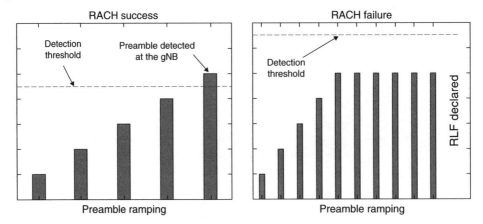

Figure 4.80 Random access success and failure. In this example, the maximum number of PRACH transmissions is 10. Once these 10 transmissions have been made without the UE receiving any response from the network, the UE declares RLF.

bad coverage conditions, it may happen that even with the maximum power allowed for random access preamble transmissions in the cell, it cannot be decoded by the gNB. It may also happen that even though gNB managed to decode a preamble, RAR message cannot be decoded by the UE. Upon reaching maximum number of allowed preamble transmissions as configured by the network, MAC layer indicates Random Access problem to RRC layer and UE declares RLF. Examples of successful and failed RACH procedures are depicted in Figure 4.80.

If an indication of RACH failure originates from MCG, before declaring RLF the UE verifies that it is not involved in RRC Connection Setup, RRC Connection Re-establishment, RRC Resume or RRC Connection Reconfiguration with sync (handover) procedures by checking that the timers related to those procedures are not running. It should be noted that RACH failure during BFR will also lead to RLF declaration.

4.4.1.3 RLC Failure

In RLC Acknowledged Mode, the receiving RLC entity provides the transmitting one with so called STATUS PDUs indicating Sequence Numbers of RLC SDUs, which were not yet successfully received (negative acknowledgements or NACKs), as described in Section 2.4.4. Based on those STATUS PDUs, the transmitting RLC entity attempts to retransmit them. In case the maximum number of allowed retransmission attempts is reached (which can be configured by the network in the range of 1 and 32), the UE declares RLF, as depicted in Figure 4.81.

There is one exception to the rule described above. When the logical channel for which the RLC failure was indicated is configured with Carrier Aggregation based duplication and with additional limitation to only transmit the data from this channel on SCells (i.e. not on PCell or PSCell), RLF is not declared. Instead the UE performs failure information procedure by which it informs the gNB about the issues it experienced, which allows the network to reconfigure the UE accordingly without the need to go through RLF procedure (as it is first assumed that its PCell or PSCell is still operational).

Figure 4.81 RLF declaration due to maximum number of RLC retransmissions.

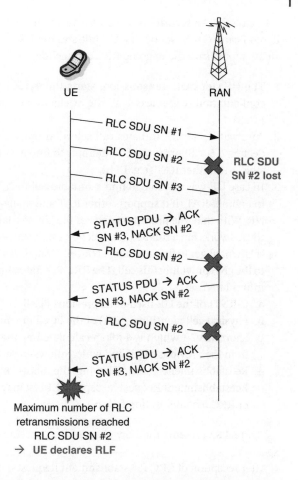

As opposed to the OOS indications from the physical layer, RLF declaration based on reaching maximum number of retransmissions in RLC layer allows to detect connectivity issues in uplink direction. Due to UE power limitations, cell coverage is often limited by signal range in uplink. It may then happen that even though downlink quality, estimated by RLM, is sufficiently good, the UE is not able to send data toward the network. The maximum RLC retransmissions criterium is used to detect and react to such situations.

4.4.2   Actions After RLF

The actions which the UE performs after RLF declaration depend on whether the RLF is declared for UE's MCG or SCG. The MCG actions are described in Section 4.4.2.1 and the SCG actions are described in Section 4.4.2.2.

4.4.2.1   RLF in MCG

The UE behavior after RLF declaration in the MCG depends on whether the Access Stratum (AS) security has been activated for the UE (see Section 2.4.6.6).

In case the AS security has been activated and the UE had SRB2 and at least one Data Radio Bearer (DRB) set up, the UE initiates the RRC re-establishment procedure, which is an attempt to save the ongoing RRC connection:

1. The UE stops UL transmission, starts timer T311 and clears most of its current RRC configuration (e.g. releases all the SCells as well as most of the parameters related to PCell).
2. Afterwards, the UE initiates cell selection process according to the rules described in Section 4.5.4. Shortly, the UE attempts to find a suitable cell either in NR or, if not possible, in another RAT (e.g. LTE).
3. In case UE was not able to find a suitable cell in NR, it moves to RRC IDLE state either in another RAT (if it supports other RAT and managed to find a suitable cell there) or in NR. With that, the procedure ends. If the UE has data to send, it may try to re-connect to the network, but a full RRC connection establishment procedure needs to be performed.
4. If UE manages to find a suitable NR cell, it sends RRC Reestablishment Request message to the gNB providing this cell. The RRC Reestablishment Request message contains such information as:
   a. C-RNTI of the UE used in its previous PCell
   b. Physical cell identities (PCIs) of the PCell in which the UE operated previously
   c. Short MAC-I which is a token calculated by the UE using its security configuration from the previous PCell and which allows to verify the UE's identity
   d. Reestablishment cause allowing the network to understand the cause of RRC Reestablishment Request (except for RLF, it may be triggered due to handover failure or RRC reconfiguration failure)

Figure 4.82 presents the flow chart of the procedure described above from the UE perspective.

After reception of RRC Reestablishment Request message, the network may reply either with RRC Setup or with RRC Reestablishment message. The former is used in case the network is not able to retrieve UE's context, which is essential for successful RRC Reestablishment. When RRC Setup message is received by the UE, it falls back to RRC connection establishment procedure as described in Section 2.4.7.6.3. However, if the gNB receiving reestablishment request is able to retrieve and verify UE's context, it replies with RRC Reestablishment message containing Next Hop Chaining Count parameter allowing the UE to update its security key and verify the integrity protection for the received message. If the message passes the check, the UE sends RRC Reestablishment Complete message and the procedure ends. The RRC connection may continue and gNB and UE are again able to exchange user and control data messages. The example of a successful RRC Reestablishment procedure is presented in Figure 4.83.

It may happen that UE is not able to identify a suitable cell before expiry of T311 timer. In such a situation UE is not allowed to initiate transmission of RRC Reestablishment Request message and goes into RRC IDLE state instead, as presented in Figure 4.84.

In case the AS security has not been activated or it has been activated, but the UE did not have SRB2 and at least one DRB established, UE goes straight to RRC IDLE state without attempting to re-establish the connection.

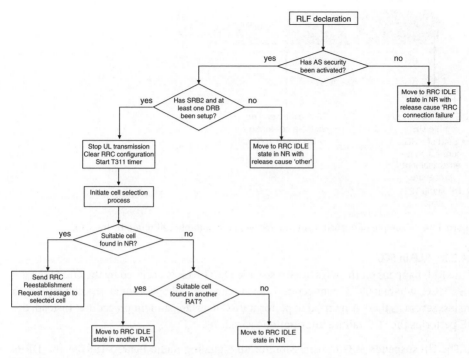

Figure 4.82 UE procedures after RLF declaration.

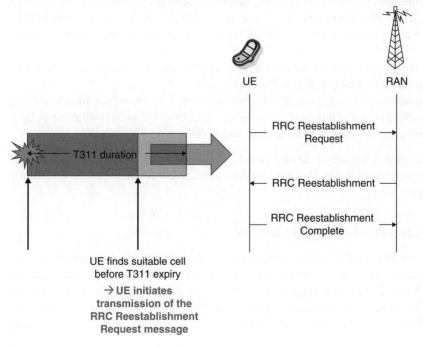

Figure 4.83 Example of a successful RLF recovery through RRC connection reestablishment procedure.

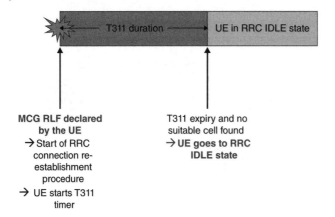

MCG RLF declared by the UE
→ Start of RRC connection re-establishment procedure
→ UE starts T311 timer

T311 expiry and no suitable cell found
→ **UE goes to RRC IDLE state**

Figure 4.84 Example of unsuccessful RLF recovery (no suitable cell found by the UE).

4.4.2.2 RLF in SCG

If the RLF happens on the UE's SCG, a so called SCG-RLF is declared by the UE. Since the UE's MCG, where its RRC connection is terminated, is still operational, the consequences are less severe as there is no need to perform RRC Reestablishment procedure. Instead, the UE performs the SCG failure information procedure:

1. The UE suspends SCG transmission for all Signaling Radio Bearers (SRBs) and DRBs and resets its MAC entity in SCG.
2. Afterwards, the UE sends SCGFailureInformation message on its primary leg, i.e. to its Master Node. Note that in case of UE operating in EN-DC mode, the SCGFailureInformation message will be sent to an eNB, i.e. to E-UTRAN. The contents of the message indicate:
   a. Failure type, i.e. whether the SCG failure occurred due to physical layer problems, random access issues, or RLC failure.
   b. Radio resource management (RRM) measurements which were obtained by the UE up to the moment of failure occurrence for the cells which were configured as UE's serving cells in SCG as well as its neighboring cells.

The SCGFailureInformation may be used by the UE's Master Node to facilitate subsequent SCG configuration. The Master Node may alternatively decide not to add an SCG for the UE, Figure 4.85.

4.4.3 Relation Between RLM/RLF and BFR

Beam recovery, discussed in Section 4.2.6, was introduced in NR to quickly overcome blockage situations that may occur in high frequencies. In beam recovery, the UE detects that the current link is no longer operational, searches for another reference signal, and tries to set up a new link based on that reference signal. On high level, the beam recovery procedure consists of four parts: BFD, new candidate beam identification, BFR request and recovery response.

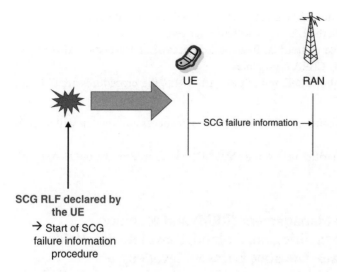

SCG RLF declared by the UE
→ Start of SCG failure information procedure

Figure 4.85 SCG failure information procedure triggered upon SCG RLF.

There are many similarities between BFI indication described in Section 4.2.6.2.2 and RLM described in Section 4.4.1.1:

- The UE estimates the link quality based on a periodic reference signal.
- The UE uses the BLER of a hypothetical PDCCH as quality measure.
- The physical layer provides indications to higher layers, and higher layers count these indications.

There are also small but fundamental differences between the two monitoring procedures:

- There is no support for explicitly configuring SS/PBCH block as a BFD RS.
- There are no in-sync indications for BFD: the absence of an OOS indication serves as an in-sync indication.
- The physical layer indications are counted by the MAC layer for BFD, and by the RRC layer for RLM.

Despite many similarities between BFD and RLM, they are different, independent procedures, and configured separately using different parameters. A physical layer indication from RLM process has no direct impact on BFD and vice versa. Also, the definition of the threshold Q_{out} is different for BFD and RLM: the tests that indirectly define these quality thresholds are slightly different [40].

After a number of consecutive OOS indications, beam recovery is directly initiated: there is no timer corresponding to T310 in the beam recovery procedure. Thus, it is quite likely that beam recovery is triggered before RLF, which is in accordance with the design objectives for beam recovery.

Both after RLF and BFD, the UE tries to re-establish the connection with the network. However, there are significant differences between the two procedures:

- There is no cell selection during beam recovery – the UE always attempts to recover to the serving cell. After RLF, the UE may select another cell.
- Beam recovery may utilize contention-free random access, and a dedicated response CORESET. After RLF, only CBRA is possible.
- Beam recovery is identical in PCell and PSCell. After RLF, the procedures are different for MCG and SCG.

Finally, if the RACH procedure initiated after BFD is unsuccessful, the UE will directly declare RLF.

Thus, there are large similarities between RLM/RLF and beam recovery, but also significant differences.

4.5 Radio Resource Management (RRM) and Mobility (Helka-Liina Määttänen, Ericsson, Finland, Dawid Koziol, Nokia Bell Labs, Poland, Claes Tidestav, Ericsson, Sweden)

4.5.1 Introduction

Mobility is a central part of any cellular system and as the name itself suggests, it is a necessity due to the nature of UEs being mobile. The mobility procedures rely on different paradigms depending on the RRC state or mode (IDLE, INACTIVE, or CONNECTED) of the UE. The RRC states and related transitions are further discussed in Section 2.4.6.

In CONNECTED mode, the mobility functionality ensures that the network can maintain service continuity, i.e. to keep the RRC connection alive, as the UE moves. The mobility in the CONNECTED state can be categorized into two types: cell level mobility and beam level (intra-cell) mobility governed by beam management. The cell level mobility requires RRC signaling to the UE. The beam management procedures are dealt with at lower layers and do not require RRC signaling other than the initial RRC configuration for the connection. RRC is also not required to know which beam is being used at a given point in time. In some cases, no signaling at all is needed for UE to switch the beam. The beam management related procedures are described in Section 4.2. In cell level mobility, as the UE moves from the coverage area of one cell to another, the network transfers the radio link so that the RRC connection is maintained during the entire connection.

From the network perspective, the handover may be intra-gNB mobility or inter-gNB mobility. In the case of intra-gNB mobility, the mobility may be either intra Distributed Unit (DU) or intra Centralized Unit (CU). The CU/DU split is discussed in Section 2.3.3. The CONNECTED mode mobility procedures, e.g. handover, are described in Section 4.5.3.2.

In the CONNECTED mode, the UE may be configured with measurements in order to provide the network input on the signal strength or interference level the UE experiences. The measurements performed by the UE give beneficial and often necessary input in order to decide whether the UE's RRC connection should be modified or whether a handover is needed. The change of RRC connection could, apart from handover, in this context be the addition or modification of the second connection (dual connectivity) or change of the secondary serving cells for the UE (carrier aggregation). These measurements that serve as an input for these procedures are referred to as RRM measurements. The basis of

RRM measurements are RSRP and Reference Signal Received Quality (RSRQ) values that are measured from serving and neighbor cells reference signals (RS) – SS/PBCH block or CSI-RS. In CONNECTED mode, the RRM measurements are explicitly RRC configured by the network. The configuration tells the UE which reference signals to measure and on which time-frequency locations those can be found. Furthermore, the configuration indicates which quantities (RSRP, RSRQ, SINR) are to be derived and how the reporting should be done. Details of the measurement configuration framework applied in the CONNECTED state can be found in Section 4.5.3.2, whereas the measurement quantities are described in Section 4.5.2.

When UE does not have an active connection to the network, i.e. when the UE is in RRC IDLE or in RRC INACTIVE mode, the UE performs RSRP or RSRQ measurements to determine a suitable cell to camp on based on specified rules for cell selection and re-selection. Cell selection and reselection processes can be influenced by the network with certain parameters, but the UEs in INACTIVE or IDLE mode do not report measurement results to the network. In those states, there is no signaling or data flow between UE and the network, other than location registration updates and when UE requests additional system information. The location registration updates are needed for the network to maintain an accurate enough information of the location of the UE to ensure that the UE can be reached, i.e. paged. Section 4.5.4 contains more information about INACTIVE and IDLE mode mobility procedures including measurement related aspects specific to these RRC states.

In order for the UE to deduce cell quality out of beam measurements, different approaches are applied.

One of the most straightforward options is that the strongest cell is determined solely based on the signal level of each cell's strongest beam. However, the measurement framework in NR also allows consideration of signal level or quality of more than just a single beam when deriving cell quality. For UEs in IDLE or INACTIVE mode, the system information may provide the UE with additional parameters assisting it to decide which of the measured cells is the strongest one. For UEs in CONNECTED mode, the cell quality derivation related parameters for RRM measurements including serving cell quality derivation are given in the RRC configuration of the UE. Further details are given in the corresponding sections.

4.5.2 UE Mobility Measurements

In NR, three types of measurements are defined for UE mobility and other RRC connection maintenance related procedures such as carrier aggregation and dual connectivity:

RSRP: This measurement is designed to capture the signal strength of the target.
RSRQ: The RSRQ measurement captures both the signal strength and the load of the target. Despite its name, RSRQ does not provide a measure of the signal quality in the target.
SINR: The SINR measurement captures the SINR in the target, and thus provides a measure of the signal quality in the target.

The measurements can be performed either on the SS/PBCH block or on the single-port CSI-RS (Section 3.7). The RSRP and RSRQ measurements are applicable in IDLE, INACTIVE and CONNECTED modes, whereas the SINR measurement is applicable only in CONNECTED mode. All CSI-RS measurements are only applicable in CONNECTED mode.

In order to enable the UE to obtain a sufficient number of physical samples to achieve the set accuracy requirements and somewhat mitigate the effect of fast fading, a physical layer measurement period is determined. This UE-internal processing is known as L1-filtering. This physical layer measurement, output from this L1-filter, serves as input to the L1-RSRP metric, discussed in Section 3.7.4.2, or to further processing in the UE in order to derive other metrics like RSRQ and SINR. For CONNECTED mode UEs, a further averaging is applied and is called Layer 3 filtering, see further detail in Section 4.5.3.2.3.

4.5.2.1 NR Mobility Measurement Quantities

The most important measurement quantity is RSRP which is designed to capture the signal strength of a beam and, when performed on the SS/PBCH block, it is called SS-RSRP. The measurement is defined as the linear average over the power contributions of the REs that carry SSSs. The UE may also use the PBCH DMRS to improve the accuracy of the SS-RSRP measurement: this is possible since the transmit power of the SSS REs and the PBCH DMRS REs are equal. This is illustrated in Figure 4.86.

The corresponding measurement for CSI-RS is called CSI-RSRP and is defined as the linear average of the power contributions of the REs that carry CSI-RS. These CSI-RS resources are separately configured for the UE which is why the CSI-RS based measurements are only applicable for CONNECTED mode UEs.

The RSRP measurement, whether performed on the SS/PBCH block or single-port CSI-RS, does not capture the load of the frequency layer where the target cell is serving the UEs; nor does it capture interference levels. Both load and interference also affect the performance in the target beam/cell, and to take also these factors into account, additional measurement quantities, RSRQ and SINR, have been defined. Since these measurements should be simple to perform, and since the configuration of the cells contributing to the interference level of the frequency layer of the target is unknown, the accuracy of load and interference estimates are not perfect, but they do provide the network with additional information.

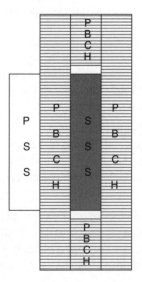

Figure 4.86 The SS-RSRP measurement is defined as the linear average of the power contributions of the REs that carry SSS. The UE may also use the REs where the PBCH DMRS is transmitted.

Both RSRQ and SINR attempt to consider the interference level of the frequency layer where the cell of the corresponding RSRP measurement is. The RSRQ is defined as

$$RSRQ = \frac{RSRP}{RSSI}$$

where RSRP is as discussed above and RSSI denotes a Received Signal Strength Indicator. The RSSI describes the linear average of the total received power in certain REs. That is, it reflects all transmissions of all neighbor cells that happen on those REs during the measurement period. A separate definition for the REs to be measured is given for the SS/PBCH block based RSRQ (SS-RSRQ) and for the CSI-RS resource based RSRQ (CSI-RSRQ).

In NR, the REs on which the RSSI is to be measured can be controlled by the network. In case of SS-RSRQ, for measurements where no measurement gaps are used, the slots to be measured can be selected by the network. Furthermore, the symbols from which the RSSI is measured can be adjusted. The RSSI is always measured starting from the first symbol of the slot and the end symbol index can be selected among {1,5,7,11}. This additional flexibility allows the focus of the measurement to certain time locations. Typically, the network would choose to configure the UE to measure the RSSI in the REs that would typically contain PDCCH/PDSCH transmissions in the neighbor cell. In particular, the network could strive to avoid REs where SS/PBCH blocks are transmitted, since these are always transmitted, resulting a constant 'load'. It is also good to note that UL/DL slot configuration is not accounted for, thus the RSSI measurement configuration may require the UE to measure from UL slots. Such a measurement should be avoided, since it would not be indicative of the load in the target: the position of a co-scheduled UL transmitter would determine the RSSI. An example of the RSSI measurement resource configuration for SS-RSRQ is shown in Figure 4.87.

For CSI-RSRQ, the RSSI is measured in the whole symbol where the CSI-RS is transmitted: the single-port CSI-RS resource only occupies 1/3 or 1/12 of the subcarriers in the OFDM symbol, and the remaining REs are available for PDCCH/PDSCH transmissions.

The SINR, which reflects the interference of the target, is defined as

$$SINR = \frac{RSRP}{I + N}$$

Note that the RSRP appears here as well, reinforcing the importance of that measurement quantity. The $I + N$ is the linear average of the interference and noise measured in the same REs where the corresponding reference signals are transmitted. Since the UE would have

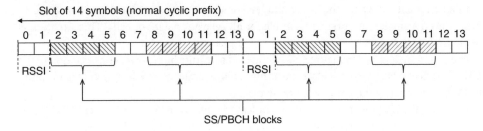

Figure 4.87 An example of RSSI measurement resources for SS-RSRQ. Note that the REs where the SS/PBCH blocks are transmitted are not included in the measurement resources.

Table 4.23 Applicability of different measurement quantities and different reference signals.

	SS/PBCH block	CSI-RS
RSRP	IDLE, INACTIVE, CONNECTED	CONNECTED
RSRQ	IDLE, INACTIVE, CONNECTED	CONNECTED
SINR	CONNECTED	CONNECTED

to estimate $I + N$ from the residuals in the channel estimation, the estimation of the SINR is more complex than estimation of RSRP and RSRQ.

As the SINR measurements takes both the RSRP and interference into account, it should be possible to predict the quality the UE would experience in the target cell when scheduled with PDSCH. Unfortunately, in tightly synchronized systems, the interference measured on the REs where the SS/PBCH block is transmitted, is generated by SS/PBCH blocks in other cells. This interference is not always representative of the interference the UE would experience when scheduled with PDSCH, which means that the SINR measurement is not an accurate prediction of the PDSCH quality in the target cell. In addition, the SINR presented here is a pre-processing SINR which means it is a signal quality measure before any receiver processing.

The RRC states where the various types of measurement quantities are applicable are summarized in Table 4.23.

4.5.2.2 SS/PBCH Block Measurement Timing Configuration (SMTC)

In LTE, the always-on reference signal that enables cell detection and measurement is the combination of PSS/SSS and CRS. Since the CRS are transmitted in every slot in every cell, the UE can collect measurements from a cell at any time after the cell is detected by the UE. Also, the PSS/SSS is in LTE transmitted in every 5th subframe. Thus, for catching PSS/SSSs of all cells on any given frequency, the UE only needs to scan the frequency for 5 ms at any given time.

In NR, the reference signals can be transmitted more sparsely: typically, the SS/PBCH block is transmitted every 20 ms, and even sparser transmissions can be configured in some cases. In an asynchronous system, there will be no coordination among the transmission timings of the SS/PBCH block in different cells. If all the cells transmit SS/PBCH blocks with a periodicity of 20 ms, the UE would have to open its receiver for more than 20 ms to collect even a single measurement sample for all relevant neighbor cells. This situation is depicted in Figure 4.88.

The situation depicted in Figure 4.88 would lead to a quite high power consumption in the UE. This effect would be particularly harmful in IDLE mode, where the power consumption is dominated by neighbor cell measurements.

To reduce the UE power consumption, for less frequent transmission of SS/PBCH blocks, NR introduced the concept of an *SS/PBCH block Measurement Timing Configuration*, or SMTC for short. The SMTC of NR shares some similarities with the discovery measurement timing configuration (DMTC) of LTE which was introduced in Release 12 for enabling

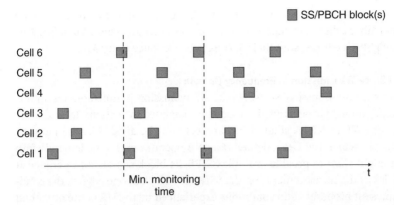

Figure 4.88 An example of an asynchronous system: the UE would have to measure during a time which is no smaller than the interval between two consecutive reference signal transmissions.

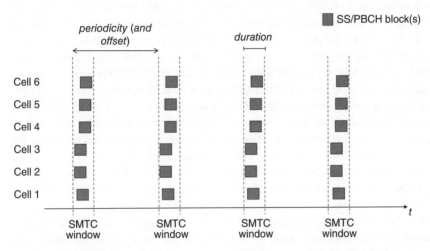

Figure 4.89 An example of a coarsely synchronized system, with an SMTC window. The UE can perform measurements on all SS/PBCH blocks by performing measurements in the SMTC window.

power savings for small cell deployments. The SMTC determines a time window with a certain period, length, and offset (versus SFN), and it is applicable in CONNECTED, INACTIVE, and IDLE modes.

There is one SMTC for a given SS/PBCH frequency location and it may be signaled to the UE in system information or via dedicated signaling. The UE may assume that the SS/PBCH blocks of the given frequency raster point of all relevant neighbors can be found inside that window: the UE is never required to search for neighbor cells outside the SMTC window. In practice, this requires that the network synchronizes the transmission of the SS/PBCH blocks to some level of accuracy: otherwise there would be neighbor cells that the UE would miss to detect even if the signal strength would be high. This would certainly be an issue as that kind of cell could cause high interference if not taken into account e.g. when a network is making handover decisions. Figure 4.89 depicts the SS/PBCH block transmissions from different cells that are aligned.

If the NW does not configure the SMTC window, the UE may assume that it can find the SS/PBCH for all neighbor cells in any measurement window of length 5 ms. In practice, this requires that all neighbor cells transmit their SS/PBCH blocks once every 5 ms.

4.5.2.3 SS/PBCH Block Transmission in Frequency Domain

While the previous section discusses SS/PBCH block transmission in time domain, in this section, the frequency domain aspects are discussed. As explained in Sections 3.4 and 2.4.5, there is a concept of BWP in NR. A single wideband cell may be divided by the network into smaller chunks in which the UEs operate. If a gNB operates with a wide bandwidth (e.g. 200 MHz), it might want to provide multiple SS/PBCH block signals for reference at different frequencies as the measurements on the SS/PBCH block at one edge of the bandwidth may not represent properly radio conditions experienced by the UE at the other end of the bandwidth. This may lead to a situation where different reference signals, being sent by the same gNB and the same sector antennas will be treated as inter-frequency measurements. Furthermore, from the perspective of the UE and its RRM measurements, they will comprise different cells, even though the network will be able to change the active BWP of the UE to be on the same frequency range as an SS/PBCH block of another physical cell ID without the need to perform handover. For this reason, a notion of cell defining Secondary Synchronization/Physical Broadcast Channel Block (CD-SSB) was introduced. The CD-SSB refers to a SS/PBCH block which is associated with SIB1 which includes the necessary system information needed to perform initial access to the cell via this SS/PBCH block and a so-called New Radio Cell Global Identity (NCGI), allowing to unambiguously identify the serving cell and the gNB in the network. When UE is pointed with a CD-SSB for its PCell, the UE knows "the home base" of the RRC connection. CD-SSB is also the one which is always used during initial access and handover to the cell. For example, if UE3 in Figure 4.90 is handed over from gNB2 to gNB1, a frequency of its CD-SSB will change from F3 to F1. Discussion on initial access can be found in Section 4.1.

Figure 4.90 below depicts a network deployment where, within a span of a single carrier bandwidth, the network transmits SS/PBCH blocks on several frequencies, see also Appendix B in TS 38.300 [10]. Two gNBs operate on the same carrier frequency, but their CD-SSBs are located on different frequencies. They also provide SS/PBCH blocks on additional frequencies. Based on the results of measurements from the UEs on different SS/PBCH blocks, gNBs may decide to do handover to a different cell or switch the BWP on which the UE operates.

4.5.3 Connected Mode Mobility

In RRC CONNECTED state, the mobility of the UE is network-controlled. The network reconfigures the UE to connect to different cells to maintain the radio link as the UE moves over a geographical area. Once the network determines that the radio link needs to be transferred to a new cell, it sends a handover command to the UE which includes the RRC configuration in the new cell including assistance information (e.g. SMTC) for the UE to synchronize to the new cell. This kind of RRC reconfiguration message is called "reconfiguration with sync" and it is used in cases where the UE needs to initiate connection

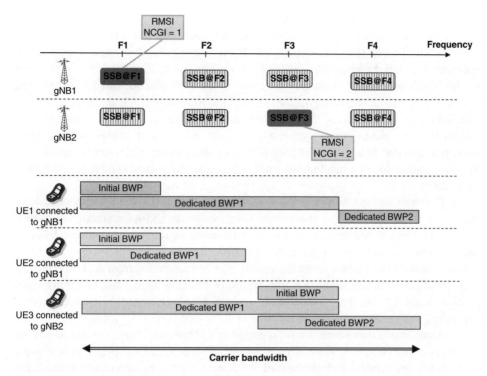

Figure 4.90 A network deployment with SS/PBCH blocks transmitted on multiple frequencies within a single carrier.

to a new cell. In addition to handover, these cases include set up of SCG (set up of dual connectivity) and modification of SCG or addition or modification of a carrier aggregation configuration in cases a new cell does not belong to the same Timing Advance Group as one of the already used cells. All these procedures can be initiated by the network without any input from the UE. However, for handover decisions, the network typically relies on measurements performed by the UE whereas configuration for an additional capacity layer, the dual connection or CA, may be feasible also without measurements.

Usually, the measurement reporting is *event-driven*, meaning that the UE provides the measurement results when certain conditions are fulfilled. The measurement events were designed to provide the network with accurate information to execute handovers that lead to competitive performance in the target cell after the handover. RRM measurements are described in the next sub-section.

In the discussion above, the mobility procedure is described from the view of the UE and the network. However, the network is not a single entity: several network nodes may be involved in the handover. For example, before a handover is carried out to a target cell served by a different gNB than the source cell, the gNB of the source cell sends a handover request message to the gNB of the target cell to check if the target gNB is willing to accept the UE in the new cell. In fact, even CN nodes can be involved in the procedure. In Section 4.5.3.2, this is described in more detail.

4.5.3.1 Overview of RRM Measurements

As explained above, the UE performs RRM measurements on two types of reference signals: SS/PBCH block(s) or CSI-RS(s). The SS/PBCH blocks can be seen as common signals meaning that the same RS is available for measurements to all UEs in a certain area while CSI-RS needs to be explicitly configured to the UE using dedicated signaling. The parameters of the CSI-RS configuration include the RE mapping, i.e. the time/frequency location of the transmissions, the periodicity and the sequence generation seed. For the SS/PBCH block, most of these parameters are fixed in the specification, and only a few need to be signaled. In addition, the SS/PBCH block includes the PSS, which facilitates easy synchronization due to a small number of candidate sequences and good auto-correlation properties.

The CSI-RSs that the UE is expected to measure are included in the RRM measurement part of the UE's overall RRC configuration. Even though the CSI-RS configuration is dedicated to a single UE, the network may reuse the same configurations for multiple UEs and this is indeed typically the case for mobility CSI-RS configurations.

Since several beamformed reference signals may be transmitted from each cell, the UE may consolidate the measurements from multiple reference signals (beams) transmitted from the same cell. The consolidated measurements are also filtered to reduce the effect of fast fading and measurement errors.

The consolidated measurements are reported to the network. There are two flavors for reporting: either periodic or event-driven, where the periodic reports are simply sent to the network at regular intervals from the start of UE receiving the measurement configuration and having performed the measurements. The event-driven reports are only delivered to the network when some conditions are fulfilled, e.g. a neighbor cell becomes stronger than the serving cell. Event-based reporting is not necessarily associated with a single report only as the UE may additionally be configured to send a certain number of periodic reports after the event if fulfilled. The event-driven reports consume less capacity on the radio interface than periodical reports and are the main tool to provide the network with adequate information about the UE movement.

We will next describe the RRC measurement configuration framework that is used for CONNECTED mode UEs. This follows with a description of the handover procedure.

4.5.3.2 Measurement Configuration

The measurement configuration itself can be included in the *RRCReconfiguration* message or the *RRCResume* message. This means that the network can update the measurement configuration for the UE while the UE is in a CONNECTED mode, resuming from INACTIVE to CONNECTED mode or provide a new measurement configuration in the handover command.

The measurement configuration is structured in such a way that the UE is provided with a list of *measurement objects (MOs), reporting configurations, measurement identities, quantity configurations*, and *measurement gap configurations*. These interrelations are depicted in Figure 4.91.

The MO provides the UE with properties of the reference signal the UE is measuring, e.g. the frequency position and SMTC. The MO may further provide a list of PCIs which may be found on the frequency position the MO is pointing to. The reporting configuration,

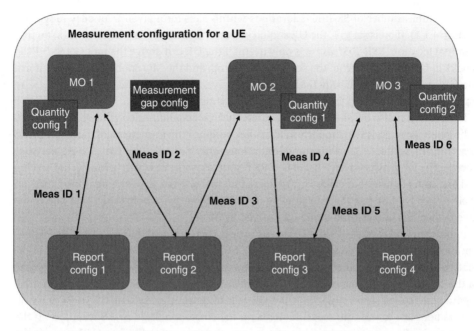

Figure 4.91 Measurement configuration.

describes the type of measurement and reporting UE should perform with respect to that MO. The measurement identity contains pointers to one reporting configuration and one MO, providing the association that defines a measurement. The measurement configuration also includes quantity configurations and a measurement gap configuration. The quantity configuration describes the filter coefficients for L3 filtering of the measurements and there can be maximum of two quantity configurations in a measurement configuration When UE reports measurement result, the measurement identity (MeasID) is added to the report so network knows with what assumptions the measurements were performed. (see Section 4.5.3.2.3).

Next, we will describe the functionality of each of the elements of the measurement configuration mentioned above.

4.5.3.2.1 Measurement Object

The MO provides the UE with information about the reference signal UE is measuring: one MO may point to an SS/PBCH block, to a CSI-RS, or both. The MO points to an SS/PBCH block by giving the center frequency as a channel raster point. Thus, the MO can flexibly point to any SS/PBCH block frequency location within NR spectrum and is not restricted to synchronization raster points. It should be noted that one MO may point only to one frequency location and that there may be only one MO per frequency location in the overall measurement configuration of the UE.

The MO also contains the SMTC for a given reference signal as well as further assistance information such as a parameter called *SSB-ToMeasure*. This parameter gives the candidate time locations of SS/PBCH blocks which the UE needs to monitor on the indicated SS/PBCH frequency location and within associated SMTC window. The parameter is a bitmap of candidate locations with L bits where L can be 4, 8, or 64 (corresponding to maximum number of SS/PBCH locations within 5 ms for a given frequency range, see Section 4.1.1). If bit is set to 1, the UE should search and measure SS/PBCH block(s) on that nominal location. *SSB-ToMeasure* is only useful if the UE can derive the index of SS/PBCH blocks in neighbor cells from the serving cell timing, and the MO contains a flag to inform the UE of this property. If *SSB-ToMeasure* is not given, the UE should attempt to measure on all candidate time locations within the SMTC window.

If the MO points to a CSI-RS to be measured, the exact configuration of the CSI-RSs needs to be given. The CSI-RS for mobility follows the configuration framework of the NZP CSI-RS described in Section 3.7.1 with some limitations: only periodic one-port CSI-RS are supported. The UE may be given a list of CSI-RS resources, some of which belong to the serving cell and some to neighbor cells. This requires that the network nodes share information on their respective CSI-RSs that are suitable for mobility measurements such that one cell is able to give this information to the UEs served by it.

In order to find the CSI-RS, the UE needs to derive a reference timing, as the CSI-RS does not include a dedicated SS like the PSS. The CSI-RS configuration needs to include information about where to obtain this timing reference. Often, it is one of the SS/PBCH blocks of the cell to which the set of CSI-RSs belongs to. This is indicated in the configuration by a parameter called *associatedSSB*. If the UE fails to detect the associated SS/PBCH block, it does not have to perform measurements on the corresponding CSI-RS resources. If the mobility CSI-RS are not configured with the parameter *associatedSSB* the UE may use the serving cell timing as the time reference.

The MO may also include a list of cells to be measured (whitelist), or list of cells not to be measured (blacklist).

4.5.3.2.2 *Reporting Configuration*

The reporting configuration describes which measurement quantities (RSRP, RSRQ, or SINR) the UE should derive after measuring the reference signals pointed in the associated MO. It further defines the reporting type the UE should perform. For mobility purposes, the UE can be configured to perform periodic or event-driven reporting. In addition, the UE may also be configured to perform reporting of Cell Global Identity (CGI) for establishment of neighbor cell relations. Thus, with the measurement configuration example given in Figure 4.91, one possible configuration could be that each MO is linked with different event triggered or CGI reporting configurations and one periodic reporting configuration linked to the MOs.

Furthermore, the reporting configuration defines whether beam specific reports should be delivered, and it gives the maximum amount of reports or cells to be reported. If the MO defines the white list, the reporting configuration has a flag whether the list should be considered with this reporting configuration as more than one reporting configuration can be associated to one MO and not all of them must consider the white list.

Periodic reporting means that the UE reports with a given periodicity without considering any thresholds for the reporting. *Event based reporting* means the configured event needs to be fulfilled in order to trigger the reporting. After an event is triggered, the reporting may continue periodically for a number of periods if so configured.

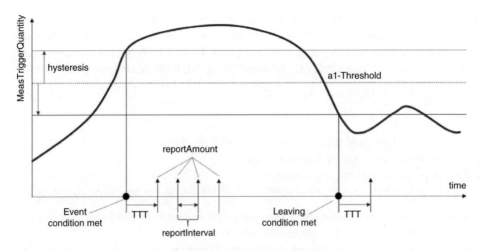

Figure 4.92 Event triggering rules taking A1 event as an example.

The events that are defined for NR do not differ much from the ones available in LTE and include:

Event A1:	Serving becomes better than absolute threshold;
Event A2:	Serving becomes worse than absolute threshold;
Event A3:	Neighbor becomes amount of offset better than PCell/PSCell;
Event A4:	Neighbor becomes better than absolute threshold;
Event A5:	PCell/PSCell becomes worse than absolute threshold1 AND Neighbor/SCell becomes better than another absolute threshold2.
Event A6:	Neighbor becomes amount of offset better than SCell.

Each of these events have a leaving and entering condition. In general, an event is triggered when the entering condition is fulfilled, e.g. for A1 when a serving cell becomes better than a threshold, and stays better than a threshold for an amount of time that is called "time to trigger" (TTT). When an event triggers, all cells that had fulfilled the entering condition for the duration of TTT are added to the "cells triggered" list which guides the reporting and prevents triggering based on the same cell consecutively. Fulfilling the leaving condition controls how cells are removed from the cells triggered list. After a cell has fulfilled leaving condition, it is removed from the cells triggered list. Once it is removed from the list, that cell can trigger the event again if the entering conditions get fulfilled. Similar to entering condition, the leaving condition needs to be fulfilled for a duration of TTT period. Each event has a hysteresis such that leaving condition can be defined to have different threshold value than entering condition to avoid ping-pong effect (i.e. UE being handed over back to the previous cell within short time), especially if TTT is short. Additionally, MO configuration may define a list of cell specific offsets which UE applies if it has detected the cell corresponding to a configured PCI. An example of A1 event scenario is presented in Figure 4.92.

4.5.3.2.3 Quantity Configuration The quantity configuration defines the layer 3 filtering coefficients which can be different for different measurement quantities, for different RS types, and for measurements per cell and per beam. Layer 3 Filtering is a memory-based averaging where each measurement performed by the UE is filtered based on the formula:

$$F_n = (1 - a) * F_{n-1} + a * M_n,$$

where:

- *Mn* is the latest received measurement result from the physical layer
- *Fn* is the updated filtered measurement result
- *Fn-1* is the old filtered measurement result
- $a = 1/2(ki/4)$, where *ki* is the filter coefficient value given in the quantity configuration

When the first measurement result from the physical layer is received to be filtered, the *F0* is set to *M1*. If k is set to 0, no layer 3 filtering is applicable.

Each MO contains a pointer to one quantity configuration.

4.5.3.2.4 Measurement Gap Configuration The UE typically needs to be configured with measurement gaps as it may not be able to receive from serving cells and measure a reference signal in an MO at the same time. The measurement gap configuration gives the UE "permission" not to listen to the serving cell during a certain time period and use that time for performing measurements in neighboring cells. Thus, it is a trade-off between performing measurements and data interruptions as UE does not receive or transmit data during the gap window. Neighbor cell measurements can thus be categorized into *gap-assisted* and *non-gap-assisted* measurements, depending on whether measurement gaps are required or not.

Similar to SMTC, measurement gap configuration contains offset, period and duration for the measurement gap. It is good to note that in NR UE may also require measurement gaps for intra-frequency measurements e.g. if the measured RS is not contained within the active BWP, as opposed to LTE where measurement gaps were required only for inter-frequency measurements.

Introduction of SS/PBCH blocks and CSI-RS signals resulted in the need to redefine what intra-frequency and inter-frequency measurements are. In LTE, the definition relied on whether the two MOs belong to the same carrier frequency expressed with so called E-UTRA Absolute Radio Frequency Channel Number (EARFCN). In NR, on the other hand, the definition depends on whether the measured signal is SS/PBCH block or CSI-RS and are as follows TS 38.300 [10]:

- SS/PBCH block based measurements are intra-frequency when the center frequency of the SS/PBCH block and the SCS of the SS/PBCH block of the serving cell is the same as that of the neighboring cell. Otherwise the measurements are treated as inter-frequency measurements.
- CSI-RS based measurements are intra-frequency when the bandwidth of the CSI-RS resource on the neighbor cell configured for measurement is within the bandwidth of the CSI-RS resource on the serving cell configured for measurement and the SCS of the two CSI-RS resources is the same.

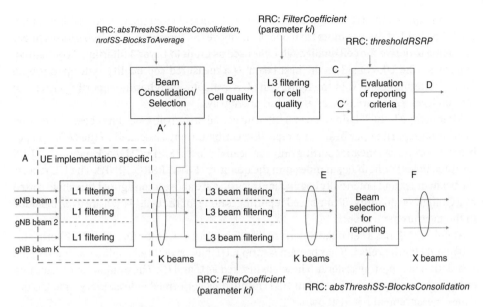

Figure 4.93 Measurement model for RRC CONNECTED.

For intra-frequency measurements, measurement gaps are not needed as long as all the configured BWPs contain the frequency domain resources of the SS/PBCH block associated with the initial BWP. Referring to Figure 4.90, measurement gaps will be required for UE1, since BWP2 does not include the frequency domain resources of the SS/PBCH block of the initial BWP. For UE2 and UE3, measurement gaps are not needed.

For inter-frequency measurements, the need for measurement gaps depend on the UE capabilities. Some UEs only support *per-UE measurement gaps*, which means that measurement gaps are always required for inter-frequency measurements. UEs may also support *per-FR measurement gaps*, in which case the UE can perform non-gap assisted measurements when none of the configured BWP frequencies is in the same frequency range as the MO.

4.5.3.3 Performing RRM Measurements

Figure 4.93 describes how the UE perform measurement filtering and consolidation. At point A, the UE has detected some of the gNB beams belonging to one cell (identified by its PCI) on a frequency pointed in the MO. The UE performs layer 1 filtering for each beam and at point A[1], layer 3 filtering is performed on the consolidated cell quality and potentially also on the individual beam qualities. Layer 1 filtering is a physical layer filtering of the measurement that the UE may do as per implementation. Layer 1 filtering is constrained by a measurement period and accuracy requirements defined in 3GPP TS 38.133 [40].

Note that we use the term 'beams' in this description. From the UE point of view, there is no notion of beams: the UE only sees a set of reference signals that it combines using certain rules. A very reasonable realization however is that the network transmits the different reference signals in different beams.

After point A[1], the UE compares the measured beams belonging to a cell with a configured threshold. If there is more than one beam above the threshold, the UE performs linear averaging to derive the cell quality, which is used as input to Layer 3 filtering, if configured. Otherwise, the UE considers the best beam as a measured cell quality. After performing Layer 3 filtering (point C in the figure), the UE evaluates whether the reporting criteria of the cell level results are met, as explained previously.

After point A[1], the UE does Layer 3 filtering per beam only in case it has been configured to perform reporting per beam in a reporting configuration associated to the MO. The per beam reporting can mean reporting only the beam indexes (SS/PBCH block index or CSI-RS resource index) or both beam index and the quantity (RSRP, RSRQ, SINR). The UE reports the beam index and potentially beam quantities that are above a configured threshold sorted by a configured quantity. Beam-based reporting is done for the cells which are to be included in the measurement report.

At point B UE has a cell quality derived which is input to L3 filtering. After L3 filtering for cell quality for the cells UE has been measuring, UE has several cell qualities as input to the Evaluation of reporting criteria. These are marked as C and C′. The evaluation of reporting criteria, for example whether Event A3 is fulfilled, is performed at least every time UE has a new measurement result at point C.

The L3 beam specific filtering is performed by the UE is the measurement configuration requests per beam measurements. After L3 filtering, the Beam selection for reporting selected which beam measurement results are added to the measurement report.

4.5.3.4 Handover Procedure

Handover is probably a procedure which could be called the essence of the cellular networks. It allows a user to move around the network and switch an ongoing connection to new cells without interrupting it. The foremost principles of the handover procedure did not change in NR as compared to LTE and the overall steps are depicted in Figure 4.94.

Step 0	While the UE establishes a connection with the network, its context is provided from the CN to RAN. Among other things, it contains mobility related information, e.g. roaming and access restrictions.
Steps 1–2	Step 1 is not a mandatory one, since the network may trigger a handover whenever it wishes to. However, in a vast majority of cases, the handover is preceded by at least one measurement report from the UE being received by the UE's serving gNB.
Step 3	Based on the UE measurement report, the source gNB identifies the cell which the UE should be moved to. In case the cell belongs to another gNB, the source gNB sends Handover Request message containing information about the UE's current configuration, its capabilities, PDU session related information (e.g. slices and Quality of Service [QoS] flows related) and may additionally include UE's latest cell and beam level measurements reported to the source gNB.

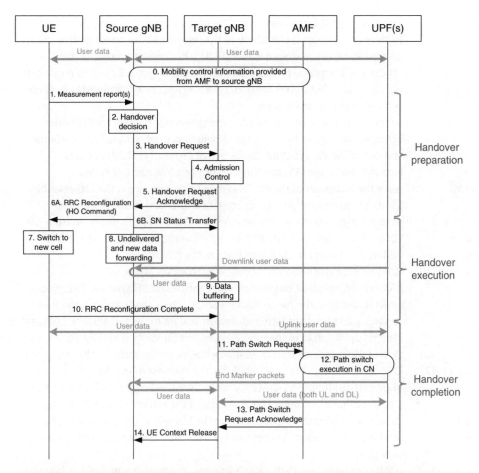

Figure 4.94 Inter-gNB handover procedure (Access Management Function (AMF) and User Plane Function (UPF) are logical entities of a 5GC network, please refer to Section 2.2 for details).

Step 4   Based on the information provided in the Handover Request message, the target gNB performs Admission Control, whose role is to decide whether the request should be accepted or rejected (e.g. due to no possibility to meet the QoS requirements of the traffic or due to no support for a specific slice indicated for the PDU session).

Step 5   In case the request is accepted, target gNB replies with the Handover Request Acknowledge message. It contains the RRCReconfiguration message to be delivered to the UE via source gNB.

Handover Request and *Handover Request Acknowledge* messages are sent either on Xn interface, in case the gNBs have different CUs, or are exchanged between the DUs of the two gNBs via CU using F1 interface, in case they share a CU.

Step 6A/6B   The source gNB forwards the *RRCReconfiguration* message (also called Handover command) contained in the reply from the target gNB to the UE transparently (i.e. it does not modify it). The message contains the information required by the UE to access the target cell such as target cell ID, UE's identifier in the new cell (C-RNTI), security configuration, target cell's system information etc.
                  At the same time, the source gNB sends Secondary Node (SN) Status Transfer message to the target gNB, which is used to transfer the uplink and downlink Packet Data Convergence Protocol (PDCP) Sequence Number and Hyper Frame Number receiver/transmitter status.

Step 7   After the reception of the *RRCReconfiguration* message, the UE detaches from the current cell and synchronizes to the new cell.

Steps 8–9   At this stage, downlink data destined for the UE is still provided from the CN to the source gNB, which forwards it to the target gNB where it is buffered, waiting for the UE to finalize the handover.

Step 10   After synchronizing to a new cell, the UE sends *RRCReconfigurationComplete* message (also called Handover Complete), which concludes the handover for the UE. Once that happens, the new serving gNB may send buffered downlink data to the UE and the UE may send its uplink data to the new gNB, which forwards it to the CN. Downlink data from the CN is still delivered to the source gNB.

Step 11–12   UE's new serving gNB sends Path Switch Request to the Access Management Function (AMF), which executes the path switch inside the CN. User Plane Function (UPF) indicates the intention to end the current data session with the source gNB for the UE subject to the handover by sending "end marker" packets and starts sending downlink user data to the new gNB.

Step 13–14   AMF acknowledges Path Switch Request by sending Path Switch Request Acknowledge message to the target gNB, which then sends UE Context Release message to the source gNB to indicate successful completion of the handover. This message is an indication to the source gNB that it may now release its radio and C-plane resources reserved for the sake of the moved UE.

It was mentioned above that the *Handover Request* message contains information assisting the target gNB in performing Admission Control. However, this is not the only piece of information included in that message by the source gNB. In fact, source gNB may provide target gNB with the current RRC configuration of the UE, including configuration of its physical data and control channels, configuration of RRM measurements as well as beam measurements etc. Thanks to the knowledge of this information, the target gNB may reuse the current configuration and only modify those parameters which require reconfiguration, e.g. the configuration of the data channel may be the same as in the source cell while the control channel properties may be different. In this situation, only the configuration of beam management would have to be modified and PDSCH/PUSCH channels' configurations could be kept. This is so called "delta signaling" and it allows to decrease the size of RRC messages being provided to the UE, saving both radio resources and UE processing resources as well as increasing the chances of successful decoding of handover message by the UE.

On high level, the steps are very similar to the ones performed in LTE, with some potential architectural differences due to introduction of CU-DU split of the gNB. However, since we deal with a beam-based system, there will be quite important differences when one inspects the procedure from lower layers perspective.

Handover Request message contains the measurement results as reported by the UE to its serving cell prior to handover triggering. Those may include both SSB and CSI-RS based measurements for the UEs currently serving cells and neighboring cells. A source gNB chooses the target gNB for the UE based on exactly those measurements, but the target gNB has some freedom to make decisions about certain aspects and knowing latest RRM measurements results of the UE is very useful for this purpose. The target gNB may, e.g. be able to decide about the following aspects:

- *Which should be the UE's primary serving cell (PCell) after the handover*: The handover is normally triggered based on the strongest cell, but if it happens that it is overloaded (source gNB may not know that) and there are strong enough cells on other frequencies, target gNB may decide to move the UE there.
- *Are there any other cells which can be configured in the UE as secondary cells (SCells)*: In case UE reported multiple strong cells belonging to the target gNB, it may decide to configure them as SCells to increase data rates achievable by the UE, if DRBs require that.
- *What are the strongest beams of the cell the UE will be connecting to*: As explained in Section 4.5.3.3 the quality of the cell is derived based on consolidating measurement results of particular beams measured by the UE. However, on top of the consolidated measurement the UE may report also signal and quality levels of particular beams identified either by the SSB index or CSI-RS resource index. This information can be used by the target gNB to deduce the most reasonable random access configuration for the UE during the handover. In particular, as explained in Section 4.1.2, random access resources (preambles and/or RACH occasions) are mapped to certain beams and the UE chooses them based on the strongest beam among those it can detect. During the handover, the gNB may assign contention-free random access (CFRA) resources for certain beams, which increases the chances for a successful handover by avoiding potential RACH collisions which can happen for CBRA. Since RACH resources are scarce and shared by many UE's performing the handover or initial cell access at a time, thanks to the knowledge about the UE's strongest beams, the gNB may assign CFRA resources only to the strongest beams as reported by the UE. Even though the UE's strongest beam may change in the meantime, it might be still beneficial to use a beam which is not the strongest one, but which has an acceptable quality and an assigned CFRA resource than to use the strongest beam, but without CFRA resource assigned to avoid the RACH collision as mentioned above. This principle is presented in Figure 4.95.

As can be seen, even though the purpose of the handover and the basic principles in NR remain similar to the ones used in the systems of previous generations, additional aspects related to beam-based air interface have to be considered.

2. Handover request including beam level measurements from the UE

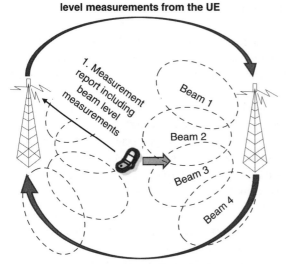

Figure 4.95 Utilization of beam level measurements for the sake of cell access during handover 4.

3. Handover request ACK including RRC message with CFRA resources for Beam 2 and Beam 3

4.5.4 Idle and Inactive Mode Mobility

4.5.4.1 Introduction

As explained in Section 2.4.7, the UE has no active connection to the network in IDLE and INACTIVE states. After cell selection, the mobility in these states relies on cell reselection procedure and occasional updates about its current location sent from the UE to the network. The UE behavior in both these states is very similar and will be elaborated further in this section. However, before proceeding to that, it is worth introducing several terms, which will be used throughout this section:

- *Camped on a cell*: The UE is camped on a cell or is camping on a cell when it has chosen this cell by executing cell selection/reselection process. The UE monitors system information and paging information on a cell it is camping on.
- Suitable cell is a cell on which the UE is allowed to camp. Such a cell must meet certain requirements:
  - o it has to belong to one of the PLMNs which the UE is allowed to use (the list is provided from Non-Access Stratum [NAS] layer of the UE)
  - o it has to fulfill a minimum signal level and quality requirement (so called cell selection criteria)
  - o the cell cannot be barred
  - o the cell cannot be part of the forbidden TAs list
- Barred cell is a cell where the network operator forbids the UEs to camp. Barring status of the cell is indicated in the System Information and can be set for numerous reasons, e.g. backhaul outage, network overload.

- Acceptable cell is a cell where the UE may camp, but only to receive emergency services, i.e. initiate emergency calls and receive notifications from Public Safety systems such as Earthquake and Tsunami Warning System (ETWS) and Commercial Mobile Alert System (CMAS). Such a cell cannot be barred and needs to meet cell selection criteria, similar as in suitable cell case. However, in contrast to suitable cell, it can belong to any available PLMN or even belong to the UE's forbidden TAs list. This is to ensure that emergency services are available for the user whenever possible.

4.5.4.2 Cell Selection and Reselection

When the UE is powered on, it executes PLMN and cell selection. The cell selection process can be executed either by leveraging the information stored in the UE, such as information about preferred frequencies, cell parameters from measurement control information elements received from the previously detected or used cells, or without using this information. Usually, the UE will begin with the cell selection based on the stored information and once that proves to be unsuccessful, it will trigger the so-called *initial cell selection*. Performing cell selection based on stored information before initial cell selection is justified by the fact that the initial cell selection is both a power and time-consuming process and the user devices tends to be utilized in the specific geographical area (e.g. neighborhood, city, country). Hence, it is natural to take advantage of the information which may speed up the cell selection process and only fall back to the more demanding initial cell selection when this is required. The initial cell selection typically happens for example when the device is powered-on after a long-haul flight in another country or even continent. The overall steps of the procedure would therefore most often be the following:

1. First, the UE scans, one by one, RF channels prioritized based on the information stored in the UE as mentioned above. If it manages to find a *suitable cell* on any of those RF channels, the cell is selected, and the procedure ends. If not, then UE continues with initial cell selection process described in step 2.
2. If no prioritized RF channel was found in step 1, the UE starts to scan all RF channels in all NR bands it supports. On each channel, the UE is only obliged to search for the strongest cell and verify whether it can be a *suitable cell* by reading SIB1 and checking the PLMNs supported by the cell. If the identified cell is suitable, it is selected, and the procedure ends. Otherwise, the UE chooses another RF channel to scan and repeats this step.

It should be remembered that the PLMN and cell selection is performed by the UE not only power-on, but also when the UE loses coverage of its current PLMN or loses network coverage completely and regains it after some time. In fact, once being out of coverage, the UE reattempts to search for a *suitable cell* on a regular basis, which is the reason for which our devices drain the battery power so quickly in such situations and where the flight mode comes in handy.

Once the UE has found a *suitable cell* where it may camp, it cannot simply rest and do nothing until it needs to establish a connection. Even when there is no immediate need to transmit or receive any data, the UE needs to constantly make sure that it camps on the best cell possible. Finding the best cell possible is done using cell ranking which is performed as part of a procedure called cell re-selection. Thanks to that, once the need to establish the connection finally comes, the UE attempts to connect on a cell which has the potential to ensure the best level of service quality. This is why the cell reselection procedure is so important and why it needs to be performed constantly while the UE is turned on and is in IDLE or INACTIVE mode. In these states, the UE is performing measurements of its serving cell (i.e. the one it is camping on) as well as of neighboring cells. However, to minimize the burden on UE's battery, there are certain pre-defined rules, which can be further affected by network configuration, allowing to make those measurements as efficient as possible:

- System Information provides the list of frequencies where the UE should perform measurements. Other frequency channels are not scanned by the UE, even though they can be supported by the UE.
- Network may provide priorities of the frequencies for UE measurements in System Information or in dedicated RRC signaling when releasing an RRC connection. The measurements on lower priority frequencies need only be performed in case the UE cannot find cells of high enough quality and signal level on its higher priority frequencies.
- The network configures also $S_{IntraSearchP}$ and $S_{IntraSearchQ}$ parameters. When the UE's serving cell's RSRP is above $S_{IntraSearchP}$ and RSRQ is above $S_{IntraSearchQ}$, it does not have to perform measurements on any neighboring cell. Only when these conditions are not fulfilled, the UE will start performing measurements on neighbor cells.

It should be noted that the "serving cell's RSRP" or "serving cell's RSRQ" used in this section refer to the representative RSRP/RSRQ value for the cell. As a default, it is the RSRP/RSRQ of the strongest SS/PBCH block of the cell, but, for intra-frequency measurements, it is also possible to configure the average of multiple SS/PBCH blocks to derive the cell's RSRP/RSRQ.

When describing cell selection and cell reselection procedures, cell selection and reselection *criteria* were mentioned several times, so it is worth clarifying what that is. The cell selection criterion is met when:

$$S_{rxlev} > 0 \text{ and } S_{qual} > 0$$

where:

$$S_{rxlev} = Q_{rxlevmeas} + (Q_{rxlevmin} - Q_{rxlevminoffset}) - P_{compensation} - Qoffset_{temp}$$
$$S_{qual} = Q_{qualmeas} + (Q_{qualmin} - Q_{qualminoffset}) - Qoffset_{temp}$$

The first expression allows verifying the minimum signal level the cell needs to provide to meet the criterion and is based on RSRP measurement ($Q_{rxlevmeas}$) while the second expression verifies cell's minimum RSRQ level ($Q_{qualmeas}$). The parameters in the formulas are provided by the network in the System Information:

$Q_{rxlevmin}$	Minimum required RX level in the cell in dBm.
$Q_{rxlevminoffset}$	Offset to the signaled $Q_{rxlevmin}$ taken into account in the Srxlev evaluation as a result of a periodic search for a higher priority PLMN while camped normally in a VPLMN (Visited PLMN, i.e. when the UE is roaming outside its service provider's country).
$Qoffset_{temp}$	Offset temporarily applied to a cell after connection establishment failure.

$Q_{qualmin}$	Minimum required quality level in the cell in dB.
$Q_{qualminoffset}$	Offset to the signaled $Q_{qualmin}$ taken into account in the Squal evaluation as a result of a periodic search for a higher priority PLMN while camped normally in a VPLMN.
$P_{compensation}$	Factor impacting the cell selection depending on the maximum transmission power levels allowed in the cell and the UE's power class ($P_{PowerClass}$).
$P_{PowerClass}$	Maximum RF output power of the UE (dBm) according to the UE power class as defined in 3GPP TS 38.101 [28, 31].

As mentioned previously, cells not meeting the cell selection criterion cannot be used as a *suitable* or an *acceptable cell* and the UE will never camp on such cells. For cell reselections, the UE additionally uses a cell ranking criterion. When the RSRP and RSRQ of the UE's serving cell drop below configured values, the UE commences neighbor cell measurements, and it starts ranking the cells meeting the cell selection criterion according to the cell ranking criterion:

$$Rs = Q_{meas,s} + Q_{hyst} - Qoffset_{temp}$$
$$Rn = Q_{meas,n} - Q_{offset} - Qoffset_{temp}$$

where:

Q_{meas}	RSRP measurement quantity used in cell reselections.
Q_{hyst}	Parameter provided by the network specifying the hysteresis value for ranking criteria
$Qoffset_{temp}$	Cell-specific offset temporarily applied to a cell after connection establishment failure.
Q_{offset}	Offset applied to the neighboring cells evaluation as signaled by the network. Its value depends also on whether the neighboring cell is on the same or different frequency as the UE's serving cell.

The *Rs* formula is used for the serving cell while the *Rn* formula is used for the neighboring cells. Once the *Rs* and *Rn* values (R values in short) of applicable cells are derived, the UE orders them starting from the highest value. If the highest value turns out to belong to a cell other than the cell the UE is currently camping on, it reselects to this cell.

In NR, there are two additional and interesting parameters, which can be used during cell ranking: "*rangeToBestCell*" and "*absThreshSS-BlocksConsolidation.*" If the *rangeToBestCell* is configured in the system information of the serving cell, the UE will not necessarily reselect to the highest ranked cell. It additionally evaluates the number of SS/PBCH blocks with signal level above the configured threshold (i.e. *absThreshSS-BlocksConsolidation*) and it performs cell reselection to the cell with the highest number of SS/PBCH blocks above this threshold among the cells whose R value is within *rangeToBestCell* of the R value of the highest ranked cell. An example of the cell reselection based on these parameters is presented in Figure 4.96.

In the example presented above, the cell with highest number of beams (SS/PBCH blocks) above the configured threshold is Cell 3. However, the cell level measurement of the Cell 3 does not fall into the range to best cell defined by the *rangeToBestCell* parameter. The second

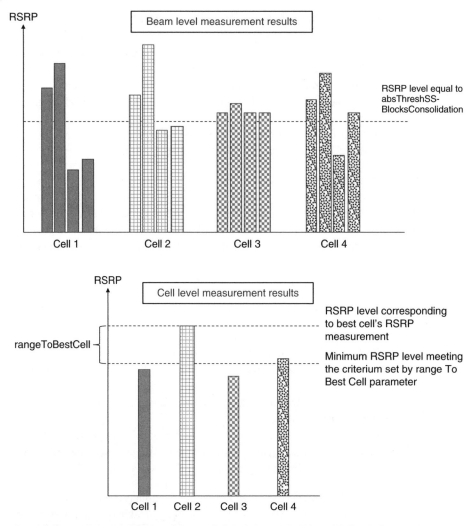

Figure 4.96 An example of usage of rangeToBestCell and absThreshSS-BlocksConsolidation parameters in cell reselection.

highest number of beams above the threshold belongs to Cell 4. Its cell level RSRP measurement result is in the range required by the *rangeToBestCell* parameter. Cell 2 has only two beams above the configured threshold and hence, a UE in this example will reselect Cell 4, even though according to the ranking criteria Cell 2 is the best one. Thanks to these additional parameters, in a deployment scenario where a cell is covered by a number of SS/PBCH block beams, situations where the UE reselects a cell with, e.g. only a single usable beam, can be avoided.

Cell selection and cell reselection procedures may lead to different outcomes. The desired result is that UE finds a *suitable cell* where it can be in a *camped normally* state. In this state, the UE can decode the system information of the cell and follow the corresponding paging channel. The UE has access to normal service which usually means the possibility to use

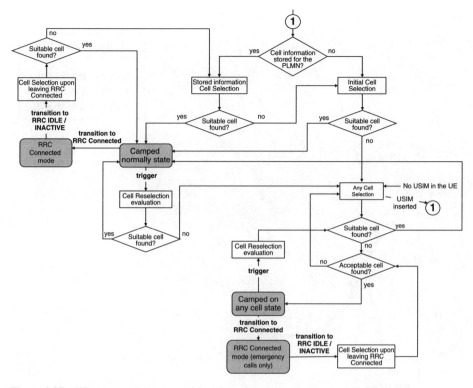

Figure 4.97   UE states and state transitions based on cell selection and reselection outcomes.

Internet, phone call, or any other services provided by the operator based on the subscription of the UE. However, if the UE does not find a suitable cell where it may camp normally, the UE attempts to find an *acceptable cell* (see explanation above in this chapter) to camp in so called "any cell state." In this state, normal services are not possible but emergency calls still are. Since, the desired state of the UE is always to be *camped normally*, when camping in *any cell state* the UE regularly attempts to find a suitable cell.

The complete view of possible cell selection and reselection related UE states and transitions between them is depicted in Figure 4.97.

4.5.4.3   Location Registration Udate

It was indicated in the beginning of the Section 4.5.4 that similar procedures are performed by the UE in RRC IDLE and RRC INACTIVE states and this is indeed true for cell reselection procedure. The main difference between those two states lies in location registration update procedures that UE is required to perform. In RRC IDLE state, the UE performs Tracking Area (TA) Update (TAU) procedure while in RRC INACTIVE the UE performs RAN-based notification area (RNA) Update (RNAU). In NR, in addition to the Tracking Area Code (TAC) defining the Core Network level area the cell belongs to (TA), a cell may also belong to a RAN area (RA) identified with a RAN area code, which is used solely in RRC INACTIVE mode. Cells can belong to one or more RAs, which cannot span across TAs as presented in Figure 4.98. Tracking Area Update (TAU) procedure is triggered when the UE enters a new

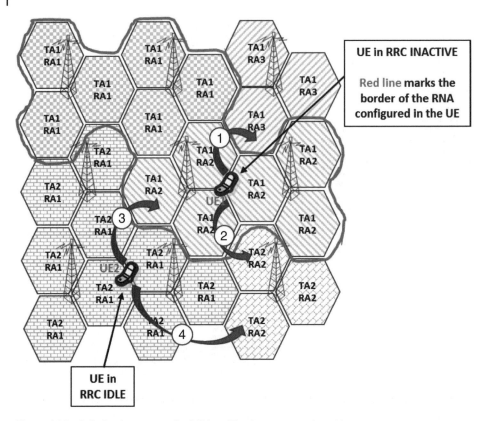

Figure 4.98 Relation between cells, RAN notification areas and tracking areas.

TA, which is not on the TA list configured in the UE by the network while UE is in RRC IDLE mode. The RNAU, on the other hand, is triggered by the UE in RRC INACTIVE state when it enters a cell which is not on the RNA as configured by the network. The RNA is configured in the UE when the network decides to move it to RRC INACTIVE state and it comprises either explicit list of cells or a list of RAs. In addition, both procedures are also executed on a periodic basis, e.g. in case the UE remains in a particular network area for a longer period of time. Periodic update is required for the network to ensure that the device is still available and, e.g. was not switched off or did not move out of the network coverage. Thanks to those updates, the network will be aware of the UE's current location in case it needs to page the UE in order to establish a connection. Figure 4.98 presents an exemplary network deployment with two UEs moving between the borders of various network areas.

In the presented example, UE1 is in RRC INACTIVE state with RNA comprising RA1 and RA2 belonging to TA1 (RNA cannot span across TAs). UE2 is in RRC IDLE and only TA2 is on its TA list:

1) UE1 moves to RA3 of TA1, which is not part of the RNA configured to it by the network. UE1 triggers RNAU procedure even though it does not cross the boundaries of the TAs.
2) UE1 moves to RA2 of TA2 and hence it crosses the boundaries of both TA and RNA. Since the UE is in RRC INACTIVE state it triggers RNAU procedure.

Table 4.24 Summary of Idle mode procedures tasks division between NAS and AS in the UE.

IDLE mode procedure	NAS layer tasks	AS layer tasks
PLMN selection	• Maintaining of the PLMN/RAT list in the priority order • PLMN/RAT selection using manual (by user) or automatic mode (based on priorities set by an operator) • Evaluation of available PLMN reports from AS layer • Maintaining a list of equivalent PLMN identities (i.e. PLMNs IDs which are treated as equivalent by the UE for cell selection, cell reselection and handover)	• Searching for available PLMNs by synchronizing to a broadcast channel • Performing measurements to support PLMN selection • Reporting of identified PLMN identities and associated RATs to NAS
Cell selection	• Controlling cell selection by providing information about the RAT(s) and associated PLMN to be searched for to AS layer • Maintaining the list of forbidden Tracking Areas and providing it to AS layer	• Performing cell selection related measurements • Similarly as for PLMN selection – synchronizing to the broadcast channel, obtaining the broadcast information and providing it to NAS layer • Identifying suitable cells and informing NAS layer on availability of such cells • Choosing the cell to camp on based on cell selection criteria
Cell reselection	• Maintaining a list of equivalent PLMN identities and providing it to AS layer • Maintaining a list of forbidden Tracking Areas and providing the list to AS layer	• Performing cell reselection related measurements • Changing a cell to camp on based on cell reselection criteria • Reception and handling of broadcast information and forwarding it to NAS layer • Changing a cell if a more suitable one is found.
Tracking area update	• Registering the UE as active in the network after power on • Registering the UE's presence in a registration area in the Core Network, e.g. when entering a new tracking area or on a periodic basis • Deregistering the UE from the network when UE is powered off • Maintaining a list of forbidden Tracking Areas	• Reporting of registration area information to NAS, e.g. upon entering new registration area
RAN notification area (RNA) update	NAS layer does not participate in this procedure	• Registering the UE's presence in RNA in the RAN network (periodically or when entering a new RNA)

3) UE2 moves from TA2 to TA1. Since only TA2 is on its TA list, the UE performs TAU procedure.

4) UE2 moves from RA1 of TA2 to RA2 of TA2. Since the UE is in RRC IDLE state, changing RAs inside the same TA does not trigger any actions. The UE2 is still in TA2 and does not trigger any location update procedure.

The location registration update procedures are required for the network to be able to find the UE when there is DL data to be delivered to it. Thanks to the knowledge of the network area where the UE currently is, the network knows where to broadcast Paging messages to trigger the UE to initiate the connection with the network. Paging is described in more detail in Section 2.4.6.6 (from procedural point of view) and in Section 3.5.3.2 (from physical layer point of view).

4.5.4.4   Division of IDLE Mode Tasks between NAS and AS Layers

When executing IDLE mode procedures, both AS and NAS layers play their role in the related tasks. For example, the UE is using both NAS layer information (e.g. eligible PLMNs) as well as AS layer parameters (e.g. cell related offsets) while performing cell selection. Both layers are also involved in the location registration tasks as explained above. Table 4.24 summarizes this aspect.

5

Performance Characteristics of 5G New Radio

Fred Vook

Nokia Bell Labs, USA

5.1   Introduction

As has been the focus in previous chapters, a key component of the New Radio air interface is the flexible and scalable framework for supporting large-scale antenna arrays at the gNB, otherwise called Massive MIMO. The two main reasons for deploying Massive MIMO are for improving system capacity and coverage reliability, both of which are important for providing the high data rates promised by 5G. For bands below 6 GHz, spectrum tends to be in short supply, so improving system capacity in areas of dense user traffic tends to be a key priority. However, the coverage performance improvements of a beam-based air interface can be of great importance for example when covering lightly populated rural areas or when deploying NR at higher carrier frequencies on site grids that were sized for operation at lower carrier frequencies. For bands above 6 GHz, large data rates can be achieved due to the use of large bandwidths, but poor propagation conditions lead to coverage reliability and range being important questions.

In this chapter, we discuss the system level performance benefits of the New Radio air interface with a focus on the MIMO performance characteristics for various antenna array configurations. The discussion is divided into two main sections: the first deals with NR performance in sub-6 GHz bands, and the second deals with NR performance in the mmWave bands. For Sub-6 GHz deployments, the focus is on the downlink, where we will show how the system performance improves as the array size increases and how the system performance is impacted by inter-site distance (ISD). We show a direct comparison between the channel state information (CSI) feedback schemes in LTE versus New Radio for the different array configurations. Next, we will show how the performance benefits of Massive MIMO are affected by the user data traffic characteristics, such as full buffer traffic versus a bursty type of traffic like FTP. Finally, we show the performance versus overhead characteristics of the various Type II CSI configurations in Rel-15 New Radio. For deployments in the mmWave bands (e.g. 28 GHz), we show what data rates and cell radius can expected to be supported in the mmWave bands. Also, we show the benefits of SU-MIMO versus MU-MIMO in the mmWave bands.

5G New Radio: A Beam-based Air Interface, First Edition. Edited by Mihai Enescu.
© 2020 John Wiley & Sons Ltd. Published 2020 by John Wiley & Sons Ltd.

5.2   Sub-6 GHz: Codebook-Based MIMO in NR

In this section, we present a performance comparison of codebook-based MIMO in NR versus LTE for various antenna array configurations in sub-6 GHz deployments. This section will highlight some of the key benefits of the NR MIMO framework with respect to the feedback of the CSI that is leveraged by the gNB when transmitting data on the downlink. This section will also quantify the impact on performance from using larger antenna array sizes and different ISDs. The impact of the user data traffic characteristics will also be explored.

5.2.1   Antenna Array Configurations

For studying NR performance in sub-6 GHz deployments, we consider antenna array configurations having 2 through 32 transceivers as shown in Figure 5.1. We consider cross-pol antenna array panel structures all having 8 rows of physical antenna elements and a variable number of columns. For arrays with 2, 4, 8, and 16 ports, the co-pol elements within each column are beamformed with an all-ones beamweight vector (of length 8) which is driven by a single transceiver. The result for the arrays having 2, 4, 8, 16 ports is a single row of 2N transceivers, where N is the number of columns in the array. For the 32-port array configuration, the top four co-pol elements in each column are beamformed with an all-ones beamweight vector (of length 4) and are driven by a single transceiver. A similar operation is done for the bottom four co-pol elements in each column, the net result of which is two rows of transceivers for a total of 32 ports in the 8-column array. An ISD-dependent mechanical downtilt is applied to the array structure to improve the coverage performance of the cell. The downtilt values and the other array parameters are summarized in Table 5.1.

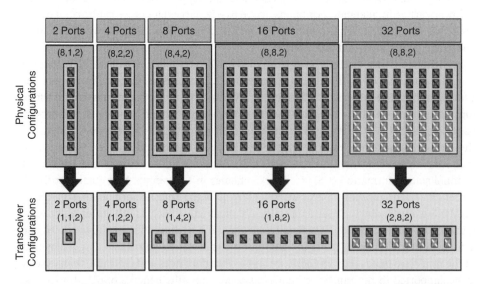

Figure 5.1 Antenna arrays at the gNB for sub-6 GHz performance comparisons: physical antenna configuration (top diagrams) and transceiver configurations (bottom diagrams), where (M,N,P) means M rows by N columns by P polarizations (P = 2 in all cases).

Table 5.1 Antenna array parameters for sub-6 GHz performance evaluation.

Antenna array parameter	Value
Number of rows of physical elements	8
Number of columns	1 (2-ports), 2 (4-ports), 4 (8-ports), 8 (16- and 32-ports)
Spacing between rows	0.8λ
Spacing between columns	0.5λ
Number of polarizations	2 (i.e. cross-pol elements)
Transceiver mapping	Full array aggregation (for 2-, 4-, 8-, and 16-ports) – single row of transceivers Subarray aggregation (for 32 ports) – two rows of transceivers
Electrical tilt	none
Mechanical tilt	ISD-dependent (6° for 200 m ISD, 2° for 1500 m ISD)

5.2.2 System Modeling

In this chapter, we will present system level simulation results to illustrate the performance of various aspects of the new radio MIMO framework. The simulations leverage the system-level methodology described in [45] for evaluating NR systems aimed at providing enhanced mobile broadband services. We consider hexagonal 3-sector cell layouts in two scenarios: the first is the Urban Micro scenario with an ISD of 200 m and a gNB height of 10 m (denoted NR-UMi-200 m), and the second is the Urban Macro scenario with an ISD of 1500 m and a gNB height of 25 m (denoted NR-UMi-1500 m). Users are assumed to be randomly placed throughout the deployment area. A 10 MHz system bandwidth used in the simulations along with the corresponding NR numerology. The performance at other operating bandwidths can be approximated from the spectral efficiency results presented in this chapter. When comparing the performance against the MIMO framework that is in LTE, the evaluation assumed both the NR and LTE systems operated with the same system overhead so as to focus on the MIMO-related performance differences. Table 5.2 summarizes the system parameters for the sub-6 GHz NR system being evaluated in this section.

5.2.3 Downlink CSI Feedback and MIMO Transmission Schemes

For transmissions on the downlink, there are numerous options for configuring the combination of CSI feedback scheme and MIMO transmission scheme. For this performance study, the downlink CSI feedback and MIMO transmission schemes are shown in Table 5.3 for the antenna array configurations under consideration. The particular transmission schemes in Table 5.3 were chosen either because those configurations were shown in previous studies (e.g. [46, 47]) to be the best performing (or essentially equivalent to the best performing) or are the configurations that are commonly deployed (as in the case of 2- and 4-port LTE deployments). Typically for larger array sizes (e.g. 8 or more transceiver ports),

Table 5.2 System modeling for performance evaluations in sub-6 GHz deployments.

Parameter	Value
Carrier frequency	2 GHz
Bandwidth	10 MHz (NR PHY numerology in all cases)
Duplexing	Frequency division duplex
Inter-site distance (ISD)	200 m, 1500 m
Scenario	NR-UMi (200 m ISD), NR-UMa (1500 m ISD)
Traffic type	Full buffer, 10 active UEs per sector (average)
	Bursty (FTP Model 1) with 0.5 MB packet sizes and varying packet arrival rates
Antenna arrays at gNB	2, 4, 8, 16, 32 ports, each with 1, 2, 4, 8, and 8 columns, respectively. See Table 5.1
UE	4 RX (2 cross-pol pairs), 80% of UEs are indoors, 20% are outdoors. 3 kmph velocity for indoor UEs, 30 kmph velocity for outdoor UEs.
Base Tx power	46 dBm (all array configurations)
Channel estimation	Non-Ideal for both CSI-RS and DM-RS
Scheduler, HARQ	Proportional fair, IR combining
MU-MIMO pairing	Greedy based on proportional-fair metrics

Table 5.3 Downlink MIMO transmission schemes.

Antenna array	LTE	NR
2-port	LTE Rel-8 codebook with SU-MIMO	NR Type I codebook with SU-MIMO (identical performance as with the LTE Rel-8 codebook)
4-port	LTE Rel-8 codebook with SU-MIMO	NR Type II codebook with MU-MIMO
8-port	LTE Rel-10 codebook with MU-MIMO	NR Type II codebook with MU-MIMO
16- and 32-port	LTE Rel-14 advanced CSI linear combination codebook with MU-MIMO	NR Type II codebook with MU-MIMO

MU-MIMO transmission can provide significant performance benefits over SU-MIMO. However, for smaller array sizes such as 2 or 4 ports, typically SU-MIMO provides the best performance, although prior studies have shown benefits to using MU-MIMO over SU-MIMO with 4 ports and the NR Type II CSI. (Using MU-MIMO with a 4-port array with either the Rel-8 LTE codebook or the NR Type I codebook tends to provide no gain over simply using SU-MIMO). The MU-MIMO transmission schemes use a regularized zero-forcing style of transmission where the channel state feedback is used to compute transmit beamforming weights that attempt to optimize the tradeoff between cross-talk suppression and beamforming gain. Prior studies have shown significant benefits to this type of transmission scheme over simply using the fed-back CSI unmodified.

5.2.4   Traffic Models and Massive MIMO

Before presenting the performance comparisons, a few points on traffic models are worth making with respect to the performance of massive MIMO systems. In this chapter, we present performance evaluations with both a full buffer traffic model and a busty traffic model (FTP Model 1) as used in 3GPP performance evaluations (e.g. [45]). Massive MIMO systems have performance that is highly dependent on the traffic model, and interesting insights into the MIMO performance can be obtained by observing system performance with both full buffer and bursty traffic models.

In a full buffer traffic model, each cell serves some number of User Equipments (UEs) on average (e.g. 10, as in the studies in this chapter). The gNB always has a packet to send to each of its attached UEs, and a new packet is generated for a UE once a packet is successfully delivered. Since the UEs attached to the cell are always in need of a downlink transmission, the gNBs in the cells are always transmitting, which can lead to deployments that are interference limited depending on the ISD. In full buffer traffic, because the attached UEs are always in need of a downlink transmission, there are plenty of opportunities for co-scheduling multiple UEs for MU-MIMO transmission as will be seen below when we present the performance comparisons.

In contrast, in the FTP model 1 traffic model considered in this chapter, UEs randomly "wake up" into the system, each needing to receive a packet from the gNB at some pre-defined arrival rate (e.g. in packets per second). A UE will go to "sleep" and disappear from the system once its packet is successfully delivered. For low arrival rates, gNB transmissions can be rare, which leads to relatively little interference between cells, in which case performance is generally path-loss limited (as opposed to interference limited) with few opportunities to co-schedule UEs for MU-MIMO transmission. As the arrival rate increases, gNB transmissions become more and more common, interference increases, and the system becomes more and more interference-limited with increasing opportunities to pair multiple UEs for MU-MIMO transmission. With bursty traffic, low arrival rates result in SU-MIMO being the predominant transmission scheme due to relatively few UEs needing to be served. With high arrival rates, MU-MIMO transmission become much more common.

However, one of the big differences between a bursty and a full buffer traffic model is the issue of the arrival rate of data into the system with respect to the spectral efficiency capabilities of the gNB. In bursty traffic, if the arrival rate is too high, the offered data load into the system can exceed the system spectral efficiency, which can cause an instability to occur since the system spectral efficiency is too low to serve the incoming packet arrival rate. This instability manifests itself as a near-zero mean UE throughput with average packet delays tending toward infinity. In contrast, there is no such instability with a full buffer traffic model since "new" data does not "arrive" into the system until the "old" data is successfully "delivered."

5.2.5   Performance in Full Buffer Traffic

Figures 5.2–5.4 show the performance of the NR and LTE configurations described above in full buffer traffic with an average of 10 active users per cell, where the details of the system

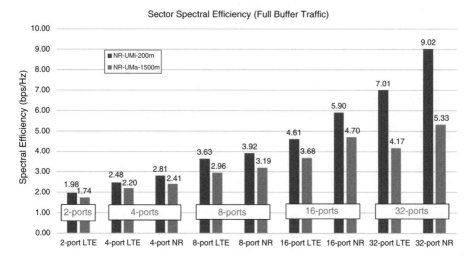

Figure 5.2 Sector spectral efficiency with full buffer traffic – 2 through 32 ports, NR vs LTE.

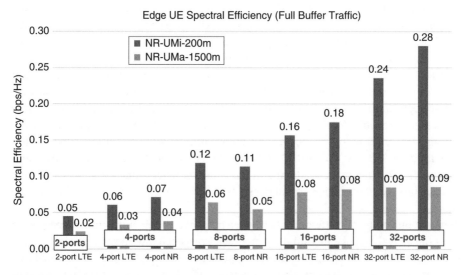

Figure 5.3 Edge user (5th-percentile) spectral efficiency with full buffer traffic – 2 through 32 ports, NR vs LTE.

level simulations have been described above. Figure 5.2 shows the sector spectral efficiency plotted as a function of the number of ports, LTE vs NR, and for two different deployment scenarios – NR-UMi-200 m and NR-UMa-1500 m. Figure 5.3 shows the edge user spectral efficiency, which is defined to be the 5th-percentile spectral efficiency (or equivalently, 95% of the users will have a user spectral efficiency greater than the value plotted in Figure 5.3). Figure 5.4 shows the mean number of users that are co-scheduled during the downlink transmission using a typical greedy-pairing algorithm based on the CSI-related feedback from the UE.

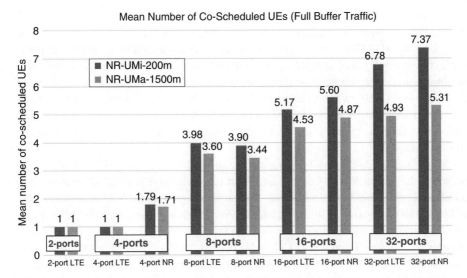

Figure 5.4 Mean number of UEs paired with a greedy pairing algorithm in full buffer traffic.

Several points are worth noting from Figures 5.2–5.4. First from Figures 5.2 and 5.3, we see a steady progression of increasing performance with increasing array size in both deployment scenarios. We also see significant gains in sector spectral efficiency of NR over LTE due to the higher accuracy of the NR Type II CSI compared to the LTE CSI. For cell edge spectral efficiency (Figure 5.3), we see higher gains in NR vs LTE for the larger array sizes (e.g. 16, 32) at the smaller ISD. With lower array sizes and higher ISDs, NR and LTE perform similarly in terms of cell edge performance. We also see slight drops in performance with the higher ISD relative to the lower ISD, which is due to the higher path losses that will be seen on average between a cell and its attached UEs. Keep in mind that in these simulations, all cells have 10 attached UEs on average regardless of the ISD, which means higher ISDs will actually have a lower UE density when measured in UEs per unit area. It is interesting to note that the 32-port LTE configuration underperforms the 16-port NR configuration in the 1500 m ISD case. For this particular scenario, the benefits of the NR Type II CSI were greater than the benefits of doubling the number of transceivers.

Following some points already made at the beginning of this chapter, a larger antenna array system provides performance benefits in two ways. The first way is to provide a higher beamforming gain by virtue of being able to coherently combine the transmitted signals from more antenna ports, thereby increasing the SNR of the signal received at the UE. The second way is to provide a higher spatial multiplexing gain by virtue of being able to use its additional degrees of freedom to pair additional users with MU-MIMO transmission. With MU-MIMO, the disadvantage of dividing the fixed transmit power amongst multiple users is typically far outweighed by multiplexing multiple users over the same bandwidth, which increases the opportunities for scheduling each user. As a result, even though the transmit power per user is decreased, the benefits of providing additional scheduling opportunities can significantly increase system performance, both in the mean and in the cell edge performance. The key to the MU-MIMO benefits is the ability of the transmit array to control the multi-user cross-talk, not only with the transmit weight computation algorithm but also with the intelligent selection of the best set of users to co-schedule. The best-performing MU-MIMO transmission schemes leverage the combination of beam steering and null-steering when computing the transmit weights for each co-scheduled user. With this combination, transmit weights are computed so that a beam is pointed in the spatial direction of a UE, while beam pattern nulls are computed to point toward the other co-scheduled UEs. Larger arrays and/or an improved accuracy/resolution of the channel knowledge at the base enable transmit weights that are more effective at controlling the multi-user cross-talk.

The greedy style of MU-MIMO pairing algorithm used in this study will pair additional users when the fed back user CSI indicates that the additional pairing will improve the spectral efficiency. From Figure 5.4, we see that larger arrays tend to co-schedule more UEs with MU-MIMO compared with the smaller arrays, which is expected in full buffer traffic since larger arrays have more degrees of freedom for suppressing the multi-user cross-talk. In Figures 5.2–5.4, the 2 and 4 port LTE cases use SU-MIMO rather than MU-MIMO, so the number of co-scheduled UEs is always one (by definition) in those configurations. Note how a higher ISD tends to result in fewer paired UEs, which is due to the tendency to avoid pairing users with relatively high path loss given how MU-MIMO must divide a fixed transmit power amongst all the paired UEs. Also note how the NR Type II CSI tends to result in more paired UEs than the LTE CSI, which is due to the higher resolution of the NR-Type II CSI, which leads to a lower predicted cross-talk by the pairing algorithm, which then favors the pairing of more users. Different pairing criteria may result in different statistics from what is shown in Figure 5.4, but we would expect similar trends to be seen with any well-designed user pairing algorithm.

It is important to note that the air interface differences (e.g. system overhead) are not taken into account in these figures so as to focus on the MIMO-oriented performance characteristics. Any advantage that NR would have over LTE in terms of system overhead will only further increase the gains of NR over LTE. It is also important to note that other implementation choices and system configuration options (e.g. pairing algorithm, transmit weight calculation, reference signal choices) will result in different performance numbers, but similar trends can be expected as the array size increases and also for the performance of NR relative to LTE. Furthermore, performance with higher system bandwidths can be approximated based on these spectral efficiency results. Finally, another key advantage for NR that is not shown in these results is any range extension benefits that result from the ability to beamform the control channels.

5.2.6 Performance in Bursty (FTP) Traffic

In this section, we focus on the impact of different traffic characteristics on the performance of massive MIMO performance. We consider the performance of two array configurations, each with an LTE configuration and an NR configuration, in both bursty and full buffer traffic. The deployment is at 2 GHz in the NR Urban Micro scenario [48] with a 200 m ISD. The two antenna configurations are the 4-column, 8-port configuration and the 8-column, 32-port configuration, both of which are show in Figure 5.1. The bursty traffic model in this section is the FTP Model 1 with 0.5 Mbyte packets with various packet arrival rates in a

10 MHz system bandwidth. For comparison, the full buffer traffic model was used with an average of 10 active UEs per cell. Unless otherwise noted, the MIMO and system details are identical to what was used with the full buffer results in the previous section.

The transmission schemes are the corresponding MU-MIMO schemes from Table 5.3, except that in this section, the CSI feedback configurations are slightly altered to improve the performance in bursty traffic. Specifically, with MU-MIMO in the full buffer results in the previous section, the UEs were configured to feed back a maximum rank of 1, which tends to optimize performance in full buffer traffic when MU-MIMO is used (the exception is the 4 port NR case with MU-MIMO, where a max rank of 2 was the better performing option in full buffer traffic). In contrast, for the results in this section, the UEs are configured for a maximum feedback rank of 2, which provides significantly improved performance in bursty traffic.

Figure 5.5 shows the mean UE spectral efficiency of the 8-port configurations and the 32-port configurations. Figure 5.6 shows the Edge UE spectral efficiency (5th percentile rate as discussed earlier). Figures 5.7 and 5.8 show the resource utilization (fraction of resources occupied on average with transmissions) and the mean number of co-scheduled UEs, respectively. Several items are worth noting from these figures and will be discussed in the following. Figure 5.9 presents the data in Figure 5.5 differently: Figure 5.9 presents the percent gain in mean UE spectral efficiency of 32 ports versus 8 ports for both NR and LTE as well as the percent gain of NR over LTE for both 8 ports and 32 ports.

First, note the general decrease in user throughput as the arrival rate increases, which is an expected behavior considering how higher arrival rates mean more users need to share the available over-the-air resources. Next, Figures 5.5 and 5.9 shows how there is a significant performance advantage in using the larger array (24–367% as shown in Figure 5.9). There are also smaller, but still significant gains in using NR rather than LTE for a fixed array size (up to 37% in Figure 5.9). As the arrival rate increases with bursty traffic, the percentage gain from using the 32-port array over the 8-port array increases significantly. Also, the gain from using NR over LTE increases with arrival rate for both the 8-port array and the 32-port array. The reason for these trends is that an increasing arrival rate provides additional opportunities for co-scheduling users, so a more capable MU-MIMO system (e.g. either a larger number of ports and/or a more accurate CSI feedback scheme) can provide higher and higher user multiplexing gains as the traffic levels increase.

From Figure 5.7, we note that the utilization of the over-the-air resources increases with the arrival rate, which is also expected given how more and more users need to share the resources. Note how the higher performing systems operate with much lower resource utilization, which is also expected since the higher-performing systems can clear users off the air faster, thereby needing less overall resource usage. The ability to clear off users faster has the advantage of lowering the interference provided to nearby cells, which helps to improve overall system performance even further. Typically, if the utilization exceeds some threshold (e.g. 70%), then the user experience will start to degrade, and operators may wish to upgrade (e.g. add another carrier, densify the deployment, add a more capable antenna array). The more capable arrays reach that utilization threshold for much higher offered loads than the less capable arrays. Also, as the utilization approaches 100%, the system will tend to go unstable where the input data rate to the system is greater than the rate at which UEs can be served. The more capable system can clear UEs faster and will reach that unstable point

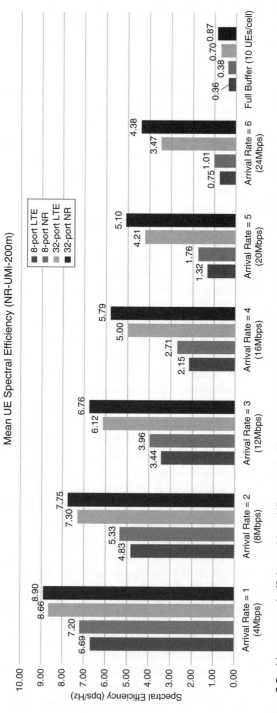

Figure 5.5 Mean spectral efficiency: Mean UE spectral efficiency with bursty traffic (arrival rates 1 through 6 packets per second). Mean sector spectral efficiency for full buffer traffic (10 UEs per cell on average). NR-UMi scenario at 2 GHz with a 200 m ISD.

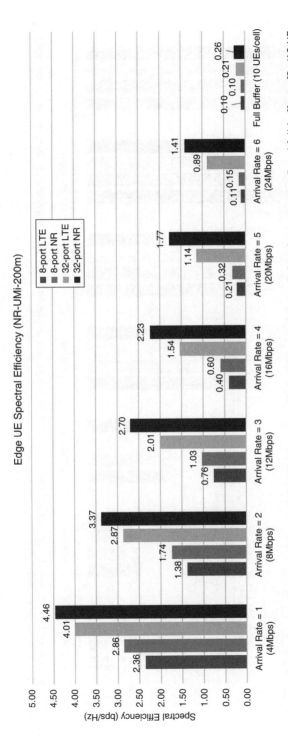

Figure 5.6 Edge UE spectral efficiency (5th percentile) with bursty traffic (arrival rates 1 through 6 packets per second) and full buffer traffic (10 UEs per cell on average). NR-UMi scenario at 2 GHz with a 200 m ISD.

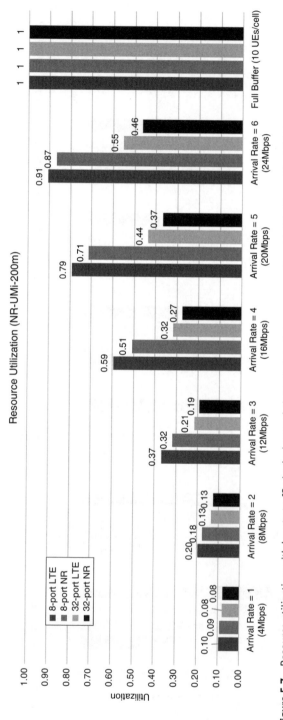

Figure 5.7 Resource utilization with bursty traffic (arrival rates 1 through 6 packets per second) and full buffer traffic (10 UEs per cell on average). NR-UMi scenario at 2 GHz with a 200 m ISD.

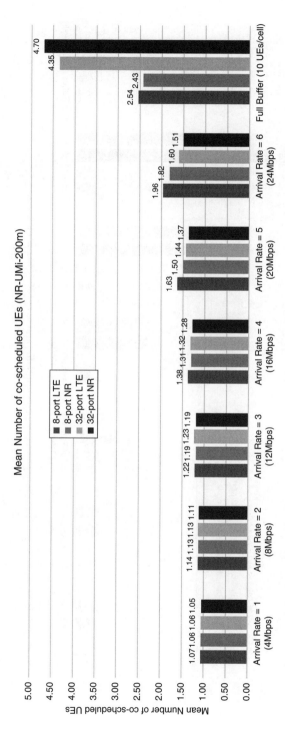

Figure 5.8 Mean number of co-scheduled users with bursty traffic and full buffer traffic. NR-UMi scenario at 2 GHz with a 200 m ISD.

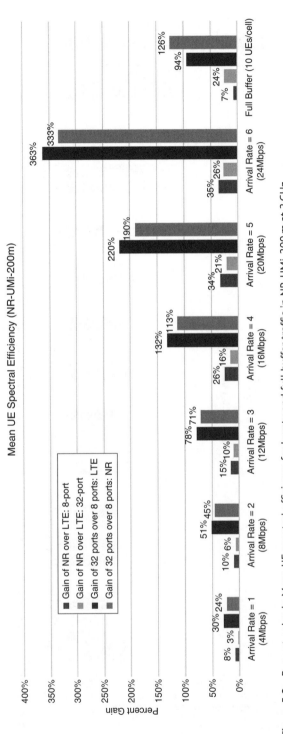

Figure 5.9 Percent gains in Mean UE spectral efficiency for bursty and full buffer traffic in NR-UMi-200 m at 2 GHz.

at higher and higher offered loads than the less capable system. Note that full buffer has utilizations of 100% which is simply due to the nature of the full buffer traffic model which was discussed earlier in this chapter.

From Figure 5.8, we note that for the arrival rates considered in this study, the average number of co-scheduled users is quite low, which is the result of relatively few opportunities to pair users given the nature of the bursty traffic. However, note that in full buffer a relatively large number of users are paired, with the higher performing arrays able to pair more users and therefore provide higher user multiplexing gains. However, in bursty, we see the opposite, namely the worse-performing arrays tend to pair more users. The explanation for this behavior is that the less capable arrays are not able to clear off the users as fast as the more capable arrays, thereby leaving more opportunities to pair users. Fundamentally however, the big arrays improve performance significantly even if there are few opportunities to pair UEs simply by virtue of the higher beamforming gain. In general, with Massive MIMO, the pairing statistic are certainly of interest, but are clearly not the final figure of merit from a performance perspective. Even if there are few opportunities to use MU-MIMO, the big arrays still have a significant performance advantage over the smaller arrays.

5.2.7 Performance of NR Type II CSI

As has been discussed already, one of the key additions into the NR-MIMO framework has been the Type II CSI, which provides the gNB with a relatively high accuracy and high resolution CSI that can significantly improve MU-MIMO performance. One of the drawbacks with the Type II CSI is the substantial feedback payload required to carry the CSI. In this section, we consider the performance of the Rel-15 Type II CSI and look at the performance versus overhead tradeoff

As discussed in Section 4.3, the Type II CSI has three main options: The first is the number of wideband beams L ($L = 2, 3,$ or 4), the second is whether to use 2-bits or 3-bits in the phase quantization, and the third is whether sub-band amplitude scaling is turned on or off. The result of these options is a total of 12 possible configurations of the Type II CSI depending on how the options are chosen. In contrast, the Type I CSI has two options: $L = 1$ or $L = 4$.

Figure 5.10 shows the relative performance of the different options for the Type I and Type II CSI feedback in New Radio Rel-15. The physical antenna array configuration was an 8 rows by 4 column cross pol array with subarray port aggregation per column for 16 transceiver ports arranged in 2 rows by 4 columns by 2 polarizations. The scenario is the NR Urban Macro channel model at a 4 GHz carrier frequency and a 200 m ISD. The traffic model was FTP Model 1 with a 0.5 MB packet size and an arrival rate of 2 packets per second, which for these configurations resulted in an average resource utilization of 0.22. MU-MIMO transmission is used, and the UEs are configured for a max transmission rank of 2. In this figure, the mean UE Spectral Efficiency and the Edge UE spectral efficiency are plotted as a percent gain over the respective spectral efficiency achieved with the Type I L = 1 CSI. Also plotted in this figure is the average size in bits of the feedback payload on the UL for delivering the CSI to the gNB. Note from this figure that there is mostly a trend of increasing performance with increased overhead. The Type II CSI can provide anywhere between 14% and 30% gain over the NR Type I L = 1 codebook, but the overhead is roughly between 8 and 30 times greater depending on the configuration. This substantial overhead is

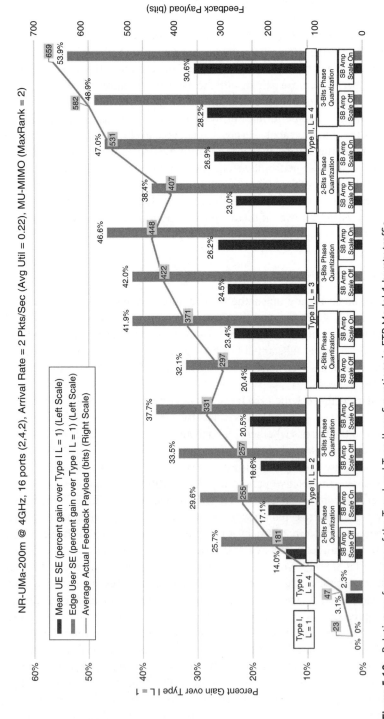

Figure 5.10 Relative performance of the Type I and Type II configurations in FTP Model 1 bursty traffic.

what prompted the development of various improvements expected to be released in Rel-16, where the reduction in the feedback overhead is expected to be on the order of 40% for roughly equivalent performance.

5.3 NR MIMO Performance in mmWave Bands

One of the major advances in the New Radio air interface was the support for operating with carrier frequencies above 6 GHz, specifically in the mmWave bands such as 28 or 39 GHz. These bands are characterized by relatively poor propagation conditions, but large user data rates can be achieved by virtue of the large bandwidths that can be used in those bands. One of the key concerns for operating in the mmWave bands is the coverage reliability since communication links can easily be disrupted by the objects in the environment due to the relative lack of diffraction as a significant propagation mechanism in those bands. One of the deployment strategies envisioned for the mmWave bands is to deploy a mmWave gNB on the existing sites that are already providing services in a sub-6 GHz band. In this strategy, the mmWave gNB becomes an additional carrier that can be used by UEs that have a good propagation link with the mmWave gNB. A UE that is close to the base station can take advantage of the high data rates from the mmWave gNB, whereas other users that do not have a good link with the mmWave gNB can connect to the sub-6 GHz gNB. Nonetheless, whether the mmWave gNB is deployed stand-alone or in conjunction with a sub-6 GHz gNB, a key question is what data rates and cell sizes can be supported in the mmWave bands.

In this section, we will evaluate the uplink (UL) and downlink performance of the new radio air interface in the mmWave bands, specifically the 28 GHz band. We consider a system operating with full buffer traffic in the Urban Macro Scenario with several ISDs (100, 200, 500, and 1000 m). In the simulations, an average of five users were active per cell in the full buffer traffic model. We assume a system bandwidth of 200 MHz and leverage the channel models in [48] and the evaluation methodology in [45]. In this evaluation, all UEs are outdoors in a fixed wireless access configuration.

The antenna array at the gNB is a cross-pol array with 16 rows and 16 columns of cross pol elements for a total of 512-physical element. The spacing between the rows and between the columns was 0.5 wavelengths. We consider three hybrid array configurations for the gNB array and assume the use of a beam refinement procedure that selects the best beam direction to use for communicating to each UE. The first configuration is aimed at supporting single user MIMO on both uplink and downlink, where the data transmission occurs with two beam ports, one formed on one polarization, and the other beam port is formed on the other polarization. The second configuration is also aimed at supporting SU-MIMO on both uplink and downlink, but the data transmission occurs with 8 beam ports. In this second configuration, the array is partitioned into four subpanels where each panel is eight rows by eight columns and two polarizations. Two beam ports are formed with each subpanel, one per polarization, for a total of eight beam ports. The third configuration uses the same subarray partitioning of the second configuration, but with the intent to support an adaptive selection between SU- and MU-MIMO on both uplink and downlink. In the third configuration, the four subpanels can transmit to (or receive from) up to four users, one

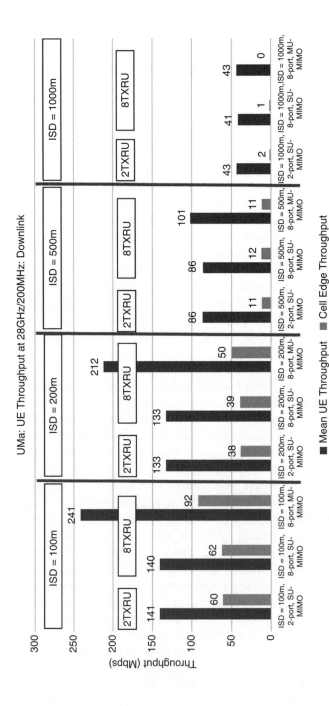

Figure 5.11 Downlink UE throughput at 28 GHz in Urban Macro.

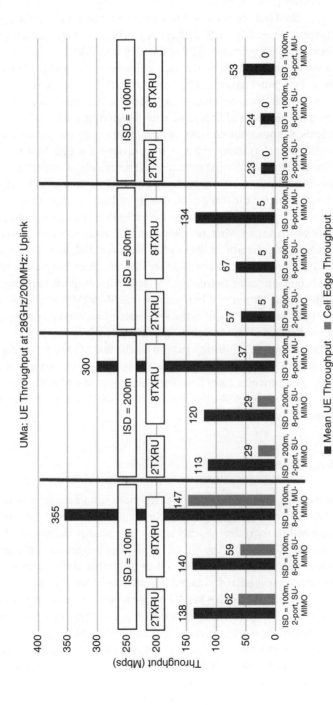

Figure 5.12 Uplink UE throughput at 28 GHz in Urban Macro.

per sub-panel. In all configurations, the downlink transmissions leverage the appropriate precoding matrix indicator (PMI) feedback based on Channel-state information reference signals (CSI-RS) transmitted out the preferred refined beam direction. For the uplink, we assume UE refinement is done at the UE so that the UE transmits and receives with the preferred beam direction. The maximum EIRP at the gNB is 60 dBm.

In this evaluation, we assume the UEs are fixed common phase errors (CPEs) having dual-panel UEs with two panel arrays facing back-to-back and randomly oriented in azimuth. We assume the UE beam refinement procedure selects the best panel and best beam per panel for transmission and reception with its serving gNB. Each UE panel is an array of 32 elements arranged in four rows by four columns by two polarizations. The maximum EIRP at the UE is 40 dBm, which is reasonable for a CPE that is in a fixed wireless access configuration, for example mounted on the outside of a residence.

Figure 5.11 shows the downlink UE throughput in the UMa scenario at 28 GHz with several values of the ISD and for the three array configurations describe above. Similarly, Figure 5.12 shows the uplink UE throughput for the same ISDs and array configurations. Several key points are worth noting: First, reasonable performance is achieved roughly out to a 200 m ISD in the NR UMa scenario in both uplink and downlink for the configurations being evaluated. For ISDs beyond 200 m, coverage performance becomes poor as evidenced by the low cell edge throughput rates. Second, SU-MIMO with two ports performs roughly the same as SU-MIMO with eightports. When restricted to SU-MIMO, the combination of the Radio Frequency (RF) beamforming with the appropriate baseband precoding provides similar performance with these hybrid array configurations. However, the 8-port configuration with MU-MIMO can provide significant gains over the two SU-MIMO configurations due to the co-scheduling of multiple users for the same reasons discussed in the previous section on sub-6 GHz performance. Note how the gains from MU-MIMO decrease with increasing ISD, which is as expected also for reasons discussed in the previous section on sub-6 GHz performance.

5.4 Concluding Remarks

In this chapter, we have provided a view into the performance characteristics of the framework for massive MIMO currently supported in 5G New Radio. We showed the performance of sub-6 GHz massive MIMO for different array sizes and looked at the impact of ISD and CSI feedback scheme. We showed that the NR CSI feedback schemes can provide roughly at 25–30% gain over the feedback schemes in LTE in full buffer traffic with MU-MIMO transmission and advanced regularized zero-forcing precoders. We also showed how the traffic characteristics can impact Massive MIMO performance. We pointed out how with full buffer traffic, there tend to be relatively frequent opportunities to pair multiple users with MU-MIMO transmission. It was observed that the higher-performing array configurations often pair more users with MU-MIMO transmission than the less capable and lower-performing array configurations. With lightly loaded traffic, there can be relatively few opportunities to pair multiple users, but larger array sizes can still provide significant gains by virtue of providing a higher beamforming gain. Finally, we looked at the performance of the NR-MIMO framework in the mmWave bands and showed reasonable coverage performance can be achieved with ISDs on the order of 200 m or so. As a final remark, it is important to note that all results shown in this chapter are for specific implementations of various aspects of the NR system (e.g. the MIMO transmission scheme, the multi-user pairing scheme & scheduler implementation). Different implementations will result in different absolute performance numbers, but in general we would expect to see similar trends and benefits from the technologies evaluated in this chapter.

6

UE Features

Mihai Enescu

Nokia Bell Labs, Finland

The 5G New Radio specification encountered the challenge of providing a unified design across all the carrier frequencies and keeping such design as optimal as possible. However, technical components had to be tailored to specific deployments where needed, like for example the choice of waveform for UL or the utilization of beams in higher carriers. Another dimension making the design more complex was the need to address all the main development cases.

All the UE functionalities used by the 3GPP 5G specifications are defined as UE features and these are being mandatory and optional, or mandatory with optional capability. These UE features are subject to UE capability signaling and are captured in TS 38.306 [49], while features that do not have capability signaling and corresponding capability definition in TS 38.306 are by definition mandatory.

In the following we are trying to give a glimpse of some of the mandatory physical layer parameters. We are keeping the parameter names as defined by the TS 38.306 [49]. While most of the parameters are applicable to both FR1 and FR2, there are situations in which parameters are applicable only to one of the frequency ranges, this is mentioned explicitly.

There is a set of mandatory without capability signaling features [50], which provide basic functionality of the UE.

CP-OFDM waveform for DL and UL, comprising of the following:

- CP-OFDM for DL
- CP-OFDM for UL

DFT-S-OFDM waveform for UL: implies the use of transform precoding for single-layer PUSCH.

DL modulation scheme, comprising of the following:

- QPSK modulation
- 16QAM modulation
- 64QAM modulation
- 264QAM modulation

UL modulation scheme, comprising of the following:

- QPSK modulation
- 16QAM modulation
- 64QAM modulation

Basic initial access channels and procedures, comprising of the following:

- Random Access Channel (RACH) preamble format
- SS block based Radio Resource Management (RRM) measurement
- Broadcast System Information Block (SIB) reception including Remaining Minimum System Information (RMSI)/Other System Information (OSI) and paging

Basic PDSCH reception: comprising of the following:

- Data RE mapping
- Single layer transmission
- Support one TCI state

Basic PUSCH transmission: comprising of the following:

- Data RE mapping
- Single layer (single Tx) transmission
- Single port, single resource Sounding reference signals (SRSs) transmission (SRS set use is configured as for codebook)
- Note: support of SRS set usage configured as for codebook does not imply UE support of codebook based PUSCH MIMO transmission

Basic DL control channel: comprising of the following:

- One configured CORESET(Control Resource Set) per Bandwidth Part (BWP) per cell in addition to CORESET0
  - CORESET resource allocation of 6 RB bit-map and duration of 1–3 OFDM symbols for FR1
  - For type 1 Common search space (CSS) without dedicated Radio Resource Control (RRC) configuration and for type 0, 0A, and 2 CSSs, CORESET resource allocation of 6 RB bit-map and duration 1–3 OFDM symbols for FR2
  - For type 1 CSS with dedicated RRC configuration and for type 3 CSS, UE specific SS, CORESET resource allocation of 6 RB bit-map and duration 1–2 OFDM symbols for FR2
  - Resource Element Group (REG)-bundle sizes of 2/3 RBs or 6 RBs
  - Interleaved and non-interleaved Control Channel Element (CCE)-to-REG mapping
  - Precoder-granularity of REG-bundle size
  - Physical Downlink Control Channel (PDCCH) demodulation reference signal (DMRS) scrambling determination
  - TCI state(s) for a CORESET configuration
- CSS and UE-SS configurations for unicast PDCCH transmission per BWP per cell
  - PDCCH aggregation levels 1, 2, 4, 8, 16
  - UP to 3 search space sets in a slot for a scheduled SCell per BWP
  - This search space limit is before applying all dropping rules.
  - For type 1 CSS with dedicated RRC configuration, type 3 CSS, and UE-SS, the monitoring occasion is within the first three OFDM symbols of a slot
  - For type 1 CSS without dedicated RRC configuration and for type 0, 0A, and 2 CSS, the monitoring occasion can be any OFDM symbol(s) of a slot, with the monitoring occasions for any of Type 1- CSS without dedicated RRC configuration, or Types 0, 0A, or 2 CSS configurations within a single span of three consecutive OFDM symbols within a slot
- Monitoring Downlink Control Information (DCI) formats 0_0, 1_0, 0_1, 1_1
- Number of PDCCH blind decodes per slot with a given sub-carrier spacing (SCS) follows Case 1-1 table
- Processing one unicast DCI scheduling DL and one unicast DCI scheduling UL per slot per scheduled CC for frequency division duplex (FDD)
- Processing one unicast DCI scheduling DL and two unicast DCI scheduling UL per slot per scheduled CC for time division duplex (TDD)

Basic UL control channel: comprising of the following:

- Physical Uplink Control Channel (PUCCH) format 0 over 1 OFDM symbols once per slot
- PUCCH format 0 over 2 OFDM symbols once per slot with frequency hopping as "enabled"
- PUCCH format 1 over 4 – 14 OFDM symbols once per slot with intra-slot frequency hopping as "enabled"
- One SR configuration per PUCCH group
- HARQ-ACK transmission once per slot with its resource/timing determined by using the DCI
- SR/HARQ (Hybrid Automatic Repeat Request) multiplexing once per slot using a PUCCH when SR/HARQ-ACK are supposed to be sent by overlapping PUCCH resources with the same starting symbols in a slot
- HARQ-ACK piggyback on PUSCH with/without aperiodic Channel State Information (CSI) once per slot when the starting OFDM symbol of the PUSCH is the same as the starting OFDM symbols of the PUCCH resource that HARQ-ACK would have been transmitted on
- Semi-static beta-offset configuration for HARQ-ACK
- Single group of overlapping PUCCH/PUCCH and overlapping PUCCH/PUSCH s per slot per PUCCH cell group for control multiplexing

Basic scheduling/HARQ operation: comprising of the following:

- Frequency-domain resource allocation
  - RA Type 0 only and Type 1 only for PDSCH without interleaving
  - RA Type 1 for PUSCH without interleaving
- Time-domain resource allocation
  - 1-14 OFDM symbols for PUSCH once per slot
  - One unicast PDSCH per slot
  - Starting symbol, and duration are determined by using the DCI
  - PDSCH mapping type A with 7-14 OFDM symbols
  - PUSCH mapping type A and type B
  - For type 1 CSS without dedicated RRC configuration and for type 0, 0A, and 2 CSS, PDSCH mapping type A with {4-14} OFDM symbols and type B with {2, 4, 7} OFDM symbols
- Transport block size (TBS) determination
- Nominal UE processing time for N1 and N2 (Capability #1)
- HARQ process operation with configurable number of DL HARQ processes of up to 16
- Cell specific RRC configured UL/DL assignment for TDD
- Dynamic UL/DL determination based on L1 scheduling DCI with/without cell specific RRC configured UL/DL assignment
- Intra-slot frequency-hopping for PUSCH scheduled by Type 1 CSS before RRC connection
- In TDD support at most one switch point per slot for actual DL/UL transmission(s)
- DL scheduling slot offset $K0 = 0$
- DL scheduling slot offset $K0 = 1$ for type 1 CSS without dedicated RRC configuration and for type 0, 0A, and 2 CSS
- UL scheduling slot offset $K2 < =12$

For type 1 CSS without dedicated RRC configuration and for type 0, 0A, and 2 CSS, interleaving for VRB-to-PRB mapping for PDSCH

Basic BWP operation with restriction: comprising of the following:

- 1 UE-specific RRC configured DL BWP per carrier
- 1 UE-specific RRC configured UL BWP per carrier
- RRC reconfiguration of any parameters related to BWP
- BW of a UE-specific RRC configured BWP includes BW of CORESET#0 (if CORESET#0 is present) and SSB for PCell/PSCell (if configured) and BW of the UE-specific RRC configured BWP includes SSB for SCell if there is SSB on SCell

Active BWP switching delay: comprising of the following:

- Support of active BWP switching delay specified in TS38.133, candidate values set: {type1, type2}. It is Mandatory to report type 1 or type 2.

Channel coding: comprising of the following:

- LDPC encoding and associated functions for data on DL and UL
- Polar encoding and associated functions for Physical Broadcast Channel (PBCH), DCI, and uplink control information (UCI)
- Coding for very small blocks

Basic power control operation: comprising of the following:

- Accumulated power control mode for closed loop
- 1 TPC command loop for PUSCH, PUCCH respectively
- One or multiple DL RS configured for pathloss estimation
- One or multiple p0-alpha values configured for open loop PC
- PUSCH power control
- PUCCH power control
- Physical Random Access Channel (PRACH) power control
- SRS power control
- PHR

6.1 Reference Signals

6.1.1 DM-RS

The support of basic DM-RS is mandatory without capability signaling, comprising of the following:

- Basic downlink DMRS for scheduling type A: this is conditioned to whether PDSCH scheduling type A is supported and it contains the following options:
  - Support 1 symbol front-loaded DM-RS without additional symbol(s)
  - Support 1 symbol front-loaded DM-RS and 1 additional DM-RS symbol
  - Support 1 symbol front-loaded DM-RS and 2 additional DM-RS symbols for at least one port.
- Basic downlink DMRS for scheduling type B: this is conditioned to whether PDSCH scheduling type B is supported and it contains the following options:
  - Support 1 symbol front-loaded DM-RS without additional symbol(s)
  - Support 1 symbol front-loaded DM-RS and 1 additional DM-RS symbol
- *Basic uplink DMRS (uplink) for scheduling type A*: this is conditioned to whether PUSCH scheduling type A is supported and it contains the following options:
  - Support 1 symbol front-loaded DM-RS without additional symbol(s)
  - Support 1 symbol front-loaded DM-RS and 1 additional DM-RS symbols
  - Support 1 symbol front-loaded DM-RS and 2 additional DM-RS symbols
- *Basic uplink DMRS for scheduling type B*: this is conditioned to whether PUSCH scheduling type B is supported and it contains the following options:
  - Support 1 symbol front-loaded DM-RS without additional symbol(s)
  - Support 1 symbol front-loaded DM-RS and 1 additional DM-RS symbol

supportedDMRS-TypeDL	Defines supported DM-RS configuration types at the UE for DL reception. Type 1 is mandatory with capability signaling. Type 2 is optional. Each of the two DM-RS types are supporting a single-symbol and a double-symbol configuration. Detailed description of the DM-RS types can be found in Section 3.8.2.1 while the main properties of the DM-RS type 1 and 2 are summarized in Table 3.13.
supportedDMRS-TypeUL	Defines supported DM-RS configuration types at the UE for UL transmission. Support both type 1 and type 2 are mandatory with capability signaling.
oneFL-DMRS-TwoAdditionalDMRS-UL	Defines support of DM-RS pattern for UL transmission with 1 symbol front-loaded (FL) DM-RS with two additional DM-RS symbols and more than 1 antenna ports. The maximum number of antenna ports is 8 for type 1 DM-RS and 12 for type 2 DM-RS.
twoFL-DMRS	Defines whether the UE supports DM-RS pattern for DL reception and/or UL transmission with 2 symbols front-loaded DM-RS without additional DM-RS symbols. The left most in the bitmap corresponds to DL reception and the right most bit in the bitmap corresponds to UL transmission.
twoFL-DMRS-TwoAdditionalDMRS-UL	Defines whether the UE supports DM-RS pattern for UL transmission with 2 symbols front-loaded DM-RS with 1 additional 2 symbols DM-RS.
oneFL-DMRS-TwoAdditionalDMRS-DL	Defines support of DM-RS pattern for DL transmission with 1 symbol front-loaded DM-RS with two additional DM-RS symbols and more than 1 antenna ports.

6.1.2 CSI-RS

maxNumberCSI-RS-SSB-CBD	Defines maximal number of different Channel-state information reference signals (CSI-RSs) [and/or SSBblocks] resources across all component's carriers, and across master cell groups (MCGs) and secondary cell group (SCG) in case of NR-dual connectivity, for new beam identifications. In release 15, the maximum value supported by the UE is up to 128. It is mandatory with capability signaling for FR2 and optional for FR1. The UE is mandated to report at least 32 for FR2. Detailed description of CSI-RS can be found in Section 3.7.
maxNumberCSI-RS-RRM-RS-SINR	Defines the maximum number of CSI-RS resources for RRM and RS-SINR measurement across all measurement frequencies per slot. If the UE supports any of csi-RSRP-AndRSRQ-MeasWithSSB, csi-RSRP-AndRSRQ-MeasWithoutSSB, and csi-SINR-Meas, the UE shall report this capability.
sp-CSI-RS	Indicates whether the UE supports semi-persistent CSI-RS.

6.1.3 PT-RS

onePortsPTRS	Defines whether the UE supports PT-RS with 1 antenna port in DL reception and/or UL transmission. It is mandatory with UE capability signaling for FR2 and optional for FR1. The left most in the bitmap corresponds to DL reception and the right most bit in the bitmap corresponds to UL transmission. Detailed description of CSI-RS can be found in Section 3.9.
ptrs-DensityRecommendationSetDL	Indicates, for each supported SCS (15, 30, 60, 120 kHz), the preferred threshold sets for determining DL PT-RS density. The reporting is done per band and it is mandatory for FR2. For each supported SCS, this field comprises: • two values of *frequencyDensity*; • three values of *timeDensity*. For each threshold set, it composes of two values each selected from {1...276} for frequency density, and three values each selected from {0...29} for time density.

6.1.4 SRS

The support of basic SRS is mandatory without capability signaling, comprising of the following: (i) Support 1 port SRS transmission, (ii) Support periodic/aperiodic SRS transmission, (iii) Support SRS Frequency intra/inter-slot hopping within BWP, (iv) At least one SRS resource per CC for aperiodic and periodic separately. Detailed description of SRS can be found in Section 3.10.

supportedSRS-Resources	Defines support of SRS resources. The capability signaling comprising indication of: • maxNumberAperiodicSRS-PerBWP indicates supported maximum number of aperiodic SRS resources that can be configured for the UE per each BWP. Candidate values from: {1, 2, 4, 8, 16}. • maxNumberAperiodicSRS-PerBWP-PerSlot indicates supported maximum number of aperiodic SRS resources per slot in the BWP. Candidate values from: {1, 2, 3, 4, 5, 6}. • maxNumberPeriodicSRS-PerBWP indicates supported maximum number of periodic SRS resources per BWP. Candidate values from: {1, 2, 4, 8, 16}. • maxNumberPeriodicSRS-PerBWP-PerSlot indicates supported maximum number of periodic SRS resources per slot in the BWP. Candidate values from: {1, 2, 3, 4, 5, 6}. • maxNumberSemiPersitentSRS-PerBWP indicate supported maximum number of semi-persistent SRS resources that can be configured for the UE per each BWP. Candidate values from: {1, 2, 4, 8, 16}. • maxNumberSP-SRS-PerBWP-PerSlot indicates supported maximum number of semi-persistent SRS resources per slot in the BWP. Candidate values from: {1, 2, 3, 4, 5, 6}. • maxNumberSRS-Ports-PerResource indicates supported maximum number of SRS antenna port per each SRS resource. Candidate values from: {1, 2, 4}.

SRS-TxSwitch

Defines whether the UE supports SRS for DL CSI acquisition as defined in clause 6.2.1.2 of TS 38.214 [22]. The capability signaling comprises of the following parameters:
- supportedSRS-TxPortSwitch indicates SRS Tx port switching pattern supported by the UE. The indicated UE antenna switching capability of "xTyR" corresponds to a UE, capable of SRS transmission on "x" antenna ports over total of "y" antennas, where "y" corresponds to all or subset of UE receive antennas, where 2T4R is two pairs of antennas; candidates are {"Not supported," "1T2R," "1T4R," "2T4R," "1T4R/2T4R," "1 T = 1R," "2 T = 2R," "4 T = 4R"}. Note: 2T4R is 2 pairs of antennas.
- txSwitchImpactToRx indicates the entry number of the first-listed band with UL in the band combination that affects this DL; Candidate value set:, {yes, no}, Note: "R" refers to a subset/set of receive antennas for PDSCH; "T" refers to the SRS antennas used for DL CSI acquisition
- txSwitchWithAnotherBand indicates the entry number of the first-listed band with UL in the band combination that switches together with this UL. Candidate value set: {yes, no}.

The UE is restricted not to include fallback band combinations for the purpose of indicating different SRS antenna switching capabilities.

6.1.5   TRS

The support of basic TRS is mandatory without capability signaling, comprising of the following: (i) Support of TRS (mandatory), (ii) All the TRS periodicities are supported. (iii) Support TRS bandwidth configuration as both "BWP" and "min (52, BWP)." Detailed description of SRS can be found in Section 3.7.4.1

csi-RS-ForTracking

Indicates support of CSI-RS for tracking (i.e. TRS). This capability signaling comprises the following parameters:
- *maxBurstLength* indicates the TRS burst length. Value 1 indicates 1 slot and value 2 indicates both of 1 slot and 2 slots. In Release 15 the UE is mandated to report value 2;
- *maxSimultaneousResourceSetsPerCC* indicates the maximum number of TRS resource sets per component carrier which the UE can track simultaneously; Candidate value set: {1 to 8}.
- *maxConfiguredResourceSetsPerCC* indicates the maximum number of TRS resource sets configured to UE per CC. It is mandated to report at least 8 for FR1 and 16 for FR2; Candidate value set: {1 to 64}.
- *maxConfiguredResourceSetsAllCC* indicates the maximum number of TRS resource sets configured to UE across CCs. It is mandated to report at least 16 for FR1 and 32 for FR2. Candidate value set: {1 to 256}.

The reporting of this UE capability is done per band.

6.1.6 Beam Management

beamCorrespondenceWithoutUL-BeamSweeping

Indicates how UE supports FR2 beam correspondence as specified in TS 38.101-2 [31], clause 6.6. The UE that fulfills the beam correspondence requirement without the uplink beam sweeping (as specified inTS 38.101-2 [31], clause 6.6) shall set the bit to 1. The UE that fulfills the beam correspondence requirement with the uplink beam sweeping (as specified inTS 38.101-2 [31], clause 6.6) shall set the bit to 0. Applicable in FR2 only

beamManagementSSB-CSI-RS

Defines support of SS/PBCH and CSI-RS based reference signal received power (RSRP) measurements. The capability comprises signaling of
- maxNumberSSB-CSI-RS-ResourceOneTx indicates maximum total number of configured one port non-zero-power (NZP) CSI-RS resources and SS/PBCH blocks that are supported by the UE for "CRI/RSRP" and "SSBRI/RSRP" reporting within a slot and across all serving cells. On FR2, it is mandatory to report $> = 8$; On FR1, it is mandatory with capability signaling to report $> = 8$.
- maxNumberCSI-RS-Resource indicates maximum total number of configured NZP-CSI-RS resources that are supported by the UE for "CRI/RSRP" reporting across all serving cells. It is mandated to report at least n8 for FR1.
- maxNumberCSI-RS-ResourceTwoTx indicates maximum total number of two ports NZP CSI-RS resources that are supported by the UE for "CRI/RSRP" reporting within a slot and across all serving cells. Candidate value set for is {0, 4, 8, 16, 32, 64}.
- supportedCSI-RS-Density indicates density of one RE per PRB for one port NZP CSI-RS resource for RSRP reporting, if supported. On FR2, it is mandatory to report either "three" or "oneAndThree"; On FR1, it is mandatory with capability signaling to report either "three" or "oneAndThree." Candidate value set: {"not supported," "one", "three," "oneAndThree"}
- maxNumberAperiodicCSI-RS-Resource indicates maximum number of configured aperiodic CSI-RS resources across all CCs. For FR1 and FR2, the UE is mandated to report at least n4. Candidate value set for is {0, 1, 4, 8, 16, 32, 64}.

For both FR1 and FR2, UE is mandated to report at least 4

periodicBeamReport

Indicates whether UE supports periodic "CRI/RSRP" or "SSBRI/RSRP" reporting using PUCCH formats 2, 3, and 4 in one slot.

aperiodicBeamReport

Indicates whether the UE supports aperiodic "CRI/RSRP" or "SSBRI/RSRP" reporting on PUSCH. The UE provides the capability for the band number for which the report is provided (where the measurement is performed).

beamReportTiming	Indicates the number of OFDM symbols X_i between the last symbol of SSB/CSI-RS and the first symbol of the transmission channel containing beam report where
	i is the index of SCS, $i = 1, 2, 3, 4$ corresponding to 15, 30, 60, 120 kHz SCS. The UE provides the capability for the band number for which the report is provided (where the measurement is performed). The UE includes this field for each supported SCS. Candidate value sets: X_1 is {2, 4, 8}, X_2 is {4, 8, 14, 28}, X_3 is {8, 14, 28}, X_4 is {14, 28, 56}.
maxNumberRxBeam	Defines whether UE supports receive beamforming switching using NZP CSI-RS resource. UE shall indicate a single value for the preferred number of NZP CSI-RS resource repetitions per CSI-RS resource set where the candidate value set is {2, 3, 4, 5, 6, 7, 8}. The support of Rx beam switching is mandatory for FR2 and the reporting is done per band.
pucch-SpatialRelInfoMAC-CE	Indicates whether the UE supports indication of PUCCH-spatialrelationinfo by a Medium Access Control (MAC) Control Element (CE) per PUCCH resource. It is mandatory for FR2 and optional for FR1.
spatialRelations	Indicates whether the UE supports spatial relations. The capability signaling comprises the following parameters. • maxNumberConfiguredSpatialRelations indicates the maximum number of configured spatial relations per CC for PUCCH and SRS. It is not applicable to FR1 and applicable to FR2 only. The UE is mandated to report 16 or higher values; candidate value set: {4, 8, 16, 32, 64, 96}. • maxNumberActiveSpatialRelations indicates the maximum number of {unique DL RS (except for aperiodic NZP CSI-RS) and SRS without spatial relation configured, and, TCI states available for DCI triggering of aperiodic NZP CSI-RS}, for indicating spatial domain transmit filter for PUCCH and SRS for PUSCH, per BWP per CC. "Unique" means RS identity. An SSB and a CSI-RS are always counted as different. Two CSI-RSs are different if they have different CSI-RS resource IDs. It is not applicable to FR1 and applicable and mandatory to report for FR2 only; • additionalActiveSpatialRelationPUCCH indicates support of one additional active spatial relation for PUCCH. It is mandatory with capability signaling if maxNumberActiveSpatialRelations is set to 1; • maxNumberDL-RS-QCL-TypeD indicates the maximum number of downlink RS resources used for Quasi co-location (QCL) type D in the active TCI states and active spatial relation information, which is optional. Candidate value set: {1, 2, 4, 8, 14}.

6.1.7 TCI and QCL

tci-StatePDSCH

Defines support of TCI-States for PDSCH. The capability signaling comprises the following parameters:
- maxNumberConfiguredTCIstatesPerCC indicates the maximum number of configured TCI-states per CC for PDSCH. For FR2, the UE is mandated to set the value to 64. For FR1, the UE is mandated to set these values to the maximum number of allowed SSBs in the supported band; candidate value set: {4, 8, 16, 32, 64, 128}.
- maxNumberActiveTCI-PerBWP indicates the maximum number of activated TCI-states per BWP per CC, including control and data. If a UE reports X active TCI state(s), it is not expected that more than X active QCL type D assumption(s) for any PDSCH and any CORESETs for a given BWP of a serving cell become active for the UE. The UE shall include this field. Candidate value set: {1, 2, 4, 8}.

Note the UE is required to track only the active TCI states.

additionalActiveTCI-StatePDCCH

Indicates whether the UE supports one additional active TCI-State for control in addition to the supported number of active TCI-States for PDSCH. The UE can include this field only if maxNumberActiveTCI-PerBWP is included in tci-StatePDSCH. Otherwise, the UE does not include this field.

multipleTCI

Indicates whether UE supports more than one TCI state configurations per CORESET. UE is only required to track one active TCI state per CORESET. UE is required to support minimum between 64 and number of configured TCI states indicated by tci-StatePDSCH. This field shall be set to 1.

*reporting is done per band

timeDurationForQCL

Defines minimum number of OFDM symbols required by the UE to perform PDCCH reception and applying spatial QCL information received in DCI for PDSCH processing as described in TS 38.214 [1] clause 5.1.5. UE shall indicate one value of the minimum number of OFDM symbols per each subcarrier spacing of 60 and 120 kHz.
- Applicable for FR2 only

6.1.8 Beam Failure Detection

maxNumberCSI-RS-BFD

Indicates maximal number of CSI-RS resources across all CCs, and across MCG and SCG in case of NR- dual connectivity, for UE to monitor PDCCH quality. In Release 15, the maximum value supported by the UE is upto 16. It is mandatory with capability signaling for FR2 and optional for FR1.

maxNumberSSB-BFD

Defines maximal number of different SSBs across all CCs, and across MCG and SCG in case of NR-dual connectivity, for UE to monitor PDCCH quality. In Release 15, the maximum value supported by the UE is up to 16. It is mandatory with capability signaling for FR2 and optional for FR1.

6.1.9 RLM

ssb-RLM
Indicates whether the UE can perform radio link monitoring (RLM) procedure based on measurement of SS/PBCH block as specified in TS 38.213 [34] and TS 38.133 [40]. This field shall be set to 1.

csi-RS-RLM
Indicates whether the UE can perform RLM procedure based on measurement of CSI-RS as specified in TS 38.213 [34] and TS 38.133 [40]. This parameter needs FR1 and FR2 differentiation. If the UE supports this feature, the UE needs to report maxNumberResource-CSI-RS-RLM.

maxNumberResource-CSI-RS-RLM
Defines the maximum number of CSI-RS resources within a slot per spCell for CSI-RS based RLM. If UE supports any of csi-RS-RLM and ssb-AndCSI-RS-RLM, UE shall report this capability.

6.1.10 CSI Framework

Basic CSI feedback is mandatory without capability signaling, comprising of the following:

- Type I single panel codebook based Precoder Matrix Indicator (PMI)
- 2Tx codebook for FR1 and FR2
- 4Tx codebook for FR1
- 8Tx codebook for FR1 when configured as wideband CSI report
- periodic-CSI on PUCCH formats over 1–2 OFDM symbols once per slot (or piggybacked on a PUSCH)
- periodic-CSI report on PUCCH formats over 4–14 OFDM symbols once per slot (or piggybacked on a PUSCH)
- aperiodic-CSI on PUSCH (at least Z value $> = 14$ symbols,)

csi-RS-IM-ReceptionForFeedback
Indicates support of CSI-RS and CSI-IM reception for CSI feedback. This capability signaling comprises the following parameters:
- maxConfigNumberNZP-CSI-RS-PerCC indicates the maximum number of configured NZP-CSI-RS resources per CC; candidate values are from {1 to 32}.
- maxConfigNumberPortsAcrossNZP-CSI-RS-PerCC indicates the maximum number of ports across all configured NZP-CSI-RS resources per CC; candidate values are from {2, 4, 8, 12, 16, 24, 32, 40, 48,..., 256}
- maxConfigNumberCSI-IM-PerCC indicates the maximum number of configured CSI-IM resources per CC; candidate values are from: {1, 2, 4, 8, 16, 32}
- maxNumberSimultaneousNZP-CSI-RS-PerCC indicates the maximum number of simultaneous CSI-RS-resources per CC; candidate values are from {5, 6, 7, 8, 9, 10, 12, 14, 16, ..., 62, 64} (includes all even numbers between 16 and 64).
- totalNumberPortsSimultaneousNZP-CSI-RS-PerCC indicates the total number of CSI-RS ports in simultaneous CSI-RS resources per CC. candidate values {8, 16, 24, ... 128}.

*reporting is done per band

csi-RS-IM-ReceptionForFeedback	For a band combination comprised of FR1 and FR2 bands, this parameter, if present, limits the corresponding parameter in MIMO-ParametersPerBand.
csi-ReportFramework	Indicates whether the UE supports CSI report framework. This capability signaling comprises the following parameters:

- maxNumberPeriodicCSI-PerBWP- ForCSI-Report indicates the maximum number of periodic CSI report setting per BWP for CSI report; candidate values: {1, 2, 3, 4}.
- maxNumberPeriodicCSI-PerBWP-ForBeamReport indicates the maximum number of periodic CSI report setting per BWP for beam report; candidate values: {1, 2, 3, 4}.
- maxNumberAperiodicCSI-PerBWP-ForCSI-Report indicates the maximum number of aperiodic CSI report setting per BWP for CSI report; candidate values {1, 2, 3, 4}.
- maxNumberAperiodicCSI-PerBWP-ForBeamReport indicates the maximum number of aperiodic CSI report setting per BWP for beam report; Component-2a candidate values {1, 2, 3, 4}.
- maxNumberAperiodicCSI-triggeringStatePerCC indicates the maximum number of aperiodic CSI triggering states in CSI-AperiodicTriggerStateList per CC; candidate values {3, 7, 15, 31, 63, 128}.
- maxNumberSemiPersistentCSI-PerBWP-ForCSI-Report indicates the maximum number of semi-persistent CSI report setting per BWP for CSI report; candidate values: {0, 1, 2, 3, 4}.
- maxNumberSemiPersistentCSI-PerBWP-ForBeamReport indicates the maximum number of semi-persistent CSI report setting per BWP for beam report; candidate values: {0, 1, 2, 3, 4}.
- simultaneousCSI-ReportsPerCC indicates the number of CSI report(s) for which the UE can measure and process reference signals simultaneously in a CC of the band for which this capability is provided. The CSI report comprises periodic, semi-persistent and aperiodic CSI and any latency classes and codebook types. The CSI report in simultaneousCSI-ReportsPerCC includes the beam report and CSI report. Candidate values: {from 1 to 8}.

* Reporting is done per band

csi-ReportFramework	For a band combination comprised of FR1 and FR2 bands, this parameter, if present, limits the corresponding parameter in MIMO-ParametersPerBand. For single CC standalone NR, it is mandatory with capability signaling to support at least 4 MIMO layers in the bands where 4Rx is specified as mandatory for the given UE and at least 2 MIMO layers in FR2.

Some relaxations to this requirement may be applicable in the future (including in Release-15).

Mandatory in all cases means mandatory with capability signaling.

It is not expected that there is a signaling change (i.e. signaling remains to be defined as {1, 2, 4, 8} in every band and every band combination, including FR1 and FR2 in all cases).

csi-RS-IM-ReceptionForFeedback-PerBandComb	Indicates support of CSI-RS and CSI-IM reception for CSI feedback. This capability signaling comprises the following parameters: • maxNumberSimultaneousNZP-CSI-RS-ActBWP-AllCC indicates the maximum number of simultaneous CSI-RS resources in active BWPs across all CCs, and across MCG and SCG in case of NR-DC. This parameter limits the total number of NZP-CSI-RS resources that the NW may configure across all CCs, and across MCG and SCG in case of NR-DC (irrespective of the associated codebook type). The network applies this limit in addition to the limits signaled in MIMO-ParametersPerBand- > maxNumberSimultaneousNZP-CSI-RS-PerCC and in Phy-ParametersFRX-Diff- > maxNumberSimultaneousNZP-CSI-RS-PerCC; • totalNumberPortsSimultaneousNZP-CSI-RS-ActBWP-AllCC indicates the total number of CSI-RS ports in simultaneous CSI-RS resources in active BWPs across all CCs, and across MCG and SCG in case of NR-DC. This parameter limits the total number of ports that the NW may configure across all NZP-CSI-RS resources across all CCs, and across MCG and SCG in case of NR-DC (irrespective of the associated codebook type). The network applies this limit in addition to the limits signaled in MIMO-ParametersPerBand- > totalNumberPorts-SimultaneousNZP-CSI-RS-PerCC and in Phy-ParametersFRX-Diff- > totalNumberPortsSimultaneous-NZP-CSI-RS-PerCC.
simultaneousCSI-ReportsAllCC	Indicates whether the UE supports CSI report framework and the number of CSI report(s) which the UE can simultaneously process across all CCs, and across MCG and SCG in case of NR-DC. The CSI report comprises periodic, semi-persistent and aperiodic CSI and any latency classes and codebook types. The CSI report in simultaneousCSI-ReportsAllCC includes the beam report and CSI report. This parameter may further limit simultaneousCSI-ReportsPerCC in MIMO-ParametersPerBand and Phy-ParametersFRX-Diff for each band in a given band combination.
maxNumberMIMO-LayersPDSCH	Defines the maximum number of spatial multiplexing layer(s) supported by the UE for DL reception. For single CC standalone NR, it is mandatory with capability signaling to support at least four MIMO layers in the bands where 4Rx is specified as mandatory for the given UE and at least two MIMO layers in FR2.
maxLayersMIMO-Indication	Indicates whether the UE supports the network configuration of maxMIMO-Layers as specified in TS 38.331 [22].

References

1 3rd Generation Partnership Project (3GPP) RP-160671, "Study on New Radio Access Technology", NTT DoCoMo.

2 3rd Generation Partnership Project (3GPP) TR 38.913, "Study on Scenarios and Requirements for Next Generation Access Technologies", 3GPP.

3 3rd Generation Partnership Project (3GPP) R1-162152, "OFDM based flexible waveform for 5G" Huawei, HiSilicon.

4 Recommendation ITU-R M.2083 - Framework and overall objectives of the future development of IMT for 2020 and beyond (September 2015).

5 Report ITU-R M.2410 "Minimum requirements related to technical performance for IMT-2020 radio interface(s)" (November 2017).

6 Report ITU-R M.2412 "Guidelines for evaluation of radio interface technologies for IMT-2020" (November 2017).

7 RP-152257 "New Study Item Proposal: Study on Scenarios and Requirements for Next Generation Access Technologies".

8 RP-160671 "New SID Proposal: Study on New Radio Access Technology".

9 3rd Generation Partnership Project (3GPP) 3GPP TS 38.213: "NR; Physical layer procedures for control".

10 3GPP TS 38.300: "NR; Overall description; Stage-2".

11 3GPP TS 38.801; "Study on new radio access technology: Radio access architecture and interfaces".

12 TS 38.401, NG-RAN; Architecture description.

13 ITU norm X.200.

14 TS 38.501, "Non-Access-Stratum (NAS) protocol for 5G System (5GS); Stage 3".

15 TS 38.321, "NR; Medium Access Control (MAC) protocol specification".

16 TS 38.322, "NR; Radio Link Control (RLC) protocol specification".

17 TS 38.323, "NR; Packet Data Convergence Protocol (PDCP) specification".

18 3rd Generation Partnership Project (3GPP) 3GPP TS 38.331: "NR; Radio Resource Control (RRC); Protocol specification".

19 ITU-T Recommendation X.680.

20 ITU-T Recommendation X.681.

21 ITU-T Recommendation X.691.

22 3rd Generation Partnership Project (3GPP) 3GPP TS 38.214: "NR; Physical layer procedures for data".

5G New Radio: A Beam-based Air Interface, First Edition. Edited by Mihai Enescu.
© 2020 John Wiley & Sons Ltd. Published 2020 by John Wiley & Sons Ltd.

23 3GPP R1-162199, "Waveform Candidates," Qualcomm Inc.

24 "Performance of FBMC Multiple Access for Relaxed Synchronization Cellular Networks", J.-B.

25 R1-162890, "5G Waveforms for the Multi-Service Air Interface below 6 GHz," Nokia, Alcatel-Lucent Shanghai Bell.

26 R4-1609225, "NR DL in-band EVM and emission requirements at BS Tx" Nokia, Alcatel-Lucent Shanghai Bell.

27 R4-168776, "On agnostic Tx and Rx unit design for NR", Nokia, Alcatel-Lucent Shanghai Bell.

28 3rd Generation Partnership Project (3GPP) 3GPP TS 38.101-1: "NR; User Equipment (UE) radio transmission and reception; Part 1: Range 1 Standalone".

29 3GPP TS38.104 "Base Station (BS) radio transmission and reception".

30 G. Berardinelli, F. M. L. Tavares, T. B. Sorensen, P. Mogensen, and K. Pajukoski, "Zero-tail DFT-spread-OFDM signals," IEEE GLOBECOM BWA'2103, pp. 1–6.

31 3rd Generation Partnership Project (3GPP) 3GPP TS 38.101-2: "NR; User Equipment (UE) radio transmission and reception; Part 2: Range 2 Standalone".

32 RP-181435: "New SID: Study on NR beyond 52.6 GHz".

33 3GPP TR38.807, "Study on requirements for NR beyond 52.6 GHz.".

34 3rd Generation Partnership Project (3GPP) 3GPP TS 38.211: "NR; Physical channels and modulation".

35 3rd Generation Partnership Project (3GPP) 3GPP TS 38.212: "NR; Multiplexing and channel coding".

36 Petrovic, D., Rave, W., and Fettweis, G. (2007). Effects of phase noise on OFDM systems with and without PLL: characterization and compensation. *IEEE Transactions on Communications* 55 (8).

37 3GPP TR38.803 "Study on new radio access technology: Radio Frequency (RF) and co-existence aspects".

38 R1-162885, "On the phase noise model for 5G New Radio evaluation", Nokia, Alcatel-Lucent Shanghai Bell, April. 2016.

39 Banerjee, D. (2006). *PLL Performance, Simulation and Design Handbook*, 4e. Dog Ear Publishing, LLC.

40 3rd Generation Partnership Project (3GPP) 3GPP TS 38.133: "NR; Requirements for support of radio resource management".

41 R1-1708927, "Discussion on PT-RS design for CP-OFDM", Nokia, Alcatel-Lucent Shanghai Bell, May, 2017.

42 V. Syrjälä, T. Levanen, M. Valkama, E. Lähetkangas, "Efficient time-domain phase noise mitigation in cm-wave wireless communications", IEEE Global conference on signal and information processing (GLOBALSIP 2016), Dec 2016.

43 3rd Generation Partnership Project (3GPP)3GPP TS 38.321: "NR; Medium Access Control (MAC) protocol specification".

44 3rd Generation Partnership Project (3GPP) 3GPP TS 38.215: "NR; Physical layer measurements".

45 3GPP TR 38.802 v14.0.0 Study on new radio access technology physical layer aspects.

46 F. W. Vook, A. Ghosh, E. Diarte, et al., 5G New Radio: Overview and Performance, 2018 52nd Asilomar Conference on Signals, Systems, and Computers.

47 F. W. Vook, W.J. Hillery, E. Visotsky, et al., System level performance characteristics of sub-6GHz massive MIMO deployments with the 3GPP New Radio, 2018 IEEE 88th Vehicular Technology Conference (VTC-Fall).

48 3GPP TR 38.901 vs14.3.0 Study on channel model for frequencies from 0.5 to 100GHz.

49 3rd Generation Partnership Project (3GPP) 3GPP TS 38.306: "NR; User Equipment (UE) radio access capabilities".

50 R1-1907862, "RAN1 NR UE features".

51 "51 38.889 Study on NR-based access to unlicensed spectrum".

Index

a

Above 52 GHz   23
Access and Mobility Management Function (AMF)   26–36, 81–82
Access Stratum (AS)   365, 388
Acknowledged mode (AM)   45, 50–51, 357, 364
ACLR. *See* Adjacent channel leakage ratio (ACLR)
ADCs. *See* Analog-to-digital converters (ADCs)
Adjacent channel leakage ratio (ACLR)   12
AF. *See* Application Function (AF)
Aggregation level   420
Always-on signals   11, 265, 374
AMF. *See* Access and Mobility Management Function (AMF)
AM. *See* Acknowledged mode (AM)
Analog beamforming   287–288
Analog front-end   210
Analog-to-digital converters (ADCs)   104
Antenna
  array   17, 105–107, 115, 317, 359, 397–400
  array architectures   108
  panels   110
  ports   221–240, 296, 307, 323–325, 345, 403
Aperiodic
  CSI-RS/IM reporting   334
  CSI-RS transmission   169–173, 300, 311, 322, 328, 426
  SRS   232–235, 244, 305–306, 424
Application Function (AF)   27
Architecture, options   18, 36–39

Area traffic capacity   4–5
ARQ. *See* Automatic repeat request (ARQ)
AS. *See* Access stratum (AS)
AUSF. *See* Authentication Server Function (AUSF)
Authentication Server Function (AUSF)   26–30
Automatic repeat request (ARQ)   45–46, 52

b

Backwards compatibility   91
Bandwidth adaptation   256–257
Bandwidth parts (BWPs)   279
Base clock   15
BCCH. *See* Broadcast Control Channel (BCCH)
BCH. *See* Broadcast Channel (BCH)
Beam
  based design   18
  determination   289, 291
  measurement   289
  reporting   289
  sweeping   18
Beam correspondence   289–290, 294
  in initial access   274
  in RACH procedure   276
Beam failure detection and recovery   312–317
  completion of BFR procedure   316
  configuration of BFD reference signals   313
  declaration   314
  detection   313

5G New Radio: A Beam-based Air Interface, First Edition. Edited by Mihai Enescu.
© 2020 John Wiley & Sons Ltd. Published 2020 by John Wiley & Sons Ltd.

Beam failure detection and recovery (*contd.*)
  instance indication   313
  new candidate beam selection   314
Beam failure/recovery RACH configuration
    280
Beam failure recovery request and response
    315
  contention based random access BFR   316
  contention free random access BFR   315
Beam Indication Framework for DL quasi
    co-location and TCI states   296–302
Beam Indication Framework for UL
    Transmission   303
Beam management during initial access   274
Beam management procedures   289
Beam sweeping   292–294
  for SS-block transmission   129
Blanked resource   11
Blocking   291
Broadcast Channel (BCH)   128
Broadcast Control Channel (BCCH)   54–55,
    91
BSD. *See* Bucket size duration (BSD)
Bucket size duration (BSD)   54
Buffer status reports   62
BWP. *See* Bandwidth parts (BWP)

C
Candidate beams   173
  selection   314–315
Carrier aggregation (CA)   71, 83, 121, 257,
    318, 364, 370
Carrier indicator   120, 253
CA. *See* Carrier aggregation (CA)
CCCH. *See* Common control channel (CCCH)
CCEs. *See* Control channel elements (CCEs)
CDM. *See* Code-division multiplexing (CDM)
Cell   70
  group   70, 265, 357, 421
  reselection   73–74, 388–389
Cell Radio-Network Temporary Identifier
    (C-RNTI)   258
Cell search   261–265
  details of PSS, SSS, PBCH   261–265
Cell-specific reference signals (CRS)   160

Cellular systems   11
Channel bandwidth   104
Channel coding   422
Channel quality indicator (CQI)   307, 325
Channel sounding   159, 228, 244
  downlink   159
Channel-state information (CSI)   22, 44, 45,
    201, 280, 298, 340, 397, 421
Channel-state-information for interference
    measurements (CSI-IM)   170, 320,
    334, 429
  resource sets   320
Channel-state-information reference signals
    (CSI-RS)   159
  antenna port mapping   167
  bandwidth   167
  CDM groups   162
  configurations   164
  interference measurement   170
  key differences with, LTE   161
  L1-RSRP measurement   173
  mapping to physical resources   162
  multiplexing with other signals   169
  reduced density   167
  resource sets   171
  sequence generation and mapping
    167
  time domain behaviour   168
  TRS   171
  use cases   159
  zero-power   170
CLI. *See* Cross link interference
Closed-loop power control   283
CMOS. *See* Complementary
    metal-oxide-semiconductor (CMOS)
CN. *See* Core Network (CN)
Codebook-based precoding   344
  single-layer uplink codebooks for case of
    four antenna ports   229
Code-division multiplexing (CDM)   161
  frequency-domain   161
  time/frequency-domain   161
"Comb" structure   230
Common control channel (CCCH)   53, 55,
    267

Common resource blocks (CRBs) 118, 194, 264

Complementary metal-oxide-semiconductor (CMOS) 211–212

Complementary SUL carrier 266

Component carriers 89

CoMP. *See* Coordinated multipoint (CoMP)

Configured grant type 54

Confined within configurable time/ frequency/spatial resources, signals and channels 11

Connected-state 46, 68, 70, 79, 80, 82, 85, 86, 270, 274, 356, 363, 370, 371, 376

Connection density 5

Connection management 32, 70

Contention
  access 12
  free access 12

Contention-free random access 265, 370, 387
  connection set up 32, 81, 86, 364

Control channel elements (CCEs) 251

Control channels 45, 68, 69, 257, 260, 312, 386, 404

Control resource sets (CORESETs) 73, 128, 251, 267, 359, 360, 428

Control signaling 18, 39, 47, 69

Control functions 44

Coordinated multipoint (CoMP) 39, 318

Core Network (CN) 25, 70, 82, 393

CORESETs. *See* Control resource sets (CORESETs)

CQI. *See* Channel quality indicator (CQI)

CRBs. *See* Common resource blocks (CRBs)

CRC. *See* Cyclic redundancy check (CRC)

C-RNTI. *See* Cell Radio-Network Temporary Identifier (C-RNTI)

Cross-carrier scheduling 120

Cross link interference (CLI) 22

CRS (Common Reference Signals) 1, 2, 6, 11, 160, 176, 186, 187, 189, 192, 209, 296, 374

CRS. *See* Cell-specific reference signals (CRS)

CSI-IM. *See* Channel state information for interference measurements (CSI-IM)

CSI-ReportConfig 318

CSI-RS 95, 271, 416, 423
  antenna port mapping 167
  differences from LTE 161
  L1-RSRP measurement 173
  mapping to physical resource elements 162
  multiplexing with other signals 169
  NZP 160, 307
  overview 159
  physical layer design 162
  resource sets 171
  sequence generation and mapping 167
  time domain behaviour 168
  TRS 171
  use cases 159
  zero power 170

ZP 160CSI. *See* Channel state information (CSI)

Cubic metric 242

Cyclic redundancy check (CRC) 15, 45, 193, 203, 250, 256, 258, 259, 267, 270, 271, 336

Cyclic shift 127, 142–146, 150, 237, 239–241

d

DACs. *See* Digital-to-analog converters (DACs)

Data
  allocation 195
  indicator 57, 120, 252
  radio bearers 28, 40, 46, 366
  transmission 249, 254

DCCH. *See* Dedicated control channel (DCCH)

DCI. *See* Downlink control information (DCI)

Dedicated control channel (DCCH) 53

Dedicated Traffic Channel (DTCH) 53, 55

Demodulation reference signals (DM-RSs) 176
  for PBCH 262

Deployment scenarios 14

DFT. *See* Discrete Fourier transform (DFT)

Digital-to-analog converters (DACs) 104

Discontinuous reception (DRX)   2, 60, 67, 70, 81, 89, 136, 313
  functionality   60–62
Discrete Fourier transform (DFT)   96, 98, 102–103, 197–200, 225, 348–355
DM-RSs. *See* Demodulation reference signals (DMRSs)
Downlink control information (DCI)   64–66, 120
  formats 0_0 and 0_1   120, 251
  formats 1_0 and 1_1   120, 251
  formats 2_0 and 2_1   2_2. 2_3, 251
Downlink multiantenna precoding   344
  type I codebook   349
  type II codebook   352
DR. *See* Dynamic range (DR)
DRX. *See* Discontinuous reception (DRX)
DTCH. *See* Dedicated Traffic Channel (DTCH)
Dual connectivity   19, 71
  multi-RAT dual connectivity   19, 93, 265
Dynamic range (DR)   104
Dynamic scheduling   13

e
eMBB. *See* Enhanced mobile broadband (eMBB)
Emission   307
  in-band emission   102, 260
  out-of-band emission   104
  requirements   12
  spectral mask emission (SEM)   104
  unwanted emission   102
Enhanced mobile broadband (eMBB)   3–4
  targets   3–4
EPC. *See* Evolved Packet Core (EPC)
Error vector magnitude (EVM)   99
EVM. *See* Error vector magnitude (EVM)
Evolved Packet Core (EPC)   18, 26, 32–37

f
Factory automation   21
Fallback format. *See* Downlink control information (DCI)—format 0_0
FDD. *See* Frequency-division duplex (FDD)

FDM. *See* Frequency domain sharing (FDM)
FDMA (frequency domain multiple access)   12
FE. *See* Front end (FE)
Figure of merit (FOM)   212
Filtering   200
  L1, L3 filtering   372, 379, 382–384
  measurements filtering   320
  spatial   173, 175–176
  or windowing   98–99
First release   10, 19
Flexible   96, 161, 243
  CSI framework   340
  CSI-RS bandwidth   162
  duplex operation   177
  flexible and dynamic HARQ-ACK feedback and retransmission timing   121
  frame structure   10
  patterns   161
  physical layer   68
  scheduling offset   341
  sequences (SRS base sequences)   241
  technology   1
FOM. *See* Figure of merit (FOM)
Forward compatibility   2, 11, 176–177
Frames   74
  half frames with SS/PBCH blocks   264
  paging frames   136
  RACH for IAB   284
  radio frames for PRACH configuration   152–153
  structure   63, 115, 135
Frequency bands   2, 6–7, 12, 14, 21, 39, 95, 104, 126, 152, 210, 265, 287, 289
Frequency bands for NR. *See* Spectrum
Frequency division duplex (FDD)   7
  bands   7
  operation   119
Frequency-domain
  mapping for DM-RS   178
  modulation scheme   110
  precoding granularity   178, 200
  signal   106
Frequency domain sharing (FDM)
Frequency error   296

Frequency ranges (FRs) 9
  FR1 9
  FR2 9
Front end (FE) 210
Front-loaded reference signals 177, 182,
    192, 210, 422–423
FRs. *See* Frequency ranges (FRs)
Full coherence 255

g

GaAs. *See* Gallium arsenide (GaAs)
Gallium arsenide (GaAs) 211–212
Gallium nitride (GaN) 139
GaN. *See* Gallium nitride (GaN)
gNB 31–42
  central unit (gNB-CU) 38–41
  distributed units (gNB-DU) 38–40
Gold sequence 127, 167, 200
Group index 243
  CDM group index 162
Guard period. *See* Guard time
Guard time 148–149, 152

h

HBT 212
Header compression 43, 46
  Robust Header compression (ROHC) 50
Higher frequency bands 2, 7, 68, 104, 127,
    287, 289
Hybrid-ARQ. *See* Hybrid Automatic Repeat
    Request (HARQ)
Hybrid Automatic Repeat Request (HARQ)
    118
  delivering, OSI 135
  mechanism 57
  resource allocation 121, 123
  supporting beamforming 135

i

Idle state mobility 33
IMT-2020 2, 3, 5, 10
  requirements 5
Indoor hotspot 4, 5, 14
Initial access 18, 20, 22, 125, 261, 265, 270
Initial beam aquisition 293

Integrated access backhaul 19
Interference
  interference mitigation 22, 124
  suppression 99, 333, 346
Interleaved CRC-to-REG mapping 251, 420
Interleaved mapping 123
International Telecommunications Union
    (ITU) 3
Interworking 3, 14, 18–19
ITU Radio Regulations (ITU-R) 3
ITU Requirements for IMT-2020 5
  average spectral efficiency 5
  bandwidth 5
  energy efficiency 5
  mobility link spectral efficiency 5
  peak data rate 5
  peak spectral efficiency 5
  reliability 5
  user experienced data rate 5
  user plane latency 5
ITU. *See* International Telecommunications
    Union (ITU)

k

Key performance indicator (KPI) 13, 15

l

LAA. *See* License-assisted access (LAA)
Latency 39, 51, 58, 68, 70, 83, 93, 96, 177,
    210, 262, 283, 326, 342–344, 430–431
  2 Step RACH 22
  URLLC 21
Layer 1 reference signal received power
    (L1-RSRP) 167, 173, 176, 243, 307,
    311, 315
  reporting of 307
  used for beam reporting 327
LBT. *See* Listen-before-talk (LBT)
LCID. *See* Logical Channel Index (LCID)
LDPC. *See* Low-density parity-check (LDPC)
License-assisted access (LAA) 20
Licensed spectrum 20, 285, 422
Listen-before-talk (LBT) 283, 285, 287
L1/L2 160
  procedure 313

Local oscillator (LO)   211, 216, 224

Logical channel(s)   43–45, 54, 56–57, 60, 62, 66, 72

Logical Channel Index (LCID)   55, 67

Logical node   26, 38

Long Term Evolution (LTE)
  bands   7
  base clock   15
  coexistence   192
  CRS   186–187
  design   2
  dual connectivity   47
  evolution   11
  spectrum   14

LO. *See* Local oscillator (LO)

Low-density parity-check (LDPC)   250, 256, 422

Lower carrier frequencies   14–16

Low frequency bands   7

Low-latency   177

L1-RSRP. *See* Layer 1 reference signal received power

LTE/NR. *See also* Long-Term Evolution (LTE)
  symbol level alignment   15

m

MAC control elements (MAC CE)   63

Machine-type communication (MTC)   3–4
  targets   3–4

Macrocells   14

MAC. *See* Medium Access Control (MAC)

Mapping to
  DRBs for downlink data   32
  Resource   45, 161–162

Massive machine type communications (mMTC)   3, 4, 6, 7, 12–15, 22, 96

Massive MIMO   14, 18, 397

Master Cell Group (MCG)   36, 357, 423

Master Information Block (MIB)   75, 119, 262

Master node   18, 25, 34, 36, 62, 262, 368

Maximum power reduction (MPR)   307

MCG. *See* Master Cell Group (MCG)

Medium Access Control (MAC)   162, 250, 267, 427

layer   19, 45, 93, 250, 313–315, 364, 369
  multiplexing operation   54
  protocol   53, 267

MIB. *See* Master Information Block (MIB)

Microcell   16

Millimeter wave   104

MIMO   14
  architectures, hybrid, analogue, digital   17
  enhancements   22
  massive MIMO support   14

Minimizing the transmission of always-on signals   11

mMTC. *See* Massive machine-type communication (mMTC)

Mobile systems   68, 80

Mobility   370

Mobility enhancements   22

Modulation   419–420

MPR. *See* Maximum power reduction (MPR)

M-sequence   127

MTC. *See* Machine-type communication (MTC)

Multi-panel   19, 245, 333–345, 352
  codebook   351

Multiple access   13

Multiple antennas   17–18

Multiple closed-loop parameter   249

Multiple RATs   15

Multiplexing capacity   240

Multiport SRS   240

Multiuser MIMO (MU-MIMO)   124, 201, 208–209, 317, 325, 333, 346, 352, 400–416

n

Narrow band Internet of Things (NB-IoT)   1, 14

NAS. *See* Non-Access Stratum (NAS)

NB-IoT. *See* Narrow band Internet of Things (NB-IoT)

Network
  energy efficiency   5–6
  response   315–316
  transmission   6

Network signaling   31

New bands   7
New candidate beam identification   312, 368
New data indicator   57, 120, 252
  ng-eNB   33–36
New Radio (NR)   1–2
  bands   389
  carrier   18, 260, 262
  coverage   15
  deployments   4, 5
  design considerations   10
  frame structure   7
  frequency ranges   7
  forward compatibility   11
  key performance indicators   15
  mobility   15
  network topology   7
  numerologies   16
  requirements and targets   2–3, 5
  single framework   10
  spectrum   7
  study item   4
  system requirements   3–7
  technical report   4
  technologies   7
  technology components   7, 10
  waveform   7, 12
NG-RAN   18, 26, 28, 31–38, 46–48
Non-Access Stratum (NAS)   26, 388
Non-codebook-based precoding   226
Non-standalone (NSA)   25, 33, 91
Non-zero-power CSI-RS (NZP-CSI-RS)   160, 170, 307, 426
NR. *See* New Radio (NR)
NR uplink multiantenna precoding   102
NSA. *See* Non standalone (NSA)
Numerologies   13, 16
  multiple   13
NZP-CSI-RS. *See* Nonzero-power CSI-RS (NZP-CSI-RS)

o

OFDM. *See* Orthogonal frequency-division multiplexing (OFDM)
OOB. *See* Out-of-band (OOB)
Open-loop power control   246–247, 283

Open-loop precoder   202
Operating bands   8–9, 126
Operators, of mobile systems   36–37, 93
Orthogonal frequency division multiplexing (OFDM)   11
  CP-OFDM   11
  DFT-s-OFDM   11
  OFDM-based waveform   96, 99, 177
Orthogonality
  among antenna ports   239
  between SRS resources   241
of subcarriers   97, 210
Out-of-band (OOB)
  emission (OBE)   104
  guarantees   260

p

power radiation   96–99
Packet Data Convergence Protocol (PDCP)   18, 38, 46, 386
  duplication   67, 93
  layer   47–48
  protocol   48
  reordering   51
  transmission   48
  window   48
Paging Channel (PCH)   54–55
Paging Control Channel (PCCH)   54–55
Paging message   55, 81–82
Paired bands   7–8
PAPR. *See* Peak-to-average power ratio (PAPR)
Partial coherence   255
PA. *See* Power amplifier (PA)
Pathloss estimate (PL estimate)   307
PBCH. *See* Physical Broadcast Channel (PBCH)
PBR. *See* Prioritized bit rate (PBR)
PCCH. *See* Paging Control Channel (PCCH)
PCell. *See* Primary cell (PCell)
PCF. *See* Policy Control Function (PCF)
PCH. *See* Paging Channel (PCH)
PCI. *See* Physical cell identity (PCI)
PDCCH. *See* Physical Downlink Control Channel (PDCCH)

PDCP. *See* Packet Data Convergence Protocol (PDCP)

PDSCH. *See* Physical Downlink Shared Channel (PDSCH)

PDU. *See* Protocol Data Unit (PDU)

Peak data rate 3–5, 15

Peak spectral efficiency 5, 15

Peak-to-average power ratio (PAPR) 23, 225, 242

   low-PAPR DM-RS design 195–200

Performance characteristics 397

Periodic CSI-RS transmission 168

Periodic reporting 380

Periodic SRS 424

Phase locked loop (PLL) 212–215

Phase noise (PN) 215

Phase noise compensation 216

Phase-tracking reference signals (PT-RS) 210

   downlink 221

   for uplink CP-OFDM 224

   for uplink DFT-s-OFDM 225

PHY. *See* Physical Layer (PHY)

Physical Broadcast Channel (PBCH) 73–74, 119, 125, 128

   association with PRACH 271

   beam failure recovery 280

   beam management during initial access 274

   block power 248

   block time pattern 262–265

   frequency band allocation 131

   initial access 261

   initial cell selection assistance information 265

   QCL associations 298

   quasi-colocation 209

   slot monitoring pattern 132

Physical cell identity (PCI) 127

Physical channel 249–250

Physical data shared channels. *See* Physical Downlink Shared Channel (PDSCH)

Physical Downlink Control Channel (PDCCH)

   basic DL control channel 420

   transmit processing 251

Physical Downlink Shared Channel (PDSCH) 250

   basic reception 420

   PDSCH/PUSCH mapping type 123

   transmit processing 250

Physical Layer (PHY) design 162

Physical random-access channel (PRACH) 265

Physical resource block groups (PRGs) 200

   narrowband 201

   wideband 202

Physical resource blocks 97

Physical Uplink Control Channel (PUCCH) 421

   basic UL control channel 421

   format 0 421

   format 1 421

   format 2 3, 4 for periodic beam reporting 426

   groups 116–117

   power control 251, 259, 422

   resource indicator 253

Physical Uplink Shared Channel (PUSCH)

   basic transmission 420

   power control 427–428

PL estimate. *See* Path-loss estimate (PL estimate)

Platooning 21

PLL. *See* Phase locked loop (PLL)

PMI. *See* Precoder matrix indicator (PMI)

PN. *See* Phase noise (PN)

Policy Control Function (PCF) 27

Positioning 22, 46

Power

   back-off 104, 122, 197

   consumption 58, 62, 68, 109, 212–215, 257

   headroom 67, 307

   ramping 282

   saving, UE 22

Power amplifier (PA) 2, 23, 102, 104, 122, 195, 242

   efficiency 104

   lower cost design 242

Power control 246, 422

LTE formula   246

NR *vs.* LTE   246

for PUSCH   246

sequence   158Power spectral density (PSD) 212–215

P1, P2, P3 beam management procedures 291–294

PRACH. *See* Physical random-access channel (PRACH)

Preamble

from pool of preambles   266

in PRACH   266

two groups: group A and group B   266

Preamble format   285, 420

for PRACH   139

Preamble transmission   139, 269–273, 283

RACH resources   20, 279–280, 284, 387

Precoder based uplink transmissions   181

Precoder codebook   323–325, 344

Precoder matrix   323, 325, 327, 328

Precoder matrix indicator (PMI)   307, 324

Preemption   251, 258

indicator   251

PRGs. *See* Physical resource block groups (PRGs)

Primary cell (PCell)   71, 259

Primary second cell (PSCell)   71

Primary synchronization sequence. *See* Primary synchronization signal (PSS)

Primary synchronization signal (PSS)/SSS 125, 128, 374

sequences   127

Prioritized bit rate (PBR)   54–57

Protocol data unit (PDU)   5, 27, 267

PSCell. *See* Primary second cell (PSCell)

PSD. *See* Power spectral density (PSD)

Pseudo-random sequence   167–168, 193

PSS. *See* Primary synchronization signal (PSS)PT-RS. *See* Phase-tracking reference signals (PT-RS)

PUCCH. *See* Physical Uplink Control Channel (PUCCH)

PUSCH. *See* Physical Uplink Shared Channel (PUSCH)

q

QCL. *See* Quasi co-location (QCL)

QFI. *See* Quality of service flow identifier (QFI)

Quadrature phase shift keying (QPSK)   168

Quality of service flow identifier (QFI)   28, 46

Quasi-colocation   173, 209

considerations for DM-RS of PDSCH   209

Quasi co-location (QCL)   296, 428

r

RACH. *See* Random-access channel (RACH)

Radio

access   2

communication   3

protocol architecture   25

Radio Access Network (RAN)   4, 5, 25, 28, 31, 357

Radio access technologies (RAT)   2

Radio frequency (RF)   261–262, 382, 416

ADC and DAC   104

Radio-interface architecture. *See also* New Radio (NR)

5G core network   16, 18, 25–26, 28, 32, 68

Radio interface technologies (RITs)   5

Radio Link Control (RLC)   38, 45, 357,

Radio link failure (RLF)   46, 53, 160, 274, 356–357

Radio Resource Control (RRC)   22, 30, 46, 65, 123

Radio resource management (RRM)   32, 160, 368, 370–396

Random-access channel (RACH)   88

2 step RACH   22

resources   387

Random-access response (RAR)   58, 63, 103, 258, 266–269

Random access RNTI   258, 267

channel   139, 263

preamble index   271–272, 283, 363–364

procedure   266, 274

response   58

Rank indicator (RI)   324

RAN. *See* Radio Access Network (RAN)

RA-RNTI. *See* Random Access RNTI
RAR. *See* Random-access response (RAR)
Rate matching   45, 120
RAT. *See* Radio-access technologies (RAT)
Receiver   211, 213, 215
    detection   141
    device   176
    gNB   145, 146, 148
    implementation   177, 209
Redundancy version (RV)   57, 252, 337
Reference signal(s)
    demodulation   95, 176
    PT-RS   210
    updates in NR Rel-16 for CP-OFDM
        low-PAPR DM-RS design   195
    updates in NR Rel-16 for DFT-S-OFDM
        waveforms   198
Reference signal received power (RSRP)
    159, 266, 327, 426
Reflective mapping   47
REGs. *See* Resource_element groups (REGs)
Remaining minimum system information
    (RMSI)   128, 420
Report configurations   66, 300–301
Resilience   147
Resource
    allocation   269
    configuration   123
    type 0   123
    type 1   123
Resource element groups (REGs)   251
Retransmission. *See also* Transmission
    fast   32
    PDCP   46
RF. *See* Radio frequency (RF)
RI. *See* Rank indicator (RI)
RITs. *See* Radio interface technologies
RLC protocol   357
RLC. *See* Radio Link Control (RLC)
RLF. *See* Radio link failure (RLF)
RMSI. *See* Remaining minimum system
    information (RMSI)
RNTI (Radio-Network Temporary Identifier)
    258
Root index   151

RRC_ACTIVE state   370
RRC_CONNECTED state   270, 370
RRC IDLE state   68–88, 274, 366–368
RRC_INACTIVE state   270, 274, 370
    state machine   68
RRC. *See* Radio Resource Control (RRC)
RRM. *See* Radio resource management
    (RRM)
RSRP. *See* Reference signal received power
    (RSRP)
RV. *See* Redundancy version (RV)

s
Sampling rate   13, 15
SCells. *See* Secondary cells (SCells)
SCG. *See* Secondary Cell Group (SCG)
Scheduler   208, 400, 417
    grants   68
    request   45, 60, 274
    transmission   283
Scrambling   45, 167, 193–194, 250, 256, 420
SDAP. *See* Service Data Application Protocol
    (SDAP)
SDL bands. *See* Supplementary Downlink
    bands (SDL bands)
SDPA. *See* Service Data Adaptation Protocol
    (SDPA)
SDU. *See* Service Data Unit (SDU)
Secondary Cell Group (SCG)   36, 357, 423
Secondary cells (SCells)   387
Secondary node   34, 37, 62, 72
Sector beam   18
Segmentation   43, 45, 50–51
Semipersistent   430
    CSI-RS/IM and CSI reporting   168, 335
    scheduling   60
Sequence
    index   141, 144, 146, 150–151, 243
    numbering   45, 46
Service Data Adaptation Protocol (SDPA)   47
Service Data Application Protocol (SDAP)
    46–47
Service data unit (SDU)   47, 49, 50
Session Management Function (SMF)   27
SFI. *See* Slot format indication/indicator (SFI)

SFN. *See* System frame number (SFN)
Short PUCCH   332
SIB1   128
SIB. *See* System Information Blocks (SIBs)
Sidelink 5   10, 21
Signaling radio bearers (SRBs)   70
Simulation   100, 216, 218–220, 399
Single-port CSI-RS   132, 358, 371–372
   format   45, 169, 262
SI-RNTI. *See* System Information RNTI
      (SI-RNTI)
Slot format indication/indicator (SFI)   169,
      232, 251
Small cells   7, 14, 39, 147
SMF. *See* Session Management Function
      (SMF)
Soft combining   53
Sounding reference signals (SRS)   95, 228,
      231–232, 305, 420
   configurations   305
   resource sets   244
   for DL CSI acquisition and
         reciprocity-based operation   244
   for UL CSI acquisition   245
   for UL beam management   246
   signalling options for UL beam
         management   306
   UE features   424
   use cases   228
Spatial filtering   173, 175–176
Spatial multiplexing   57, 105, 287, 309, 317,
      403, 431
Spectrum   7
   for low-to-mid frequency (FR1)   8
   for high frequency (FR2)   9
Split bearers   46, 50, 66, 71
SRBs. *See* Signaling radio bearers (SRBs)
SRI. *See* SRS resource indicator (SRI)
SRS. *See* Sounding reference signals (SRS)
SRS resource indicator (SRI)   229, 245, 246,
      248, 249
SS block. *See* Synchronization Signal block
      (SS block)
Subcarrier spacing   115
Subframe number   152

SUL. *See* Supplementary uplink (SUL)
Supplementary Downlink bands (SDL bands)
      8
Supplementary uplink (SUL)   7–9, 166, 272
   bands   8
Symmetric utilization, of DL/UL   12
Synchronization raster   129, 261–262, 265,
      379
Synchronization Signal block (SS block)   20,
      73, 293, 330, 383, 391, 392
System frame number (SFN)   73, 152
System Information Blocks (SIBs)   40, 123
System Information RNTI (SI-RNTI)
      195–196, 259

t
TCI. *See* Transmission Configuration Index
      (TCI)
TDD. *See* Time division duplex (TDD)
TDM. *See* Time-domain sharing (TDM)
Technical specifications (TS)   70
TF. *See* Transport format (TF)
Time division duplex (TDD)   7
   bands   7
Time domain
   behaviour   168–169
   behaviour for CSI report   308–332
   behaviour for SRS   230, 306
   locations for DM-RS   177
   locations for SS/PBCH block   262
   mapping for DM-RS   181
   patterns for PT-RS   223
   PRACH   276
   response   199
Time-domain sharing (TDM)
Time multiplexed   130, 135
TM. *See* Transparent mode (TM)
Tracking reference signal (TRS)   95, 171,
      298, 328
Traffic
   bandwidth   96, 122, 210, 283
   channel   53, 55
   steering   19
Transmission Configuration Index (TCI)
      296, 297

Transmission Reception Point (TRP)   130, 289

Transmission scheme   149, 399, 401, 417

Transparent mode (TM)   45, 50–51

Transport block(s)   44, 45, 51
  payload   128, 129
  sizes   51, 57, 421
  single   325

Transport channels   42, 43, 45, 53, 55
  processing in downlink   250, 251

Transport format (TF)   282, 283

TRP. *See* Transmission Reception Point (TRP)

TRS. *See* Tracking reference signal (TRS)

TS. *See* Technical specifications (TS)

u

UCI. *See* Uplink control information (UCI)

UDM. *See* Unified Data Management (UDM)

UE power class   391

UE registration area   395

UE. *See* User equipment (UE)

UL-SCH (Uplink Shared Channel)   55, 270

UL/SUL indicator   272

Ultra Reliable Low Latency and communication (URLLC)   3–4, 21–23. 93, 96,
326

Unacknowledged mode (UM)   45, 50, 51

Unified Data Management (UDM)   27

Unlicensed spectra, operation in   21

Unpaired bands   7–8

Unwanted emissions   102

UPF. *See* User Plane Function (UPF)   27,
385, 386

Uplink and downlink shared channels   45

Uplink carrier   260, 267

Uplink channel   239, 243
  and downlink channel   228
  multiport SRS   240
  SRS resource set   63, 233

Uplink control information (UCI)   21, 44,
325, 422
  commands 45   249, 251

Urban Macro   5, 14, 399, 411,
413–415

Urban Macro-URLLC   5

URLLC. *See* Ultra Reliable Low Latency and
communication (URLLC)

Usage scenarios   4–5
  for IMT-2020   4–5

User equipment (UE)
  multiple panels   306

User experienced data rate   5

User Plane Function (UPF)   27, 385,
386

User plane protocols   28
  physical layer PHY   38, 42, 45, 82
  RLC   38, 42, 44, 50, 85, 93

User specific bandwidth, waveform,
  numerology   11

Uu interface   28, 81, 82

U1, U2, U3 beam management procedures
295

v

Vehicle-to-everything communication
  (V2Xcommunication)   21, 23

Vehicle-to-vehicle communication
  (V2Vcommunication)   21

Virtual resource blocks   123

w

Waveform   12

Wi-Fi   19, 20, 22

Wireless   41
  access configuration   413, 416
  backhaul   284
  channels   177, 201–209
  modem   27

Wireless backhaul   19, 284

World Radio-communication Conference
  (WRC)
  WRC-15   14
  WRC-19   14
  WRC 2015   9

X

Xn interface   26, 32–36

Z

Zadoff-Chu sequences (ZC sequences)   140,
    198

Zero-correlation zone parameter   143
Zero-power
    channel state   160
    CSI-RS   170